MATHEMATICS IN ENGINEERING AND SCIENCE

Books are to be returned on or before
the last date below.

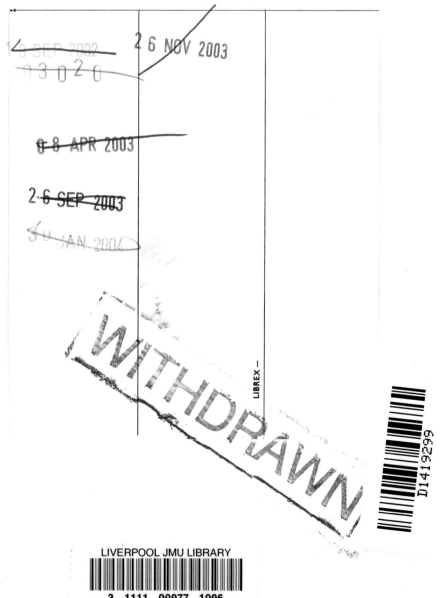

MATHEMATICS IN ENGINEERING AND SCIENCE

L R Mustoe
Loughborough University, Loughborough

M D J Barry
University of Bristol, Bristol

JOHN WILEY & SONS

Chichester · New York · Weinheim · Brisbane · Singapore · Toronto

Other Wiley Editorial Offices

John Wiley & Sons, Inc., 605 Third Avenue,
New York, NY 10158-0012, USA

Wiley-VCH GmbH, Pappelallee 3,
D-69469 Weinheim, Germany

Jacaranda Wiley Ltd, 33 Park Road, Milton,
Queensland 4064, Australia

John Wiley & Sons (Asia) Pte Ltd, 2 Clementi Loop #02-01,
Jin Xing Distripark, Singapore 129809

John Wiley & Sons (Canada) Ltd, 22 Worcester Road,
Rexdale, Ontario M9W 1L1, Canada

British Library Cataloguing in Publication Data

A catalogue record for this book is available from the British Library

ISBN 0 471 97093 X (pbk); 0 471 97095 6 (cloth)

Typeset by Techset Composition Ltd, Salisbury
Printed and bound in Great Britain by Bookcraft (Bath) Ltd
This book is printed on acid-free paper responsibly manufactured from sustainable forestation,
for which at least two trees are planted for each one used for paper production.

To the Memory of
My Mother, Father and Grandmother

(L. R. Mustoe)

To Angus and Katharine

(M. D. J. Barry)

CONTENTS

PREFACE

In today's world, technology plays an increasingly important rôle. At the same time, mathematics is finding ever wider areas of application as we seek to understand more about the way in which nature works. Traditionally, engineering and science have relied on mathematical models for design and for the prediction of the behaviour of many phenomena.

Although widespread availability of computers and pocket calculators has reduced the need for long, tedious calculations to be carried out 'by hand', it is still important to be able to perform simple calculations in order to have a feel over the processes involved.

This book starts with a detailed synopsis of the authors' *Foundation Mathematics*. It then expands the material in the areas of trigonometry, solution of equations and algebra. Vectors are covered next, then the calculus is taken forward into geometrical applications. Matrix algebra and uncertainty follow before deeper analysis in the chapters on integer variable, differential equations and complex numbers lead towards an appendix on mathematical modelling.

Each chapter opens with a list of learning objectives and ends with a summary of key points and results. A generous supply of worked examples incorporating motivational applications is designed to build knowledge and skill. Drill and practice is essential and the exercises vary in difficulty for reading and revision; the answers at the end of each chapter include helpful hints. Use of a pocket calculator is encouraged where appropriate (indicated by calculator icon). Many of the exercises can be validated by computer algebra (indicated by a computer icon) and its use is strongly recommended where higher algebraic accuracy can be achieved and drudgery removed. Asterisks indicate difficult exercises. Lecturers' Resource Guides/Teachers' Manuals will be available in due course; please contact the publisher for details.

The authors have between them a wide experience of teaching mathematics to a broad spectrum of students: mature students, those with an arts background who wish to convert to science and engineering, those who need to advance from GCSE to first year university level, in addition to more traditional undergraduates and to secondary level pupils. They are concerned at the growing gap between secondary and tertiary level mathematics. They have many contacts with the educational scene in continental Europe and are aware that the UK has fallen behind in its mathematics provision. They believe that there is a need for a no-nonsense textbook to take students across the interface between secondary and tertiary level and have striven to provide one.

The text aims to provide a modern mathematical education which matches closely the interface between Core Zero and Level One of the Core Curriculum prepared by the Mathematics Working Group of SEFI (*Société Européene pour la Formation des Ingénieurs*). The core of Advanced Level GCE mathematics and its equivalent in GNVQ are embraced, as are parts of Further Mathematics, although there is no direct link to any one particular syllabus. The main purpose of the book is to build upon the aims and contents of *Foundation Mathematics* in order to provide a systematic post-GCSE mathematical education for both users of mathematics and mathematics specialists. Its concise and focused approach will help the student reader to meet the challenges of mathematics in a course at university level.

Foundation Mathematics and *Mathematics in Engineering and Science* are written to be both complementary and independent; students may follow both books consecutively or may use just one, depending on their previous mathematical experience and the level of mathematical development that they wish to achieve.

L R Mustoe
M D J Barry

FOUNDATION
MATHEMATICS

ARITHMETIC

1. An integer, or whole number, can be expressed uniquely as the product of prime number factors or it is prime. For example

$$300 = 2 \times 2 \times 3 \times 5 \times 5 = 2^2 \times 3 \times 5^2$$
$$1999 = 1 \times 1999 \quad \text{(prime)}$$

This property is known as the **fundamental theorem of arithmetic**.

The **highest common factor** (HCF) of two integers is the largest whole number that divides exactly into them both. For example

$$300 = 2 \times 2 \times 3 \times 5 \times 5$$

$$180 = 2 \times 2 \times 3 \times 3 \times 5$$

To find the HCF identify common prime factors in the decomposition of each.

$$\text{HCF } (300, 180) = 2 \times 2 \times 3 \times 5 = 60$$

The **lowest common multiple** (LCM) of two integers is equal to their product divided by their HCF.

$$\text{LCM } (300, 180) = \frac{54\,000}{60} = 900$$

It follows that the product of two integers is therefore equal to the product of their LCM and HCF.

$$300 \times 180 = 900 \times 60 = 54\,000$$

2. **Integer powers** of the form N^M can be determined for any integers N and M. 0^0 cannot be found. For non-zero N, $N^0 = 1$. When powers of the same number are

multiplied or divided, the powers are added or subtracted respectively. When raising a power to a power the two powers are multiplied. Examples are

$$5^4 = 5 \times 5 \times 5 \times 5 = 625$$

$$5^2 \times 5^3 = 5^5 = 25 \times 125 = 3025$$

$$5^6 \times 5^{-4} = \frac{15\ 125}{625} = 25 = 5^2$$

$$5^{-2} = \frac{1}{5^2} = \frac{1}{25}$$

$$(5^2)^3 = (5^3)^2 = 5^6 = 15\ 625$$

$$(5^2)^{-2} = \frac{1}{5^4} = \frac{1}{625}$$

$$(-3)^2 = (-1)^2 3^2 = 9$$

$$(-4)^{-3} = (-1)^{-3} 4^{-3} = \frac{1}{(-1)^3 4^3} = -\frac{1}{64}$$

3. A **fraction** or **rational number** is a ratio of integers $\dfrac{N}{M}$ where $M \neq 0$. It is positive if N and M have the same sign and negative if they have opposite signs. The fraction is **proper** if $|N| \leq |M|$ and **vulgar** if $|N| > |M|$. If all common factors of N and M are cancelled the fraction is in its **lowest form**. Examples are

$$\frac{-2}{3} = \frac{2}{-3} = -\frac{2}{3} \quad \text{(proper)}$$

$$\frac{-6}{-4} = \frac{6}{4} = \frac{3}{2} \quad \text{(vulgar and in lowest form)}$$

A decimal number, or **decimal**, is a fraction whose denominator is a power of 10.

4. The **precedence of operators** for both written mathematics and calculators is:

raise to a power   II   multiply or divide   II   add or subtract

$$2 \times 3 - 1 \div 3 + 5^2 + 5 = (2 \times 3) - \frac{1}{3} + (5^2) + 5$$

$$= 6 - \frac{1}{3} + 25 + 5 = 35\frac{2}{3}$$

Brackets are used to modify the order. For example

$$(2 \times 3 - 1) \div 3 + (5 \times 5)^2 = \frac{5}{3} + 25^2 = 626\frac{2}{3}$$

5. A general number, i.e. an integer, a rational number or an irrational number is called a **real number**. Real numbers in decimal form can be expressed to a specified number of **significant figures** (s.f.) or to an appropriate number of **decimal places** (d.p.). By decimal places we mean the number of digits after the decimal point and by significant figures we mean the number of digits relevant to its required level of accuracy. Using standard notation

   671.618 725 613 7 . . . becomes
   670 (2 s.f.)
   671.6 (4 s.f.)
   671.62 (2 d.p.)
   671.619 (3 d.p.)

   The last figure is rounded up if the one that follows is 5 or more.
   With **scientific notation** the same number is, for example

   $$6.7162 \times 10^2 \ (5 \text{ s.f.}) \quad \text{or} \quad 6.7 \times 10^2 \ (2 \text{ s.f.})$$

   Decimal places do not have any significance in this context.

6. Real powers of real numbers of the form x^y exist for any y and any non-negative x. The rules for manipulating powers are

   $0°$ cannot be found $a^0 = 1$
   $a^x \times a^y = a^{(x+y)}$ $a^x \div a^y = a^{(x-y)}$
   $(a^x)^y = a^{xy}$ $a^{-y} = 1/a^y$

   For rational x and y, e.g. $x = \frac{2}{3}, y = \frac{1}{4}$ the meaning is

   $$\left(\frac{2}{3}\right)^{1/4} \times \left(\frac{2}{3}\right)^{1/4} \times \left(\frac{2}{3}\right)^{1/4} \times \left(\frac{2}{3}\right)^{1/4} = \frac{2}{3}$$

   so $\left(\frac{2}{3}\right)^{1/4}$ is a positive number whose fourth power is $\frac{2}{3}$. Sometimes negative numbers have rational powers, e.g. $(-27)^{-2/3} = \frac{1}{9}$, though this is not always the case.

   Calculators are programmed to find roots of positive numbers only, so they will generate an error message when asked to calculate roots of negative numbers.

For $x \geq 0$ and y real, e.g. $x = 11.574$ (5 s.f.) and $y = 2.23$ (3 s.f.) the meaning of

$$(11.574)^{2.23} \quad \text{is} \quad (11.574)^{1/100} \times \cdots \times (11.574)^{1/100} \qquad \text{(223 times)}$$

7. **Surds** are expressions involving the sum or difference of rational numbers and their square roots, e.g.

$$\sqrt{6}, \qquad 2 + \sqrt{3}, \quad \sqrt{5} - \sqrt{7}.$$

Arithmetic operations involving sum, difference, product ratio and integer powers of surds usually produce another surd, e.g.

$$(2 + 3\sqrt{5})^2 = 4 + 12\sqrt{5} + 45 = 49 + 12\sqrt{5}$$

$$\frac{\sqrt{2} + 3}{\sqrt{2} - 1} = \left(\frac{\sqrt{2} + 3}{\sqrt{2} - 1}\right)\left(\frac{\sqrt{2} + 1}{\sqrt{2} + 1}\right) = \frac{2 + 3\sqrt{2} + \sqrt{2} + 3}{2 - 1} = 4\sqrt{2} + 5$$

Surds can sometimes be expressed in simpler forms, e.g.

$$\sqrt{75} = \sqrt{5^2 \times 3} = 5\sqrt{3}, \qquad \sqrt{\frac{16}{27}} = \sqrt{\frac{4^2 \times 3}{9^2}} = \frac{4}{9}\sqrt{3}$$

8. In the relationship $10^3 = 1000$ we see that 10 must be raised to the power 3 to obtain the number 1000. We say that 3 is the **logarithm** of 1000 **to the base** 10. In symbolic form this is written $\log_{10} 1000 = 3$.

   The general rules for dealing with logarithms are

$$\log_b n = x \text{ if } b^x = n \qquad \log n^m = m \log n$$

$$\log(nm) = \log n + \log m \qquad \log\left(\frac{n}{m}\right) = \log n - \log m$$

Note that $\log 1 = 0$, any base, and

$$\log\left(\frac{1}{\text{number}}\right) = -\log\ (\text{number})$$

9. The **modulus** or absolute value of x, written $|x|$, is the magnitude of x. For example

$$|3.241| = 3.241, \qquad |-6.832| = 6.832$$

For two numbers x and y

$$|x + y| \leq |x| + |y| \qquad \text{(the triangle inequality)},$$

$$|xy| = |x| \times |y| \qquad \left|\frac{x}{y}\right| = \frac{|x|}{|y|}$$

10. The error in an approximation to a number is defined by the relationship

Error = approximate value − true value

or $\varepsilon = x - a$

$|\varepsilon| = |x - a|$ is the **absolute error**

$\dfrac{|\varepsilon|}{|a|} = \dfrac{|x - a|}{|a|}$ is the **relative error**

The absolute error in the sum or the difference of two appropriate values is

$$|\varepsilon| \leq |\varepsilon_1| + |\varepsilon_2|$$

The relative error in multiplying or dividing two numbers is at most the sum of the relative errors in each. Consider 20.0 ± 0.1 and 30.0 ± 0.3. For addition or subtraction

Absolute error $= 0.1 + 0.3 = 0.4$

For multiplication or division

$$\text{Relative error} = \frac{0.1}{20} + \frac{0.3}{30} = 0.015 \quad \text{or} \quad 1.5\%$$

BASIC ALGEBRA

1. An algebraic **expression** is a combination of constants, variables and arithmetic operators. In the expression $a - 3b + 2c$ the **coefficients** of a, b and c are respectively 1, −3 and 2.

   Algebraic expressions are evaluated according to the following order of precedence: **Brackets, Exponentiation, Division/Multiplication, Addition/Subtraction**.

   The **expansion of expressions** is carried out using the distributive laws

   $$a(b + c) = ab + ac \quad \text{and} \quad (a + b)c = ac + bc$$

Note the results

$$(a + b)^2 = a^2 + 2ab + b^2 \qquad (a - b)^2 = a^2 - 2ab + b^2$$

$$(a - b)(a + b) = a^2 - b^2$$

2. A **formula** relates one variable quantity, the subject, to other variable quantities. In the formula $T = 2\pi\sqrt{\dfrac{l}{g}}$, T is the subject, 2π and g are constants. We can rearrange the formula as $l = \dfrac{gT^2}{4\pi^2}$ to make l the subject.

3. An **identity** states that two algebraic expressions are always equal to each other; an **equation** is true for one or more values of the unknown(s).
    Examples of identities are $a^2 - b^2 \equiv (a - b)(a + b)$ and $a^2 - 4 \equiv (a - 2)(a + 2)$. On the other hand, $a^2 - 4 = (a + 2)(a + 2)$ is an equation with solution $a = -2$.

4. In a **linear equation** the **unknown variable** occurs to the **power one**. A linear equation can be solved for an unknown variable by making that variable the subject of a formula. For example $3x + 2 = 5 - 2x$ becomes $5x = 3$ or $x = \dfrac{3}{5}$

5. Two linear simultaneous equations in two unknowns are **inconsistent** if they have no solution; if they are consistent they either have a **unique solution** or **infinitely many solutions**. For example

$$\left.\begin{array}{l} x + y = 3 \\ 2x - y = 3 \end{array}\right\} \quad \text{have the unique solution } x = 2, y = 1$$

$$\left.\begin{array}{l} x + y = 1 \\ x + y = 2 \end{array}\right\} \quad \text{is an inconsistent set}$$

$$\left.\begin{array}{l} x + y = 1 \\ 2x + 2y = 2 \end{array}\right\} \quad \text{is a repeated set with infinitely many solutions.}$$

Similar principles apply to **three equations** in **three unknowns**, and to higher orders too. When the equations can be solved, the unknowns are eliminated one by one.

6. **Proportionality**

$$y \text{ is \textbf{directly proportional} to } x \text{ if } y = Cx$$

$$y \text{ is \textbf{inversely proportional} to } x \text{ if } y = \frac{C}{x}$$

$$z \text{ is \textbf{jointly proportional} to } x \textbf{ and } y \text{ if } z = Cxy$$

C is a constant.

Suppose that p is directly proportional to x, inversely proportional to the square of y and proportional to z to the power $\frac{3}{2}$, then

$$p = \frac{Cxz^{3/2}}{y^2}$$

7. Possible factors of an expression can be tested by setting them equal to zero. In such a case the whole expression is zero. For example, consider the expression

$$14x^2 + 65x - 25 = (14x - 5)(x + 5)$$

Setting $x = \frac{5}{14}$ or -5 will make the quadratic vanish. Similarly, $(2p + q)(q - p) + q^3 - p^3$ has a factor $q - p$, which can be checked by setting $q - p = 0$, i.e. by putting $q = p$.

8. The rules for dealing with **inequalities** are as follows:

$$a > b \quad \text{implies} \quad a + c > b + c$$
$$a < b \quad \text{implies} \quad a + c < b + c$$
$$\text{if } a > b, \lambda > 0 \quad \text{then} \quad \lambda a > \lambda b, \qquad \text{if } \lambda < 0 \quad \text{then} \quad \lambda a < \lambda b$$
$$ab > 0 \quad \text{implies} \quad a > 0, b > 0 \quad \text{or} \quad a < 0, b < 0$$
$$ab < 0 \quad \text{implies} \quad a > 0, b < 0 \quad \text{or} \quad a < 0, b > 0.$$

Inequalities can be represented by intervals on the real line.

9. In dealing with inequalities involving $|x|$ we use the following results:

$$|x| < a \text{ is equivalent to } -a < x < a$$
$$|x| > a \text{ is equivalent to } x < -a \text{ or } x > a$$
$$|x| = a \text{ is equivalent to } x = -a \text{ or } x = a.$$

For example, the inequality $|2x - 1| < 4$ is equivalent to the inequalities $-4 < 2x - 1 < 4$, $-3 < 2x < 5$, $-\frac{3}{2} < x < \frac{5}{2}$ and $\left| x - \frac{1}{2} \right| < 2$.

STRAIGHT LINES

1. A **graph** is a scaled picture which shows the relationship between two variables. Typically, the variables extend horizontally and vertically and their magnitudes are

measured along axes, x horizontally and y vertically. The point (x, y) is a distance x from the y-axis and y from the x-axis. The values x and y are the Cartesian **coordinates** of the point (Figure 1). The coordinate plane is split into four **quadrants**, I, II, III and IV.

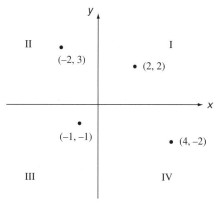

Figure 1 Coordinates

2. An equation of the **first power** or **degree** in both variables x and y forms a **straight-line graph**. Such an equation is therefore called **linear**.

   The standard form for the equation of the line is

   $$y = mx + c$$

   where m is the **gradient** of the line and the point at which the y-axis is crossed, $(0, c)$, is called the **intercept** (Figure 2).

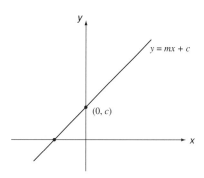

Figure 2 Straight line

3. The equation of a straight line can take one of several forms:

   (a) $y - y_1 = m(x - x_1)$ gradient m, passing through (x_1, y_1)

(b) $\dfrac{y - y_1}{y_2 - y_1} = \dfrac{x - x_1}{x_2 - x_1}$ passing through (x_1, y_1) and (x_2, y_2)

(c) $\dfrac{x}{a} + \dfrac{y}{b} = 1$ passing through $(a, 0)$ and $(0, b)$

(d) $lx + my + n = 0$ general equation.

4. Two lines $y = m_1 x + c_1$ and $y = m_2 x + c_2$ are **parallel** if their gradients are equal, i.e. $m_1 = m_2$. Two lines are **perpendicular** if the product of their gradients equals -1, i.e. $m_1 m_2 = -1$.

5. Two straight lines with equations

$$a_{11} x + a_{12} y = b_1$$
$$a_{21} x + a_{22} y = b_2$$

meet whenever they are not parallel, i.e. when

$$a_{11} a_{22} - a_{21} a_{12} \neq 0$$

or in determinant form (Chapter 8 of this volume)

$$\begin{vmatrix} a_{11} & a_{12} \\ a_{21} & a_{22} \end{vmatrix} \neq 0$$

The coordinates (x, y) of the meeting point satisfy both equations simultaneously.

If $a_{11} a_{22} - a_{12} a_{12}$ is much smaller in magnitude than either of the two products of which it is the difference, then the lines are nearly parallel and the meeting point may vary widely with small fluctuations of slope. If this is the case the equations are said to be **ill-conditioned**.

6. A line or curve separates the coordinate plane into disjoint zones of inequality, as shown in Figure 3. For the straight line $y = mx + c$.

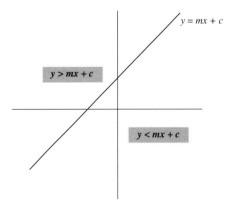

Figure 3   Linear inequalities

7. **Simultaneous inequalities** define regions of the $x-y$ plane; in these regions the coordinates of the points satisfy all the inequalities. For example, in Figure 4 the shaded region is where

$$3x + y \leq 8, \qquad x + y \leq 3, \qquad x + 3y \leq 8, \qquad x \geq 0, \qquad y \geq 0.$$

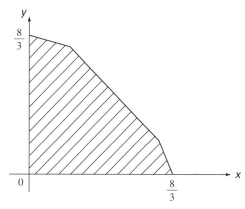

Figure 4   Simultaneous inequalities

8. Sometimes a relationship can be rewritten in a form from which a straight line graph can be drawn. This is known as **reduction to linear form**. Examples are

(a) $y = ax^n$ plot $\log y$ against $\log x$

(b) $y = ba^x$ plot $\log y$ against x

(c) $\dfrac{1}{x} + \dfrac{1}{y} = a$ plot $\dfrac{1}{y}$ against $\dfrac{1}{x}$

(d) $y = ax^2 + bx$ plot $\dfrac{y}{x}$ against x.

9. By taking logarithms, the relationship $y = ka^x$ becomes $\log y = \log k + x \log a$, which is known as a **log-linear relationship**. If we plot $\log y$ against x, the gradient of the line is $\log a$ and the intercept is $\log k$.

By taking logarithms, the relationship $y = ax^n$ becomes $\log y = \log a + n \log x$, which is known as a **log-log relationship**. If we plot $\log y$, against $\log x$, the gradient of the line is n and the intercept is $\log a$.

Special graph paper can be used for direct plots of log-linear and log-log relationships.

QUADRATICS AND CUBICS

1. A **quadratic expression** is an expression of the form $ax^2 + bx + c$ where a, b and c are constants. Its graph can be obtained from the graph of x^2 by a combination of scaling, translation and (if $a < 0$) reflection in the x-axis.

2. The **graph** of a **quadratic expression** $y = ax^2 + bx + c$ is a **parabola**. The quadratic curve has its vertex at $x = -\dfrac{b}{2a}$; this is the lowest point if $a > 0$, the highest point if $a < 0$.

   The graphs of $y = -x^2 + x + 6$, $y = (x + 1)^2$, $y = x^2 - 2x + 1.5$ and $y = 1 - 0.5x$ are shown in Figure 5.

3. The **quadratic equation** $ax^2 + bx + c = 0$ has no roots if $b^2 - 4ac < 0$, a repeated root if $b^2 - 4ac = 0$ and two roots if $b^2 - 4ac > 0$. The quantity $\Delta = b^2 - 4ac$ is called the **discriminant** of the equation. If $(x - \alpha)$ is a **factor** of a quadratic, cubic or higher **polynomial**, then $x = \alpha$ is a **root** of the corresponding equation and vice versa. The value of the expression is **zero** at a root $x = \alpha$.

   If the real roots are α and β, the equation can be expressed as $a(x - \alpha)(x - \beta) = 0$, as in Figure 5(a).

   If there is a repeated root α, the equation can be expressed as $a(x - \alpha)^2 = 0$, as in Figure 5(b).

   If real roots α and β exist, the maximum or minimum is at $s = \dfrac{1}{2}(\alpha + \beta)$, as in Figure 5(a).

4. The main methods for **solution of a quadratic equation** are as follows:

   Factorisation
   $$x^2 + bx + c \equiv (x - \alpha)(x - \beta)$$
   so that the roots are $\quad x = \alpha, x = \beta$

   Completing the square
   $$x^2 + bx + c \equiv \left(x + \frac{b}{2}\right)^2 - \frac{b^2}{4} + c, \text{ gives the formula}$$
   $$x = -\frac{b \pm \sqrt{b^2 - 4ac}}{2a}$$

5. The **sum of the roots** of the quadratic equation $ax^2 + bx + c = 0$ is $\alpha + \beta = -\dfrac{b}{a}$, the **product of the roots** is $\alpha\beta = \dfrac{c}{a}$.

   Use can be made of identities like $\alpha^2 + \beta^2 \equiv (\alpha + \beta)^2 - 2\alpha\beta$ to determine quadratics whose roots are α^2 and β^2 or $\dfrac{1}{\alpha}$ and $\dfrac{1}{\beta}$, etc., without actually finding α and β directly.

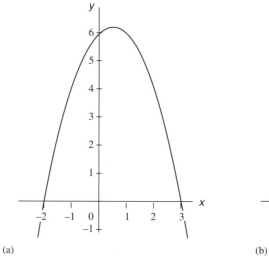

(a)

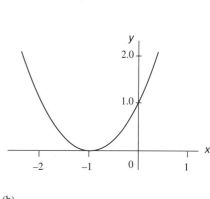

(b)

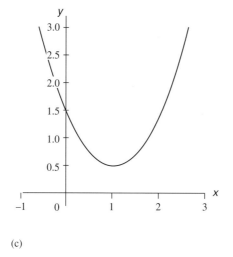

(c)

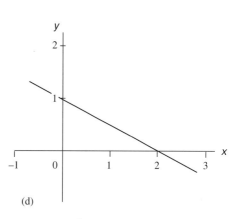

(d)

Figure 5   Graphs of (a) $y = -x^2 + x + 6$, (b) $y = (x + 1)^2$, (c) $y = x^2 - 2x + 1.5$ and (d) $y = 1 - 0.5x$ ($a = 0$)

6. Certain **non-linear equations** can be solved by forming a quadratic equation. For example, to solve the equations

$$x + y = 9$$
$$xy = 18$$

substitute $y = \dfrac{18}{x}$ in the first equation.

7. Sometimes only one solution of a quadratic equation may be relevant, perhaps when modelling a physical situation. For example, a rectangle of area $8\,\mathrm{m}^2$ which has its longer side $2\,\mathrm{m}$ longer than the adjacent side has sides of length $x\,\mathrm{m}$ and $(x+2)\,\mathrm{m}$. Then $x(x+2)=8$, with solutions $x=-4, x=2$. Only the second solution is meaningful.

8. **Quadratic inequalities**. Examples of quadratic inequalities are $y > x^2 + 4$, shown in Figure 6(a) and $x^2 + x - 2 < y < 6 - x$, shown in Figure 6(b).

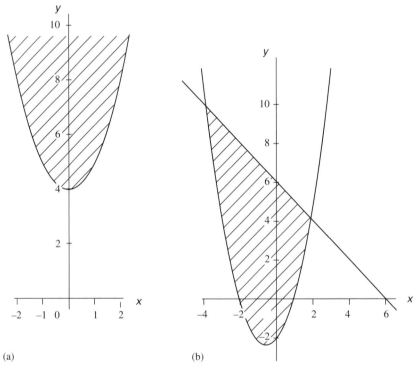

(a)   (b)

Figure 6   Graphs to show (a) $y > x^2 + 4$ and (b) $x^2 + x - 2 < y < 6 - x$

9. A **cubic expression** has the form $ax^3 + bx^2 + cx + d, a \neq 0$.

10. The **graph** of $y = x^3$ is shown in Figure 7. However the graph of a general cubic is not like the graph of $y = x^3$.

   If $a > 0$ the graph of $y = ax^3 + bx^2 + cx + d$ starts in the south-west quadrant and ends in the north-east quadrant. If $a < 0$ the graph starts in the north-west quadrant and ends in the south-east quadrant. Expressed in terms of the quadrants in Figure 1, $a > 0$ implies a curve from III to I and $a < 0$ implies a curve from II to IV.

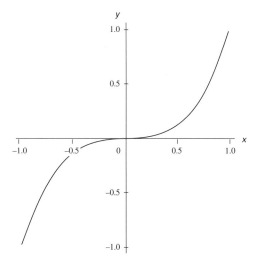

Figure 7 Graph of $y = x^3$

Travelling from left to right along the graph of a general cubic expression with $a > 0$ we meet in turn a local **maximum**, then a **point of inflection** and then a **local minimum**. If $a < 0$ we meet a local minimum, then a point of inflection and then a local maximum.

Figure 8(a) shows the graph of $y = x(x - 1)(x + 2)$, and Figure 8(b) shows the graph of $y = -x^3 - 3x^2 - x - 1$.

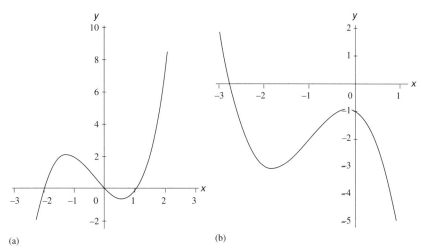

(a)   (b)

Figure 8 Graphs of (a) $y = x(x - 1)(x + 2)$ and (b) $y = -x^3 - 3x^3 - x - 1$

GEOMETRY

1. Angles between $0°$ and $90°$ are **acute**, angles between $90°$ and $180°$ are **obtuse** and angles between $180°$ and $360°$ are **reflex**. An angle of $90°$ is a **right angle**, an angle of $180°$ is a **straight angle** and an angle of $360°$ represents a **full turn**.

2. When a **transversal** cuts parallel lines, it generates three sets of equal angles: **vertically opposite**, **corresponding** and **alternate**. Angles a in Figure 9(a) are vertically opposite angles and angles b are corresponding angles. In Figure 9(b) they form pairs of alternate angles.

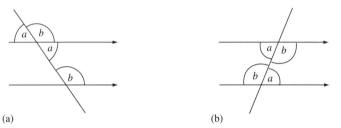

(a)   (b)

Figure 9   Equal angles: (a) vertically opposite and corresponding and (b) alternate.

3. There are three main types of **triangle**. **Scalene** triangles have three sides of unequal length. **Isosceles** triangles have two sides equal. **Equilateral** triangles have all three sides equal. If a triangle has one angle equal to $90°$ it is called a **right-angled triangle**.

   The **sum of the interior angles** of a triangle is $180°$. All three interior angles of an equilateral triangle are $60°$.

4. Two triangles are **congruent**, i.e. of identical size and shape, if in both triangles there is equality between

   (a) two sides and the included angle   (b) two angles and the side joining them
   (c) three sides   (d) right angle, hypotenuse and another side.

5. Two triangles are **similar** if corresponding angles are equal. The length of corresponding sides are in the same ratio. One triangle is an enlargement of the other, whereas congruent triangles have corresponding sides equal. The concept of similarity applies to shapes in general.

6. The following properties of the triangle are useful:

   (a) An **exterior angle** is equal to the sum of the two opposite **interior** angles.

   (b) The **bisectors of the angles** of a triangle are **concurrent**, i.e. meet at a point I, the **incentre**, which is equidistant from each side.

(c) The **perpendicular bisectors of the sides** of a triangle meet at a point O, the **circumcentre**.

(d) The **altitudes** of a triangle meet at a point H, the **orthocentre**.

(e) The **medians**, i.e. the lines joining the vertices of a triangle to the midpoints of the opposite sides, meet at a point G, the **centroid**, i.e. the point about which a cardboard cut-out of the triangle would balance.

(f) The points O, H and G all lie on a straight line, the **Euler line**.

The **area of a triangle** is equal to its height times half the length of its base. Any side can be taken as the base for this purpose.

7. A polygon with all n sides equal is called **regular**. Its interior angles are all equal. In general the sum of the interior angles of a polygon is $(2n - 4) \times 90°$.

   A quadrilateral with two sides parallel is a **trapezium**. When the quadrilateral has opposite sides parallel, or a pair of sides equal and parallel, it is a **parallelogram**. If all the sides are equal it is called a **rhombus**.

   A quadrilateral with all angles equal to $90°$ is a **rectangle**. A regular quadrilateral with all angles and sides equal, i.e. a rhombus with right-angles, is called a **square**. A square is therefore a rectangle with all sides equal.

   The area of a polygon is equal to the total area of its constituent triangles. For example, a regular hexagon can be subdivided into six congruent equilateral triangles.

8. One **radian**, written 1^c, is the angle subtended at the centre of a circle by an arc whose length is equal to the radius of the circle. To convert from radians to degrees, or vice versa, use the formula $\pi^c \equiv 180°$. The **arc length** of a sector which subtends an angle θ at the centre of a circle of radius r is $r\theta$ and the **sector area** is $\frac{1}{2}r^2\theta$.

9. The key parts of a circle are shown in Figure 10. The **area of a circle** of radius r is πr^2, the **circumference** is $2\pi r$.

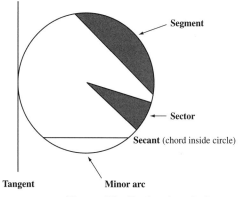

Figure 10 Parts of a circle

10. **Volumes of solids**

Cuboid	base area × height
Tetrahedron	$\frac{1}{3}$ × base area × height
Pyramid	$\frac{1}{3}$ × base area × height
Sphere	$\frac{4}{3}\pi r^3$
Cylinder	$\pi r^2 h$
Cone	$\frac{1}{3}\pi r^2 h$

Surface areas of solids

Sphere	$4\pi r^2$
Closed cylinder	$2\pi r^2 + 2\pi rh$
Closed cone	$\pi r^2 + \pi rl$

PROOF

1. An **axiom** is a fundamental truth justified by deductive reasoning at a basic level. A **theorem** is a higher-order result that emerges from deductive mathematical reasoning at a lower level. A **corollary** is a further result which follows as a consequence of a theorem.

2. **Pythagoras' theorem** may be derived using Figure 11. In the right-angled triangle PQR, the square on the hypotenuse is equal to the sum of the squares on the other two sides, i.e.

$$PQ^2 + QR^2 = PR^2$$

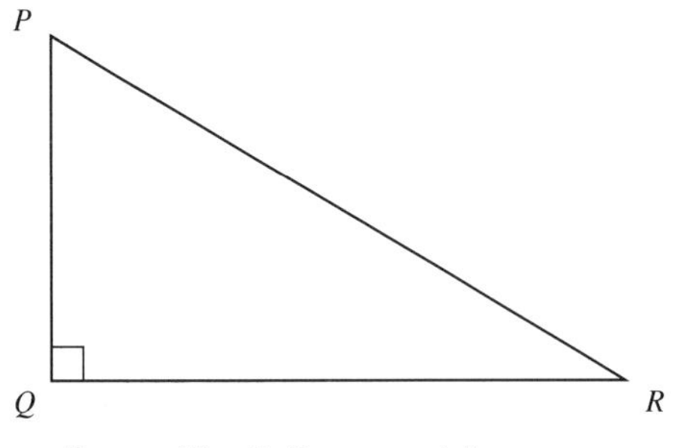

Figure 11   Pythagoras' theorem

3. **Theorems in geometry** may be categorised by shape. Triangle theorems are covered in item 6 of the section on geometry. Here are some circle theorems.

(a) **Angle in a segment theorem**

$$A\hat{C}B = A\hat{D}B = \frac{1}{2}A\hat{O}B \qquad \text{(Figure 12)}$$

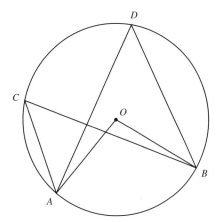

Figure 12 Angle in a segment theorem

(b) **Cross-chord theorem**

$$RX \times XQ = PX \times XS \qquad \text{(Figure 13)}$$

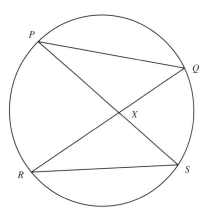

Figure 13 Cross-chord theorem

(c) **Alternate segment theorem**: the angles α are equal (Figure 14).

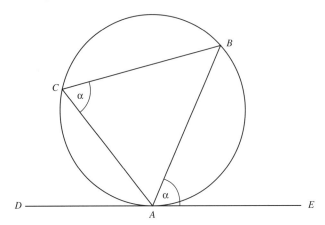

Figure 14 Alternate segment theorem

(d) **Chord–tangent theorem**

$$CP \times CQ = CT^2 \qquad \text{(Figure 15)}$$

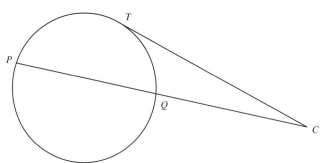

Figure 15 Chord–tangent theorem

4. A proposition p is a **sufficient condition** for a proposition q if q is true when p is true. We write $p \Rightarrow q$, read as p *implies* q.

A proposition p is a **necessary condition** for a proposition q if p is true when q is true. We write $q \Rightarrow p$ or $p \Leftarrow q$.

The notation $p \Leftrightarrow q$, means that p and q both hold, or neither hold, and that both are **necessary and sufficient** to imply one another.

Examples are $x = 4 \Rightarrow x^2 = 16$, but $x^2 = 16 \nRightarrow x = 4$, i.e. $x = 4$ is sufficient for $x^2 = 16$, but not necessary. However, $x = 2 \Leftrightarrow x^3 = 8$.

5. **Methods of proof**

(a) **Reductio ad absurdum**, or reduction to the absurd, is a method of proof based upon the principle that a proposition is false. The ensuing argument based upon the falsehood then leads to a contradiction.

(b) If $p \Rightarrow q$ and $q \Rightarrow r$ then $p \Rightarrow r$. The chain of implication leads to the **method of induction**, which involves a sequence of propositions $\{p_n\}$ with the property that $p_n \Rightarrow p_{n+1}$, for all n. If the specific proposition p_1 is true then so is p_2, hence all of them are true. This method of proof establishes the truth of a formula dependent on n, e.g.

$$1 + 2 + \cdots + n = \frac{1}{2}n(n+1)$$

(c) $\bar{p}$ is the **negation** of proposition p, i.e. if p is true then $\bar{p}$ is not true. For example, if $p\colon x = 5$ then $\bar{p}\colon x \neq 5$. To prove that a proposition is true we can prove that its negation is false. To prove that a proposition is false we can find a **counter-example**.

TRIGONOMETRY

1. In a right-angled triangle (Figure 16) one of the smaller angles is θ; we define the **trigonometric ratios** sine, cosine and tangent as

$$\sin\theta = \frac{\text{opposite}}{\text{hypotenuse}} \qquad \cos\theta = \frac{\text{adjacent}}{\text{hypotenuse}} \qquad \tan\theta = \frac{\text{opposite}}{\text{adjacent}}$$

Some useful results are

$$\sin(90° - \theta) = \cos\theta \qquad \cos(90° - \theta) = \sin\theta$$
$$\sin(180° - \theta) = \sin\theta \qquad \sin(360° - \theta) = \sin\theta$$

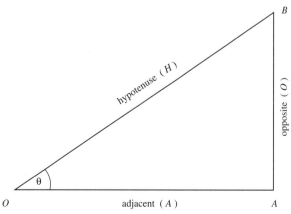

Figure 16 Defining the trigonometric ratios

Note the results in radians

$$\sin\left(\frac{\pi}{2} - \theta\right) = \cos\theta \qquad \cos\left(\frac{\pi}{2} - \theta\right) = \sin\theta$$

Some special values are shown in the following table.

θ	0	30°($\pi/6$)	45°($\pi/4$)	60°($\pi/3$)	90°($\pi/2$)
$\sin\theta$	0	$1/2$	$1/\sqrt{2}$	$\sqrt{3}/2$	1
$\cos\theta$	1	$\sqrt{3}/2$	$1/\sqrt{2}$	$1/2$	0
$\tan\theta$	0	$1/\sqrt{3}$	1	$\sqrt{3}$	–

2. The ratios of angles in quadrants other than the first can be found by noting that $\sin(180° - \theta)$, $\sin(180° + \theta)$, $\sin(360° - \theta)$ are all of the form $k\sin\theta$ where k is either 1 or -1. Similar results apply for the other ratios. The value of k can be found from the **CAST rule** (Figure 17). For example, $\cos 330° = \cos(360° - 30°) = \cos 30°$ and $\sin 240° = \sin(180° + 60°) = -\sin 60°$.

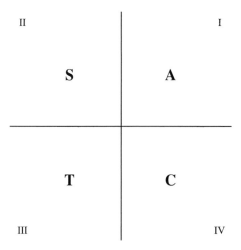

Figure 17 CAST rule

3. The **fundamental trigonometric identity** is $\sin^2\theta + \cos^2\theta \equiv 1$.

4. In any triangle ABC whose sides are a, b and c, opposite angles A, B and C, the following results are true:

$$\frac{a}{\sin A} = \frac{b}{\sin B} = \frac{c}{\sin C} \qquad \text{(sine rule)}$$

$$a^2 = b^2 + c^2 - 2bc\cos A$$
$$b^2 = c^2 + a^2 - 2ca\cos B \qquad \text{(cosine rule)}$$
$$c^2 = a^2 + b^2 - 2ab\cos C$$

The **area** of a triangle is

$$\frac{1}{2}bc\sin A = \frac{1}{2}ca\sin B = \frac{1}{2}ab\sin A$$

5. The **solution of a triangle** can be accomplished if we know (i) three sides, (ii) two sides and the included angle, (iii) two angles and one side. The ambiguous case can occur if two sides and a non-included angle are given.

For example, suppose we know $b = 6.2$, $c = 4.8$, $B = 30°$ (Figure 18). Then we use the sine rule

$$\frac{\sin C}{c} = \frac{\sin B}{b} \quad \text{i.e.} \quad \frac{\sin C}{4.8} = \frac{0.5}{6.2}$$

Since $c < b$ then $C < B$, so C is an acute angle. Therefore $C = 22.6°$ and we could find the third angle as $A = 180° - 30° - 22.6° = 127.4°$. The third side could be found from the sine rule or via the cosine rule, $a^2 = b^2 + c^2 - 2bc\cos A = 98.64$; hence $a = 9.93$.

Note that if b had been greater than c than we might have found two triangles satisfying the given data, one with C acute and one with C obtuse, both having the same sine.

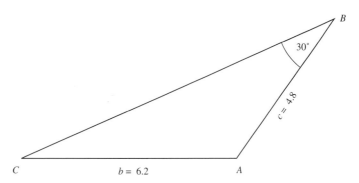

Figure 18 Using the sine and cosine rules

6. A **periodic** function has a graph that repeats its pattern regularly; the **period** is the smallest interval over which the pattern repeats.

The sine, cosine and tangent of an angle of any magnitude can be found by using the periodic nature of these functions. Since $\sin(360° + \theta) = \sin\theta$ for any angle θ, we deduce that in radians the result is $\sin(2\pi + \theta) = \sin\theta$ and we see that sine has a period of 2π. Similarly, $\cos(2\pi + \theta) = \cos\theta$ and cosine has a period of 2π. However, $\tan(\pi + \theta) = \tan\theta$ so tangent has a period of π.

The graphs of $\sin x$ and $\cos x$ are shown in Figure 19 and the graph of $\tan x$ is shown in Figure 20.

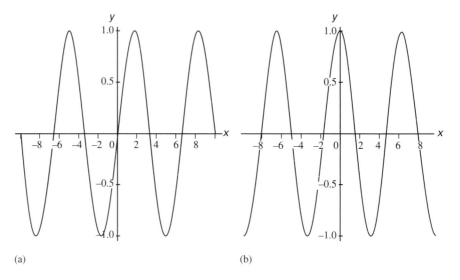

(a)   (b)

Figure 19 Graphs of (a) $\sin x$ and (b) $\cos x$

7. The general solution of $\tan\theta = p$ is $\theta_0 \pm n\pi$ and the general solution of $\cos\theta = p$, $|p| \leq 1$ is $2n\pi \pm \theta_0$ where θ_0 is the value obtained from a calculator. This is known as the **principal value**.

Example   Find the solutions in $0° \leq x \leq 360°$ of (a) $\tan x = 1$, (b) $\tan x = -1$, (c) $\cos x = 0.5$, (d) $\cos x = -0.5$, (e) $\sin x = 0.5$ and (f) $\sin x = -0.5$.

Solution

(a) A calculator gives the value $x = 45°$; the second solution is $x = 180° + 45° = 225°$.

(b) A calculator gives the value $x = -45°$; the first solution is $x = 180° - 45° = 135°$ and the second solution is $x = 180° + 135° = 315°$.

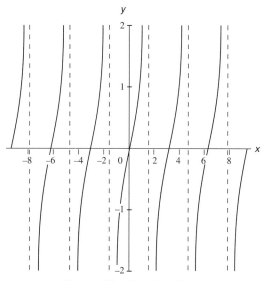

Figure 20 Graph of tan x

(c) A calculator gives the value $x = 60°$; the second solution is $x = 360° - 60° = 300°$.

(d) A calculator gives the value $x = 120°$; the second solution is $x = 360° - 120° = 240°$.

(e) A calculator gives the value $x = 30°$; the second solution is $x = 180° - 30° = 150°$.

(f) A calculator gives the value $x = -30°$; the first solution is $x = 180° + 30° = 210°$ and the second solution is $x = 360° - 30° = 330°$.

9. The **angle between a line and a plane** is the angle between the line and its projection on the plane.

   Two planes meet in a **common line** or are parallel. The angle between them is equal to the angle between the normals to the planes.

FURTHER ALGEBRA

1. An **arithmetic progression** (AP) is a sequence in which each term is obtained from its predecessor by adding a constant amount, the **common difference**. It takes the form $a, a + d, a + 2d, \ldots$, where a is the first term and d is the common difference.

   The sum S of an arithmetic progression (AP) can take one of the forms

   $$S = \frac{n}{2}(2a + (n - 1)d), \qquad S = \frac{n}{2}(a + l)$$

   where n is the number of terms, and l is the last term.

2. A **geometric progression** (GP) is a sequence in which each term is obtained from its predecessor by multiplying it by a constant amount, the **common ratio**. It takes the form $a, ar, ar^2, \ldots$, where a is the first term and r is the common ratio.

The sum S of a geometric progression takes the form

$$S = \frac{a(1 - r^n)}{1 - r} = \frac{a(r^n - 1)}{r - 1} \qquad (r \neq 1)$$

or $\qquad S = na \qquad (r = 1)$

where n is the number of terms. If $|r| < 1$, $r^n \to 0$ as $n \to \infty$ and $S \to \dfrac{a}{1 - r}$

3. For two numbers a and b, the following definitions apply:

$$\text{Arithmetic mean} = \frac{1}{2}(a + b)$$

$$\text{Geometric mean} = \sqrt{ab}$$

$$\text{Harmonic mean} = \frac{1}{\dfrac{1}{a} + \dfrac{1}{b}}$$

If $a, b > 0$ the geometric mean is never greater than the arithmetic mean.

4. The Greek capital S, or $\sum$ (called sigma) is used as a shorthand for summation, i.e.

$$\sum_{r=1}^{n} t_r = t_1 + t_2 + \cdots + t_n$$

For example $\quad \displaystyle\sum_{r=1}^{4} r^5 = 1^5 + 2^5 + 3^5 + 4^5$

A constant multiplier k can be taken inside or outside the $\sum$ sign, i.e.

$$\sum_{r=1}^{n} kt_r = kt_1 + \cdots + kt_n = k(t_1 + \cdots + t_n) = k\sum_{r=1}^{n} t_r$$

If the sequence is infinite then

$$\sum_{r=1}^{\infty} t_r = t_1 + t_2 + \cdots$$

Some special sums are

$$\sum_{r=1}^{n} r = 1 + 2 + \cdots + n = \frac{n}{2}(n+1)$$

$$\sum_{r=0}^{n-1} a^r = \frac{1 - a^n}{1 - a} = \frac{a^n - 1}{a - 1} \qquad (a \neq 1)$$

If $|a| < 1$ then the sum to infinity is $\dfrac{1}{1 - a}$

5. A **polynomial of degree** n in the variable x, written $p(x)$, has the form

$$a_0 + a_1 x + a_2 x^2 + \cdots + a_n x^n$$

A polynomial can be evaluated by **nested multiplication**. For example, $5x^3 - 2x^2 + 3x + 4 \equiv [(5x - 2)x + 3]x + 4$ and we evaluate the result from the inside outwards.

6. The equation

$$y = L(x) = \frac{(x - x_2)}{(x_1 - x_2)} y_1 + \frac{(x - x_1)}{(x_2 - x_1)} y_2$$

represents the straight line through the points (x_1, y_1) and (x_2, y_2).
   The line $y = L(x)$ is the line of linear interpolation between the points (x_1, y_1) and (x_2, y_2).

7. If a polynomial is written as $p(x) \equiv (x - a)q(x) + R$ then $q(x)$ is the **quotient** and R is the remainder. If we put $x = a$ then the remainder when $p(x)$ is divided by $(x - a)$ is $R = p(a)$. If the remainder is zero then $(x - a)$ is a linear **factor** of the polynomial.

8. A **rational function** is the ratio of two polynomials, $r(x) = \dfrac{p(x)}{q(x)}$. Assume that common factors have been cancelled top and bottom. Where $p(x) = 0$, $r(x)$ has a **zero**; where $q(x) = 0$, $r(x)$ has a **vertical asymptote**.
   If both $p(x)$ and $q(x)$ are linear then $r(x)$ has one zero and one vertical asymptote. The graph of $r(x) = \dfrac{3x - 5}{x + 3}$ is shown in Figure 21, as $|x| \to \infty$; $r(x) \to 3$.

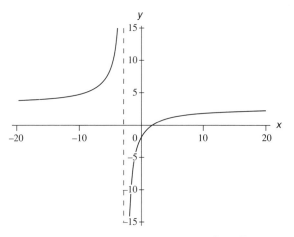

Figure 21   Graph of $r(x) = \dfrac{3x - 5}{x + 3}$

COORDINATE GEOMETRY

1. The **distance between the points** (x_1, y_1) and (x_2, y_2) is given by

$$\sqrt{(x_1 - x_2)^2 + (y_1 - y_2)^2}$$

The point which **divides the line segment joining the points** (x_1, y_1) and (x_2, y_2) in the ratio $m:n$ is

$$\left(\frac{mx_2 + nx_1}{m + n}, \frac{my_2 + ny_1}{m + n} \right)$$

The coordinates of the midpoint of the line segment are

$$\left(\frac{1}{2}(x_1 + x_2), \frac{1}{2}(y_1 + y_2) \right)$$

2. The **angle** θ **between the lines** $y = m_1x + c_1$ and $y = m_2x + c_2$ with gradients m_1 and m_2, respectively, is given by

$$\tan \theta = \frac{m_2 - m_1}{1 + m_2 m_1}$$

If $m_1 = m_2$ then the lines are **parallel**. If $1 + m_1 m_2 = 0$, i.e. $m_1 m_2 = -1$ then the lines are **perpendicular** to each other. For example, if the lines are $y = 2x + 5$ and $3x + y = 4$ then $m_1 = 2$, $m_2 = -\dfrac{1}{3}$ and $\tan \theta = -1$. Then the angle between the lines is $135°$; the acute angle is $45°$.

3. The **distance of the point** (x_1, y_1) **from the line** $ax + by + c = 0$ is given by

$$d = \frac{|ax_1 + by_1 + c|}{\sqrt{a^2 + b^2}}$$

If $ax_1 + by_1 + c = 0$ then $d = 0$, so the point (x_1, y_1) lies on the line.

4. The **area of the triangle** with vertices (x_1, y_1), (x_2, y_2) and (x_3, y_3) is given by the absolute value of

$$\frac{1}{2}\{x_1(y_2 - y_3) + x_2(y_3 - y_1) + x_3(y_1 - y_2)\}$$

5. A **locus** is a set of points which satisfy a given condition. For example, the set of points which are equidistant from two given points form the perpendicular bisector of the line segment joining the given points.

   The equation $(x - x_0)^2 + (y - y_0)^2 = a^2$. represents a **circle** centre (x_0, y_0) and radius a. The equation $x^2 + y^2 + 2gx + 2fy + c = 0$ is the **general equation of a circle**; it has centre $(-g, -f)$ and radius $(g^2 + f^2 - c)^{1/2}$.

   A **parabola** is the locus of points which are equidistant from a **focus**, a fixed point, and a **directrix**, a given line. If the focus is $(a, 0)$ and the y-axis is the directrix then the **standard equation** of the parabola is $y^2 = 4ax$.

6. A curve in the x–y plane may be specified by a pair of **equations** expressing x and y in terms of a third variable, the **parameter**. For example, the circle radius a and centre the origin can be represented as $x = a \cos \theta$, $y = a \sin \theta$ where the parameter θ satisfies the condition $0 \leq \theta < 2\pi$.

7. A point can be represented in **polar coordinates** (r, θ) where $x = r \cos \theta$, $y = r \sin \theta$. Note the result $x^2 + y^2 = r^2$.

   The polar form $r = 4 \cos \theta$ can be multiplied by r to give the equation $r^2 = 4r \cos \theta$. Converting to Cartesian form we obtain $(x - 2)^2 + y^2 = 4$, which is the equation of a circle with centre $(2, 0)$ and radius 2.

8. The **equation of a straight line in three dimensions** which passes through the point (x_1, y_1, z_1) with direction ratios (a, b, c) is

$$\frac{x - x_1}{a} = \frac{y - y_1}{b} = \frac{z - z_1}{c}$$

The equation

$$\frac{x-1}{2} = \frac{y-2}{3} = \frac{z-3}{1}$$

can be replaced by the component equations $x = 1 + 2t, y = 2 + 3t, z = 3 + t$ by putting the fractions equal to the parameter t. When $t = -3$ the line passes through the plane $z = 0$ at the point $(-5, -7, 0)$.

Two lines in three dimensions can be parallel, skew or intersecting. The lines

$$\frac{x-1}{2} = \frac{y+1}{1} = \frac{z}{-2} \quad \text{and} \quad \frac{x-1}{-3} = \frac{y-2}{0} = \frac{z-2}{4}$$

can be rewritten as

$$x = 1 + 2t, y = -1 + t, z = -2t \quad \text{and} \quad x = 1 - 3u, y = 2, z = 2 + 4u$$

with parameters t and u. The zero in the denominator indicates that the second line lies in the plane $y = 2$. When $t = 3, u = -2$ the equations give the same coordinates $(7,2,6)$, which is the point where the lines intersect. In three dimensions it is rare for two lines to meet.

9. The **general equation of a plane** is $ax + by + cz = d$. The **normal**, i.e. perpendicular, has direction ratios (a, b, c). If $d = 0$ the origin lies in the plane. If $b = c = d = 0$ and $a \neq 0$ then the plane is $x = 0$, i.e. the y–z plane.

The straight line

$$\frac{x-1}{2} = \frac{y+1}{1} = \frac{z}{-2} = t$$

meets the plane $x + y + 2z = 2$ where $(1 + 2t) + (-1 + t) + 2(-2t) = 2$, so $t = -2$. The point of intersection is therefore $(-3, -3, 4)$.

FUNCTIONS

1. A **function** is a rule which associates each member of one set, the **domain**, with a unique member of a second set, the **co-domain**. The set of values taken by the function is its **range**. We write $y = f(x)$ where x is the **independent variable** and y is the **dependent variable**.

For example, the function $f : x \mapsto 2x + 3$ doubles the input x and adds 3 to it. An alternative notation is $f(x) = 2x + 3$. When $x = 2, y = f(2) = 7$. Both the domain and the range are the set of all real numbers, denoted by $\mathbb{R}$. Some standard functions are shown in the following table.

Function	Domain	Range
$\sqrt{x}$	$x \geq 0$	$y \geq 0$
$\sin x$	$\mathbb{R}$	$-1 \leq y \leq 1$
$\dfrac{1}{x}$	$x \neq 0$	$y \neq 0$
$|x|$	$\mathbb{R}$	$y \geq 0$

Sometimes a function requires a piecewise approach:

$$f(x) = \begin{cases} x & x \leq 3 \\ 3 & 3 < x \leq 7 \\ 10 - x & x > 7 \end{cases}$$

2. The function $y = f(x)$ may be **transformed** as follows:

Translation $f(x - a)$ represents a move to the right by a
$f(x) + a$ represents a move upwards by a

Reflection $-f(x)$ represents a reflection in the x-axis
$f(-x)$ represents a reflection in the y-axis

Scaling $af(x)$ represents a scaling in the y-direction,
an enlargement if $a > 1$, a contraction if $0 < a < 1$
$f(ax)$ represents a scaling in the x-direction,
an enlargement if $0 < a < 1$, a contraction if $a > 1$

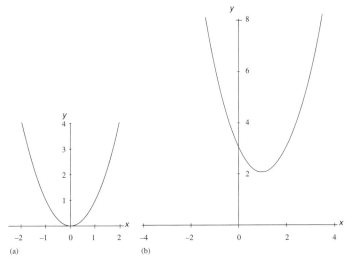

Figure 22   Graphs of (a) $y = x^2$ and (b) $y = (x - 1)^2 + 2$

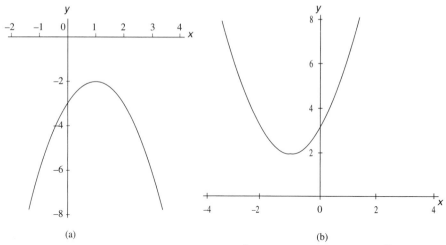

(a)   (b)

Figure 23   Graphs of (a) $y = -(x-1)^2 - 2$ and (b) $y = (-x-1)^2 + 2$

Figure 22(a) shows the graph of $y = x^2$ and Figure 22(b) shows the graph of $y = (x-1)^2 + 2$, indicating a translation of 1 unit to the right and a translation of 2 units upwards. The translations can be carried out in either order.

Figure 23(a) shows the graph of $y = -(x-1)^2 - 2$; it is a reflection of Figure 22(b) in the x-axis. Figure 23(b) shows the graph of $y = (-x-1)^2 + 2$; it is a reflection of Figure 22(b) in the y-axis.

Figure 24(a) shows the graph of $y = (x-1)^2$; Figure 24(b) shows the graph of $y = 2(x-1)^2$, a vertical stretching by a factor of 2. Figure 24(c) shows the graph of $y = (2x-1)^2$, a horizontal contraction by a factor of 2.

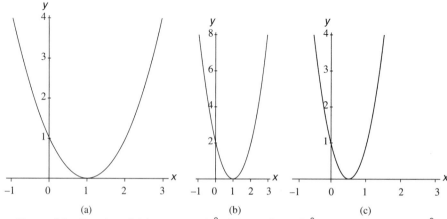

(a)   (b)   (c)

Figure 24   Graphs of (a) $y = (x-1)^2$, (b) $y = 2(x-1)^2$ and (c) $y = (2x-1)^2$

A function which satisfies $f(-x) = f(x)$, i.e. is its own reflection in the y-axis is said to be **even**, e.g. $f(x) = x^2$, $f(x) = \cos x$. Likewise a function which satisfies $f(-x) = -f(x)$, $f(0) = 0$, i.e. an upside-down reflection in the y-axis, is said to be **odd**, e.g. $f(x) = x$, $f(x) = \sin x$.

3. The output of one function may become the input of a second function; this is known as **function of a function**. The notation $f(g(x))$ implies that the function f is applied to the output of the function g. The domain of the composite function $f(g(x))$ will depend on the domains of the two functions. If f is the square root function then g can send only non-negative outputs, so we must limit the inputs to g.

   The order of applying the functions usually makes a difference. Let $g(x) = x^2$ and $f(x) = x + 3$, then $f(g(x)) = x^2 + 3$ but $g(f(x)) = (x + 3)^2$.

4. The **inverse** $f^{-1}(x)$ reverses the action of the function $f(x)$; its graph is obtained by reflecting the graph of $f(x)$ in the line $y = x$ (Figure 25). The inverse $f^{-1}(x)$ is not always a function, but it will be a function if $f(x)$ is such that no two inputs are mapped to the same output. The domain of an inverse function is the range of the original function and vice versa.

   If $f(x) = 2x + 3$ then $f^{-1}(x) = \dfrac{x - 3}{2}$, also a function. To obtain this formula, put $y = 2x + 3$. Rearranging gives $x = \dfrac{y - 3}{2}$, then we interchange x and y.

   To obtain the inverse function to $f(x) = x^2$ it is necessary to restrict the domain to $x \geq 0$, then $f^{-1}(x) = \sqrt{x}$. On the other hand, if $f(x) = x^3$ then $f^{-1}(x) = x^{1/3}$ and both functions have as domain and range the set of all real numbers.

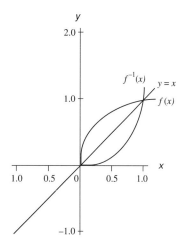

Figure 25 Inverse of a function

5. The **exponential function** is $f(x) = e^x$, where $e \approx 2.718$. It is sometimes written $\exp(x)$. Figure 26 shows the graphs of $y = e^x$ and $y = e^{-x}$. The domain of each function is the set of all real numbers and the range of each function is the set of all non-negative real numbers.

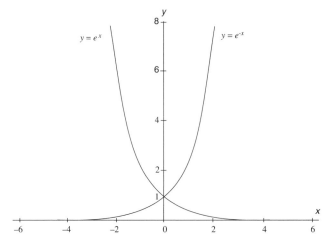

Figure 26 Graphs of $y = e^x$ and $y = e^{-x}$

6. The inverse of the exponential function is the **natural logarithmic function**, $\ln x$. Its domain is all positive real numbers and its range is all real numbers. The graph of $y = \ln x$ is shown in Figure 27.

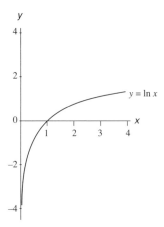

Figure 27 Graph of $y = \ln x$

From the laws of indices, if $y_1 = e^{x_1}, y_2 = e^{x_2}$ then $y_1 y_2 = e^{x_1 + x_2}$. It follows that $x_1 = \ln y_1, x_2 = \ln y_2$ and therefore $\ln(y_1 y_2) = x_1 + x_2 = \ln y_1 + \ln y_2$.
Other laws for the logarithmic function can be derived in a similar way.

7. The domains of the trigonometric functions are restricted to an interval of length π in order that their inverses are functions. The function $\cos x$ is restricted to the domain $(0, \pi)$; the functions $\sin x$ and $\tan x$ are restricted to the domain $\left(-\dfrac{\pi}{2}, \dfrac{\pi}{2} \right)$. The **inverse trigonometric functions** are written $\cos^{-1} x$, $\sin^{-1} x$ and $\tan^{-1} x$. The function $\sin^{-1} x$ has domain $-1 \leq x \leq 1$ and is shown in Figure 28.

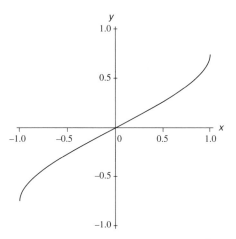

Figure 28   Graph of $y = \sin^{-1} x$

DIFFERENTIATION

1. A function $f(x)$ is **differentiated** to give the **derived function** $f'(x)$. The value $f'(a)$ is the **derivative** of $f(x)$ at $x = a$. The derivative measures the instantaneous rate of change of the function. In formal terms

$$f'(a) = \lim_{h \to 0} \frac{f(a+h) - f(a)}{h}.$$

Consider the function $f(x) = x^2$ over the interval $[a, a + h]$, i.e. $a \leq x \leq a + h$. The **average rate of change** of the function over the interval is

$$\frac{(a + h)^2 - a^2}{(a + h) - a} = \frac{2ah + h^2}{h} = 2a + h$$

As h tends to zero this average rate tends to the limit $2a$. This is the derivative of the function at $x = a$. Since the result is true for any a, we say that the derived function of $f(x) = x^2$ is $f'(x) = 2x$.

An alternative notation is to replace a by x, h by δx, y by $f(x)$ and $y + \delta y$ by $f(a + h)$. Then

$$\frac{dy}{dx} = \lim_{\delta x \to 0} \frac{f(x + \delta x) - f(x)}{\delta x}$$

2. **Some derived functions**

$f(x)$	x^n	e^{kx}	$\sin kx$	$\cos kx$	$\ln x$
$f'(x)$	nx^{n-1}	ke^{kx}	$k \cos kx$	$-k \sin kx$	$\dfrac{1}{x}$

3. To differentiate a **linear combination** of functions we use the result

$$\frac{d}{dx}\{\alpha f(x) + \beta g(x)\} = \alpha f'(x) + \beta g'(x)$$

For example

$$\frac{d}{dx}(6x^{2/3} + 11x^{-7}) = 4x^{-1/3} - 77x^{-2}$$

Finite polynomial $p_n(x) = a_0 + a_1 x + a_2 x^2 + \cdots + a_n x^n$ may be differentiated to give

$$p'_n(x) = a_1 + 2a_2 x + \cdots + na_n x^{n-1}$$

4. In kinematics, if s is the displacement of an object and v is its velocity then

$$v = \frac{ds}{dt}$$

For a projectile moving vertically under gravity, we have

$$s = u - \frac{1}{2}gt^2 \qquad \frac{ds}{dt} = u - gt$$

The object is stationary when $v = 0$, i.e. when $t = \dfrac{u}{g}$

5. The **gradient of the curve** $y = f(x)$ is given by

$$\frac{dy}{dx} = f'(x)$$

The **equation of the tangent** to the curve $y = f(x)$ at the point (x_0, y_0) is

$$y - y_0 = m(x - x_0)$$

The **equation of the normal** to the curve at that point is

$$y - y_0 = -\frac{1}{m}(x - x_0)$$

For example, we find the equations of the tangent and normal to the curve $y = x^2 - 4x + 3$ at the point $(4, 3)$. First, $\frac{dy}{dx} = 2x - 4$ and at $x = 4$ this has value 4.

The equation of the tangent is

$$y - 3 = 4(x - 4) \quad \text{or} \quad y = 4x - 13$$

and the equation of the normal is

$$y - 3 = -\frac{1}{4}(x - 4) \quad \text{or} \quad 4y + x = 16$$

6. The function $f(x)$ is **increasing** where $f'(x) > 0$; $f(x)$ is **decreasing** where $f'(x) < 0$. Where $f'(x) = 0$ the function has a **stationary point**.

The function has a **local minimum** at $x = a$ if $f(x) \geq f(a)$ for all x near a. The function has a **local maximum** at $x = a$ if $f(x) \leq f(a)$ for all x near a.

The stationary point is a local minimum at $x = a$ if $f'(x) < 0$ for $x < a$ and $f'(x) > 0$ for $x > a$. The stationary point is a local maximum at $x = a$ if $f'(x) > 0$ for $x < a$ and $f'(x) < 0$ for $x > a$. If $f'(x)$ has the same sign on both sides of $x = a$ the stationary point is a **point of inflection**. Local maxima and minima are called **turning points** of the function.

For example, if

$$y = x^4 - 6x^2 + 8x + 9 \quad \text{then} \quad \frac{dy}{dx} = 4x^3 - 12x + 8 \equiv 4(x - 1)^2(x + 2)$$

$$\frac{dy}{dx} = 0 \quad \text{where} \quad x = -2 \quad \text{or} \quad x = 1$$

We investigate the values of $\dfrac{dy}{dx}$ at $x = -3, 0, 2$.

At $x = -3$ $\qquad \dfrac{dy}{dx} < 0$

At $x = 0$ $\qquad \dfrac{dy}{dx} > 0$

At $x = 2$ $\qquad \dfrac{dy}{dx} > 0$

Hence the function has a local minimum at $x = -2$ and a horizontal point of inflection at $x = 1$. The graph is shown in Figure 29.

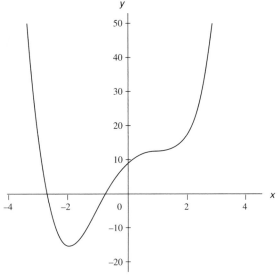

Figure 29 Graph of $y = x^4 - 6x^2 + 8x + 9$

7. The **second derivative** of a function is the derivative of the first derivative, written

$$f''(x) \quad \text{or} \quad \frac{d^2y}{dx^2}$$

The **second derivative test** for a stationary point at $x = a$ is as follows:

If $f''(a) > 0$ the point is a local minimum

If $f''(a) < 0$ the point is a local maximum

If $f''(a) = 0$ we need further investigation.

For the function $y = x^4 - 6x^2 + 8x + 9$, $\dfrac{d^2y}{dx^2} = 12(x^2 - 1)$

At $x = -2$ $f''(x) = \dfrac{d^2y}{dx^2} > 0$ (a local minimum)

At $x = 1$ $f''(x) = 0$ (a horizontal point of inflection)

8. If $f''(a) = 0$ then when $f'(x) \neq 0$ the function has a **point of inflection** at $x = a$; if $f'(x) = 0$ we need to investigate further, as we have said. For the function $y = x^4 - 6x^2 + 8x + 9$, $\dfrac{d^2y}{dx^2} = 0$ when $x = 1, x = -1$. The first of these we know and the second is a point of inflection with a non-zero gradient.

INTEGRATION

1. **Integration** is the reverse of differentiation. **Indefinite integration** produces a function, incorporating an **arbitrary constant**; **definite integration** produces a numerical value.

    If $f(x) = 2x$ then the indefinite integral is $F(x) = x^2 + C$ where C is an arbitrary **constant of integration**. We write $F(x) = \int f(x)dx$. An alternative notation: if $\dfrac{dy}{dx} = 2x$ then $y = x^2 + C$.

2. Rules for integration follow the rules for differentiation; for example

    $$\int \{\alpha f(x) + \beta g(x)\}dx = \alpha \int f(x)dx + \beta \int g(x)dx$$

    Hence $\int \{3 \cos x - 2x\}dx = 3 \sin x - x^2 + C$.

3. The **area under the curve** $y = f(x)$ can be calculated as a definite integral; areas under the x-axis are assigned a negative value. The **fundamental theorem of calculus** states that the area under the curve $y = f(x)$ between the ordinates $x = a$ and $x = b$ is given by

    $$\int_a^b f(x)dx = F(b) - F(a)$$

    where $F'(x) = f(x)$. Hence $y = 3x^2$, so $\int_1^2 3x^2 \, dx = [x^3]_1^2 = 8 - 1 = 7$

4. Here are some useful results:

(a) If $a \le c \le b$ then $\int_a^b f(x)dx = \int_a^c f(x)dx + \int_c^b f(x)dx$

Hence $\displaystyle\int_0^\pi \cos x \, dx = [\sin x]_0^\pi = 0 - 0 = 0$

but $\displaystyle\int_0^{\pi/2} \cos x \, dx = 1$ and $\displaystyle\int_{\pi/2}^\pi \cos x \, dx = -1$

(b) If $m \le f(x) \le M$ then $m(b-a) \le \int_a^b f(x)dx = M(b-a)$

(c) As shown in Figure 30, the curves $y = x^2$ and $y = x^3$ meet at the points $(0,0)$ and $(1,1)$. The area enclosed by the curves is

$$\int_0^1 (x^2 - x^3)dx = \frac{1}{12}$$

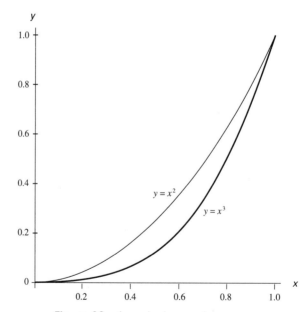

Figure 30 Area between two curves

5. The **mean value of a function** in the interval $a \leq x \leq b$ is

$$\frac{1}{b-a} \int_a^b f(x)dx$$

The mean value is the height of a rectangle of base $(b-a)$ and area $\int_a^b f(x)dx$.
The **root mean square of a function** in the interval $a \leq x \leq b$ is

$$\sqrt{\frac{1}{b-a} \int_a^b \{f(x)\}^2 \, dx}$$

6. The **volume of a solid of revolution** formed by rotating the graph of $y = f(x)$ in the interval $a \leq x \leq b$ about the x-axis through an angle 2π is $\pi \int_a^b y^2 \, dx$. The formula is derived by dividing the volume into thin discs of radius y and thickness δx; the total volume, $\sum \pi y^2 \, \delta x$ tends to $\pi \int_a^b y^2 \, dx$ as $\delta x \to 0$

7. The **moment of the plane area** bounded by the curve $y = f(x)$, the x-axis and the ordinates $x = a, x = b$ about the y-axis is $\int_a^b xy \, dx$ and about the x-axis is $\frac{1}{2} \int_a^b y^2 \, dx$

8. The **centre of gravity of the plane area** bounded by the curve $y = f(x)$, the x-axis and the ordinates $x = a, x = b$ has coordinates $(\bar{x}, \bar{y})$ where

$$\bar{x} = \frac{\displaystyle\int_a^b xy \, dx}{\displaystyle\int_a^b f(x)dx}, \qquad \bar{y} = \frac{\dfrac{1}{2}\displaystyle\int_a^b y^2 \, dx}{\displaystyle\int_a^b f(x)dx}$$

9. The **centre of gravity of a solid of revolution** lies on the axis of revolution, hence $\bar{y} = 0$ and

$$\bar{x} = \frac{\displaystyle\int_a^b xy^2 \, dx}{\displaystyle\int_a^b y^2 \, dx}$$

10. According to the **trapezium rule**, the integral $\int_a^b f(x)dx$ is approximated by

$$\frac{h}{2}(f_1 + 2f_2 + \cdots + 2f_{n-1} + f_n) \quad \text{where} \quad h = x_i - x_{i-1} \quad \text{and} \quad f_i = f(x_i)$$

For example, if $a = 0, b = 0.4, n = 4, h = 0.1$ then $\int_0^{0.4} x^2 \, dx$ is approximated by

$$\frac{1}{2} \times 0.1 \times (0 + 2 \times 0.01 + 2 \times 0.04 + 2 \times 0.09 + 0.16) = 0.022$$

This compares favourably with the analytical value of 0.0213 (4 d.p.).

1 TRIGONOMETRY

Introduction

Many phenomena in science and engineering are periodic in nature and they can be modelled by trigonometric functions. Problems can be simplified using trigonometric identities to simplify expressions, so that equations can be solved more readily.

Objectives

After working through this chapter you should be able to

- Use the reciprocal functions secant, cosecant and cotangent
- Simplify expressions by using identities based on Pythagoras' theorem
- Recognise and use the addition identities
- Recognise and use the double-angle identities
- Know the derived functions and indefinite integrals of the functions $\sin kx$ and $\cos kx$
- Know and apply the result $\lim\limits_{x \to 0} \dfrac{\sin x}{x} = 1$
- Rewrite the expression $a \cos \theta + b \sin \theta$ in the form $R \sin(\theta + \phi)$
- Define the frequency, amplitude and period of a wave
- Calculate wave quantities from the trigonometric function which models the wave
- Explain the phenomenon of beats.

1.1 SEC, COSEC AND COT

In Foundation Mathematics we defined tangent, sine and cosine in terms of the right-angled triangle of Figure 1.1.

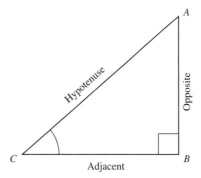

Figure 1.1 Right-angled triangle

In the same way we define the three reciprocal functions

$$\sec C = \frac{\text{hypotenuse}}{\text{adjacent}} = \frac{CA}{CB} = \frac{1}{\cos C}$$

$$\operatorname{cosec} C = \frac{\text{hypotenuse}}{\text{opposite}} = \frac{CA}{AB} = \frac{1}{\sin C}$$

$$\cot C = \frac{\text{adjacent}}{\text{opposite}} = \frac{CB}{AB} = \frac{1}{\tan C}$$

where **sec**, **cosec** and **cot** are abbreviated forms for **secant**, **cosecant** and **cotangent** respectively.

The reciprocal trigonometric functions are in common use and we need to be thoroughly familiar with them. They satisfy the obvious complementary rules, e.g.

$$\operatorname{cosec}(90° - \theta) = \sec \theta \qquad\qquad \sec(90° - \theta) = \operatorname{cosec} \theta$$

and equivalent identities as we shall see in Section 1.2. They also must follow the sign conventions of their parent functions in each of the four quadrants; for example, $\sec 120° = -\sec 60°$ in the same way that $\cos 120° = -\cos 60°$. We can display $\sin x$ and $\operatorname{cosec} x$ on the same graph (Figure 1.2); $\sin x$ is shown dashed.

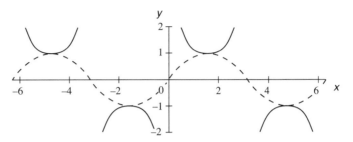

Figure 1.2 Graphs of sin x and cosec x

The function $\sec x = \dfrac{1}{\cos x}$ always has the same sign as $\cos x$ and is infinite when $\cos x = 0$, i.e. at all odd multiples of $x = \pi/2$. It also has maxima and minima at the same points, $(\pi, -1)$, etc., where $\cos x$ has minima and maxima. The graph of $y = \sec^{-1} x$ has a branched structure similar to $y = \sec x$, with the role of the x- and the y-axes interchanged. It has no maxima or minima but each branch is contained within an interval of width π between odd multiples of $\pi/2$ on the y-axis. Note that the depiction of $y = \sec^{-1} x$ in Figure 1.3 does not represent a function of x as such, because an infinite number of y's can be defined for any given x that satisfies $|x| \geq 1$.

To evaluate $\sec^{-1} x$ you must choose a suitable branch, usually the upper part of the principal branch. A further point to note is that cosec x, sec x and cot x, like sin x, cos x and tan x must be respectively odd, even and odd functions.

We can also use the principle of reflection in the line $y = x$ to depict $y = \sec x$ and $y = \sec^{-1} x$ on the same axes; the graph of $y = \sec^{-1} x$ is dashed in Figure 1.3.

As with the inverse of $y = \cos x$, we must restrict the domain of $y = \sec x$ to $0 \leq x < \pi$. This time we exclude $x = \dfrac{\pi}{2}$.

Examples

1. Write down the following, using your calculator as necessary.

(a) (i) sec 60°   (ii) cosec(−30°)   (iii) cot 216°

(b) (i) $\operatorname{cosec}\left(-\dfrac{\pi}{3}\right)$ (ii) $\sec \dfrac{19\pi}{6}$ (iii) $\cot\left(-\dfrac{45\pi}{4}\right)$

(c) $\sec^{-1} 2$, principal value

Solution

(a) (i) 2   (ii) −2   (iii) 1.3764

(b) (i) −1.1547   (ii) −1.1547   (iii) −1

(c) $\dfrac{\pi}{3}$

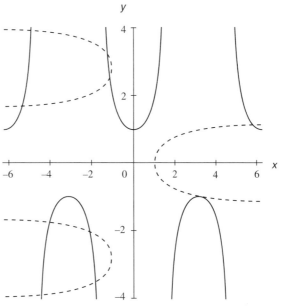

Figure 1.3 Graph of $y = \sec^{-1} x$

2. Prove that in any triangle ABC

$$\frac{a - b \cos C}{c - b \cos A} = \frac{\sin C}{\sin A} \quad \text{and} \quad \frac{a \sec C - b}{c \sec A - b} = \frac{\tan C}{\tan A}$$

where A, B, C and a, b, c have their usual meanings. Refer to Figure 1.4.
Now

$$c = AB = AD + DB = b \cos A + a \cos B$$

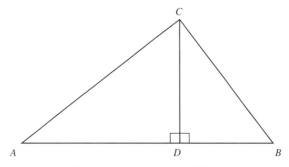

Figure 1.4 Triangle ABC

so that

$$c - b \cos A = a \cos B$$

Similarly, by drawing AE perpendicular to CB to meet CB at E, we have

$$a - b \cos C = c \cos B$$

so that

$$\frac{a - b \cos C}{c - b \cos A} = \frac{c}{a} = \frac{\sin C}{\sin A} \qquad \text{(by the sine rule)}$$

Dividing the numerators by $\cos C$ and the denominators by $\cos A$ gives

$$\frac{a \sec C - b}{c \sec A - b} = \frac{\tan C}{\tan A}$$

You should satisfy yourself this identity is true even if the foot of the perpendicular from C is outside AB.

3. Prove that the equation of the tangent to the circle $x^2 + y^2 = a^2$ at the point $(a \cos \theta, a \sin \theta)$ is given by

$$y + x \cot \theta = a \ \text{cosec} \ \theta$$

Referring to Figure 1.5, the normal at P has gradient $\dfrac{a \sin \theta}{a \cos \theta} = \tan \theta$. The tangent,

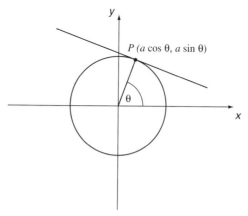

Figure 1.5 Tangent to a circle

being perpendicular to the normal, must have gradient $-\dfrac{1}{\tan\theta} = -\cot\theta$. Its equation is therefore

$$(y - a\sin\theta) = -\cot\theta(x - a\cos\theta)$$

i.e. $(y + x\cot\theta) = a(\sin\theta + \cos\theta\cot\theta)$

But

$$\sin\theta + \cos\theta\cot\theta = \sin\theta + \frac{\cos^2\theta}{\sin\theta}$$

$$= \frac{\sin^2\theta + \cos^2\theta}{\sin\theta} = \frac{1}{\sin\theta} = \operatorname{cosec}\theta$$

therefore the required equation follows.

4. Determine the general solution in radians of the equations

(a) $\sec x = 2$

(b) $\cot\left(2x - \dfrac{\pi}{4}\right) = \sqrt{3}$

Solution

(a) The given equation is equivalent to the equation $\cos x = \dfrac{1}{2}$ so that

$$x = 2n\pi \pm \frac{\pi}{3}; \; n = 0, \pm 1, \pm 2$$

The first positive solutions are

$$x = \frac{\pi}{3}, \frac{5\pi}{3}, \frac{7\pi}{3}, \frac{11\pi}{3}$$

(b) $\tan\left(2x - \dfrac{\pi}{4}\right) = \dfrac{1}{\sqrt{3}}$,

so $2x - \dfrac{\pi}{4} = n\pi + \dfrac{\pi}{6}$ (*n* integer)

Hence $2x = n\pi + \dfrac{5\pi}{12}$

$$\text{then} \quad x = \frac{n\pi}{2} + \frac{5\pi}{24}$$

■

Exercise 1.1

 1 Write down the following:

(a) (i) sec 30°   (ii) cosec 45°   (iii) cot 120°

     (iv) cot(−45°)   (v) sec 153°   (vi) cosec 217°

     (vii) cot 334°   (viii) cosec 495°   (ix) sec(−315°)

     (x) cosec(−508°)

(b) (i) $\operatorname{cosec} \dfrac{\pi}{3}$ (ii) $\sec \dfrac{5\pi}{6}$ (iii) $\cot \dfrac{11\pi}{4}$

     (iv) $\sec \dfrac{19\pi}{2}$ (v) $\cot\left(-\dfrac{6\pi}{5}\right)$ (vi) $\sec\left(-\dfrac{11\pi}{3}\right)$

2 On the same axes draw the following graphs.

(a) $y = \cos x$ (continuous)   (b) $y = \sec x$ (dashed)

3 On the same axes draw the following graphs:

(a) $y = \cot x$ (b) $y = \tan x$

Then draw the graph of $y = \cot x$, reflect it in the line $y = x$ and dash in the graph of $y = \cot^{-1} x$.

4 Draw the straight line whose polar equation is

$$r = 3 \sec\left(\theta - \frac{\pi}{6}\right)$$

5 Triangle ABC has incircle centre I, radius r. Prove that the area of the triangle is

$$r^2 \left(\cot \frac{A}{2} + \cot \frac{B}{2} + \cot \frac{C}{2} \right)$$

where $\cot \dfrac{A}{2}$ is the cotangent of the bisected angle at the vertex A, etc.

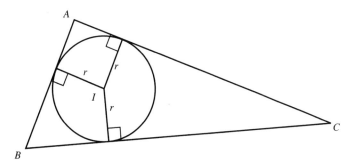

6 P is a point on the ellipse $\dfrac{x^2}{a^2} + \dfrac{y^2}{b^2} = 1$

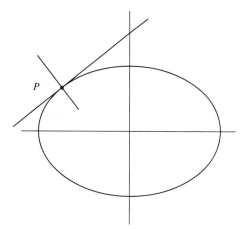

The coordinates of P can be written in the form $(a\cos\phi, b\sin\phi)$.

(a) Prove that the tangent at P has equation

$$\frac{x}{a}\cos\phi + \frac{y}{b}\sin\phi = 1,$$

given that its gradient is $\dfrac{-b}{a}\cot\phi$

(b) Prove that the equation of the normal is

$$ax\sec\phi - by\cos\mathrm{ec}\,\phi = a^2 - b^2$$

(c) For an ellipse with $a = 4$, $b = 3$, what is the equation of the normal when $\phi = \dfrac{\pi}{3}$

7 Determine the general solution in radians of the following equations:

(a) $\cos x = \dfrac{\sqrt{3}}{2}$ (b) $\tan x = -1$

(c) $\cot x = \sqrt{3}$ (d) $\cos\left(2x - \dfrac{\pi}{2}\right) = \dfrac{1}{2}$

(e) $\cot\left(3x + \dfrac{\pi}{4}\right) = -1$ (f) $\operatorname{cosec}\left(4x - \dfrac{2\pi}{3}\right) = -\dfrac{2}{\sqrt{3}}$

8 Use the cosine rule to prove that in any triangle ABC, with the usual notation

$$\frac{\cos A}{a} + \frac{\cos B}{b} + \frac{\cos C}{c} = \frac{a^2 + b^2 + c^2}{2abc}$$

9 Prove that the functions $\operatorname{cosec} x$ and $\cot x$ are odd, and the function $\sec x$ is even.

10* Show that $\cot A + \cot B + \cot C = \cot A \cot B \cot C$ whenever $A + B + C = \dfrac{\pi}{2}$

1.2 IDENTITIES AND EQUATIONS

As we have seen, an algebraic identity must hold all the time, whereas an algebraic equation is satisfied only for certain specific values which can be determined. Trigonometric identities and equations are no exception, though trigonometric equations often have an infinite number of possible solutions.

Pythagoras' theorem is

$$\sin^2 \theta + \cos^2 \theta \equiv 1 \tag{1.1}$$

and we know that

$$\tan \theta \equiv \frac{\sin \theta}{\cos \theta} \tag{1.2}$$

Dividing identity (1.1) by $\sin^2 \theta$ and $\cos^2 \theta$ in turn gives the further identities

$$1 + \operatorname{cosec}^2\theta \equiv \cot^2 \theta \tag{1.3}$$

and $$1 + \tan^2 \theta \equiv \sec^2 \theta \tag{1.4}$$

whereas (1.2) gives

$$\cot\theta = \frac{\cos\theta}{\sin\theta}$$

as we know.

Examples

1. Prove the identity

$$\frac{\cos\theta - 1}{\sec\theta + \tan\theta} + \frac{\cos\theta + 1}{\sec\theta - \tan\theta} = 2(1 + \tan\theta)$$

First we write the left-hand side (LHS) as a single fraction:

$$\frac{(\cos\theta - 1)(\sec\theta - \tan\theta) + (\cos\theta + 1)(\sec\theta + \tan\theta)}{\sec^2\theta - \tan^2\theta}$$

$$= \frac{1 - \sec\theta - \sin\theta + \tan\theta + 1 + \sec\theta + \sin\theta + \tan\theta}{1}$$

by multiplying out and using (1.4). Hence LHS $= 2 + 2\tan\theta = $ right-hand side (RHS).

2. Solve the equations

(a) $2\sin^2\theta - 3\cos\theta = 3$ (b) $12\sin\theta\cos\theta - 8\sin\theta + 3\cos\theta = 2$

(c) $2(\tan\theta - \cot\theta) + \sec\theta = 0$ (d) $3\sin^2\theta = 5\sin\theta + 2$

In each case give all the roots between $0°$ and $360°$.

Solution

(a) Using (1.1)

$$2(1 - \cos^2\theta) - 3\cos\theta = 3$$

i.e. $2\cos^2\theta + 3\cos\theta + 1 = 0$

which factorises to give

$$(2\cos\theta + 1)(\cos\theta + 1) = 0$$

so that $\cos\theta = -\dfrac{1}{2}$ or -1

The solutions are therefore

$$180° \pm 60° \quad \text{and} \quad 180°$$

i.e. $120°, 180° \quad \text{and} \quad 240°$

(b) $4 \sin \theta (3 \cos \theta - 2) + 3 \cos \theta - 2 = 0$

Hence $(4 \sin \theta + 1)(3 \cos \theta - 2) = 0$

so that

$$\sin \theta = -\frac{1}{4} \quad \text{or} \quad \cos \theta = \frac{2}{3}$$

i.e. $\theta = 180° + 14.48°, 360° - 14.48°$

or $\theta = 48.19°, 360° - 48.19°$

The solutions are therefore

$$48.19°, 194.48°, 311.81°, 345.52°$$

(c) The equation can be written as

$$\frac{2 \sin \theta + 1}{\cos \theta} = \frac{2 \cos \theta}{\sin \theta}$$

which, after multiplying up by the product $\sin \theta \cos \theta$, gives
$2 \sin^2 \theta + \sin \theta = 2 \cos^2 \theta$, and since $2 \cos^2 \theta = 2(1 - \sin^2 \theta)$ we obtain the equation

$$4 \sin^2 \theta + \sin \theta - 2 = 0$$

hence

$$\sin \theta = \frac{-1 \pm \sqrt{33}}{8} = -0.8431 \quad \text{or} \quad 0.5931$$

From the negative value of $\sin \theta$

$$\theta = 180° + 57.47°, 360° - 57.47°$$

and from the positive value

$$\theta = 36.38°, 180° - 36.38°$$

Hence the required solution are

$$\theta = 36.38°, 143.62°, 237.47°, 302.53°$$

(d) The equation can be written

$$3\sin^2\theta - 5\sin\theta - 2 = 0$$

so that $\qquad (3\sin\theta + 1)(\sin\theta - 2) = 0$

Hence

$$\sin\theta = -\frac{1}{3} \text{ or } 2$$

Only the first root is valid, because it is impossible for $\sin\theta$ to equal 2, so that

$$\theta = 199.47° \quad \text{or} \quad 340.53° \qquad\qquad \blacksquare$$

We know that a simple trigonometric equation, e.g. $\sin\theta = p$, where $|p| \le 1$, has an infinite number of solutions for θ but if θ is restricted to a single period of $360°$ or $2\pi^c$, exactly two solutions exist when $|p| < 1$; refer to Figure 1.6.

For example, if $\sin\theta = -\frac{3}{5}$, then $\cos\theta = \pm\frac{4}{5}$. We need therefore to specify **both** $\sin\theta$ **and** $\cos\theta$ for uniqueness. The same applies to other combinations, e.g. cosec θ and $\tan\theta$, with a single solution guaranteed provided that individual values of both $\sin\theta$ and $\cos\theta$ can be separately identified.

Example   Determine the value of θ between $0°$ and $360°$ which satisfies

$$\text{cosec } \theta = -\frac{37}{12} \qquad \tan\theta = \frac{12}{35}$$

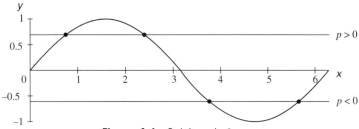

Figure 1.6 Solving $\sin\theta = p$

Note that

$$\sin\theta = -\frac{12}{37} \quad \text{so} \quad \cos\theta = \frac{\sin\theta}{\tan\theta} = -\frac{35}{37}.$$

The correct value must lie in the third quadrant. However, a calculator solution obtained using the $\sin^{-1}$ and $\cos^{-1}$ functions can appear ambiguous and misleading, e.g.

$$\sin\theta = -\frac{12}{37} \Rightarrow \theta = -18.92°$$

$$\text{and} \quad \cos\theta = \frac{35}{37} \Rightarrow \theta = 161.08°$$

Neither of these answers is correct because the calculation is programmed to select specified **principal values** which are uniquely defined in the range $[-90°, 90°]$ for $\sin\theta$ and $[0°, 180°]$ for $\cos\theta$, as depicted in Figure 1.7. The correct answer can be found by taking $\tan^{-1}\left(\frac{12}{35}\right) = 18.92°$ and adding $180°$ to obtain $\theta = 198.92°$. ∎

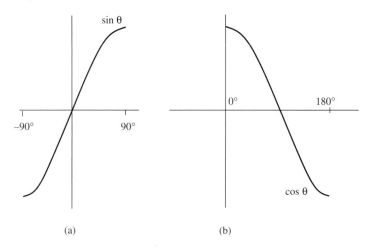

(a)   (b)

Figure 1.7 Principal values for (a) $\sin^{-1}\theta$ and (b) $\cos^{-1}\theta$

Exercise 1.2

1 Express in terms of $\sin\theta$

(a) $\sin^2\theta - \cos^2\theta$ (b) $2\cos^2\theta - 4\sin\theta$

2 If $\tan \theta = t$, express the following in terms of t:

(a) $3 \sec^2 \theta - \tan^2 \theta$

(b) $(1 + 2 \sin^2 \theta)/ \cos^2 \theta$

3 Simplify

(a) $\dfrac{\sin^2 \theta + \cos^2 \theta}{\cos^2 \theta}$

(b) $\dfrac{\sin^3 \theta + \sin \theta \cos^2 \theta}{\cos \theta}$

4 If $2 \sin^2 \theta - \cos^2 \theta = 1$ show that $\sin^2 \theta = 2 \cos^2 \theta$, hence find the possible values of $\tan \theta$.

5 Angle θ satisfies the equation $\sec^2 \theta + \tan^2 \theta = 5$; find the possible values of

(a) $\tan \theta$

(b) $\cos \theta$

(c) $\sin \theta$

6 Solve the following equations for θ in the range $0 \le \theta \le 180°$:

(a) $\cos^2 \theta - \sin^2 \theta = 0$

(b) $3 \tan^2 \theta = 2 \sec^2 \theta$

(c) $2 \cos^2 \theta + \sin \theta = 1$

(d) $2 \operatorname{cosec}^2 \theta = 3 \cot \theta + 1$

(e) $\cos^2 \theta + \cos \theta = \sin^2 \theta$

(f) $\tan^2 \theta = \sec \theta + 1$

(g) $1 + \sin \theta \cos^2 \theta = \sin \theta$

7 Solve the following equations for x in the range $0 \le x \le 360°$:

(a) $2 \sin^2 x + 5 \sin x - 3 = 0$

(b) $\cos^2 x - 1 = 0$

(c) $2 \sin^2 x - \sin x - 1 = 0$

8 Establish the following identities:

(a) $\dfrac{\cot A + \tan B}{\cot B + \tan A} \equiv \cot A \tan B$

(b) $\dfrac{1 + \sin \theta}{\cos \theta} \equiv \dfrac{\cos \theta}{1 - \sin \theta}$

(c) $\tan \theta + \cot \theta = \sec \theta \operatorname{cosec} \theta$

9 Given that r is positive and $-180° < \theta \le 180°$ find r and θ such that $r \cos \theta = -4$ and $r \sin \theta = 2.5$.

10 Solve the following equations for $0° \le \theta < 360°$:

(a) $\sin 2\theta = 2 \sin 11.5°$

(b) $\cos 2\theta = 0.7665$

(c) $\sin 2\theta = -0.3636$

(d) $\tan 2\theta = 1.9883$

11 Determine (to 4 s.f.) the sole angles between $0°$ and $360°$ which satisfy the properties given in each of the following cases:

(a) $\sin \theta = \dfrac{3}{5}, \cos \theta = -\dfrac{4}{5}$

(b) $\sin \theta = -\dfrac{8}{17}, \cot \theta = \dfrac{15}{8}$

(c) $\operatorname{cosec} \theta = \dfrac{25}{7}, \sec \theta = -\dfrac{24}{7}$

12   Illustrate graphically the general solution to Question 11(a) and determine its value in degrees.

1.3   FUNCTIONS OF COMPOUND ANGLES

Compound angles

We now prove that if A and B are acute angles then

$$\sin(A + B) \equiv \sin A \cos B + \cos A \sin B$$

From Figure 1.8

$$\sin(A + B) = \frac{RP}{OP} = \frac{RX + XP}{OP}$$

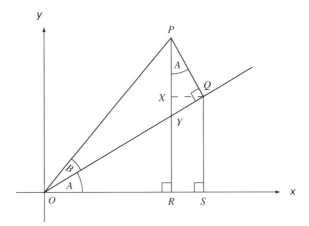

Figure 1.8   Diagram to prove $\sin(A + B) \equiv \sin A \cos B + \cos A \sin B$

Observe that triangles OYR and PYQ are similar (both contain a right angle and $O\hat{Y}R = P\hat{Y}Q$, vertically opposite angles) so that $Y\hat{P}Q = A$. Now

$$\frac{RX + XP}{OP} = \frac{RX}{OQ} \times \frac{OQ}{OP} + \frac{XP}{PQ} \times \frac{PQ}{OP} = \frac{SQ}{OQ} \times \frac{OQ}{OP} + \frac{XP}{PQ} \times \frac{PQ}{OP}$$

$$= \sin A \cos B + \cos A \sin B$$

The same figure can be used to prove that

$$\cos(A + B) \equiv \cos A \cos B - \sin A \sin B$$

In fact, both results are true whatever the values of A and B.

By replacing B by $-B$ and using the evenness and oddness properties of cosine and sine respectively we arrive at the **addition theorems**

$$\sin(A \pm B) \equiv \sin A \cos B \pm \cos A \sin B$$
$$\cos(A \pm B) \equiv \cos A \cos B \mp \sin A \sin B$$

(Notice that $\mp$ means that the sign on the right-hand side of the last identity is opposite to that on the left-hand side.)

Example   Determine exact values for

(a) $\sin 15°$

(b) $\tan 75°$

Solution

(a) $\sin 15° = \sin(45° - 30°) = \sin 45° \cos 30° - \cos 45° \sin 30°$

$$= \frac{1}{\sqrt{2}} \times \frac{\sqrt{3}}{2} - \frac{1}{\sqrt{2}} \times \frac{1}{\sqrt{2}} = \frac{\sqrt{3} - 1}{2\sqrt{2}}$$

(b) $\tan 75° = \dfrac{\sin 75°}{\cos 75°} = \dfrac{\sin 75°}{\sin 15°}$

Writing $75° = 45° + 30°$ we see that

$$\sin 75° = \sin 45° \cos 30° + \cos 45° \sin 30° = \frac{\sqrt{3} + 1}{2\sqrt{2}}$$

whence

$$\tan 75° = \frac{\sqrt{3}+1}{2\sqrt{2}} \times \frac{2\sqrt{2}}{\sqrt{3}-1} = \frac{(\sqrt{3}+1)}{(\sqrt{3}-1)} \times \left(\frac{\sqrt{3}+1}{\sqrt{3}+1}\right) = 2 + \sqrt{3}$$ ■

Putting $B = A$ in the compound angle formulae and combining with the identity $\cos^2 A + \sin^2 A = 1$ gives the **double-angled formulae**

$$\sin 2A \equiv 2 \sin A \cos A$$
$$\cos 2A \equiv 2 \cos^2 A - 1$$
$$\equiv \cos^2 A - \sin^2 A$$
$$\equiv 1 - 2 \sin^2 A$$

Also note that

$$\cos^2 A \equiv \frac{1}{2}(1 + \cos 2A)$$

$$\sin^2 A \equiv \frac{1}{2}(1 - \cos 2A)$$

Furthermore

$$\sin 3A \equiv \sin(2A + A)$$
$$\equiv \sin 2A \cos A + \cos 2A \sin A$$
$$\equiv 2 \sin A \cos^2 A + \sin A \cos 2A$$
$$\equiv 2 \sin A(1 - \sin^2 A) + \sin A(1 - 2 \sin^2 A)$$
$$\equiv 3 \sin A - 4 \sin^3 A$$

A similar identity for $\cos 3A$ is left as an exercise, as are the tangent identities

$$\tan(A + B) \equiv \frac{\tan A + \tan B}{1 - \tan A \tan B}$$

$$\tan(A - B) \equiv \frac{\tan A - \tan B}{1 + \tan A \tan B}$$

Extensions to the addition theorems

Adding two of the compound angle formulae gives

$$\sin(A + B) + \sin(A - B) \equiv 2 \sin A \cos B$$

so, by putting $A + B = P$ and $A - B = Q$, we find

$$\sin P + \sin Q \equiv 2 \sin\left(\frac{P + Q}{2}\right) \cos\left(\frac{P - Q}{2}\right)$$

In a similar way we find that

$$\sin P - \sin Q \equiv 2 \cos\left(\frac{P + Q}{2}\right) \sin\left(\frac{P - Q}{2}\right)$$

$$\cos P + \cos Q \equiv 2 \cos\left(\frac{P + Q}{2}\right) \cos\left(\frac{P - Q}{2}\right)$$

$$\cos Q - \cos P \equiv 2 \sin\left(\frac{P + Q}{2}\right) \sin\left(\frac{P - Q}{2}\right)$$

Examples

1. Illustrations of the extension to the addition theorems are

   (a) $\sin 60° + \sin 40° = 2 \sin 50° \cos 10°$

   (b) $\sin 120° - \sin 60° = 2 \cos 90° \sin 30° = 0$

   (c) $\cos 60° \cos 30° = \dfrac{1}{2}(\cos 90° + \cos 30°) = \dfrac{1}{2} \cos 30° = \dfrac{\sqrt{3}}{4}$

2. Prove the identity

$$\frac{\sin A + 2 \sin 3A + \sin 5A}{\sin 3A + 2 \sin 5A + \sin 7A} = \frac{\sin 3A}{\sin 5A}$$

Group together the outer terms in both numerator and denominator of the left-hand side to get

$$\frac{2 \sin 3A \cos 2A + 2 \sin 3A}{2 \sin 5A \cos 2A + 2 \sin 5A}$$

i.e.
$$\frac{2 \sin 3A(1 + \cos 2A)}{2 \sin 5A(1 + \cos 2A)}$$

Cancelling top and bottom by $1 + \cos 2A$, we see that

$$\text{left-hand side} \equiv \frac{\sin 3A}{\sin 5A} \equiv \text{right-hand side}$$

Note that if $\cos 2A = -1$, i.e. $2A = (2n + 1)\pi$ or $A = (2n + 1)\frac{\pi}{2}$, the left-hand side is indeterminate, having the form $\frac{0}{0}$, whereas the right-hand side is equal to -1. It requires calculus to prove that, for these values of A, the identity is still valid.   ∎

Simplifying functions

We are now in a position to resolve trigonometric expressions which incorporate both a function and its inverse.

Examples   1. Using principal values, simplify

(a) $\cos\left(2\cos^{-1}\left(\frac{1}{2}\right)\right)$ (b) $\cot^{-1}\left(\cos\left(\frac{2\pi}{3}\right)\right)$ angles in $(0, \pi)$

Also using principal values, find the following as functions of x:

(c) $\sin(2\sin^{-1} x)$ (d) $\tan(\cot^{-1} x)$

Solution

(a) $\cos^{-1}\left(\frac{1}{2}\right) = \frac{\pi}{3}$ and $\cos\left(\frac{2\pi}{3}\right) = -\frac{1}{2}$. Answer is $-\frac{1}{2}$.

(b) $\cos\left(\frac{2\pi}{3}\right) = -\frac{1}{2}$ and $\cot^{-1}\left(-\frac{1}{2}\right) = \tan^{-1}(-2) = \pi - \tan^{-1} 2 = 116.57°$

(c) Figure 1.9(a) may help. Here $\sin\theta = x$, and using the identity $\sin 2A \equiv 2\sin A \cos A$, we obtain

$$\sin(2\sin^{-1} x) = 2\sin(\sin^{-1} x)\cos(\sin^{-1} x)$$

The angle shown is $\theta = \sin^{-1} x$ and its cosine is $(1 - x^2)^{1/2}$, so that

$$\sin(2\sin^{-1} x) = 2x(1 - x^2)^{1/2}$$

(d) Using another right-angled triangle, Figure 1.9(b), $\theta = \cot^{-1} x$ is the angle and we obtain

$$\tan(\cot^{-1} x) = \frac{1}{x}$$

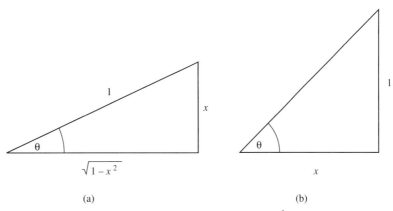

(a)   (b)

Figure 1.9   Finding the principal value of (a) $\sin(2\sin^{-1} x)$ and (b) $\tan(\cot^{-1} x)$

2. Determine the equation of the straight line passing through the point $(1, 2)$ at $60°$ to the line $y = 2x + 1$, rotated clockwise (Figure 1.10). We work in radians. The line $y = 2x + 1$ has angle of slope $\tan^{-1} 2$.

The angle of slope of the required straight line is $(\tan^{-1} 2) - \dfrac{\pi}{3}$. The tangent of this angle is $\tan\left(\tan^{-1} 2 - \dfrac{\pi}{3}\right)$

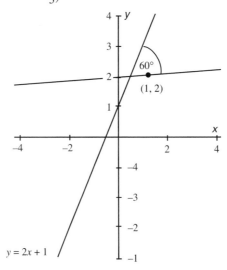

Figure 1.10   Finding the equation of the straight line through $(1, 2)$ at $60°$ to $y = 2x + 1$

Using the formula for $\tan(A - B)$, we get

$$\tan\left(\tan^{-1} 2 - \frac{\pi}{3}\right) = \frac{2 - \sqrt{3}}{1 + 2\sqrt{3}} = \frac{2 - \sqrt{3}}{1 + 2\sqrt{3}} \times \frac{1 - 2\sqrt{3}}{1 - 2\sqrt{3}}$$

$$= \frac{2 - \sqrt{3}}{2\sqrt{3} + 1} \times \frac{2\sqrt{3} - 1}{2\sqrt{3} - 1}$$

$$= \frac{5\sqrt{3} - 8}{11} \approx 0.0600$$

The equation of the required line is

$$y - 2 = m(x - 1) \quad \text{where} \quad m \approx 0.06$$

i.e. $y = mx + 2 - m$ or $y \approx 0.06x + 1.94$ ■

Calculus and trigonometry

Compound angles can be used to determine the derivatives of trigonometric functions. Consider Figure 1.11.

BA is a short arc of a unit circle centre O of radius 1, subtending a small angle δx at O, i.e. $BA = \delta x$. If δx is very small, it can reasonably be assumed that $BH < BA < CA$, i.e. that $\sin(\delta x) < \delta x < \tan(\delta x)$.

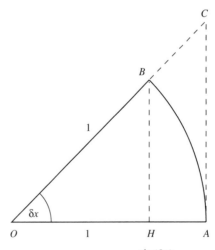

Figure 1.11   Limit of $\dfrac{\sin(\delta x)}{\delta x}$

Dividing by $\sin(\delta x)$ we obtain

$$1 < \frac{\delta x}{\sin(\delta x)} < \frac{1}{\cos(\delta x)}$$

As $\delta x \to 0$, $\cos(\delta x) \to 1$, so the fraction $\dfrac{\delta x}{\sin(\delta x)}$ is sandwiched between 1 and a fraction $\dfrac{1}{\cos(\delta x)}$ which has a limit of 1. Hence $\dfrac{\delta x}{\sin(\delta x)} \to 1$, so that $\dfrac{\sin(\delta x)}{\delta x} \to 1$ as $\delta x \to 0$.

You should be aware this is not a rigorous proof.

We now find the derived function of $\sin x$ from first principles. If $y = \sin x$ and x changes to $x + \delta x$, where δx is small, then y changes to

$$y + \delta y = \sin(x + \delta x)$$

so that

$$\delta y = \sin(x + \delta x) - \sin x = 2 \cos\left(x + \frac{1}{2}\delta x\right) \times \sin\left(\frac{1}{2}\delta x\right)$$

Since

$$\lim_{\delta x \to 0} \frac{\sin(\delta x)}{\delta x} = 1, \quad \text{then} \quad \frac{\delta x}{\sin(\delta x)} \to 1 \text{ as } \delta x \to 0.$$

Hence

$$\frac{\delta y}{\delta x} = \frac{2 \cos(x + \frac{1}{2}\delta x) \times \sin(\frac{1}{2}\delta x)}{\delta x} = \cos\left(x + \frac{1}{2}\delta x\right) \times \frac{\sin(\frac{1}{2}\delta x)}{\frac{1}{2}\delta x}$$

As $\delta x \to 0$, $\cos\left(x + \dfrac{1}{2}\delta x\right) \to \cos x$ and, since $\dfrac{1}{2}\delta x \to 0$

$$\frac{\sin(\frac{1}{2}\delta x)}{\frac{1}{2}\delta x} \to 1$$

Therefore

$$\frac{\delta y}{\delta x} \to \frac{dy}{dx} = \cos x \times 1 = \cos x$$

It is left as an exercise for you to prove that $\dfrac{d}{dx}(\cos x) = -\sin x$ from first principles

using the arguments above. These can be extended to the function $\sin kx$ and $\cos kx$ where

k is a constant, and extended to their integrals by reversing the differentiation process. The summary is

$$\frac{d}{dx}(\sin kx) = k \cos kx$$

$$\frac{d}{dx}(\cos kx) = -k \sin kx$$

$$\int \sin kx \, dx = -\frac{1}{k}\cos kx + C$$

$$\int \cos kx \, dx = \frac{1}{k}\sin kx + C$$

where C is the constant of integration.

Exercise 1.3

1 If $\sin A = \dfrac{3}{5}$, $\cos B = -\dfrac{4}{5}$ where $0 < A < \dfrac{\pi}{2}$ and $O < B < \pi$, then without the use of a calculator, determine the values of

(a) $\sin(A + B)$ (b) $\cos(B - A)$ (c) $\tan(A + B)$

(d) $\tan 2A$ (e) $\sin 3A$

2 Prove the identities

(a) $\cos 3A \equiv 4\cos^3 A - 3\cos A$ (b) $\tan(A \pm B) \equiv \dfrac{\tan A \pm \tan B}{1 \mp \tan A \tan B}$

and deduce that

(c) $\tan 2A \equiv \dfrac{2\tan A}{1 - \tan^2 A}$

3 Express each of the following as half the sum of two trigonometric functions:

(a) $\cos 42° \sin 38°$ (b) $\cos 80° \cos 40°$

(c) $\sin 52° \sin 38°$

(d) $\sin 60° \cos 40°$

(e) $\sin 160° \cos 140°$

(f) $\cos(2x + d) \cos(2x - d)$

4   Without the use of a calculator find the values of

(a) $\dfrac{\sin 45° \cos 15° + \sin 15° \cos 45°}{\sin 45° \cos 15° - \sin 15° \cos 45°}$

(b) $\sin^2 15°$

(c) $\tan 15°$

(d) $\cos 22.5°$ given that $\cos 45° = \dfrac{1}{\sqrt{2}}$

5   If $\sin B = -\dfrac{4}{5}$ where $\pi < B < \dfrac{3\pi}{2}$, and $\cos C = -\dfrac{24}{25}$ where $\dfrac{\pi}{2} < C < \pi$ calculate the values of

(a) $\sin(B - C)$

(b) $\cos(B + C)$

(c) $\cos 2B$

(d) $\tan(B - C)$

6   Prove the identities

(a) $\dfrac{\sin \theta}{1 + \cos \theta} \equiv \tan \dfrac{\theta}{2}$

(b) $\cot(A + B) \equiv \dfrac{\cot A \cot B - 1}{\cot A + \cot B}$

7   Solve the following equations for $0° \le \theta \le 360°$:

(a) $\cos 2\theta - \cos \theta = 0$

(b) $3 \cos 2\theta + \sin^2 \theta + 2 \sin \theta = 0$

(c) $\cos 2\theta + 2 \cos \theta - 3 = 0$

(d) $\cos 2\theta - 2 \cos \theta - 3 = 0$

(e) $4 \sin \theta \cos \theta - 2 \sin \theta - 2 \cos \theta + 1 = 0$ (*Hint*: group in pairs and factorise.)

(f) $2 \cos^2 \theta + 3 \sin \theta - 3 = 0$

8   Solve for values of θ in the range of $0° \le \theta \le 360°$:

(a) $\cos \theta - \cos 3\theta = \sin 2\theta$

(b) $\cos(\theta - 30°) - \cos(\theta + 30°) = \dfrac{1}{2} \tan \theta$

(c) $\cos 3\theta + \cos \theta = \sin 3\theta + \sin \theta$

9   Prove the identities

(a) $\dfrac{\sin 2m\theta + \sin 2n\theta}{\cos 2m\theta + \cos 2n\theta} \equiv \tan(m + n)\theta$

(b) $\dfrac{\sin A + \sin 3A + \sin 5A}{\cos A + \cos 3A + \cos 5A} \equiv \tan 3A$

10 Simplify the following:

(a) $\sin\left(2\sin^{-1}\left(\frac{1}{2}\right)\right)$

(b) $\tan(\sec^{-1}(\sqrt{2}))$

(c) $\cot^{-1}\left(\sin\left(-\frac{\pi}{3}\right)\right)$ (give the answer in degrees)

and in terms of x find

(d) $\cos(3\cos^{-1}x)$

(e) $\cos(2\sin^{-1}x)$

(f) $\sin(\tan^{-1}x)$

(g) $\sin(2\tan^{-1}x)$

(The principal values of $\sin^{-1}$, $\tan^{-1}$, cosec^{-1} and $\cot^{-1}$ are taken between $-\frac{\pi}{2}$ and $\frac{\pi}{2}$ and those of $\cos^{-1}$ and $\sec^{-1}$ between 0 and π.)

11 A horizontal railway tunnel runs from A to B underneath a ridge which is at a height h above the tunnel. The highest point of the ridge C is at an angle of elevation α above A and β above B.

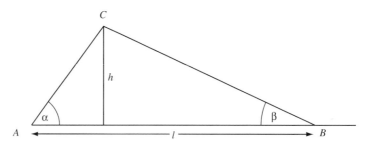

Prove that

$$h = \frac{l\sin\alpha\sin\beta}{\sin(\alpha+\beta)}$$

If $\alpha = 12°$, $\beta = 5°$ and $l = 1\,\text{km}$, what height is C above the tunnel (3 s.f.)?

12 In the triangle ABC, the angles and sides are denoted in the usual way, i.e. A, B, C, a, b, c, with the semiperimeter of the triangle given by

$$s = \frac{1}{2}(a+b+c)$$

Starting with the cosine rule for A prove the following results:

(a) $\cos\frac{A}{2} = \sqrt{\frac{s(s-a)}{bc}}$

(b) $\sin\frac{A}{2} = \sqrt{\frac{(s-b)(s-c)}{bc}}$

(c)　$\sin A = \dfrac{2}{bc} \sqrt{s(s-a)(s-b)(s-c)}$

(d)　$\dfrac{1}{a}\cos^2\left(\dfrac{A}{2}\right) + \dfrac{1}{b}\cos^2\left(\dfrac{A}{2}\right) + \dfrac{1}{c}\cos^2\left(\dfrac{C}{2}\right) = \dfrac{s^2}{abc}$

(e)　(i) $\dfrac{\sin B - \sin C}{\sin B + \sin C} = \dfrac{b-c}{b+c}$　　(ii) $\tan\dfrac{1}{2}(B-C) = \left(\dfrac{b-c}{b+c}\right)\cot\dfrac{A}{2}$

13 Determine the equations of the following straight lines:

(a)　At $45°$ anticlockwise to $y = 6x - 7$ passing through the origin.

(b)　At $\tan^{-1} 2$ clockwise to $y + 4x = 5$ passing through the point $(1, 1)$.

14 Solve the simultaneous equations

$$\sin x + \sin y = 1 \qquad \operatorname{cosec} x + \operatorname{cosec} y = 4$$

giving the values of x and y between $0°$ and $180°$.

15 Find all the angles between $0°$ and $90°$ which simultaneously satisfy

$$\cos x \cos y = 0.5 \qquad \sin x \sin y = 0.4$$

Are there any solutions between $90°$ and $180°$?

16 (a)　Prove that $\tan^{-1} x + \tan^{-1} y \equiv \tan^{-1}\dfrac{x+y}{1-xy}$

(b)　Determine the acute angle between the straight lines

$$3x + 5y + 6 = 0 \quad \text{and} \quad y = 5x - 4$$

(c)*　If $\tan^{-1} x + \tan^{-1} y + \tan^{-1} z = \dfrac{\pi}{2}$, deduce that

$$xy + yz + zx = 1$$

(*Hint*: take $\tan^{-1} x + \tan^{-1} y$ first of all, then include $\tan^{-1} z$; note that $\tan^{-1}\left(\dfrac{\pi}{2}\right) = \infty$, which implies a zero denominator.)

17 (a)　Prove from first principles that

$$\dfrac{d}{dx}(\cos x) = -\sin x$$

(b)   Use first principles to prove that

$$\frac{d}{dx}(\sin kx) = k \cos kx$$

18   By using the extensions to the addition theorems, determine the derivatives of the following functions:

(a) $\sin 5x \sin 3x$ (b) $\cos 2x \cos 4x$ (c) $\sin x \cos 3x$

19   Determine the following indefinite integrals:

(a) $\displaystyle\int \sin 5x \sin 3x \, dx$ (b) $\displaystyle\int \sin^2 x \, dx$ (c) $\displaystyle\int \sin^3 x \, dx$

and the following definite integrals:

(d) $\displaystyle\int_0^{\pi/2} (2\cos^2 x + 3\sin^2 x)dx$ (e) $\displaystyle\int_0^{\pi/3} 2\sin 3x \cos x \, dx$

1.4   WAVES AND FURTHER APPLICATIONS

The form $a\cos\theta + b\sin\theta$

The graphs of $y = \sin\theta$ and $y = \cos\theta$ are of wave forms, separated horizontally by $90°$ or $\dfrac{\pi}{2}$ radians. We call the gap of $90°$ or $\dfrac{\pi}{2}$ radians the **phase difference**.

Consider the expression $a\cos\theta + b\sin\theta$; Figure 1.12 shows the graphs of $a\cos\theta$, $b\sin\theta$ and their sum, for $\theta > 0$.

The resultant, i.e. the sum, is the dashed curve. We prove this is a sinusoidal wave of the same period as the component functions, namely 2π.

We start with the identity

$$R\sin(\theta + \phi) \equiv R\sin\theta\cos\phi + R\cos\theta\sin\phi \equiv R\sin\phi\cos\theta + R\cos\phi\sin\theta$$

and try to find R and ϕ such that

$$R\sin\phi = a \qquad R\cos\phi = b$$

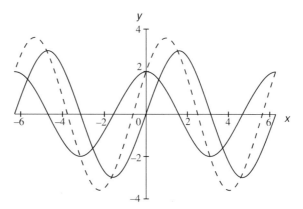

Figure 1.12 Graphs of $a\cos\theta$, $b\sin\theta$ and $a\cos\theta + b\sin\theta$

so that

$$R\sin(\theta + \phi) \equiv a\cos\theta + b\sin\theta$$

If we choose $R^2 = a^2 + b^2$ and $\tan\phi = \dfrac{a}{b}$ where ϕ is an acute angle, we can achieve

our aim since $\sin\phi = \dfrac{a}{\sqrt{a^2 + b^2}} = \dfrac{a}{R}$ and $\cos\phi = \dfrac{b}{\sqrt{a^2 + b^2}} = \dfrac{b}{R}$.

Note that

$$\sin^2\phi + \cos^2\phi = \frac{a^2}{a^2 + b^2} + \frac{b^2}{a^2 + b^2} = 1$$

and that $R\cos\phi = b$, $R\sin\phi = a$.

In fact, we can express $a\cos\theta + b\sin\theta$ in any of the forms

$$R\sin(\theta + \phi_1) \qquad R\sin(\theta - \phi_2) \qquad R\cos(\theta + \phi_3) \qquad R\cos(\theta - \phi_4)$$

by equating the coefficients of $\cos\theta$ and $\sin\theta$ to a and b respectively. The angles ϕ_i, $i = 1(1)4$, will differ in each case, though $R = (a^2 + b^2)^{1/2}$ throughout.

Examples

1. Express $4\cos\theta + 3\sin\theta$ in the forms

    (a) $R\sin(\theta + \phi_1)$ (b) $R\sin(\theta - \phi_2)$

    (c) $R\cos(\theta + \phi_3)$ (d) $R\cos(\theta - \phi_4)$

Solution

For all cases $R^2 = 3^2 + 4^2$, so that $R = 5$.

(a) $\tan \phi_1 = \dfrac{4}{3}$, therefore $\phi_1 = 53.13°$

We take the angle in the first quadrant since $\cos \phi_1, \sin \phi_1 > 0$.

(b) $\tan \phi_2 = -\dfrac{4}{3}$, therefore $\phi_2 = 306.87°$

We take the angle in the fourth quadrant since $\cos \phi_2 > 0, \sin \phi_2 < 0$.

(c) $\cot \phi_3 = -\dfrac{4}{3}$, therefore $\phi_3 = 323.13°$

We take the angle in the fourth quadrant since $\cos \phi_3 > 0, \sin \phi_3 < 0$.

(d) $\cot \phi_4 = \dfrac{4}{3}$, therefore $\phi_4 = 36.87°$

We take the angle in the first quadrant since $\cos \phi_4 > 0, \sin \phi_4 > 0$.

2. Solve the equation

$$4 \cos \theta + 3 \sin \theta = 2.5 \qquad (0 \le \theta \le 360°)$$

We can use any of the four forms above to solve $3 \sin \theta + 4 \cos \theta = 2.5$. In this solution we use the form $R \sin(\theta + \phi_1)$, so

$$5 \sin(\theta + 53.13°) = 2.5$$

hence $\quad \sin(\theta + 53.13°) = \dfrac{1}{2}$

therefore $\theta + 53.13° = 30°, 150°, 390°, \ldots$

$$\text{or} = -210°, -330°, \ldots$$

Hence $\theta = 96.87°, 336.87°$ are the appropriate solutions. ∎

In solving problems of the form

$$a \sin \theta + b \cos \theta = c$$

you can choose any of the four forms. R is always the same. The appropriate solutions of $\sin(\theta + \phi_1) = p$, etc., need to be chosen so that values of θ satisfying $0 \le \theta \le 360°$ can be found.

Waves and applications

Waveforms occur frequently in engineering and science; for example, they occur in the areas of mechanical vibration and electrical oscillation. Alternating electric current is designed to be sinusoidal. In such cases the waveforms are time dependent, i.e. $\theta = \omega t$, where ω is called the **angular velocity**, measured in radians per second and t is the time in seconds. Refer to Figure 1.13.

Sometimes **frequency** f is used, measured in s^{-1} or Hz, particularly in the case of electric current generation. The **period** of oscillation, i.e. the time for a complete cycle, is found by putting $\theta = 2\pi$, i.e. $t = \dfrac{2\pi}{\omega} = \dfrac{1}{f}$, where $\omega = 2\pi f$.

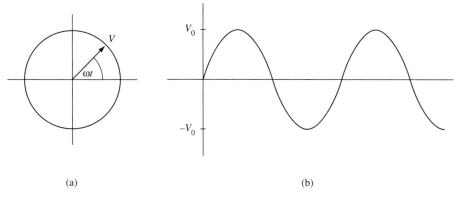

(a)   (b)

Figure 1.13 Waveforms: (a) circular generation of voltage $V = V_0 \sin \omega t \Leftrightarrow$ (b) sinusoidal voltage $V = V_0 \sin \omega t$

Example

Produce a graph of the function $\sin \omega t + \sin 3\omega t$ and comment. The extension to the addition theorems gives

$$\sin \omega t + \sin 3\omega t = 2 \sin 2\omega t \cos \omega t$$

There is an oscillation of period $\dfrac{2\pi}{2\omega}$ taking place within a slower oscillation of period $\dfrac{2\pi}{\omega}$.

This gives rise to the phenomenon of **beats**; see Figure 1.14. ∎

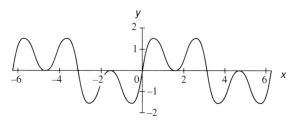

Figure 1.14 Graph of $\sin \omega t + \sin 3\omega t$

Some items of heavy-duty electrical equipment operate on the principle of aggregated a.c. voltages of different **phases**. There are often three or more of these phases, but for simplicity in the next example we consider only two.

Example

Determine the aggregate of two similar voltages which differ by a phase of $\dfrac{\pi}{4}$.

Let the voltages be $V \sin \omega t$ and $V \sin\left(\omega t - \dfrac{\omega}{4}\right)$. Their aggregate is

$$V\left(\sin \omega t + \sin\left(\omega t - \frac{\pi}{4}\right)\right) = 2V \sin\left(\omega t - \frac{\pi}{8}\right) \cos \frac{\pi}{8} \approx 1.848V \sin\left(\omega t - \frac{\pi}{8}\right)$$

i.e. a sinusoidal voltage of 1.848 times each of the original ones with an exactly intermediate phase angle; see Figure 1.15.

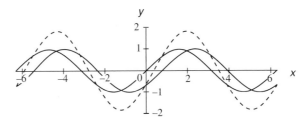

Figure 1.15 Aggregate of two voltages which differ in phase by $\pi/4$

Note that the amplitude of each component voltage is V and that the aggregate has amplitude $1.848V$. The aggregate wave is shown dashed. ∎

Further applications in calculus

We know that a function such as $5 \cos x + 12 \sin x$ can be written as $13 \sin(x + \phi)$, since $5^2 + 12^2 = 13^2$, so the maximum and minimum values are ± 13. Differentiation of the function yields

$$-5 \sin x + 12 \cos x$$

and this derivative is zero when $\tan x = \dfrac{12}{5}$. Hence, either

$$\sin x = \frac{12}{13} \quad \text{and} \quad \cos x = \frac{5}{13}$$

or

$$\sin x = -\frac{12}{13} \quad \text{and} \quad \cos x = -\frac{5}{13}.$$

The maximum and minimum values are therefore respectively

$$5 \times \left(\frac{5}{13}\right) + 12 \times \left(\frac{12}{13}\right) = \frac{169}{13} = 13 \quad \text{and} \quad 5 \times \left(-\frac{5}{13}\right) + 12 \times \left(-\frac{12}{13}\right)$$

$$= -\frac{169}{13} = -13 \text{ (as above)}$$

You can then see without any differentiation that a function such as

$$f(x) = \frac{33}{5 \cos x + 12 \sin x + 20}$$

is bounded between $\frac{33}{33} = 1$ and $\frac{33}{-13 + 20} = \frac{33}{7}$. On the other hand, were you to differentiate $f(x)$ directly to find its maxima and minima, you would obtain the same result but with greater difficulty.

For integration much use is made of the t substitution $t = \tan\frac{\theta}{2}$. It follows that

$$\sin \theta = 2 \sin\frac{\theta}{2}\cos\frac{\theta}{2} = \frac{2\tan\frac{\theta}{2}}{\sec^2\frac{\theta}{2}} = \frac{2t}{1 + t^2}$$

$$\cos \theta = \frac{1 - t^2}{1 + t^2} \qquad \text{(using Pythagoras' theorem)}$$

$$\tan \theta = \frac{2t}{1 - t^2} \qquad \text{(dividing the first result by the second)}$$

We will explore its uses for integration in Chapter 6.

Example   Express the function $f(x) = \dfrac{1 + \cos x}{2 + \sin x}$ in terms of $t = \tan\dfrac{x}{2}$ and use differentiation to obtain its maximum and minimum values.

First observe that

$$0 \leq 1 + \cos x \leq 2 \qquad 1 \leq 2 + \sin x \leq 3$$

so that

$$0 \le f(x) \le 2.$$

We can refine these limits as follows. With the t substitution we obtain the replacement function

$$g(t) = \frac{1 + \left(\dfrac{1 - t^2}{1 + t^2} \right)}{2 + \left(\dfrac{2t}{1 + t^2} \right)} = \frac{(1 + t^2) + 1 - t^2}{2(1 + t^2) + 2t} = \frac{1}{1 + t + t^2}$$

Differentiation gives

$$g'(t) = -\frac{(1 + 2t)}{(1 + t + t^2)^2}$$

This is zero when $t = \dfrac{1}{2}$, so that

$$\sin x = \frac{-1}{1 + \dfrac{1}{4}} = -\frac{4}{5} \qquad \cos x = \frac{3}{5} \qquad f(x) = \frac{4}{3}$$

Hence $0 \le f(x) \le \dfrac{4}{3}$ is the true bound.

(Strictly speaking, we should show that $f''\left(-\dfrac{1}{2}\right) < 0$. This is left for you to verify.)

■

Exercise 1.4

 1 Transform each of the following into the form indicated:

    (a) $3 \sin \theta - 4 \cos \theta$ $R \cos(\theta + a)$

    (b) $7 \cos \theta + 24 \sin \theta$ $R \cos(\theta - a)$

    (c) $\cos \theta + \sin \theta$ $R \sin(\theta - a)$

    (d) $3 \sin \theta - \cos \theta$ $R \sin(\theta + a)$

    (e) $5 \cos \theta - 12 \sin \theta$ $R \cos(\theta - a)$

    (f) $2 \cos \theta + 4 \sin \theta$ $R \cos(\theta + a)$

2 Solve the following equations for values of θ from $0°$ to $360°$:

(a) $3 \sin \theta + 4 \cos \theta = 2$ (b) $2 \cos \theta - \sin \theta = 0.5$

(c) $5 \sin \theta + 12 \cos \theta = 6.5$ (d) $\cos \theta - \sin \theta = 1$

(e) $7 \sin \theta - 24 \cos \theta = 10$

3   Express $8 \cos 2\theta - 15 \sin 2\theta$ in the form $R \cos(2\theta + \phi)$. Hence solve the equation

$$8 \cos 2\theta - 15 \sin 2\theta = 17 \qquad \text{for } 0 < \theta < 360°$$

4   Express $\cos \omega t + \cos\left(\omega t - \dfrac{\pi}{3}\right)$ in the form

(a) $R \cos(\omega t + \phi)$ (b) $R \cos(\omega t - \phi)$

5   Resolve the following waveforms into components in $\sin \omega t$ and $\cos \omega t$:

(a) $25 \sin(\omega t + 73.74°)$ (b) $\sqrt{40} \cos(\omega t + 18.43°)$

6   Three different alternating voltages of the same magnitude are applied across a resistor. Each of them is phased $\dfrac{\pi}{3}$ apart, i.e.

$$V_1 = V \sin\left(2\pi ft - \frac{\pi}{3}\right) \qquad V_2 = V \sin 2\pi ft \qquad V_3 = V \sin\left(2\pi ft + \frac{\pi}{3}\right)$$

where f is the frequency of the voltage, measured in hertz (i.e. s^{-1}) and t the time in seconds (s). Show that the total voltage is

$$V = V_1 + V_2 + V_3 = 2V \sin 2\pi ft$$

(*Hint*: start by setting $\theta = 2\pi ft$ and adding V_1 and V_3.)

7   An amplitude-modulated signal has waveform

$$y = A(B + \sin \omega_m t) \sin \omega_c t$$

where A and B are constraints and ω_m and ω_c are respectively the modulating and carrier frequencies, *both constant*. Show that the signal contains three angular frequencies ω_c, $\omega_c - \omega_m$ and $\omega_c + \omega_m$.

     (*Hint*: multiply out and reduce the number of symbols, i.e. write out as

$$A \sin \theta + C \sin \phi \sin \theta, \qquad \text{where } \theta = \omega_c t, \phi = \omega_m t, C = AB.$$

The answer involves sines and cosines of θ, $\theta - \phi$, $\theta + \phi$.)

8   An alternating electric current generated at 50 Hz with an RMS current of 2 amps through an inductive coil of resistance 50 ohms and inductance 2 henries takes the form

$$V = 2(50 \sin 100\pi t - 200\pi \cos 100\pi t)$$

(a)   Express V in the form $2A \sin(100\pi t - \phi)$.

(b)   For the general case of I amps, frequency f Hz, resistance R ohms and inductance L henries,

$$V = I(R \sin 2\pi f t - 2\pi f L \cos 2\pi f t).$$

Express V in the form

$$IZ \sin(2\pi f t - \phi)$$

Z is called the **impedance** and ϕ the **phase lag**.

9   If $t = \tan \dfrac{\phi}{2}$, prove that $\sin \theta = \dfrac{2t}{1 + t^2}$, $\cos \theta = \dfrac{1 - t^2}{1 + t^2}$. What is $\tan \theta$?

Prove further that

$$\frac{1 + \sin \theta}{4 + 3 \cos \theta} = \frac{(1 + t)^2}{7 + t^2}$$

Deduce, by differentiating with respect to t that

$$0 \le \frac{1 + \sin \theta}{4 + 3 \cos \theta} \le \frac{8}{7}$$

10   (a)   Without using differentiation determine the maximum and minimum, values of

$$4 \sin \theta - 3 \cos \theta$$

(b)   Hence determine the maximum and minimum values of

$$\frac{1}{4 \cos \theta - 3 \sin \theta + 6}$$

(c)   Verify the result in (b) by substituting $t = \tan \dfrac{\theta}{2}$ in the expression and using differentiation.

11 (a) Expand $\sin(ax + b)$ in terms of $\sin ax$, $\cos ax$, etc. Differentiate with respect to x, assuming a and b to be constant, and show that

$$\frac{d}{dx}(\sin(ax + b)) = a\cos(ax + b)$$

(b) What is the derivative of $\cos(ax + b)$?

12 Determine the equation of the tangent to the curve

$$y = \sin\left(2x - \frac{\pi}{3}\right)$$

at the first point to the right of the y-axis where the slope is equal to 1.

13 In the figure $B\hat{A}C = \theta$ and the shaded area in the circle is twice that of $\triangle ABC$.

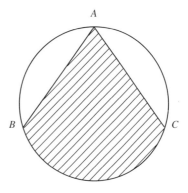

(a) Prove that θ must satisfy the equation

$$\theta = \sin\theta + \sin 2\theta$$

(b) Draw the graphs of $y = \theta$ and $y = \sin\theta + \sin 2\theta$ and use them to estimate the solution of the equation in (a) which lies between $\frac{\pi}{3}$ and $\frac{\pi}{2}$ (3 s.f.).

14 (a) Write down the indefinite integrals of

(i) $\sin(ax + b)$ (ii) $\cos(ax + b)$

(b)   Determine

$$\int_0^{\pi/2} \sin\left(2x - \frac{\pi}{3}\right) dx \qquad \int_0^p \cos\left(3x + \frac{\pi}{4}\right) dx$$

(c)   What is the smallest value of p for the second integral to be zero?

15   A cone is inscribed in a sphere of radius a; θ is the semi-angle of the cone.

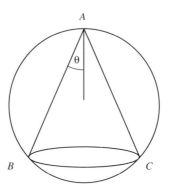

(a)   Prove that the volume of the cone is

$$\frac{8}{3} \pi a^3 \sin^2 \theta \cos^4 \theta$$

(b)   Put $x = \cos \theta$ and determine the value of x, hence θ, for which the volume is a maximum. Write down the volume.

16   Write down each of the following sequences of numbers, $n = 1\ (1)\ 10$, whose nth terms are

(a) $\sin n\pi$ (b) $\cos n\pi$ (c) $\sin\left(\dfrac{n\pi}{2}\right)$

(d) $\sqrt{2}\sin\left(\dfrac{n\pi}{4}\right)$ (e) $2\cos(n\pi/3)$ (f) $\sin\left[\dfrac{(2n-1)\pi}{2}\right]$

Which of the above can be replaced by $(-1)^n$?

SUMMARY

- **Reciprocal ratios**

$$\sec\theta = \frac{1}{\cos\theta} \qquad \mathrm{cosec}\,\theta = \frac{1}{\sin\theta} \qquad \cot\theta = \frac{1}{\tan\theta}$$

- **Identities**

$$\cos^2\theta + \sin^2\theta \equiv 1$$

$$1 + \tan^2\theta \equiv \sec^2\theta$$

$$\cot^2\theta + 1 \equiv \mathrm{cosec}^2\theta$$

- **Addition theorems**

$$\sin(A \pm B) \equiv \sin A \cos B \pm \cos A \sin B$$

$$\cos(A \pm B) \equiv \cos A \cos B \mp \sin A \sin B$$

$$\tan(A \pm B) \equiv \frac{\tan A \pm \tan B}{1 \mp \tan A \tan B}$$

- **Double-angle formulae**

$$\sin 2A \equiv 2\sin A \cos A$$

$$\cos 2A \equiv \cos^2 A - \sin^2 A \equiv 2\cos^2 A - 1 \equiv 1 - 2\sin^2 A$$

$$\tan 2A \equiv \frac{2\tan A}{1 - \tan^2 A}$$

- **Product formulae**

$$\sin P + \sin Q \equiv 2\sin\left(\frac{P+Q}{2}\right)\cos\left(\frac{P-Q}{2}\right)$$

$$\sin P - \sin Q \equiv 2\cos\left(\frac{P+Q}{2}\right)\sin\left(\frac{P-Q}{2}\right)$$

$$\cos P + \cos Q \equiv 2\cos\left(\frac{P+Q}{2}\right)\cos\left(\frac{P-Q}{2}\right)$$

$$\cos Q - \cos P \equiv 2\sin\left(\frac{P+Q}{2}\right)\sin\left(\frac{P-Q}{2}\right)$$

- **An important limit**

$$\lim_{x\to 0}\frac{\sin x}{x} = 1$$

- **Derivatives and integrals**

$$\frac{d}{dx}(\sin kx) = k\cos kx \qquad \frac{d}{dx}(\cos kx) = -k\sin kx$$

$$\int \sin kx \, dx = -\frac{1}{k}\cos kx + C \qquad \int \cos kx \, dx = \frac{1}{k}\sin kx + C$$

- **$a\cos\theta + b\sin\theta$** can be expressed as $R\sin(\theta + \phi)$ where

$$R = (a^2 + b^2)^{1/2} \qquad \sin\phi = \frac{a}{R} \qquad \cos\phi = \frac{b}{R}$$

- **Waveform** $A\cos\omega t$ has amplitude A, angular frequency ω, frequency $f = \frac{\omega}{2\pi}$ and period $T = \frac{2\pi}{\omega} = \frac{1}{f}$

Answers

Exercise 1.1

1 (a) (i) $\dfrac{2}{\sqrt{3}}$ (ii) $\sqrt{2}$ (iii) $-\dfrac{1}{\sqrt{3}}$

  (iv) -1 (v) -1.1223 (vi) -1.6616

  (vii) -2.0503 (viii) $\sqrt{2}$ (ix) $\sqrt{2}$

  (x) -1.8871

 (b) (i) $\dfrac{2}{\sqrt{3}}$ (ii) $-\dfrac{2}{\sqrt{3}}$ (iii) -1

  (iv) infinite (v) -1.3764 (vi) 2

2

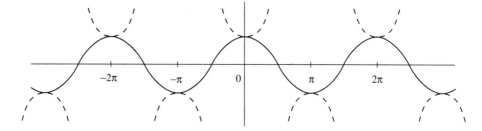

3

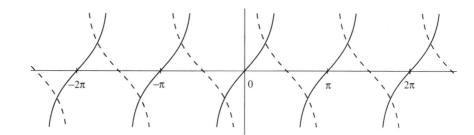

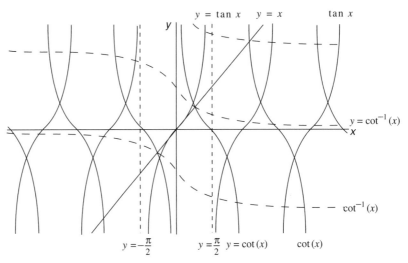

4

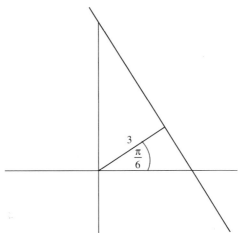

6 (c) $y = \dfrac{-1}{2\sqrt{3}}(8x + 7)$

7 (a) $x = 2n\pi \pm \dfrac{\pi}{6}$

(b) $x = n\pi + \dfrac{3\pi}{4}$

(c) $x = n\pi + \dfrac{\pi}{6}$

(d) $x = n\pi + \dfrac{\pi}{12}, n\pi + \dfrac{5\pi}{12}$

(e) $x = \dfrac{n\pi}{3} + \dfrac{\pi}{6}$

(f) $x = \dfrac{n\pi}{2}, \dfrac{n\pi}{2} + \dfrac{\pi}{12}$

Exercise 1.2

1 (a) $2\sin^2\theta - 1$ (b) $2 - 2\sin^2\theta - 4\sin\theta$

2 (a) $3 + 2t^2$ (b) $1 + 3t^2$

3 (a) $\sec^2\theta$ (b) $\tan\theta$

4 $\tan\theta = \pm\sqrt{2}$

5 $\tan\theta = \pm\dfrac{1}{\sqrt{2}},\ \cos\theta = \pm\sqrt{\dfrac{2}{3}},\ \sin\theta = \pm\dfrac{1}{\sqrt{3}}$

6 (a) $45°, 135°$ (b) $54°.73, 125°.27$
   (c) $90°$ (d) $45°, 63.43°$
   (e) $60°, 180°$ (f) $60°, 180°$
   (g) $90°$

7 (a) $30°, 150°$ (b) $0, 180°, 360°$
   (c) $90°, 210°, 330°$

9 $r = 4.717, \theta = 148.0°$

10 (a) $11.75°, 191.75°, 78.25°, 258.25°$
   (b) $19.98°, 160.02°, 199.98°, 340.02°$
   (c) $100.66°, 169.34°, 280.66°, 349.340°$
   (d) $31.65°, 121.65°, 211.65°, 310.65°$, etc.

11 (a) $143.1°$ (b) $208.1°$ (c) $163.7°$

12

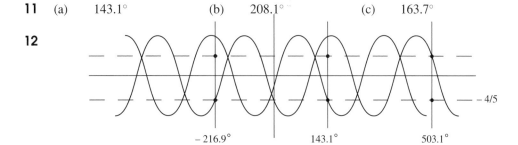

General solution $143.1° + n.360°$

Exercise 1.3

1 (a)   0   (b) $-\dfrac{7}{25}$ (c)   0

    (d) $\dfrac{24}{7}$ (e) $\dfrac{117}{125}$

3 (a) $\dfrac{1}{2}(\sin 80° - \sin 4°)$ (b) $\dfrac{1}{2}(\cos 120° + \cos 40°)$

    (c) $\dfrac{1}{2}(\cos 14° - \cos 90°)$ (d) $\dfrac{1}{2}(\sin 100° + \sin 20°)$

    (e) $\dfrac{1}{2}(\sin 300° + \sin 20°)$ (f) $\dfrac{1}{2}(\cos 4x + \cos 2d)$

4 (a) $\sqrt{3}$ (b) $\dfrac{2 - \sqrt{3}}{4}$ (c) $2 - \sqrt{3}$

    (d) $\left(\sqrt{2 + \sqrt{2}}\right)\Big/ 2$

5 (a) $\dfrac{117}{125}$ (b) $\dfrac{4}{5}$ (c) $-7/25$ (d) $117/44$

7 (a)   0°, 120°, 240°, 360°   (b)   90°, 216.87°, 323.13°

    (c)   0°, 360°   (d)   180°

    (e)   30°, 60°, 150°, 300°   (f)   30°, 90°, 150°

8 (a)   0°, 30°, 90°, 150°, 180°, 270°, 360°

    (b)   0°, 60°, 180°, 300°, 360°

    (c)   22.5°, 90°, 112.5°, 202.5°, 270°, 292.5°

10 (a) $\sqrt{3}/2$ (b)   1   (c)   49.11°

     (d) $4x^3 - 3x$ (e) $1 - 2x^2$ (f) $\dfrac{x}{\sqrt{1 + x^2}}$

     (g) $\dfrac{2x}{1 + x^2}$

11   62.0 m

13 (a) $5y + 7x = 0$ (b) $y = \dfrac{1}{7}(6x + 1)$

14 $x = y = 30°$, $x = y = 150°$

15 $x = 55.05°, y = 29.21°$ or vice versa; also $124.95°, 150.79°$

16 (b) $\tan^{-1}\dfrac{14}{5} = 70.35°$

18 (a) $4\sin 8x - \sin 2x$ (b) $-(3\sin 6x + \sin 2x)$

(c) $2\cos 4x - \cos 2x$

19 (a) $\dfrac{1}{4}\left(\sin 2x - \dfrac{1}{4}\sin 8x\right) + C$ (b) $\dfrac{1}{2}\left(x - \dfrac{1}{2}\sin 2x\right) + C$

(c) $\dfrac{1}{4}\left(\dfrac{1}{3}\cos 3x - \cos x\right) + C$ (d) $\dfrac{5\pi}{4}$

(e) $\dfrac{9}{8}$

Exercise 1.4

1 (a) $5\cos(\theta + 216.87°)$ (b) $25\cos(\theta - 73.73°)$

(c) $\sqrt{2}\sin(\theta - 315°)$ (d) $\sqrt{10}\sin(\theta + 314.57°)$

(e) $13\cos(\theta - 37.62°)$ (f) $2\sqrt{5}\cos(\theta + 296.57°)$

2 (a) $103.28°, 330.45°$ (b) $50.52°, 256.35°$

(c) $82.62°, 322.62°$ (d) $0°, 270°, 360°$

(e) $97.32°, 230.15°$

3 $17\cos(2\theta + 61.93°)$; $149.03°, 329.03°$

4 (a) $\sqrt{3}\cos\left(\omega t + \dfrac{11\pi}{6}\right)$ (b) $\sqrt{3}\cos\left(\omega t - \dfrac{\pi}{6}\right)$ (in radians)

5 (a) $7\sin\omega t + 24\cos\omega t$ (b) $6\cos\omega t - 2\sin\omega t$

7 $A\left(\sin\omega_c t + \dfrac{B}{2}(\cos(\omega_c - \omega_m)t - \cos(\omega_c + \omega_m)t)\right)$

8 (a) $2(50(1 + 16\pi^2)^{1/2})\sin 100\pi t$, $\phi = \tan^{-1} 4\pi = 0$

(b) $Z = (R^2 + 4\pi^2 f^2 L^2)^{1/2}$, $\phi = \tan^{-1}(2\pi f L/R)$

10 (a) $-5, 5$ (b) $\dfrac{1}{11}, 1$

11 (b) $-a\sin(ax + b)$

12 $\quad y = x - \dfrac{\pi}{2} + \dfrac{\sqrt{3}}{2}$

13 (b) $\quad$ 1.37

14 (a) $\quad -\dfrac{1}{a}\cos(ax+b) + C, \quad \dfrac{1}{a}\sin(ax+b) + C$

$\quad$ (b) $\quad \dfrac{1}{2}, \quad \dfrac{\sqrt{2}}{6} - \dfrac{\cos(3p + \pi/4)}{3}$ $\qquad$ (c) $\quad \dfrac{\pi}{2}$

15 (b) $\quad \theta = \cos^{-1}\sqrt{\dfrac{2}{3}};$ max volume $= \dfrac{32\pi a^3}{81}$

16 (a) $\quad 0, 0, 0, 0, 0, 0, 0, 0, 0, 0$

$\quad$ (b) $\quad -1, 1, -1, 1, -1, 1, -1, 1, -1, 1$

$\quad$ (c) $\quad 1, 0, -1, 0, 1, 0, -1, 0, 1, 0$

$\quad$ (d) $\quad 1, \sqrt{2}, 1, 0, -1, -\sqrt{2}, -1, 0, 1, \sqrt{2}$

$\quad$ (e) $\quad 1, -1, -2, -1, 1, 2, 1, -1, -2, -1$

$\quad$ (f) $\quad 1, -1, 1, -1, 1, -1, 1, -1, 1, -1$

$\qquad$ (b) only is $(-1)^n$ $\qquad$ (f) is $(-1)^{n-1}$ $\quad$ or $\quad (-1)^{n+1}$

2 SOLVING EQUATIONS

Introduction

Equations lie at the heart of mathematical models in science and engineering. Methods of solution must therefore be accurate and efficient. Many equations can be solved algebraically and our aim is to present some of the more common types. But most equations cannot be solved algebraically, therefore numerical methods are needed to obtain good approximations to their roots.

Objectives

After working through this chapter you should be able to

- Understand the nature of the roots of polynomial equations
- Solve equations involving simple surd expressions
- Reduce suitable equations to equivalent quadratic equations and identify which of their roots are valid
- Know the formulae for the sum and product of the roots of a quadratic equation
- Use the sum and product to obtain the quadratic equation whose roots are simple combinations of the roots for a given quadratic
- Solve equations involving logarithmic and exponential functions and identify which of the solutions are valid
- Use logarithms to solve equations where the unknown is an index
- Use equations to help solve inequalities
- Recognise hyperbolic functions and their inverses
- Locate roots by noting sign changes in a function
- Use simple iterative methods such as successive bisection
- Use the Newton–Raphson method and the secant method.

2.1 POLYNOMIAL EQUATIONS

The practical solution of equations by simple direct methods is generally the exception rather than the rule as equations become more intricate. A seemingly innocuous equation like $x = e^{-x}$ cannot be solved directly but we will see later that it has only one root and a way to find that root will be described.

However, for the moment we will concentrate on classically solvable equations and give an overview of some of the methods available. Equations that can be solved classically fall largely into one of the following types:

(a) Polynomial equations that can be factorised into linear and quadratic factors.

(b) Rational equations that can be reduced into a form of the type (a).

(c) Simple functional equations, e.g. surds, that can be reduced either to type (a) or to type (b) and then to type (a).

(d) Trigonometric equations reducible to type (a).

(e) Exponential or logarithmic equations that by suitable variable change or by taking logs or exponents converts them to type (a).

In this section we look systematically at polynomials, their roots and rational functions before moving on in the next section to exponential and logarithmic equations, some of which reduce to polynomial form. We then consider irreducible equations such as $x = e^{-x}$ and non-direct methods of solution.

Surd properties and equations

We know that the positive integer N possesses two square roots, $\sqrt{N}$ and $-\sqrt{N}$. These are irrational unless N is a perfect square but both are expressible as the roots of the **synthetic**, i.e. artificial, quadratic equation

$$x^2 - N = (x - \sqrt{N})(x + \sqrt{N}) = 0$$

However, if we seek a **synthetic factorisation**, i.e. artificial factorisation, of the quadratic $x^2 - p - \sqrt{q}$ in the form

$$(x^2 - p - \sqrt{q}) \equiv (x - a - \sqrt{b})(x + a + \sqrt{b}) = 0$$

where p, q, a, b are all rational, then separating the rational part of the equation from the square root part, we require

$$p + \sqrt{q} = (a + \sqrt{b})^2 = a^2 + b + 2a\sqrt{b}$$

i.e. $$p = a^2 + b, \qquad q = 4a^2b$$

Eliminating a^2 gives $q = 4b(p - b)$, which is a quadratic equation in b. This must therefore have rational roots for the synthetic factorisation to be possible. As this will only be true for suitably chosen values of p and q, we reason thsat $\sqrt{p + \sqrt{q}}$ is **not** generally expressible as $\pm(a \pm \sqrt{b})$, though it may be in special cases, as we see in the exercises.

Putting it another way, rational numbers and irrational square roots cannot be equated with one another, so any surds possessing rational and irrational elements must satisfy the following properties:

(a) $a + \sqrt{b} = 0 \iff a = 0$ and $b = 0$

(b) $a + \sqrt{b} = c + \sqrt{d} \iff a = c$ and $b = d$

These properties are most important in solving surd equations. Note that $\iff$ means **implies and is implied by**.

Examples

1. Solve the equation

$$2 + \sqrt{3} = x + 1 + \frac{1}{2}\sqrt{(x + 2)(x + 3)}$$

Equating the rational and square root parts, we get $x = 1$ directly, and squaring the surd parts

$$3 = \frac{1}{4}(x + 2)(x + 3) \quad \text{i.e.} \quad x = 1, -6$$

we see that $x = 1$ is the only possible solution.

2. Solve $\sqrt{x + 5} - \sqrt{x + 2} = \sqrt{2x + 3}$.

Solving a surd equation very often involves squaring both sides of the equation twice over, following the transposing of terms.
Square both sides to obtain

$$x + 5 + x + 2 - 2\sqrt{(x + 5)(x + 2)} = 2x + 3$$

i.e. $$\sqrt{(x + 5)(x + 2)} = 2$$

Square again to give

$$x^2 + 7x + 10 = 4$$

or $x^2 + 7x + 6 = 0$, i.e. $x = -1$ or -6

Substituting these values into the original equation, where positive square roots are needed, shows $x = -1$ to be the only admissible root. The other root $x = -6$ was introduced by squaring the equation. ∎

Equations reducible to a quadratic

Surd equations of the form given above reduce to a quadratic equation which is readily solved. It may happen that only one of the two roots is acceptable.

A simple rational equation such as

$$\frac{(x-1)^2}{x-2} = \frac{9}{2}$$

reduces via cross-multiplication to

$$2(x^2 - 2x + 1) = 9(x - 2)$$

i.e. $2x^2 - 13x + 20 = 0$ which solves to give

$$x = \frac{5}{2} \quad \text{or} \quad x = 4$$

We are equipped to solve quadratic equations only, not equations of higher order. Sometimes, however, a transformation or substitution can make solution possible.

Example   Solve $x(x+2) + 1 = \dfrac{12}{x(x+2)}$

Cross-multiplication directly would lead to a quartic equation. Instead, set $y = x(x+2)$ to obtain

$$y + 1 = \frac{12}{y}$$

so that $y(y+1) = 12$ or $y^2 + y - 12 = 0$

hence $(y-3)(y+4) = 0$,   i.e. $y = 3$ or -4

For $y = 3$ we must solve

$$x(x + 2) = 3$$

i.e. $x^2 + 2x - 3 = 0$

hence $(x - 1)(x + 3) = 0,$ giving $x = 1$ or -3

For $y = -4$ we must solve

$$x^2 + 2x + 4 = 0$$

But $x^2 + 2x + 4 \equiv (x + 1)^2 + 3$ has no roots, so $x = 1$ and $x = -3$ are the only real roots of the original equation.   ■

Properties and roots of polynomials

The general quadratic equation is usually written in the form $ax^2 + bx + c = 0$. If $a = 0$ the quadratic becomes linear, i.e. $bx + c = 0$ to which the solution is simply $x = -b/c$.

If we wish therefore to examine the root properties of a quadratic, we can assume the coefficient of x^2 is not zero and *without loss of generality* make it equal to 1. For example, the equations $3x^2 - 6x + 9 = 0$ and $x^2 - 2x + 3 = 0$ have identical solutions. Polynomials with their leading coefficients equal to 1 are said to be *monic*.

Assume the monic quadratic $x^2 + px + q$ has roots α and β. These roots may be rational or irrational. Then

$$x^2 + px + q \equiv (x - \alpha)(x - \beta) = 0$$

Comparing coefficients yields

$$\alpha + \beta = -p$$
$$\alpha\beta = q$$

In other words, the middle-term coefficient p is equal to **minus the sum of the roots** and the constant term q is equal to the **product of the roots**.

Example   Call the roots of the following quadratic equation α and β:

$$x^2 - 5x + 2 = 0$$

Without solving this equation, determine the quadratic equations with integer coefficients whose roots are

(a) α^2 and β^2

(b) $\alpha(\alpha + 1)$ and $\beta(\beta + 1)$

(c) $\dfrac{\alpha}{\beta}$ and $\dfrac{\beta}{\alpha}$
(d) $\dfrac{\alpha}{\alpha+\beta}$ and $\dfrac{\beta}{\alpha+\beta}$

Solution

Observe first of all that

$$\alpha + \beta = 5$$
$$\alpha\beta = 2$$

(a) We require first the quadratic expression

$$(x - \alpha^2)(x - \beta^2) \equiv x^2 - (\alpha^2 + \beta^2)x + (\alpha\beta)^2$$

The constant term, $(\alpha\beta)^2 = 4$, and since $\alpha^2 + \beta^2 = (\alpha + \beta)^2 - 2\alpha\beta$ then $\alpha^2 + \beta^2 = 25 - 4 = 21$, so the new quadratic equation is

$$x^2 - 21x + 4 = 0$$

(b) With $\alpha(\alpha + 1)$ and $\beta(\beta + 1)$, we have

$$\text{'minus the sum' of the roots} = \alpha^2 + \alpha + \beta^2 + \beta = \alpha^2 + \beta^2 + \alpha + \beta$$
$$= 21 + 5 = 26$$

$$\text{product of the roots} = \alpha\beta(\alpha\beta + \alpha + \beta + 1)$$
$$= 2(2+5+1) = 16$$

Hence $x^2 - 26x + 16 = 0$

(c)
$$\text{minus the sum} = \frac{\alpha}{\beta} + \frac{\beta}{\alpha} = \frac{\alpha^2 + \beta^2}{\alpha\beta} = \frac{21}{2}$$

$$\text{product} = \frac{\alpha}{\beta}\frac{\beta}{\alpha} = 1$$

The quadratic equation with integer coefficients is therefore

$$2x^2 - 21x + 2 = 0$$

(d)
$$\text{minus the sum} = \frac{\alpha}{\alpha+\beta} + \frac{\beta}{\alpha+\beta} = 1$$

$$\text{product} = \frac{\alpha\beta}{(\alpha+\beta)^2} = \frac{2}{25}$$

The quadratic equation is thus

$$25x^2 - 25x + 2 = 0$$ ∎

Suppose we have a cubic equation in the form

$$x^3 + px^2 + qx + r \equiv (x - \alpha)(x - \beta)(x - \gamma) = 0$$

then comparing coefficients gives

$$\alpha + \beta + \gamma = -p$$
$$\alpha\beta + \beta\gamma + \gamma\alpha = q$$
$$\alpha\beta\gamma = -r$$

This is more awkward to manipulate, but the sum and product of the roots are once again simply related to the coefficients. A polynomial of degree n, in monic form, can be written as

$$P_n(x) = x^n + a_{n-1}x^{n-1} + \cdots + a_0$$

It is reasonable that a quadratic has a maximum of two roots, that a cubic has a maximum of three roots, etc. The fundamental theorem of algebra states that a polynomial of degree n possesses n roots exactly when repeated roots are counted as many times as they are repeated and any complex roots are also included.

Therefore

$$P_n(x) = x^n + a_{n-1}x^{n-1} + \cdots + a_0$$
$$\equiv (x - \alpha_1)(x - \alpha_2)\cdots(x - \alpha_n)$$
$$\equiv x^n - (\alpha_1 + \alpha_2 + \cdots + \alpha_n)x^{n-1} + \cdots + (-1)^n\alpha_1\alpha_2\alpha_3\ldots\alpha_n = 0$$

We therefore have the general result

> **minus** sum of roots $=$ coefficient of x^{n-1}
>
> $(-1)^n$ product of roots $=$ constant term

The precise location of the roots is a more difficult task and beyond our scope, except for the quadratic polynomial. Note that many cubic and quartic equations can be solved by algebraic methods but not quintic equations and those of higher order. Computer algebra suites are pre-programmed to use algebraic methods. However, we can quite easily determine **bounds** within which all roots of a polynomial must lie.

Example Prove that the equation $x^5 + 4x^4 - 19x^3 + 21x^2 + 100x - 219 = 0$ has no root larger than 8 in magnitude.

Recall that

$$|a + b| \le |a| + |b|$$

and that

$$|a + b + c| \le |a + b| + |c|$$
$$\le |a| + |b| + |c|, \text{ etc.}$$

If $|x| = 8$ then

$$|4x^4 - 19x^3 + 21x^2 + 100x - 219|$$
$$\le 4|x|^4 + 19|x|^3 + 21|x|^2 + 100|x| + 219 = 28\,475$$

whereas $|x|^5 = 32\,768$. The quintic cannot possibly be zero for $|x| = 8$ or any larger value of $|x|$.

Note that the sum of the roots is -4 and the product 219. This information alone is insufficient to tell whether or not roots are real or complex, or what magnitude they have.

■

Composite equations

The roots of composite equations comprising polynomial, exponential, logarithmic and trigonometric functions in different combinations can very often be found by examining the roots of the constituent parts.

Example (a) Determine the roots of $f(x) = 0$ for the functions $f(x)$ given below.

(i) $\dfrac{x^2 - x - 12}{x^2 + 1}$ (ii) $(2x - 5) \sin \pi x$

(iii) $e^x(x^2 - 5x + 6)$ (iv) $\ln(x - 3) + \ln\left(x - \dfrac{1}{3}\right)$

(b) When is (i) negative? Sketch its graph.

Solution

(a) (i) $x^2 + 1 \ge 1$ and the roots of $f(x)$ are the roots of $x^2 - x - 12 = 0$. These are $x = -3$ and 4.

(ii) $2x - 5 = 0$ when $x = \dfrac{5}{2}$ and $\sin \pi x = 0$ when $x = n$, integer. The roots are therefore $x = \dfrac{5}{2}$ and all integers.

(iii) In Foundation Mathematics we assumed that $e^x > 0$ for all x, having plotted its graph. We shall later prove this rigorously, but for now observe that $e^x > x$ if $x > 0$, so $e^x > 0$ if

$x > 0$. Also if $x > 0$ then $-x < 0$, and since $e^{-x} = \dfrac{1}{e^x}$, we have $e^{-x} > 0$ if

$e^x > 0$, thus $e^x > 0$ for all $x < 0$, and $e^0 = 1$. Hence $f(x) = e^x(x^2 - 5x + 6) = 0$ only when $x^2 - 5x + 6 = 0$, i.e. $x = 2$ or $x = 3$.

(iv) $\ln(x - 3) + \ln\left(x - \dfrac{1}{3}\right) = \ln\{(x - 3)\left(x - \dfrac{1}{3}\right)\} = \ln\left(x^2 - \dfrac{10}{3}x + 1\right)$ which is

zero when $x^2 - \dfrac{10}{3}x + 1 = 1$, so $x = 0$ or $\dfrac{10}{3}$. However, $f(x)$ is defined only for

$x > 3$, so the only valid root is $x = \dfrac{10}{3}$.

(b) The function $x^2 - x - 12$ is negative for $-3 < x < 4$. The graph of $\dfrac{x^2 - x - 12}{x^2 + 1}$ is shown in Figure 2.1. ∎

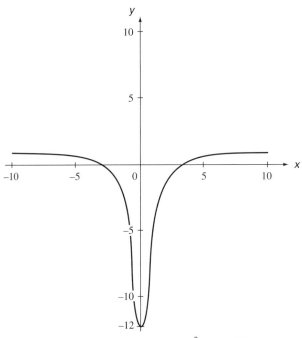

Figure 2.1 Graph of $\dfrac{x^2 - x - 12}{x^2 - 1}$

Inequalities

The roots of polynomial equations play a key part in resolving algebraic inequalities as the following example shows.

Example

Determine the values of x for which the following inequalities hold:

(a) $\dfrac{x+1}{x+2} < \dfrac{x}{x+1}$

(b) $\dfrac{x}{x+4} > 2$

Solution

(a) The inequality is equivalent to

$$\frac{x}{x+1} - \frac{x+1}{x+2} > 0$$

i.e.

$$\frac{x(x+2) - (x+1)^2}{(x+1)(x+2)} > 0$$

or

$$\frac{-1}{(x+1)(x+2)} > 0$$

Hence

$$(x+1)(x+2) < 0 \quad \text{so that} \quad -2 < x < -1.$$

(b) If we multiply the inequality by a positive quantity, the direction of the inequality is maintained. Hence we multiply by $(x+4)^2$ to obtain

$$x(x+4) > 2(x+4)^2$$

i.e. $(x+4)(x - 2(x+4)) > 0$

or $(x+4)(-x-8) > 0$

Then

and hence $(x+4)(x+8) < 0$

$$-8 < x < -4 \qquad \blacksquare$$

In the next chapter we will look closely at graphs with asymptotes and with limited domains of existence.

Closely related to polynomial equations are trigonometric and exponential equations which can be reduced to quadratic and similar forms. We will be looking at exponential and logarithmic equations in the next section, and in order to set out the ground rules, we shall focus upon related ideas that we have already met in solving trigonometric equations.

Examples

1. Determine all solutions of the equation $\sin 3\theta + \sin \theta = 0$ by writing $\sin 3\theta$ as a polynomial in $\sin \theta$. Recall that $\sin 3\theta \equiv 4 \sin^3 \theta - 3 \sin \theta$, so that $4 \sin^3 \theta - 2 \sin \theta = 0$ hence $\sin \theta = 0$ or $\pm \dfrac{1}{\sqrt{2}}$.

   The solutions are therefore $\theta = n\pi$ or $\theta = (2n - 1)\dfrac{\pi}{4}$, n integer. In other words, the solutions are all integer multiples of π or all odd integer multiples of $\dfrac{\pi}{4}$.

2. Prove that

   (a) $\tan^{-1}\left(\dfrac{1}{3}\right) + \sin^{-1}\left(\dfrac{1}{\sqrt{5}}\right) = \dfrac{\pi}{4}$

   (b) $\tan^{-1} x + \cot^{-1} x = \dfrac{\pi}{2}$, x acute

   (c) Show there are only two values of x which satisfy the equation

   $$\tan^{-1}(2x + 1) + \tan^{-1}(2x - 1) = \tan^{-1} 2$$

Solution

(a) Construct a right-angled triangle, as in Section 1.3. If $\alpha = \tan^{-1}\left(\dfrac{1}{3}\right)$ and $\beta = \sin^{-1}\left(\dfrac{1}{\sqrt{5}}\right)$ then we see that $\beta = \tan^{-1}\left(\dfrac{1}{2}\right)$. Refer to Figure 2.2(a).

Then $\quad \tan(\alpha + \beta) = \dfrac{\dfrac{1}{3} + \dfrac{1}{2}}{1 - \dfrac{1}{3} \times \dfrac{1}{2}} = 1 \quad$ and $\quad \tan^{-1}(1) = \dfrac{\pi}{4}$

Hence $\alpha + \beta = \dfrac{\pi}{4}$ as required.

(b) Figure 2.2(b) demonstrates the result since

$$x = \tan \alpha = \cot \beta \quad \text{and} \quad \alpha + \beta = \dfrac{\pi}{2}$$

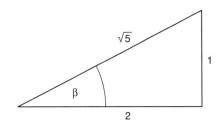

(a)

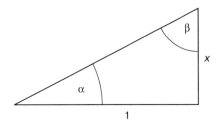

(b)

Figure 2.2 Diagrams to prove (a) $\tan^{-1}\left(\dfrac{1}{3}\right) + \sin^{-1}\left(\dfrac{1}{\sqrt{5}}\right) = \dfrac{\pi}{4}$ and

(b) $\tan^{-1} x + \cot^{-1} x = \dfrac{\pi}{2}$, x acute

$$\left(\text{Also } \tan(\tan^{-1} x + \cot^{-1} x) = \frac{x + \dfrac{1}{x}}{1 - x\dfrac{1}{x}} = +\infty, \text{ and } \tan^{-1}(\infty) = \frac{\pi}{2}\right)$$

(c) Take tangents of both sides to get

$$\frac{2x + 1 + 2x - 1}{1 - (2x + 1)(2x - 1)} = 2$$

i.e. $$\frac{4x}{2 - 4x^2} = 2$$

and hence $$2x^2 + x - 1 = 0$$

so that $x = \dfrac{1}{2}$ or $x = -1$. These are the two values required. ∎

Exercise 2.1

1   Find the square roots of the following numbers in the form shown in brackets.

(a) $11 + 6\sqrt{2}$ $(a + \sqrt{b})$ (b) $4(2 - \sqrt{3})$ $(\sqrt{a} - \sqrt{b})$

(c)   If $\sqrt{a + \sqrt{b}} = \pm(\sqrt{x} + \sqrt{y})$
prove that x and y satisfy the simultaneous equations

$$x + y = a \qquad 4xy = b$$

Find the positive square root of $2(4 + \sqrt{15})$ in the above form.

2   Determine in simplest rational form the quadratic equation whose roots are

$$\frac{\sqrt{2}}{(\sqrt{2} \pm \sqrt{7})}$$

3   (a)   Solve the equation

$$\sqrt{x + 4} + \sqrt{x + 7} = \sqrt{15 + 2x}$$

(b)   Determine the two values of x which satisfy

$$\sqrt{x + 2} + \sqrt{11 + 2x} = 2\sqrt{2(x + 3)}$$

4   Expand the following as a quartic equation:

$$x(x + 2) + 1 = \frac{12}{x(x + 2)}$$

Sketch its graph, verifying there are only two real roots, $x = -1$ and -3.

5   Solve the equations

(a) $3x(x + 3) - 19 = -\dfrac{28}{x(x + 3)}$

(b) $x(x + 2) + \dfrac{3}{x(x + 2)} = 4$

(c)   Represent the sixth-order polynomial

$$(x(x+1)(x+2))^2 + 4x(x+1)(x+2) + 5$$

as a quadratic and explain why it has no real roots.

6   Solve the simultaneous equations

$$x + y = 5, \qquad x^2 y^2 + 3xy = 28$$

Start by putting $z = xy$ to solve the second equation for xy.

7   Solve the simultaneous equations

$$\frac{7}{x+y} + \frac{1}{x-y} = 2, \qquad x^2 - y^2 = 7$$

8   Label the roots of the equations $x^2 - 7x + 5 = 0$ as α and β.

(a)   Write down the values of

(i) $\alpha + \beta$ (ii) $\alpha\beta$

(b)   Without solving the quadratic equation above, write down the quadratic equations with integer coefficients whose roots are

(i) α^2 and β^2 (ii) α^3 and β^3

(iii) α/β and β/α (iv) $\dfrac{\alpha}{\alpha+\beta}$ and $\dfrac{\beta}{\alpha+\beta}$

Do not attempt to find α and β.

9   Find the value of K given that the equation

$$x^2 + 2(3+K)x + K^2 = 0$$

has equal roots.

10   The monic quadratic equation

$$x^2 + px + q = 0$$

has two real roots, one of which is twice as large as the other. Prove that

$$2p^2 = 9q$$

11   The cubic equation

$$x^2(x+3) = \frac{(x+15)}{4}$$

has roots α, β, γ. Determine the cubic equation with integer coefficients whose roots are $2\alpha, 2\beta, 2\gamma$.

12   A monic polynomial of degree n has two roots α and $\beta \neq 1$. The $n-2$ remaining roots are all equal to 1.

(a)   Write down

     (i)   the coefficient of x^{n-1}

     (ii) the constant term.

(b)   If $n = 4$ and the quartic is

$$x^4 - x^3 + ax^2 + bx - 6$$

determine the other two roots together with a and b.

13   (a)   Use the property

$$|x+y| \leq |x| + |y|$$

to obtain, without differentiation, a maximum bound for

$$f(x) = |x| + |3+x| + |6-x|, \qquad 0 \leq x \leq 3$$

(b)   For $p_6(x) = x^6 - 5x^5 + 4x^3 + 3x^2 + 17x + 3$ prove that

$$3 + 17R + 3R^2 + 4R^3 + 5R^5 \leq R^6$$

when $|x| = R$.

     What is the smallest positive integer value of R for which the inequality is strict? What conclusion do you reach?

14 If $f(x)$ is of the form $f(x) = \dfrac{N(x)}{D(x)}$ where $N(x)$ and $D(x)$ are numerator and denominator functions, roots of $f(x) = 0$ occur whenever $N(x) = 0$, provided $N(x)$ and $D(x)$ do not have zeros in common. Identify the roots of the following functions:

(a) $\dfrac{(x^2 - 5x + 6)}{x^2 + 1}$

(b) $\dfrac{(x - 1)\sin \pi x}{x^2 + 4x + 4}$

(c) $\dfrac{x - 1}{(x + 3)(x - 4)} - \dfrac{x}{x + 3}$

(d) $e^x(x^2 + ax + b)$

In (d) consider all values of a and b.

15 Determine the range of values of x for which the following inequality holds:

$$\frac{x}{8 - x} < 11$$

16 Prove that the expression

$$y = \frac{x^2 + x + 1}{x + 1}$$

takes no values between -3 and 1.

(*Hint*: Cross-multiply and obtain an inequality for y in the discriminant of the resulting quadratic equation.)

17 (a) If $x = \alpha$ is a root of the quartic equation

$$x^4 + cx^2 + d = 0$$

prove that $x = -\alpha$ is also a root of the equation.

(b) Solve the equation

$$x^4 + 22x^2 + 49 = 0$$

for x^2 first of all. Use the result of Question 1 to determine in surd form the only two real roots.

18 (a) Solve the equation

$$\tan^3 \theta - \tan^2 \theta = 3(\tan \theta - 1) \qquad 0 \le \theta < 360°$$

(b)   If $x\cos\theta + y\sin\theta = a$ and $x\sin\theta - y\cos\theta = b$, prove that

$$\tan\theta = \frac{bx + ay}{ax - by}, \qquad x^2 + y^2 = a^2 + b^2$$

(c)   If $\theta = \tan^{-1}\left(\frac{1}{3}\right)$ and $\phi = \tan^{-1}\left(\frac{1}{2}\right)$, prove that $\theta + \phi = \frac{\pi}{4}$ (θ and ϕ acute).

(d)   Prove that

$$\cot^{-1}\left(\frac{1}{3}\right) = \cot^{-1}3 + \cos^{-1}\left(\frac{3}{5}\right)$$

(e)   Determine x from the equation

$$\tan^{-1}(2x) + \tan^{-1}(3x) = \frac{\pi}{4}, \quad x \text{ acute}$$

(f)   If $x = \cos\beta$ and β is an acute angle, prove that

$$\sin^{-1}x + \cos^{-1}x = \text{constant}$$

Draw a right-angled triangle to help prove the result and determine the constant.

19   Write down expressions for $\cos 2\theta$ and $\cos 3\theta$ in terms of $\cos\theta$. If θ happens to be very small so that $\cos\theta = 1 - x$, where x is small and x^2 can be neglected, prove that $\cos 2\theta \approx 1 - 4x$ and $\cos 3\theta \approx 1 - 9x$.

20*   The equation $\sin x = \lambda x$ cannot be solved by conventional means. There is, however, one trivial solution. On the same axes draw $y = \sin x$ and a selection of lines $y = \lambda x$ of different gradients both positive and negative.

(a)   Write down the trivial solution.

(b)   Prove that if $x = \alpha$ is a root of the above equation, so is $x = -\alpha$, whatever λ. Replace α by $-\alpha$ in the equation and illustrate this property graphically.

(c)   Use the illustration in (b) to demonstrate how the equation will always possess an odd number of roots. Do not attempt to solve it.

(d)   How many roots does the equation have when

(i) $\lambda > 1$ (ii) $\lambda = 0.3$

(iii) $\lambda = 0.1$ (iv) λ very small and positive.

As $\lambda \to 0$ what do the roots of $\sin x = \lambda x$ become?

2.2 EXPONENTIAL AND LOGARITHMIC EQUATIONS

Exponential equations

In Foundation Mathematics we defined the exponential and logarithmic functions. The number $e \simeq 2.718$ is a **natural constant** (like π) with the unique property that when it is raised to the variable power x, i.e. e^x, the gradient of e^x at the point, satisfies

$$\frac{d}{dx}(e^x) = e^x$$

If $f: x \mapsto e^x$ or $\exp(x)$ defines the exponential function, then the inverse function $f^{-1}: x \mapsto \ln x$ or $\exp^{-1}(x)$ defines the logarithmic function. The functions are plotted on the same axes in Figure 2.3.

We recall that $y = e^x$ and $y = \ln x$, i.e. $f(x)$ and $f^{-1}(x)$ are reflections of each other in the line $y = x$, and from Foundation Mathematics we have

$$\frac{d}{dx}(\ln x) = \frac{1}{x}$$

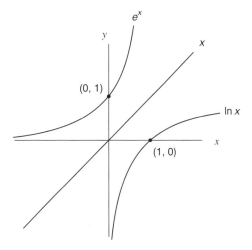

Figure 2.3 The logarithmic function and the exponential function

Also the domain of f is the set of all real numbers and its range the set of all positive real numbers. For f^{-1} they are reversed, so neither f nor f^{-1} is defined in the third quadrant.

Both e^x and $\ln x$ increase with x because their gradients are positive, so the equations $e^x = a$ ($a > 0$) and $\ln x = b$ possess the unique solutions $x = \ln a$ and $x = e^b$ respectively. This is shown in Figure 2.4.

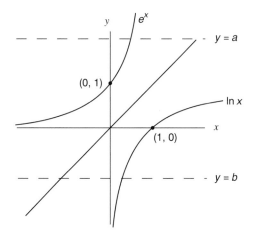

Figure 2.4 Unique solutions to $e^x = a(a > 0)$ and $\ln x = b$

In other words, we can exponentiate or take logarithms to solve equations as necessary.

One immediate consequence of the unique relationship is that all exponential and logarithmic functions can be related to **natural logarithms**, i.e. equivalents based on e. For instance, given $a > 0$, we can write $a = e^b$, so that

$$a^x = e^{bx}$$

Taking natural logarithms gives

$$x \ln a = bx, \quad \text{i.e. } b = \ln a$$

Also if $d = \log_b c$, exponentiating with respect to b, we get

$$b^d = c$$

so taking natural logarithms

$$d \ln b = \ln c,$$
$$\text{i.e. } d = \frac{\ln c}{\ln b}$$

Summarising the results, we have

$$a^x = e^{x \ln a} \quad \text{if } a > 0$$

$$\log_b c = \frac{\ln c}{\ln b} \quad \text{if } b, c > 0$$

Example

(a) If $2^x = e^{bx}$ determine b and solve the equation $2^x = 3$.

(b) If $2^x = 5^p$, determine p in terms of x.

(c) Solve the equation $3^x + 3^{-x} = 5$.

Solution

(a) $b = \ln 2 \approx 0.6931$, and if $e^{x \ln 2} = 3$, then

$$x = \frac{\ln 3}{\ln 2} = 1.5850$$

(b) $2^x = e^{x \ln 2} = 5^p = e^{p \ln 5}$

so $x \ln 2 = p \ln 5$, i.e. $p = \dfrac{\ln 2}{\ln 5} x \qquad \approx 0.4307x$

(c) Let $z = 3^x$, so that

$$z + \frac{1}{z} = 5$$

$$\text{or} \quad z^2 - 5z + 1 = 0$$

Solving gives

$$z = \frac{5 \pm \sqrt{21}}{2}$$

Since $z = 3^x$, only the positive root is acceptable, so $z = \frac{1}{2}(5 + \sqrt{21})$. But $3^x = e^{x \ln 3}$, whence

$$x = \frac{\ln\left(\frac{1}{2}(5 + \sqrt{21})\right)}{\ln 3} \approx 1.4262 \qquad \blacksquare$$

Logarithmic equations

Equations involving logarithms can be solved by experimentation.

Example   Solve the equations

(a) $\log 2x - \log(x^2 + 2x - 2) = \log 2$ (log to any base)

(b) $\log_3 x = \log_9(x + 6)$

Solution

(a) First, regroup using the property that $\log a - \log b = \log \frac{a}{b}$. This gives

$$\log \frac{2x}{x^2 + 2x - 2} = \log 2$$

Exponentiate both sides (i.e. remove logs):

$$\frac{2x}{x^2 + 2x - 2} = 2$$

i.e. $x^2 + x - 2 = 0$

so that $x = 1$ or -2

But $x = 1$ is the only viable answer because $x = -2$ gives $\log 2x = \log(-4)$, which cannot be found.

(b) Note first that $9 = 3^2$ and the logarithmic property that if $y = \log_3 x$, then $3^y = x$,

i.e. $9^{y/2} = x$ or $9^y = x^2$

So substituting in the equation above

$$\log_9 x^2 = \log_9(x + 6)$$

Exponentiating both sides gives

$$x^2 = x + 6 \quad \text{or} \quad x^2 - x - 6 = 0$$

Therefore $(x - 3)(x + 2) = 0$; hence $x = 3$ or -2. However $x = 3$ is the only viable answer, as we see by checking

$$\log_3 3 = \log_9(3 + 6) = 1 \qquad \blacksquare$$

Hyperbolic functions and equations

In Foundation Mathematics we introduced the hyperbolic sine and cosine functions together with their inverses by way of exercises. These are **hyperbolic sine** of x, defined to be

$$\sinh x = \frac{1}{2}(e^x - e^{-x}), \text{ pronounced } \textit{shine } x.$$

and **hyperbolic cosine** of x, defined to be

$$\cosh x = \frac{1}{2}(e^x + e^{-x})$$

They are exponential in nature, not trigonometric, so they are not periodic. But they do satisfy similar identities, the more important of which are summarised below.

$$\cosh^2 x - \sinh^2 x \equiv 1$$
$$\sinh 2x \equiv 2 \sinh x \cosh x$$
$$\cosh 2x \equiv 2 \cosh^2 x - 1$$

You can prove them for yourself, along with the key properties

$\sinh x = 0$ when $x = 0$ only

$\cosh x \geq 1$

$\sinh x$ is an odd function

$\cosh x$ is an even function

Also in common use is the **hyperbolic tangent** of x:

$$\tanh x = \frac{\sinh x}{\cosh x} = \frac{e^x - e^{-x}}{e^x + e^x}$$

an odd function whose values must lie between ± 1.

You can determine hyperbolic function values on your calculator by pressing

hyp   cos

etc. For example check that $\cosh 2 = 3.7621$.

Note also the reciprocal functions

$$\operatorname{cosech} x = \frac{1}{\sinh x} \qquad \operatorname{sech} x = \frac{1}{\cosh x} \qquad \coth x = \frac{1}{\tanh x}$$

The graphs of $\sinh x$, $\cosh x$ and $\tanh x$ are shown in Figure 2.5.

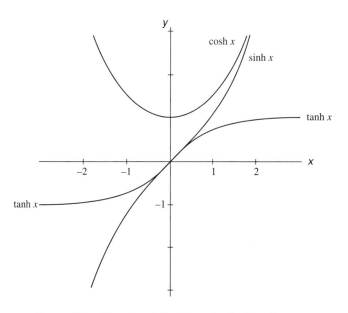

Figure 2.5 Graphs of the hyperbolic functions

Directly related to the hyperbolic functions are the **inverse hyperbolic functions** $\sinh^{-1} x$, $\cosh^{-1} x$, $\tanh^{-1} x$, which are logarithmic in character. By reflecting the original hyperbolic functions in the line $y = x$, their graphs can be drawn as shown in Figure 2.6.

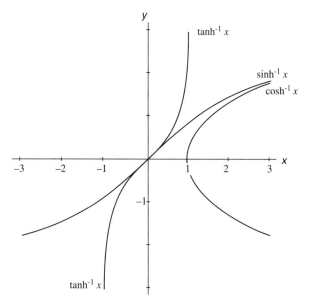

Figure 2.6   Graphs of the inverse hyperbolic functions

These functions can be written in the 'logarithmic equivalent' form

$$\sinh^{-1} x = \ln(x + \sqrt{x^2 + 1})$$

$$\cosh^{-1} x = \ln(x \pm \sqrt{x^2 - 1}) = \pm \ln(x + \sqrt{x^2 - 1})$$

$$\tanh^{-1} x = \frac{1}{2} \ln\left(\frac{1 + x}{1 - x}\right)$$

The derivation of these results is described in the exercises to Chapter 10 of *Foundation Mathematics*.

Both $\sinh^{-1} x$ and $\tanh^{-1} x$ are single-valued and $\tanh^{-1} x$ is defined for $-1 < x < 1$ only, whereas $\cosh^{-1} x$ is double-valued. The values of $\sinh^{-1}$ and $\tanh^{-1}$ and the upper branch of $\cosh^{-1}$ can be determined by using

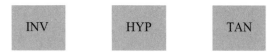

etc. on the calculator, Check for example that $\tanh^{-1}(0.5) = 0.5493$.

Example   (a) Solve the equation $5 \cosh x - 3 \sinh x = 5$.

(b) Solve the simultaneous equations

$$\cosh x = 4 \sinh y, \qquad 12 \cosh y + 3 \sinh x = 17$$

Solution

(a) Express the hyperbolic functions in terms of e^x, e^{-x}, whence

$$\frac{5}{2}(e^x + e^{-x}) - \frac{3}{2}(e^x - e^{-x}) = 5$$

Put $e^x = z$, then

$$5\left(z + \frac{1}{z}\right) - 3\left(z - \frac{1}{z}\right) = 10.$$

or $z^2 - 5z + 4 = 0$

Therefore $z = 1$ or 4, so that $x = 0$ or $\ln 4$. (We could equally well have eliminated $\cosh x$ by squaring up and using $\cosh^2 x = 1 + \sinh^2 x$, thereby solving for $\sinh x$.)

(b) Eliminate x by setting

$$\cosh x = 4 \sinh y, \qquad \sinh x = \frac{17}{3} - 4 \cosh y$$

Since $\cosh^2 x - \sinh^2 x = 1$, we have

$$(4 \sinh y)^2 - \left(\frac{17}{3} - 4 \cosh y\right)^2 = 1$$

$$16(\sinh^2 y - \cosh^2 y) + \frac{136}{3} \cosh\, y - \frac{17}{9} = 1$$

which reduces to

$$\cosh y = \frac{13}{12}, \qquad \text{so that } \sinh y = \frac{5}{12}$$

$$\text{Hence } \cosh x = \frac{5}{3}, \ \sinh x = \frac{4}{3}.$$

$$\text{Thus} \quad x = \ln\left(\frac{4}{3} + \left(1 + \left(\frac{4}{3}\right)^2\right)^{1/2}\right) = \ln 3$$

$$\text{and} \quad y = \ln\left(\frac{5}{12} + \left(1 + \left(\frac{5}{12}\right)^2\right)^{1/2}\right) = \ln\frac{3}{2} \qquad \blacksquare$$

Integer solution and incrementation

If we do not have to solve an equation directly but instead need to satisfy an inequality involving the integers we can **step forward**, i.e. increment, so that we find the smallest or largest integer that works.

Example

An industrial cleansing process removes 85% of a synthetic dye from a contaminated nylon sample. How many cleansings are required to ensure that 99% of the dye is removed?

The amount of dye remaining after a cleansing is 0.15 of what was present before. After two cleansings it is $(0.15)^2$ proportionately and after n cleansings it is $(0.15)^n$. We require $(0.15)^n \leq 0.01$.

Taking logs on the equality, $n\ln(0.15) = \ln 0.01$, so

$$n \geq \frac{\ln 0.01}{\ln 0.15} = 2.42$$

Therefore three cleansings are required. We also see that $(0.15)^3 < 0.01$. $\qquad \blacksquare$

Sometimes there is no alternative but to increment forwards, as the following example shows.

Example

Data is transmitted in blocks of m bits. Extra parity bits are required to provide for error correction, making for a total of n bits overall. The value of n needs to be chosen large enough to ensure that

$$2^{n-m} \geq n + 1$$

for a single error-correcting code. Given that $m = 8$, how large in n?

We need $2^{n-8} \geq n + 1$. Stepping forward, we start with $n = 9$ and see that $n = 12$ is the smallest value. $\qquad \blacksquare$

Exercise 2.2

 1   By taking natural logarithms solve the equations

(a) $2^x = 10$ (b) $e^{2x} = 5$

(c) $(5.612)^x = 12$ (d) $3^{-x} = \dfrac{1}{100}$

 2   Solve the following equations in terms of the natural logarithm. Convert to a quadratic equation as necessary

(a) $e^{2x} - 6e^x + 5 = 0$ (b) $5^x + 3 \times 5^{1-x} = 8$

Determine also the integers x and y which satisfy the simultaneous equations

(c) $2^x + 3^y = 17, \qquad 2^{2x} - 5 \times 3^y = 19$

3   For the simultaneous equations

$$2^{x+y} = 8, \qquad 3^x + 3^y = 12$$

determine $x + y$ from the first equation then substitute into the second equation, solving a quadratic.

4   (a)   Determine x when

$$\log_{100}((x^2 + 2x + 4)^2) - \log_{1000} x^3 = \log_{10} 6$$

(b)   Solve the following simultaneous equations for x and y:

$$\log_{10} x - \log_{10} y = \log_{10} 2, \qquad 2\log_{10} x + \log_{10} y = 2 + \log_{10} 5$$

(c)   Given that $2\log_2 N = 3\log_8 N^2$, determine N.

(d)   Solve $\log_3 x = \log_{27}(2x^2 - x + 2)$.

5   (a)   Solve $2x = \ln(3e^x - 2)$.

(b)   Given that $\sqrt{xy} = \ln(e^{2\sqrt{xy}} - 2)$ and $x + y = 2\ln 2$, determine x and y.

6   (a)   For $\sinh x$ and $\cosh x$ establish the identities

(i) $\cosh^2 x - \sinh^2 x \equiv 1$

(ii) $\sinh 2x \equiv 2 \sinh x \cosh x$

(iii) $\cosh 2x \equiv 2\cosh^2 x - 1$

$$\equiv \cosh^2 x + \sinh^2 x$$

(b)   If $z = \sinh^{-1} x$ then $x = \sinh z = \dfrac{1}{2}(e^z - e^{-z})$. Solve a quadratic equation for e^z then take logs to establish that

$$\sinh^{-1} x = \ln(x + \sqrt{x^2 + 1}), \quad \text{for all } x$$

(c)   Prove that

$$\tanh^{-1} x = \frac{1}{2}\ln\left(\frac{1+x}{1-x}\right)$$

(d)   Establish that $x - \sqrt{x^2 - 1} \equiv \dfrac{1}{x + \sqrt{x^2 - 1}}$ and prove that

$$\cosh^{-1} x = \pm\ln(x + \sqrt{x^2 - 1}) \equiv \ln(x \pm \sqrt{x^2 - 1})$$

7 Establish the identities

(a) $\sinh(x \pm y) \equiv \sinh x \cosh y \pm \cosh x \sinh y$

(b) $\cosh(x \pm y) \equiv \cosh x \cosh y \pm \sinh x \sinh y$

Establish identities for

(c) $\tanh(x \pm y)$

In terms of the tanh function only.

8 Solve for x and y the simultaneous equations

(a) $\cosh x = 3\sinh y,$ $2\sinh x + 6\cosh y = 5$

(b) $3\cosh x + 5\sinh y = 17,$ $\sinh x \cosh y = \dfrac{52}{15}$

expressing the answers in logarithms.

9* Simplify

(a) $\sinh(\sinh^{-1} 2 + \sinh^{-1} 3)$ (b) $\cosh\left(\tanh^{-1}\left(\dfrac{1}{5}\right)\right)$

(c) $\tanh\left(\dfrac{1}{2}\sinh^{-1} 2\right)$

In terms of x determine

(d) $\sinh(2\sinh^{-1}x)$ (do not expand)

(e) $\tanh(x+\tanh^{-1}x)$ (f) $\cosh(3\ln x+2)$

10 Solve the equations

(a) $\sinh(\ln x)=2$ (b) $\tanh^{-1}2x=\tanh^{-1}x+\tanh^{-1}\left(\dfrac{1}{2}\right)$

(c) $\sinh^{-1}x=\ln 5$

11 In a binomial expansion the ratio of the $(n+1)$th term to the nth term is

$$\frac{27}{16}\left(\frac{13-n}{n}\right)$$

where n is a positive integer. Which term will have the largest positive value?

12 A fair coin is tossed until a head appears. The probability this occurs by the nth toss can be

proved to be $1-\left(\dfrac{1}{2}\right)^{n}$

(a)   Increment forwards to determine the value of n which ensures this probability exceeds 99%.

(b)   If the coin is further tossed until a second head appears, the probability of that happening by the nth toss is $1-(1+n)\left(\dfrac{1}{2}\right)^{n}$. Continue to increment forwards to find a larger value of n than (a) which ensures the probability again exceeds 99%.

2.3   ROOTS BY SIGN CHANGE

Figure 2.7 shows graphs of $y=x$ and $y=e^{-x}$. Is the point of intersection P unique?

If $x<0$, $e^{-x}>1$ so no meeting points.
If $x>1$, $e^{-x}<1/e$ so no meeting points.

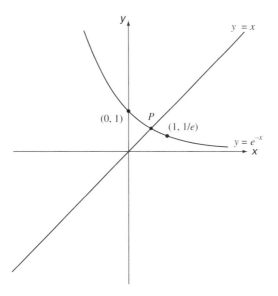

Figure 2.7 Intersection of $y = x$ and $y = e^{-x}$

Note also as x increases from left to right then e^{-x} decreases and any meeting point must be in the range $0 \leq x \leq 1$.

The strictly increasing nature of $y = x$ and strictly decreasing nature of $y = e^{-x}$ appear to rule out more than one meeting point, but how do we prove it?

Assume there are two points x_1 and x_2 where $x = e^{-x}$. Take $x_2 > x_1$, then $x_2 - x_1 = e^{-x_2} - e^{-x_1} > 0$, a contradiction as e^{-x} is a strictly decreasing function, so $x_2 = x_1 = \alpha$, say, the root. The value of α seems close to $x = 0.5$ and this will serve as a good starting-point for its determination using sign change.

Examples

1. Take $f(x) = x - e^{-x}$ and investigate its *sign change*. Tabulating $f(x)$ gives

0	0.2	0.4	0.5	0.6	0.8	1.0
-1.0000	-0.6187	-0.2703	-0.1065	$+0.0512$	0.3507	0.6321

The root seemingly lies about two-thirds of the way between $x = 0.5$ and $x = 0.6$, so taking a second decimal place we determine $f(x)$ over ranges to include such values.

0.50	0.55	0.56	0.57	0.58	0.60
-0.1065	-0.0269	-0.0112	$+0.0045$	0.0201	0.0512

The root is between 0.56 and 0.57, again about two-thirds of the way along, so we can take a third decimal place for $f(x)$.

0.560	0.566	0.567	0.568	0.570
-0.0112	-0.0018	-0.0002	$+0.0013$	0.0045

Now it seems that the root is very close to 0.567, much nearer than it is to 0.566 or 0.568. Were you to find the root this way you would doubtless try 0.5671, 0.5672, etc., so that four significant figures could be obtained. However we can reasonably conclude that the root is 0.567 to three decimal places. Systematically, we can alternatively use the **method of bisection**, in which the interval size of 'root' containment is halved each time, see the next example.

2. Show graphically that the equation $2x - 3 = \ln x$ has a root just to the right of the origin and another root near $x = 1.8$. Drawing $y = \ln x$ and $y = 2x - 3$, as in Figure 2.8, shows this quite well, the roots being at the intersection points. Note that both roots must be positive because we cannot take the logarithm of a negative number.

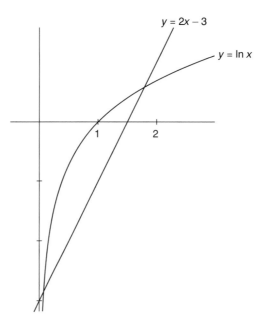

Figure 2.8   The equation $2x - 3 = \ln x$ has a root just to the right of the origin and another root near $x = 1.8$

For the upper root near $x_0 = 1.8$, the initial approximation, take $f(x) = 2x - 3 - \ln x$, whence $f_0 = f(x_0) = 0.0122$.

The root α is obviously a little lower. Try a lower interval value $x_1 = 1.75$, whence $f_1 = f(x_1) = -0.0596$, too low.

Halve the interval size, so take $x_2 = 1.775$, $f_2 = f(x_2) = -0.0238$, too low again.

The root therefore lies between 1.775 and 1.8. Successively repeating, the process, we obtain a sequence of approximations for the containment of α together with the value of $f(x)$, hopefully getting closer to zero:

$x_0 = 1.8$	$f_0 = 0.012\ 2$	α within
$x_1 = 1.75$	$f_1 = -0.059\ 6$	(1.75, 1.80)
$x_2 = 1.775$	$f_2 = -0.023\ 8$	(1.775, 1.80)
$x_3 = 1.787\ 5$	$f_3 = -0.005\ 8$	(1.787 5, 1.80)
$x_4 = 1.793\ 75$	$f_4 = 0.006\ 8$	(1.787 5, 1.793 75)
$x_5 = 1.790\ 625$	$f_5 = -0.001\ 3$	(1.790 625, 1.793 75)

There is obviously a root $\alpha \approx 1.791$ but the method of bisection approaches it relatively slowly. In the next section we will develop numerical methods of root determination that are also efficient and generate convergence to the root. ■

Exercise 2.3

 1 Given $f(x) = x^4 - 5x + 2$.

(a) Prove that $f(x) > 0$ if $x < 0$.

(b) Prove that $x^4 + 2 > 5x$ if $x > 2$.

(c) Draw the graph of $f(x)$ for $x = 0\ (0.25)\ 2$, i.e. from $x = 0$ to $x = 2$ in steps of 0.25, and show that $f(x)$ has two roots between $x = 0$ and $x = 2$. Deduce there are no further roots.

(d) Refine the step length of the graphical tabulation of $f(x)$ and obtain a realistic estimate of the upper root to 2 d.p.

(e) Use the method of bisection to achieve the result in (d) for the larger root to 2 d.p.

2 The lower root of the equation in Question 1, $2x - 3 = \ln x$, is known to lie close to 0.05. Increment (i.e. step up or down) in steps of 0.01 to locate the sign change in $f(x)$, then use the method of bisection to estimate the root to 3 d.p.

3 The method of bisection halves the width of the interval of root containment. Find how many bisections are required to reduce the interval width by a factor of

(a)   1000   (b)   10 000   (c) 10^6

4   If an equation $f(x) = 0$ is supposed to have a root α in the interval (a, b), which of the following conditions is sufficient for this to be true?

(a) $f(a) f(b) > 0$ (b) $f(a) f(b) = 0$

(c) $f(a) f(b) < 0$

5   A cubic is of the form

$$f(x) = x^3 + ax^2 + bx + c$$

Assuming the term in x^3 totally dominates $f(x)$ if x is large, what do you deduce about the sign of $f(x)$ when x is large, positive or negative? How many roots must $f(x)$ have at least?

6   Prove that $f(x) = x^3 + 3x^2 + 2x + 1$ has no positive roots. Plot $f(x)$ for $-4 \le x \le 0$ and estimate the root α in that range to 2 d.p. using change of sign methods. Differentiate $f(x)$ and determine the location and nature of the turning points. Deduce that α is the only root.

7   Write down the value of x^x for $x = 1 (1) 4$ and use the sign change method to determine the solution of $x^x = e^5$ to 2 d.p. (Taking logarithms will simplify the problem.)

8   Draw a graph to show that the curves $y = 4 - x^2$ and $y = \dfrac{1}{x}$ meet in three points. Write down the cubic equation whose roots coincide with x-coordinates of these points. Guess an approximation to the only negative root and use the bisection method to find this root to 2 d.p.

2.4   THE CONCEPT OF ITERATION

To solve the equation $f(x) = 0$ we understand that a direct and explicit determination of a root α is the exception rather than the rule. The existence of α on the other hand is manifestly clear from the plot and sign change of $f(x)$ wherever this occurs. So we must start by taking an approximation to α, say x_0, and do what we can to refine it progressively until we are tolerably close to α in numerical terms. The process of refining x_0 via a sequence $\{x_1, x_2, x_3 \ldots\}$ of values which converge to α is called **iteration** and the values themselves are called **iterates**. The method of bisection is an example of such a process, as we now see.

Example

Construct a sequence of iterates to solve $f(x) = x - e^{-x} = 0$ using the method of bisection. We know that the root α is close to $x_0 = 0.5$ from the graph, and that $0.5 < \alpha < 0.6$. Take

$$x_0 = 0.5, \quad e^{-x_0} = 0.606\,53 \quad \text{(too low)}$$
$$x_1 = 0.6, \quad e^{-x_1} = 0.548\,81 \quad \text{(too high)}$$

Split the difference, i.e. halve the internal width and take

$$x_2 = 0.55, \quad e^{-x_2} = 0.576\,94 \quad \text{(too low)}$$

The root is between 0.55 and 0.60 rather than between 0.50 and 0.55. Again, split the difference.

$$x_3 = 0.575, \quad e^{-x_3} = 0.562\,70 \quad \text{(too high)}$$

Continuing,

$$x_4 = 0.562\,5, \quad e^{-x_4} = 0.597\,83$$
$$x_5 = 0.568\,75, \quad e^{-x_5} = 0.566\,23$$
$$x_6 = 0.565\,13, \quad e^{-x_6} = 0.568\,29$$
$$x_7 = 0.566\,69, \quad e^{-x_7} = 0.567\,40$$
$$x_8 = 0.567\,47, \quad e^{-x_8} = 0.566\,96$$

The actual root $\alpha = 0.5671$ to 4 d.p. Its determination is an example of an **algorithm**, i.e. a method which solves a problem in a finite number of steps, even if the solution is approximate. Regarding root determination, the method of bisection is known to be crude and rather slow in taking 10 bisections to achieve 3 d.p.; see Question 3 of Exercise 2.3. It is, however, usually assured of success in finding a root to a prescribed accuracy. ∎

A common iteration procedure, but simple to devise, is the **method of transposition**, an example of a **fixed point method**. The idea is to rewrite $f(x)$ in the form $f(x) = x - g(x)$, so when $f(\alpha) = 0$, $\alpha - g(\alpha) = 0$, i.e. $\alpha = g(\alpha)$. In practice we determine x_0 by looking for a sign change in $f(x)$ and construct a sequence of iterates as follows:

$$x_1 = g(x_0)$$
$$x_2 = g(x_1)$$
$$\vdots$$
$$x_{n+1} = g(x_n)$$
$$\vdots$$

so that $\alpha = g(\alpha)$ in the limit.

Provided the sequence converges, we terminate the iteration when two successive iterates agree to within a specified tolerance.

Example

Rework the previous example for $f(x) = x - e^{-x}$. An obvious choice for $g(x)$ is e^{-x}, so that $x_1 = e^{-x_0}$, etc. Write

$$
\begin{aligned}
x_0 &= 0.5 \\
x_1 &= e^{-0.5} &&= 0.606\ 53 \\
x_2 &= e^{-0.60653} &&= 0.545\ 24 \\
x_3 &= e^{-0.54524} &&= 0.579\ 70 \\
x_4 &= e^{-0.57970} &&= 0.560\ 06 \\
&\ \ \vdots \\
x_{12} &= e^{-0.56728} &&= 0.567\ 07 \\
x_{13} &= e^{-0.56707} &&= 0.567\ 18 \\
x_{14} &= e^{-0.56718} &&= 0.567\ 12 \\
&\ \ \vdots
\end{aligned}
$$

The pattern of convergence is evident, and after about 20 iterations it can be seen that $\alpha \approx 0.567\ 14$ to 5 d.p. Observe too that

$$ x_1 > x_3 > x_5 \ldots > \alpha \quad \text{and that } x_2 < x_4 < x_6 \ldots < \alpha. $$

This feature is peculiar to the iterative sequence we have chosen and occurs for reasons we demonstrate in Figure 2.10. What you must know is that transposition iterative sequences converge to a root $x = \alpha$ only when $|g'(\alpha)| < 1$. We will not prove the result but the diagrams below demonstrate why the slope of $y = g(x_1)$ has to be less than the slope of $y = x$ i.e. ± 1 (angle $\pm 45°$). Figures 2.9 to 2.12 show the possible cases.

If $|g'(x)| > 1$ successive iterates are repelled away from the root. The diagrams show the importance of $|g'(\alpha)| < 1$ for convergence, and how important is its magnitude for the speed of convergence. The closer it is to zero, the faster the convergence will be. In practice $|g'(x_0)| < 1$ usually tells you whether the iterative scheme will converge.

Example

Solve the equation $2x - 3 = \ln x$ using the schemes

(a) $\quad x_{n+1} = \dfrac{1}{2}(\ln x_n + 3)$ \hspace{2cm} (b) $\quad x_{n+1} = \exp(2x_n - 3)$

for the roots near $x = 1.80$ and $x = 0.10$; retain 4 d.p. Determine $|g'|$ in each case to verify the observed behaviour.

Solution

(a) $\quad g_1(x) = \dfrac{1}{2}(\ln x + 3)$

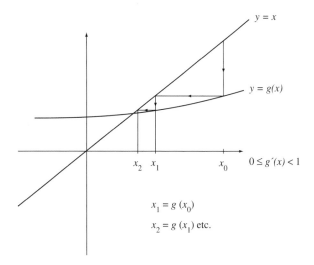

$y = x$

$y = g(x)$

x_2 x_1

x_0

$0 \le g'(x) < 1$

$x_1 = g(x_0)$

$x_2 = g(x_1)$ etc.

Figure 2.9 Convergence with positive scale factor

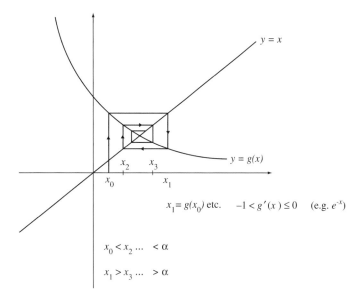

$y = x$

x_2 x_3

x_0

x_1

$y = g(x)$

$x_1 = g(x_0)$ etc. $-1 < g'(x) \le 0$ (e.g. e^{-x})

$x_0 < x_2 \ldots < \alpha$

$x_1 > x_3 \ldots > \alpha$

Figure 2.10 Convergence with negative scale factor

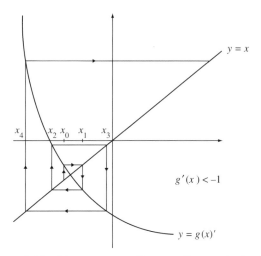

Figure 2.11   Divergence with negative scale factor

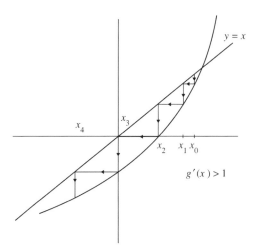

Figure 2.12   Divergence with positive scale factor

x_0	1.8000	0.1000	$g_1'(x) = \dfrac{1}{2x}$
x_1	1.7939	0.3487	
x_2	1.7922	0.9732	$|g_1'(1.8)| \approx 0.278$
x_3	1.7922	1.4864	$|g_1'(0.1)| \approx 5.0$
x_4	1.7916	1.6982	
x_5	1.7916	1.7648	
$\vdots$		1.7840	
		1.7894	
$\alpha \approx 1.7916$		1.7909	
		1.7914	
		$\vdots$	

(b) $g_2(x) = \exp(2x - 3)$

x_0	0.1000	1.8000	$g_2'(x) = 2\exp(2x - 3)$
x_1	0.0608	1.8221	
x_2	0.0562	1.9045	$|g_2'(1.8)| \approx 3.64$
x_3	0.0557	2.2458	
x_4	0.0557	4.4443	$|g_2'(0.1)| \approx 0.122$
x_5	0.0556	360.90	
$\beta \approx 0.0556$			

$x = g_1(x)$ and $x = g_2(x)$ make for convergence to the upper and lower roots, α and β respectively. Note that the approximation 0.1 to β is turned away by $g_1(x)$ into a sequence converging to α. ■

Exercise 2.4

1   The equation $\ln x = x^2 - x - 1$ has two roots near the values $x = 0.4$ and $x = 1.8$. Three possible transposition iteration methods are

(a) $x_{n+1} = x_n^2 - \ln x_n - 1$ (b) $x_{n+1} = \dfrac{\ln x_n + x_n + 1}{x_n}$

(c) $x_{n+1} = \exp(x_n^2 - x_n - 1)$

Test each method for convergence to a root, hence solve to 4 d.p.

2   Choose a suitable transposition iteration method to solve the following equation to 4 d.p.:

$$\frac{1}{e^x} - \sqrt{x} = 0$$

3   The only positive root of $x^3 - x^2 - x - 1 = 0$ is to be obtained by iteration. Assume the root is near 1.75.

(a)   Show the iteration $x_{n+1} = x_n^3 - x_n^2 - 1$ cannot be used.

(b)   Show the iteration $x_{n+1} = 1 + \dfrac{1}{x_n} + \dfrac{1}{x_n^2}$ converges for $\dfrac{10}{6} \le x \le 2$. Hence find this root to 4 d.p.

4   A way to ensure relatively quick convergence of a transposition iteration to solve $f(x) = x - g(x) = 0$ is to choose

$$g(x) = x - \frac{f(x)}{m}$$

where $m = f'(x_0)$ and x_0 is a good approximation to the root α.

(a)   Prove that $g'(x_0) = 0$.

(b) $f(x) = x^5 - 3x - 2$ has a root close to $x_0 = 1.45$. Use the above method to find the root of $f(x)$, recording each iterate to monitor the speed of the convergence; retain 5 s.f.

5   The function

$$f(x) = x^3 - x - 1$$

is known to have a root to $x = 1$. Consider the convergence properties of the transposition schemes $x_{n+1} = g_i(x_n)$, $i = 1, 2, 3$ where

$$g_1(x) = x^3 - 1$$
$$g_2(x) = 1 + 2x - x^3$$
$$g_3(x) = (x + 1)^{1/3}$$

and choose the one which best assures quick convergence to find the root; retain 5 s.f.

6 Given

$$P_6(x) = x^6 - 12x^5 - 3x^4 + 175x^3 - 132x^2 - 33x + 4$$

(a) Tabulate $P_6(x)$ and remove two of the roots which are known to be small positive integers. Determine the remaining quartic polynomial factor, $P_4(x)$.

(b) Tabulate $P_4(x)$ for $x = -4\,(1)\,12$ and determine the intervals where sign changes take place. Use nested multiplication.

(c) Construct a convergent iterative scheme for the two larger (positive and negative) roots; retain 4 s.f.

(d) Observe that the two remaining roots lie close together near $x = 0$, but on either side. Choose suitable iterative schemes for them, noting that you need a good initial starting approximation in each case. (4 sf)

(e) Verify your answers with respect to the full expanded factors of $P_6(x)$ given in the answer.

7 Draw the graphs of $y = \sin x$ and $y = e^{-x}$. Observe there is a countable infinity of roots to the equation $e^{-x} = \sin x$.

(a) Determine the smallest root using the iteration scheme

$$x_{n+1} = x_n + \frac{e^{-x_n} - \sin x_n}{e^{-x_n} + \cos x_n}$$

starting with $x_0 = 0.6$ and returning iterates to 6 d.p. What do you observe?

(b) What are the approximate values of the larger roots (to 4 d.p.) for $n = 10$.

8 We wish to compute the smallest positive roots of the equation $x = a - bx^2$, $(a, b > 0)$ by using the iterative method $x_{n+1} = a - bx_n^2$. What is the condition for convergence satisfied by a and b?

9 On the same graph draw the curves

(a) $y = \cos x$

(b) $y = e^{-x}$ $\left(0 \leq x \leq \dfrac{\pi}{2}\right)$

and estimate the common area between the two curves immediately to the right of the y-axis.

(*Hint*: the scheme $x_{n+1} = \cos^{-1}(e^{-x_n})$ converges near the root close to $x_0 = 1.2$.)

10　The convergence of a tranposition iteration of the form $x = g(x)$ to solve $f(x) = 0$ can be speeded up using **Aitken's acceleration**. If x_n, x_{n+1}, x_{n+2} are three successive iterates take

$$\bar{x}_{n+2} = x_n - \frac{(x_{n+1} - x_n)^2}{(x_{n+2} - 2x_{n+1} + x_n)}$$

as the accelerated iterate. Apply this to Question 9 with $x_0 = 1.2$ and compute $\bar{x}_3$.

2.5　SYSTEMATIC ITERATIVE METHODS

The Newton–Raphson method

In order to derive the formula on which this method is based, we draw a tangent to the graph of $y = f(x)$ near a root α of the equation $f(x) = 0$ and take the intercept on the x-axis as the next approximation. Refer to Figure 2.13.

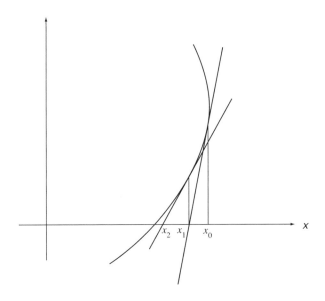

Figure 2.13　Newton–Raphson method

At $(x_0, f(x_0))$, where x_0 is an approximation to α, the equation of the tangent is given by

$$y - f(x_0) = f'(x_0)(x - x_0)$$

where $f'(x_0)$ is the gradient of $f(x)$ at $x = x_0$. This tangent line intercepts the x-axis where

$$x = x_0 - \frac{f(x_0)}{f'(x_0)}$$

So take $x_1 = x_0 - \dfrac{f(x_0)}{f'(x_0)}$ as the next approximation and in general take

$$x_{n+1} = x_n - \frac{f(x_n)}{f'(x_n)}$$

The Newton–Raphson method converges *much faster* than the bisection method or any transposition method and it is *usually reliable* given $f'(x)$ and a good approximation x_0 to α known in advance.

Examples

1. Solve the equation $x - e^{-x} = 0$ using the Newton–Raphson method starting with $x_0 = 0.5$. Firstly,

$$f(x) = x - e^{-x}$$

and

$$f'(x) = 1 + e^{-x}$$

The iterative scheme is thus

$$x_{n+1} = x_n - \frac{(x_n - e^{-x_n})}{1 + e^{-x_n}}$$

where $x_0 = 0.5$.

We therefore have

$$x_1 = 0.5 - \frac{(0.5 - e^{-0.5})}{1 + e^{-0.5}}$$
$$= 0.566\,31$$

And in like manner,

$$x_2 = 0.567\ 14$$
$$x_3 = 0.567\ 14 \quad \text{(unchanged to 5 d.p.)}$$

(Notice how fast the iteration is! The methods of bisection and transposition converge to a root at a fixed rate, e.g. at two iterations per decimal place. The Newton–Raphson method has the property that if x_0 is a good approximation to α, then each iteration tries to double the number of significant figures in convergence. In other words, if x_0 is accurate to 1 d.p., x_1 is accurate to 2 d.p., x_3 to 4 d.p., x_4 to 8 d.p., etc.)

2. Solve $x^2 - 2 = 0$ using the Newton–Raphson method to determine $\sqrt{2}$.

$$f(x) = x^2 - 2$$
$$f'(x) = 2x$$
$$x_{n+1} = x_n - \frac{(x_n^2 - 2)}{2x_n}$$
$$= \frac{1}{2}\left(x_n + \frac{2}{x_n}\right)$$

Taking $x_0 = 1.4$ and retaining 6 d.p., we have

$$x_0 = 1.400\ 000$$
$$x_1 = 1.414\ 286$$
$$x_2 = 1.414\ 214$$
$$x_3 = 1.414\ 214$$

Full accuracy to 3 d.p. occurs in one iteration and accuracy to 6 d.p. is there in two iterations.   ∎

The secant method

The **secant method** is an adapted form of the classical *regula falsi* method. The root to an equation $f(x)$ is obtained by taking two approximations to a root, drawing a chord between them, choosing the axis intercept as a third approximation, eliminating max $|f(x)|$ between the first two, then repeating the process again iteratively.

Any of the chord constructions in Figure 2.14 are appropriate; x_0 and x_1 are approximations to α on either the same side, or opposite sides of the root, and x_2 is the intercept on the axis. The line joining $P(x_0, f(x_0))$ and $Q(x_1, f(x_1))$ is

$$\frac{y - f(x_0)}{f(x_1) - f(x_0)} = \frac{x - x_0}{x_1 - x_0}$$

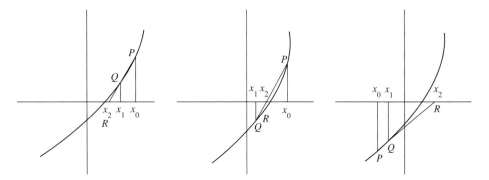

Figure 2.14 Secant method: three possible constructions using approximations x_0 and x_1 to obtain intercept x_2, which becomes a new approximation

The intercept R is at $(x_2, 0)$, so take

$$x_2 = x_0 - \frac{(x_1 - x_0) f(x_0)}{f(x_1) - f(x_0)}$$

With x_2 as the next approximation, eliminate either x_0 or x_1, depending on which of $|f(x_0)|$ or $|f(x_1)|$ is the larger, then start again to get

$$x_3 = x_1 - \frac{(x_2 - x_1) f(x_1)}{f(x_2) - f(x_1)}$$

assuming x_0 to have been eliminated.
In general we obtain a sequence of iterates, $\{x_n\}$ where

$$x_{n+1} = x_{n-1} - \frac{(x_n - x_{n-1}) f(x_{n-1})}{f(x_n) - f(x_{n-1})}$$

Examples

1. Solve the equation $x - e^{-x}$ using the secant method starting with $x_0 = 0.5$, $x_1 = 0.6$ and retaining 5 d.p.

$$x_2 = 0.5 - \frac{(0.6 - 0.5)(0.5 - e^{-0.5})}{(0.6 - e^{-0.6} - 0.5 + e^{-0.5})}$$

$$= 0.567\ 54$$

$$x_3 = 0.567\ 14 \quad \text{(accurate to 5 d.p.)}$$

2. To find $\sqrt{2}$, we set up the scheme

$$x_{n+1} = x_{n-1} - \frac{(x_n - x_{n-1})(x_{n-1}^2 - 2)}{x_n^2 - 2 - x_{n-1}^2 + 2}$$

$$= x_{n-1} - \frac{(x_{n-1}^2 - 2)}{x_n + x_{n-1}}$$

Taking

$$x_0 = 1.40, x_1 = 1.42$$

$$x_2 = 1.40 - \frac{(1.40^2 - 2)}{2.82}$$

$$= 1.414\ 184$$

$$x_3 = 1.414\ 214 \quad \text{(good to 6 d.p.)} \qquad \blacksquare$$

This method seems just about as efficient as the Newton–Raphson method given good approximations x_0 and x_1. In fact, it is not just as fast in convergence. It increases the number of significant figures per iteration by a factor of about 1.6, so if x_2 is accurate to 3 d.p. x_3 will be good to 5 d.p. and x_4 to 8 d.p.

Nevertheless, the secant method is the thoroughbred method of practical computation where iteration is involved. Should x be a computer input and $f(x)$ its associated computer output, then $f(x)$ can never be algebraically known, though its value can always be calculated. Therefore $f'(x)$ can never be determined exactly, and this would be needed for the Newton–Raphson method. We would thus need to approximate $f'(x)$ and the best approximation would lead to the secant method.

You should be fully aware that there is some element of risk in all iterative methods and that pitfalls await in real life. Close roots, multiple roots, and so on, can present difficulties as can poor initial approximations. Broadly speaking, the methods described work well within their prescribed limitations, but take care!

Exercise 2.5

1 Devise the Newton–Raphson iterative scheme to solve the equation $e^{-x} = \sin x$.

2 Find two roots of $x \tan x - 3x + 1 = 0$ for $0 \le x \le \dfrac{\pi}{2}$ using

(a) Newton–Raphson with $x_0 = 0$ and 1.

(b) Secant method with $x_0 = 0, x_1 = 0.5$ and $x_0 = 1, x_1 = 1.5$. Retain 4 d.p.

3 By sketching $y = 2^x$ and $y = 5x - 2$ estimate a root of the equation $2^x - 5x + 2 = 0$. Hence, using the secant method find this root to 4 d.p.

4 Derive an iterative process for finding the pth root of a number $c > 0$ by applying Newton's method to the equation

$$x^p - c = 0$$

Use this process to find the fifth root of 21.

5 In the design of high voltage tubular electrical insulators, the equation

$$Q = \frac{\pi q^2 (x^2 - 1)}{(\ln x)^2} \quad \text{occurs.}$$

Assuming q to be constant and $x > 1$, find the value of x (to 4 d.p.) for which Q is a minimum. [$x_0 = 2$].

6 The equation $x = 0.5 + 0.2 \sin x$ is a type of Kepler equation for predicting the orbit of satellites. Using the secant method, find the solution to 4 d.p.

7 Draw the graph of

$$f(x) = \sin(x - \sin x) \quad \text{for} \quad -4 \le x \le 4$$

Determine $f'(x)$ and solve $f'(x) = 0$ using iteration to determine the smallest positive value of x for which $f(x)$ is a maximum.

8 If $f(x) = (x - a)^k h(x)$, $h(\alpha) \ne 0$, we say that α is a k-fold root of $f(x)$. If $k > 1, f(x)$ will touch the x-axis at $x = \alpha$, as illustrated.

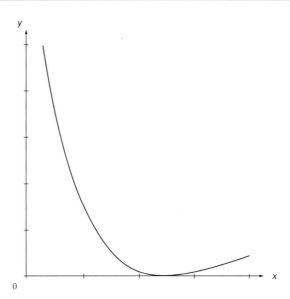

The Newton–Raphson method then needs to be modified to

$$x_{n+1} = x_n - \frac{kf(x_n)}{f'(x_n)}$$

Now

$$f(x) = x^5 - 7x^4 + 10x^3 + 10x^2 - 7x + 1$$

has a double root ($k = 2$) close to $x_0 = 0.3$. Use the modified formula to find this root to 3 d.p.

9   The curves $y = 2\sin x$ and $y = \ln x - C$ touch close to $x = 8$.

(a)   Sketch a graph which shows this happening.

(b)   Determine the equation that x must satisfy for the touching condition to be fulfilled. Solve this equation to 3 d.p.

(c)   Hence determine C to 3 d.p.

10   This displacement $x(t)$ has the analytic form

$$x(t) = (t - 1)e^{-(t-2)} - (t - 2) \qquad t \geq 0$$

(a)   Tabulate $x(t)$ for $t = 0\,(0.5)\,3$ to show that $x = 0$ at two values of t between 0 and 3. Retain 3 s.f.

(b)   Investigate the convergence of the iterative scheme

$$t_{n+1} = (t_n - 1)e^{-(t_n-2)} + 2$$

in the vicinity of each zero, and determine the value of t for the zero in the convergent case $\left(\text{assume } \dfrac{d}{dt}(te^{-t}) = (t - 1)e^{-t}\right)$. Retain 5 s.f.

(c)   For the divergent case in (b) use the Newton–Raphson method to find the value of t for the zero.

SUMMARY

- **Quadratic equation:** roots a and b of $x^2 + px + q = 0$ have sum $-p$ and product q

- **Polynomial equation:** $x^n + a_{n-1}x^{n-1} + \cdots + a_0 = 0$ has roots whose sum is $-a_{n-1}$ and whose product is $(-1)^n a_0$

- **Exponential function:** if $a > 0$ then $a^x = e^{x \ln a}$

- **Logarithmic function:** $\log_b c = \dfrac{\ln c}{\ln b}$ if $b, c > 0$

- **Hyperbolic functions**

$$\sinh x = \frac{1}{2}(e^x - e^{-x}) \qquad \cosh x = \frac{1}{2}(e^x + e^{-x}) \qquad \tanh x = \frac{\sinh x}{\cosh x}$$

- **Properties of hyperbolic functions**

$$\cosh^2 x - \sinh^2 x \equiv 1 \qquad \sinh 2x \equiv 2 \sinh x \cosh x$$
$$\cosh 2x \equiv \cosh^2 x + \sinh^2 x$$
$$\sinh x = 0 \quad \text{when} \quad x = 0 \qquad \cosh x \geq 1, \cosh x > \sinh x$$
$$\sinh x \text{ is odd} \qquad \cosh x \text{ is even}$$

- **Inverse hyperbolic functions**

$$\sinh^{-1} x \equiv \ln(x + \sqrt{x^2 + 1}) \qquad \cosh^{-1} x \equiv \ln(x \pm \sqrt{x^2 - 1})$$
$$\tanh^{-1} x \equiv \frac{1}{2}\ln\left(\frac{1+x}{1-x}\right)$$

- **Location of roots:** if $f(x)$ has a continuous graph in $a \leq x \leq b$ and $f(a) \times f(b) < 0$ then $f(x) = 0$ has at least one root in the interval $a < x < b$.

- **Successive bisection method:** evaluates $x_m = \dfrac{1}{2}(a + b)$ and $f(x_m)$; if $f(a) \times f(x_m) < 0$ then put $b = x_m$; if $f(a) \times f(x_m) > 0$ put $a = x_m$.

- **Transposition method:** to solve $x = g(x)$ make an initial guess x_0; the sequence $x_{n+1} = g(x_n)$ will converge to a root if $|g'(x)| < 1$ near the root.

- **Newton–Raphson method:** to solve $f(x) = 0$ make an initial guess x_0; the sequence $x_{n+1} = x_n - \dfrac{f(x_n)}{f'(x_n)}$ may converge to a root.

- **Secant method:** initial guesses x_0 and x_1, then

$$x_{n+1} = x_{n-1} - \frac{(x_n - x_{n-1})f(x_{n-1})}{f(x_n) - f(x_{n-1})}$$

Answers

Exercise 2.1

1   (a) $3 + \sqrt{2}$ (b) $\sqrt{6} - \sqrt{2}$ (c) $\sqrt{3} + \sqrt{5}$

2 $5x^2 + 4x - 2$

3   (a) -3 (b) $-1, -\dfrac{33}{17}$

4 $x^4 + 4x^3 + 5x^2 + 2x - 12 = 0$

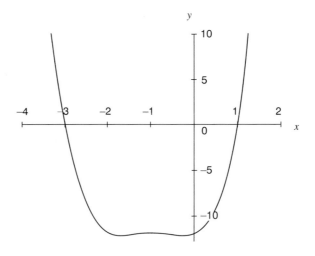

5   (a) $1, -4, \dfrac{-3 \pm \sqrt{55/3}}{2}$

    (b) $1, -3, \sqrt{2} - 1, -\sqrt{2} - 1$

    (c)   If $y = x(x + 1)(x + 2)$, $y^2 + 4y + 5 = (y + 2)^2 \ \ + 1 \geq 1$

6 $x = 1, y = 4$ and vice versa and

$$x = \frac{5 + \sqrt{53}}{2}, \qquad y = \frac{5 - \sqrt{53}}{2} \text{ and vice versa.}$$

7 $x = 4, y = 3$

8 (a) (i) 7 (ii) 5

(b) (i) $x^2 - 39x + 25$
(ii) $x^2 - 238x + 125$
(iii) $5x^2 - 39x + 5$
(iv) $49x^2 - 49x + 5$

9 $-\dfrac{3}{2}$

11 $x^3 + 6x^2 - x - 30$

12 (a) (i) $-(\alpha + \beta + n - 2)$ (ii) $(-1)^n \alpha \beta$
Note that $((-1)^{n-2} = (-1)^n)$

(b) $\alpha = 2, \beta = -3, a = -7, b = 13$

13 (a) $|f(x)| \le 3 + 6 + 6 = 15$
(b) $R = 6$; the magnitude of all the roots, real or complex, cannot exceed 6.

14 (a) 2, 3 only; denominator never zero.

(b) $x = n$, integer $(\sin \pi x = 0)$; $x = 1$ is also a root; denominator never zero.

(c) $x^2 - 5x + 1 = 0$, i.e. $x = \dfrac{5 \pm \sqrt{26}}{2}$; denominator is zero when $x = -3, 4$ but not when the numerator is zero.

(d) $x = \dfrac{-a \pm (4b - a)^{1/2}}{2}$, $4b > a$, but no root if $4b < a$; one repeated root if $4b = a$.

15 All values except $\dfrac{87}{11} < x < 8$

17 (a) $(-\alpha)^4 + c(-\alpha)^2 + d = \alpha^4 + c\alpha^2 + d = 0$
(b) $\pm(3 + \sqrt{2})$

18 (a) $\tan \theta = 1, \pm\sqrt{3}$;
$\theta = 45°, 60°, 120°, 225°, 240°, 300°$.

(e) $\dfrac{1}{6}$

(f)

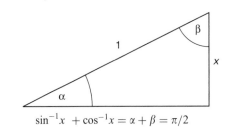

$$\sin^{-1}x + \cos^{-1}x = \alpha + \beta = \pi/2$$

20 $x = 0$

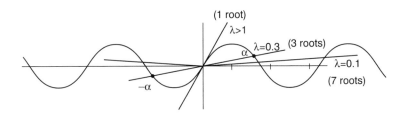

(b) If $\sin \alpha = \lambda\alpha$, α a root
$\sin(-\alpha) = \lambda(-\alpha)$, $-\alpha$ also a root.

(c) Number of roots: 1, 7, 11, 15, etc. The most extreme roots may coincide if λx is a tangent to $\sin x$.

(d) (i) One root.

(ii) Three roots: $y = 0.3$ passes below $\left(\dfrac{\pi}{2}, 1\right)$, which is the first positive maximum of $\sin x$.

(iii) Seven roots: $y = 0.1x$ passes below $\left(\dfrac{5\pi}{2}, 1\right)$, which is the second positive maximum but above $\left(\dfrac{9\pi}{2}, 1\right)$, the next maximum.

(iv) As $\lambda \to 0$, the roots become those of $\sin x$, i.e. $x = n\pi$; $x = 0$ is always a root.

Exercise 2.2

1 (a) $\dfrac{\ln 10}{\ln 2} \approx 3.322$

(b) $\dfrac{1}{2}\ln 5 \approx 0.8047$

(c) $\dfrac{\ln 12}{\ln(5.612)} \approx 1.4406$

(d) $\dfrac{\ln 100}{\ln 3} \approx 4.1918$

2 (a)   0, ln 5

(b) $1, \dfrac{\ln 3}{\ln 5} \approx 0.6826$

(c) $x = 3, y = 2$

3 $x = 2, y = 1$ or $x = 1, y = 2$

4 (a)   2

(b) $x = 10, y = 5$

(c)   64

(d)   2

5 (a)   0, ln 2

(b) $x = y = \ln 2$

7 (c) $\tanh(x \pm y) \equiv \dfrac{\tanh x \pm \tanh y}{1 \pm \tanh x \tanh y}$

8 (a) $x = \ln(1/2), \quad y = \ln(3/2)$

(b) $x = \ln 3, \quad y = \ln 5$

9 (a) $\sqrt{5}(2\sqrt{2} + 3)$

(b) $\dfrac{5}{2\sqrt{6}}$

(c) $\dfrac{1}{2}(\sqrt{5} - 1)$

(d) $\dfrac{1}{2}\left(\left(x + \sqrt{1 + x^2}\right)^2 - \left(x + \sqrt{1 + x^2}\right)^{-2}\right)$

(e) $\dfrac{x + \tanh x}{1 + x \tanh x}$

(f) $\dfrac{1}{2}\left(\left(x^3 + \dfrac{1}{x^3}\right)\cosh 2 + \left(x^3 - \dfrac{1}{x^3}\right)\sinh 2\right)$

10 (a) $2 + \sqrt{5}$

(b) $\dfrac{1}{2}(-1 \pm \sqrt{3})$

(c)   2.4

11 9th term.

12 (a) $n = 7$ (b) $n = 12$

Exercise 2.3

1 (c)

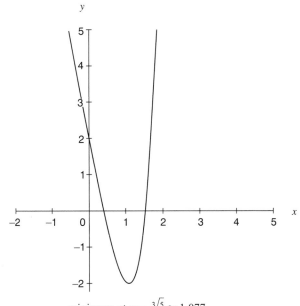

minimum at $x = \sqrt[3]{\frac{5}{4}} \simeq 1.077$

x	0	0.25	0.5	0.75	1	1.25	1.5	1.75	2
$f(x)$	2	0.754	-0.438	-1.434	-2	-1.809	-0.438	2.629	8

(d) $x = 1.55, f(x) = 0.022$
       $x = 1.548, f(x) = 0.002$

(e)   (1.5, 1.75)
       (1.5, 1.625)
       (1.5, 1.5625)
       (1.531 25, 1.5625)
       (1.546 875, 1.5625)
       (1.5469, 1.5547) (retaining 4 d.p.)
       Root is 1.55 to 2 d.p.

2

x	0.04	0.05	0.06	0.07
$f(x)$	0.299	0.096	-0.066	-0.201

Successive intervals are
        (0.05, 0.06)
        (0.055, 0.060)
        (0.055, 0.0575)
        (0.055, 0.056 25)
        (0.055 63, 056 25)
        Root is 0.056 to 3 d.p.

3   (a)   10 $(2^{10} = 1024)$

     (b)   14 $(2^{14} = 16\,384)$

     (c)   20 $(2^{20} = 1\,048\,576)$

4   (a)   No.

     (b)   Yes if a or b is the root.

     (c)   Yes.

5 $f(x) > 0$ if $x \gg 0$ (i.e. large positive).
     $f(x) < 0$ if $x \ll 0$ (i.e. large negative).
     Axis must be crossed, so there is one root at least; this is true for any odd-order polynomial.

6   As all coefficients are positive, $f(x) > 1$ if $x > 0$, so no positive roots.

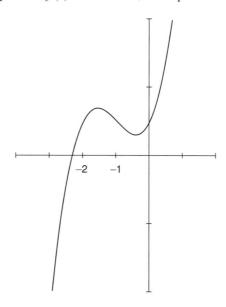

$f'(x) = 3x^2 + 6x + 2 = 0$ when $x = \pm\dfrac{1}{\sqrt{3}} - 1 = -1.58$ (max), -0.42 (min)

Check max and min with $f''(x) = 6(x + 1)$. The coordinates are $(-1.58, 1.38)$ and $(-0.42, 0.62)$. $f'(x) > 0$ for $x < -1.58$, so $f(x)$ decreases as x becomes more negative. The axis can only be crossed once.

7 $x \ln x = 5$; take $f(x) = x \ln x - 5$

x	1	2	3	4	5
$f(x)$	-5	-3.614	-1.704	0.545	3.047

x	3.7	3.8	3.76	3.77	5
$f(x)$	-0.159	0.073	-0.020	0.003	3.047

Root $= 3.77$ to 2 d.p.

8

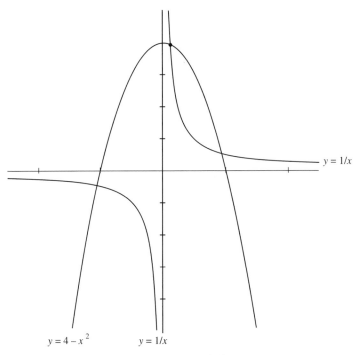

$y = 4 - x^2$ $y = 1/x$

$y = 1/x$

Root $= -2.11$ to 2 d.p.

Exercise 2.4

1 (a) $g_1(x) = x^2 - \ln x - 1$ $g_1'(0.4) = -1.7, \quad g_1'(1.8) = 3.04$

    (b) $g_2(x) = \dfrac{\ln x + x + 1}{x}$ $g_2'(0.4) = 5.7, \quad g_2'(1.8) = -0.18$

    (c) $g_3(x) = \exp(x^2 - x - 1)$ $g_3'(0.4) = -0.058, \quad g_3'(1.8) = 4.03$

    (b) converges to $\alpha = 1.8696$

    (c) converges to $\beta = 0.2984$

2   0.4263

3 (b)   1.8393

4 (b) $x_1 = 1.446\ 85$

        $x_2 = 1.446\ 86$

        $x_3 = 1.446\ 86 \ (= \alpha$ to 5 d.p.)

5   1.3247, $g_1'(1) \approx 3,$ $g_2'(1) \approx -1,$ $g_3'(1) \approx 0.21;$ use $x = g_3(x)$

6 (a) $P_6(x) = x^6 - 12x^5 - 3x^4 + 175x^3 - 132x^2 - 33x + 4 \uparrow$ denotes a sign change.

-4	-3	-2	-1	0	1	2	3	4
2440	-2408	-1458	-260	4	0	442	1012	0
	↑				↑	↑		↑
5	6	7	8	9	10	11		12
		$-37\,908$		$-80\,240$		$11\,620$		
-5336	$-17\,690$		$-62\,468$		$-68\,526$			$220\,792$
							↑	

    (b) $P_4(x): x^4 - 7x^3 - 42x^2 - 7x + 1$

-4	-3	-2	-1	0	1	2	3	4
61	-86	-81	-26	1	-54	-221	-506	-891

5	6	7	8	9	10	11	12
-334	-1769	-2106	-2231	-2006	-1269	166	2509

    (c)   Upper root, near $x = 11$ $(x_0 = 11)$

$$x_{n+1} = x_n - \frac{f(x_n)}{m} \qquad (\alpha = 10.91)$$

$$m = f'(x_0) = 1852$$

Lower root, near $x = -3.8$ $(x_0 = -3.8)$

$$x_{n+1} = x_n - \frac{f(x_n)}{m}$$ $(m = f'(x_0) = -210.5)$

$(\alpha = -3.732)$

(d)   positive root $x_0 = 0.1, f'(x_0) = -15.606, \ \alpha = 0.092$

       negative root $x_0 = -0.25, f'(x_0) = 12.625, \ \alpha = -0.267$

(e) $P_6(x) = (x-1)(x-4)(x^2 - 11x + 1)(x^2 + 4x + 1)$

7

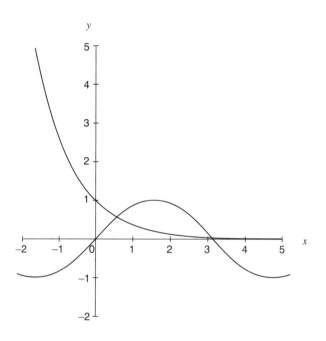

(a) $x_0 = 0.6$

      $x_1 = 0.5885$

      $x_2 = 0.588\,532 \ (\alpha_0 \text{ to 6 d.p.})$

(b) $\alpha_n \approx n\pi \quad (\sin x = 0)$

      to 4 d.p., $n > 10$

      $e^{-10} < 10^{-4}$

8 $ab < 3$

9

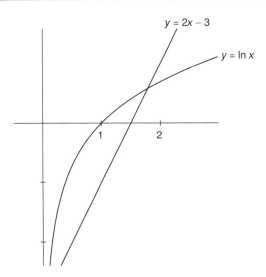

$y = 2x - 3$

$y = \ln x$

Iterates
1.2
1.2649
1.2846
1.2904
1.2920
1.2925
1.2926
1.2927   Area $= 0.2361$

10 $\bar{x}_3 = 1.2928$ (to within 10^{-4})

Exercise 2.5

1   See Question 8 of Exercise 2.4

2   Roots 0.3855 and 1.1290

3   Root lies between 0.5 and 0.75; secant method converges to 0.7322

4 $x_{n+1} = \dfrac{1}{p}\left\{(p-1)x_n + \dfrac{c}{x_n^{p-1}}\right\}$

Root $= 1.8384$

5   2.2185

6   0.6155

7 $f'(x) = \cos(\sin x - x)(1 - \cos x)$

$= 0$ if $x = \sin x + \dfrac{\pi}{2}$

Root $= 2.3099$

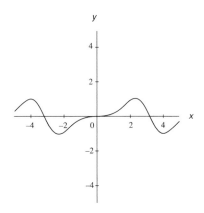

8   Root $= 0.268$

9   (a)

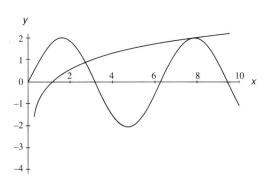

(b) $x = 7.790$

(c) $C = 0.057$

10   (a)

t	0	0.5	1.0	1.5	2.0	2.5	3.0
$x(t)$	-5.39	-0.74	1.00	1.32	1.00	0.41	-0.26

(b)   For $t_{n+1} = g(t_n)$, $g'(t) = (2 - t)e^{-(t-2)}$; $g'(0.7) \approx 4.77$, $g'(2.8) \approx -0.36$
Upper root ≈ 2.8064

(c)   Newton–Raphson method gives lower root ≈ 0.6500

3 ALGEBRA

Introduction

Selection from a set of objects finds application in probability theory, which models situations that involve uncertainty. The binomial theorem can be used to study how small perturbations affect a system in equilibrium. A rational function may be split into partial fractions so it becomes easier to manipulate and its graph can be sketched more straightforwardly. The techniques in this chapter therefore help us to understand the behaviour of many more systems in engineering and science.

Objectives

After working through this chapter you should be able to

- Use factorial notation
- Calculate permutations and combinations
- Use Pascal's triangle to find binomial expansions with positive integer indices
- Find the binomial expansion when the index is any rational number
- Examine the effect of small perturbations on some simple systems
- Obtain the partial fractions of a given proper rational function where the denominators are simple or repeated linear factors or quadratic factors
- Find the partial fractions of a given improper rational function
- Sketch the curves of simple rational functions and identify their main features.

3.1 PERMUTATIONS AND COMBINATIONS

Factorial notation

The expression $1 \times 2 \times 3 \times \cdots \times n$ occurs regularly and is given a special notation. We write it as $n!$ and refer it to as **factorial** n, or n factorial. By adding the definition that $0! = 1$ we can say that $n!$ is a function with domain the non-negative integers.

Example

(a) List the values of $n!$ for $n = 1\,(1)\,10$.

(b) Show that $n! + 1$ must either be a prime number or have a factor greater than n.

Solution

(a) $1! = 1, \quad 2! = 1 \times 2 = 2, \quad 3! = 1 \times 2 \times 3 = 2! \times 3 = 6,$
$4! = 1 \times 2 \times 3 \times 4 = 24, \quad 5! = 120, \quad 6! = 720, \quad 7! = 5040, \quad 8! = 40\,320,$
$9! = 362\,880, \quad 10! = 3\,628\,880$

(b) If $n = 1$ then $n! + 1 = 2$ and is prime. If $n > 1$ then $n!$ is divisible by all integers from 2 to n inclusive; therefore $n! + 1$ cannot be divisible by any of these numbers. Hence the factors of $n! + 1$ must be greater than n, unless it has no factors and is therefore a prime number. ■

Notice how rapidly the values of $n!$ grow as n increases; $n!$ is quite large even for relatively small n.

Permutations

Consider the following problems.

Examples

1. There are four boxes in a row and four objects A, B, C and D. In how many ways can we fill the boxes to make a different arrangement of objects if each box is to contain exactly one object?

   We have four choices of object for the first box, three for the second, two for the third and only one for the last box (i.e. the choice is automatic). The possible arrangements are

$$
\begin{array}{cccccc}
ABCD & ABDC & ACBD & ACDB & ADBC & ADCB \\
BACD & BADC & BCAD & BCDA & BDAC & BDCA \\
CABD & CADB & CBAD & CBDA & CDAB & CDBA \\
DABC & DACB & DBAC & DBCA & DCAB & DCBA.
\end{array}
$$

The number of arrangements is 24, which is the value of $4!$. This is no coincidence. For example, without listing all the possibilities, we can say the number of arrangements of six objects into six boxes is $6!=720$.

2. Suppose that we still have four boxes in a row but now six objects A, B, C, D, E, F. We therefore have six choices of object for box 1, five for box 2, four for box 3 and three for box 4. The total number of arrangements is $6 \times 5 \times 4 \times 3 = 360$. Note that

$$6 \times 5 \times 4 \times 3 = \frac{6 \times 5 \times 4 \times 3 \times 2 \times 1}{2 \times 1} = \frac{6!}{2!}$$

The fraction is written as $^6P_4$ and is an example of a **permutation**; it represents the number of ways in which a specifically ordered set of objects can be chosen. Later in this section we give a formal definition. ∎

Combinations

Suppose we are interested only in which objects are chosen, not in the order of their selection.

Example

A team of four is to be selected from six people A, B, C, D, E, F. Then, for example, $CAFB$ is the same **selection** as $ABCF$, $BAFC$, etc., since the same team will be selected, irrespective of the order of selection. There are fewer selections than permutations. In the first example the table of arrangements shows that the selection of A,B,C and D can be arranged in $4! = 24$ ways. This will be true of each selection of four objects. Hence the number of team selections (using the result of the second example) is $360/24 = 15$.

Note that this fraction is $\dfrac{6!}{2! \times 4!}$ and this is an example of a **combination**. Two notations are in common use: $^6C_4$ and $\dbinom{6}{4}$.

It is important to note that the selection of A, B, C and F is effectively the same as the non-selection of D and E. The number of ways of achieving these results must be the same, i.e.

$$^6C_2 = \frac{6!}{4! \times 2!} = \frac{6!}{2! \times 4!} = {}^6C_4 = 15$$

Note also that a selection is a ratio of factorials. ∎

The general case

The number of distinct permutations (or ordered arrangements) of m objects from a set of n different objects is

$$n \times (n-1) \times (n-2) \times \cdots \times (n-m+1) = \frac{n!}{(n-m)!} \qquad \text{written} \quad {}^nP_m \qquad (3.1)$$

The number of distinct combinations (where the selection is unordered) of m objects from a set of n different objects is

$$\frac{n!}{m! \times (n-m)!} \quad \text{written} \quad {}^nC_m \tag{3.2}$$

When computing this quantity it is easier to cancel out the larger factorial on the denominator, making use of the result

$${}^nC_m = {}^nC_{n-m}$$

Example   Evaluate ${}^8P_3, {}^8C_5, {}^8C_6$.

$$^8P_3 = \frac{8 \times 7 \times 6 \times 5 \times 4 \times 3 \times 2 \times 1}{5 \times 4 \times 3 \times 2 \times 1} = 8 \times 7 \times 6 = 336$$

$$^8C_5 = \frac{8 \times 7 \times 6 \times 5 \times 4 \times 3 \times 2 \times 1}{3 \times 2 \times 1 \times 5 \times 4 \times 3 \times 2 \times 1} = \frac{8 \times 7 \times 6}{3 \times 2 \times 1} = 56$$

$$^8C_6 = \frac{8 \times 7}{1 \times 2} = 28$$ ∎

Note that in the simplified version there are as many numbers in the numerator as in the denominator; in the numerator we start at n and step *down* by 1, whereas in the denominator we start at 1 and step *up* by 1.

Examples   1. A telephone exchange holds six-digit numbers. Numbers starting with 00 are excluded.

   (a) How many numbers are available?

   (b) What proportion of these have all six digits different?

   (c) What proportion have no adjacent repeated digits?

Solution

   (a) With no restrictions each digit could be any of 10, giving 10^6 possibilities. The numbers 000000 to 009999 are excluded, leaving $10^6 - 10^4 = 990\,000$.

   (b) If we insist that all six digits must be different, there are
   $^{10}P_6 = 10 \times 9 \times 8 \times 7 \times 6 \times 5$ numbers. The required proportion is

$$\frac{10 \times 9 \times 8 \times 7 \times 6 \times 5}{990\,000} = 0.1527 \quad \text{(4 d.p.)}$$

(c) There are 10 options for the first digit, 9 for the second digit, 9 for the third and so on, since each successive digit must differ from its predecessor only. The required proportion is

$$\frac{10 \times 9^5}{990\ 000} = 0.5965 \quad (4\ \text{d.p.})$$

2. Find how many arrangements of the letters are possible in the following:

(a) LARGE   (b) SMALL   (c) ILLITERATE

Solution

(a) The letters are all different; hence the number of arrangements is $5! = 120$.

(b) L is repeated, therefore we divide 5! by 2! to allow for the L's to be interchanged, giving 60 arrangements.

(c) There are 2 L's, 2 I's, 2 T's and 2 E's. The 10! arrangements are grouped into sets of $(2!)^4$ which are indistinguishable. Hence there are $\dfrac{10!}{(2!)^4} = 226\ 800$ distinct arrangements. ■

Combinations of similar groupings

In the last example we had a set of 10 letters of which there were four sets of two identical letters and two separate letters. More generally, suppose that we have a set of n objects which can be partitioned into k subsets such that the objects in each subset are identical to each other. Let n_1 be the number of objects in the first subset, n_2 be the number of objects in the second subset, and so on.
Then

$$n_1 + n_2 + n_3 + \cdots + n_k = n$$

The number of distinct arrangements of the n objects is then

$$\frac{n!}{n_1! \times n_2! \times \cdots \times n_k!}$$

Example

Twelve new students arriving on campus are allocated accommodation with five going to block A, three to block B and four to block C. The allocation can be carried out in
$$\frac{12!}{5! \times 3! \times 4!} = 27\ 720 \text{ ways.}$$ ■

Note that $\dbinom{n}{r} = {}^nC_r = \dfrac{n!}{r! \times (n-r)!} = \dfrac{n(n-1)\ldots(n-r+1)}{1 \times 2 \times \cdots \times r}$ is always an integer.

An important result is

$$\binom{n}{r} + \binom{n}{r-1} = \binom{n+1}{r} \tag{3.3}$$

We prove a specific example: $\dbinom{8}{6} + \dbinom{8}{5} = \dbinom{9}{6}$. The proof of the general result follows a similar approach.

The left-hand side is

$$\frac{8!}{6! \times 2!} + \frac{8!}{5! \times 3!} = \frac{8!}{6 \times 5! \times 2!} + \frac{8!}{5! \times 3 \times 2!}$$

$$= \frac{3 \times 8!}{6 \times 5! \times 3 \times 2!} + \frac{6 \times 8!}{6 \times 5! \times 3 \times 2!} = \frac{9 \times 8!}{6 \times 5! \times 3 \times 2!}$$

$$= \frac{9 \times 8!}{6! \times 3!} = \frac{9!}{6! \times 3!}$$

which is the right-hand side.

Exercise 3.1

 1   (a)   Find the smallest value of n for which $n! > 10^6$?

      (b)   List the factors, if any, of $n! + 1$ for $n = 1$ to $n = 12$.

2   (a)   Five pictures are hung in a row. In how many ways can they be arranged?

      (b)   Five people sit around a circular table. Why is the number of arrangements less than that in (a)?

      (c)   In how many ways can n people sit around a circular table in any rotation?

3   (a)   What are the values of ${}^6P_2, {}^{11}P_4, {}^nP_3$ (as a function of n)?

      (b)   In a class of 74 foundation level students each student is given a unique ranked position after the first set of examinations. How many possibilities are there for the first three positions, in that specific order?

4 Each of the four digits of a bicycle 'combination' lock can be 1, 2, 3, 4, 5 or 6.

    (a)   How many possible settings of the lock are there?

    (b)   If the digits in the setting are to be all different, how many settings are possible?

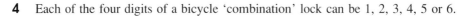

5   (a)   When UK car registration 'numbers' were introduced, most numbers were of the form two capital letters followed by up to four digits. The letters did not include I, Q or Z and the second part was a number from 1 to 9999. How many possible registrations were there?

    (b)   In the 1930s a new scheme was introduced in order to cater for the increase in the number of cars. The format was three capital letters followed by three digits in which the letters Q and Z were omitted and the second part went from 1 to 999. How many cars could this new scheme cope with?

6 A bar code has a maximum of 10 digits. If a code of all zeros is excluded find the number of possible codes

    (a)   with no further restrictions

    (b)   if all digits must be different

    (c)   if no two adjacent digits can be equal.

7 Three items are chosen from a large number n in a specific order. How big does n have to be to ensure there are more than a million possible choices?

8   (a)   Find the values of

        (i) $^6C_3$ (ii) $^{15}C_{12}$ (iii) $^{11}C_7$

    (b)   In how many ways can we short-list three candidates from a total of nine? If we decide instead to short-list six candidates, how many possibilities are there? Compare the two answers and comment.

9 How many distinct rearrangements are there for the following sets?

    (a)   (i) 1, 7, 5 and 2   (ii) 3, 1, 1 and 5   (iii) 1, 2, 1, 3, 1 and 7

    (b)   (i) A, B, C, X, Y and Z   (ii) A, B, R, A, C, A, D, A, B, R and A

10 In how many ways can each of the following seatings be arranged?

    (a)   Six men and six women sit around a non-circular table so that no two people of the same gender sit next to each other.

(b) At a long rectangular table the host and hostess sit at opposite ends and the four pairs of guests sit so that each man sits between two women, each woman sits between two men and the members of each pair of guests sit apart.

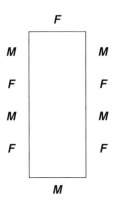

11 (a) Write expressions for

(i) $\dbinom{n}{3}$ (ii) $\dbinom{n}{n-4}$

(iii) $\dbinom{n+1}{n-1}$ (iv) $\dbinom{2n}{n}$

(b) The probability of holding three aces in a hand of 13 cards in a round of bridge or whist is

$$\frac{\dbinom{4}{3}\dbinom{48}{10}}{\dbinom{58}{13}}$$

Evaluate this expression.

12 Six girls at a summer school are allocated duties with three being given task A, one task B and two task C. They are allocated three double rooms as accommodation. How many possibilities are there of (a) allocating tasks, (b) allocating accommodation? (in case (b) it may be easier to list the different possibilities.)

13   In how many ways can a team of five players be chosen from six men and four women if there are to be more men than women in the team? How many ways are there if the team must contain more women than men?

14   Two cards are selected at random from a set of seven black and five red cards.

(a)   Find the number selections which contain

(i) two red cards   (ii) one card of each colour   (iii) two black cards.

(b)   The probability that two black cards are selected is the ratio of the number of possible selections of two red cards to the number of selections of any two cards. Find this probability.

3.2 THE BINOMIAL THEOREM

Expansion of $(1 + x)^n$ and Pascal's triangle

The following expansions, known as **binomial expansions**, can be verified by directly multiplying each expansion in turn by $(1 + x)$; 'binomial' means two terms, in this case 1 and x.

$$(1 + x)^1 = 1 + x$$

$$(1 + x)^2 = 1 + 2x + x^2$$

$$(1 + x)^3 = 1 + 3x + 3x^2 + x^3$$

$$(1 + x)^4 = 1 + 4x + 6x^2 + 4x^3 + x^4$$

$$(1 + x)^5 = 1 + 5x + 10x^2 + 10x^3 + 5x^4 + x^5$$

$$(1 + x)^6 = 1 + 6x + 15x^2 + 20x^3 + 15x^4 + 6x^5 + x^6$$

What if we wanted the expansion of $(1 + x)^{19}$ or merely the coefficient of x^8 in that expansion? It would be very tedious to work steadily forwards, increasing the power of $(1 + x)$ by one each time. We therefore look for a pattern in the results so far. Note that the number of different terms in each expansion is one more than the power of $(1 + x)$ and that the coefficients start with 1, rise to a maximum and decrease symmetrically back to 1. If the power of $(1 + x)$ is even, there is a single largest coefficient.

Consider the expansion of $(1 + x)^4 \equiv (1 + x)(1 + x)(1 + x)(1 + x)$. To get the terms in x^3 we need to take x from three of the four brackets and 1 from the fourth. There are four choices of bracket to provide the number 1, the three others having automatically to provide x. Therefore there will be four terms x^3 and hence the term $4x^3$ in the expansion.

In a similar way terms in x^2 are obtained by selecting x from two brackets, which can be done in $^4C_2 = 6$ ways, the two other brackets automatically providing the number 1; hence the term $6x^2$ in the expansion.

These principles apply to all expansions of the form $(1 + x)^n$ where n is a positive integer. Note that the sums of coefficients in the expansion of $(1 + x)^4$ is $1 + 4 + 6 + 4 + 1 = 16 = 2^4$, the sum of the coefficients in the expansion of $(1 + x)^5$ is $1 + 5 + 10 + 10 + 5 + 1 = 32 = 2^5$, and so on.

The general result, known as the **Binomial theorem**, can be stated as follows:

$$(1 + x)^n \equiv 1 + {}^nC_1 x + {}^nC_2 x^2 + \cdots + {}^nC_r x^r + \cdots + {}^nC_{n-1} x^{n-1} + x^n$$

$$\equiv 1 + \binom{n}{1}x + \binom{n}{2}x^2 + \cdots + \binom{n}{r}x^r + \cdots + \binom{n}{n-1}x^{n-1} + x^n \qquad (3.4)$$

A more practical version is

$$(1 + x)^n \equiv 1 + \frac{n}{1}x + \frac{n \times (n-1)}{1 \times 2}x^2 + \cdots + \frac{n \times (n-1) \times \cdots \times (n-r+1)}{1 \times 2 \times \cdots \times r}x^r$$
$$+ \cdots + \frac{n}{1}x^{n-1} + x^n \qquad (3.5)$$

Examples

1. In the expansion of $(1 + x)^{19}$ the coefficient of x^3 is $^{19}C_3 = \dfrac{19 \times 18 \times 17}{1 \times 2 \times 3} = 969$. The coefficient of x^{17} is $^{19}C_{17} = {}^{19}C_2 = \dfrac{19 \times 18}{1 \times 2} = 171$. Notice how we use the symmetry property of the coefficients to make the second evaluation easier.

2. If we put $x = 1$ in the expansion of $(1 + x)^n$, we obtain the result $\sum_{r=0}^{n}\binom{n}{r} = 2^n$, i.e. the sum of the coefficients is equal to 2^n as we had suggested earlier.

    If we put $x = -1$ in the expansion, we obtain the result $\sum_{r=0}^{n}(-1)^r\binom{n}{r} = 0$. In the case of $n = 3$ this means that $1 - 3 + 3 - 1 = 0$, which is a feature of the symmetry of the coefficients. In the case of $n = 4$ the result means $1 - 4 + 6 - 4 + 1 = 0$, which is not a feature of symmetry. ∎

Pascal's triangle

The relationships between the coefficients of the expansions $(1 + x)^n$ can be illustrated by a diagram known as **Pascal's triangle**. Figure 3.1 shows the first seven rows

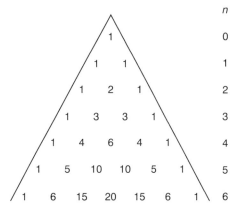

Figure 3.1 Pascal's triangle

Each row starts and finishes with the number 1. Other coefficients in the row are related by the process depicted in Figure 3.2: the coefficient is the sum of the two numbers in the previous row which are closest to it, i.e. above left and above right. This relationship, which may be symbolised $\binom{n}{r-1} + \binom{n}{r} = \binom{n+1}{r}$ was mentioned in the previous section.

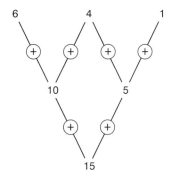

Figure 3.2 Relationship between the coefficients

Expansions of powers of expressions other than $(1+x)$ can be obtained by a modification of the method we have developed. We quote a more general form of the binomial expansion:

$$(a+b)^n \equiv a^n + \frac{n}{1}a^{n-1}b + \frac{n \times (n-1)}{1 \times 2}a^{n-2}b^2$$
$$+ \cdots + \frac{n \times (n-1) \times \cdots \times (n-r+1)}{1 \times 2 \times \cdots \times r}a^{n-r}b^r + \cdots + \frac{n}{1}ab^{n-1} + b^n$$

$$(3.6)$$

Examples   1. Obtain the expansions of

(a) $(2 + x)^4$ (b) $(4x + y)^5$ (c) $(3x - 2y)^3$

Solution

(a) We put $a = 2$, $b = x$ and $n = 4$ in equation (3.6) to obtain

$$2^2 + 4 \times 2^3 \times x + 6 \times 2^2 \times x^2 + 4 \times 2 \times x^3 + x^4$$
$$= 16 + 32x + 24x^2 + 8x^3 + x^4$$

Alternatively, we could write first $(2 + x)^4 = 2^4(1 + x/2)^4$ and expand $(1 + x/2)^4$ by equation (3.6), replacing x by $x/2$ and putting $n = 4$. In expansions where the power is not a positive integer this is often the best option.

(b) Here we put $a = 4x$, $b = y$ and $n = 5$ to obtain

$$(4x)^5 + 5(4x)^4 y + 10(4x)^3 y^2 + 10(4x)^2 y^3 + 5(4x)y^4 + y^5$$
$$= 1024x^5 + 1280x^4 y + 640x^3 y^2 + 160x^2 y^3 + 20xy^4 + y^5$$

(c) This time, $a = 3x$, $b = -2y$ and $n = 3$. Hence

$$(3x - 2y)^3 = (3x + (-2y))^3$$
$$= (3x)^3 + 3(3x)^2(-2y) + 3(3x)(-2y)^2 + (-2y)^3$$
$$= 27x^3 - 54x^2 y + 36xy^2 - 8y^3$$

2. Obtain the coefficient of x^3 in the expansion of $(2 + x - x^2)^6$

This is a much harder problem. First we regroup the terms in the brackets as $(2 + x(1 - x))^6$ and expand in powers of 2 and $x(1 - x)$, i.e. we put $a = 2$, $b = x(1 - x)$ and $n = 6$. Then we obtain

$$2^6 + 6 \times 2^5 x(1 - x) + 15 \times 2^4 x^2(1 - x)^2 + 20 \times 2^3 x^3(1 - x)^3 + \cdots$$

All further terms contain at least x^4 and can therefore be ignored.

The two terms in x^3 are $15 \times 2^4 x^2(-2x)$ and $20 \times 2^3 x^3(1)$, so the coefficient required is $15 \times 2^4(-2) + 20 \times 2^3(1) = -320$.   ∎

The general binomial theorem

The series we have met so far are all finite, i.e. they have a finite number of terms. This is a direct consequence of the power n being a positive integer so that at some stage the quantity $n - r + 1$ is zero. Once a quantity appears in the numerator of one of the coefficients it appears in the numerator of all the subsequent terms, so the series terminates. This is *not* the case if the power is a fraction or a negative number.

We state without proof the result that

$$(1+x)^p = 1 + \frac{p}{1}x + \frac{p(p-1)}{1 \times 2}x^2 + \frac{p(p-1)(p-2)}{1 \times 2 \times 3}x^3$$

$$+ \cdots + \frac{p(p-1)(p-2)\cdots(p-r+1)}{1 \times 2 \times 3 \times \cdots \times r}x^r + \cdots \qquad (3.7)$$

provided that $|x| < 1$.

Remember that when the power was a positive integer the series expansion was valid for all real values of x.

Examples

1. Obtain the binomial series for

   (a) $(1+x)^{-1}$ (b) $(1-x)^{-1}$

   Solution

   (a) $(1+x)^{-1} = 1 + \frac{(-1)}{1}x + \frac{(-1)(-2)}{1 \times 2}x^2 + \frac{(-1)(-2)(-3)}{1 \times 2 \times 3}x^3$

   $$+ \cdots + \frac{(-1)(-2)(-3)\cdots(-r)}{1 \times 2 \times 3 \times \cdots \times r}x^r + \cdots$$

   $$= 1 - x + x^2 - x^3 + \cdots + (-1)^r x^r + \cdots$$

   (b) Replacing x by $-x$ we obtain

   $$(1-x)^{-1} = 1 + x + x^2 + x^3 + \cdots + x^r + \cdots$$

   (These binomial series are geometric series.)

2. Expand $(1+x)^{1/2}$ as a binomial series to $O(x^3)$ defined opposite, and examine the convergence of the series when $x = 1$.

   Use the series to obtain approximations to 3 d.p. to (a) $\sqrt{1.01}$ and (b) $\sqrt{3.96}$

   Now $(1+x)^{1/2} = 1 + \frac{\frac{1}{2}}{1}x + \frac{\frac{1}{2}\left(-\frac{1}{2}\right)}{1 \times 2}x^2 + \frac{\frac{1}{2}\left(-\frac{1}{2}\right)\left(-\frac{3}{2}\right)}{1 \times 2 \times 3}x^3 - \cdots$

   $$= 1 + \frac{1}{2}x - \frac{1}{8}x^2 + \frac{1}{16}x^3 - \cdots$$

   provided that $|x| < 1$.

When $x = 1$ the expansion becomes

$$1 + \frac{1}{2} - \frac{1}{8} + \frac{1}{16} - \cdots$$

Since the terms alternate in sign, we obtain successively under- and over-estimate by adding on the next term to the approximation we have so far. These estimates are 1, 1.5, 1.375, 1.4375, ... Convergence to $\sqrt{2}$ is slow because we are at the threshold with $x = 1$.

(a) Putting $x = 0.01$ gives

$$\sqrt{1.01} = (1.01)^{1/2} = 1 + \frac{1}{2}(0.01) - \frac{1}{8}(0.01)^2 + \frac{1}{16}(0.01)^3 + \cdots$$

$$= 1.005 \quad (3 \text{ d.p.})$$

(b) First, $\sqrt{3.96} = \sqrt{4 - 0.4} = \sqrt{4(1 - 0.01)} = 2\sqrt{1 - 0.01}$. With $x = -0.01$ we have

$$\sqrt{3.96} = 2(0.99)^{1/2} = 2\left(1 - \frac{1}{2}(0.01) - \frac{1}{8}(0.01)^2 - \frac{1}{16}(0.01)^3 - \cdots\right)$$

$$= 1.990 \quad (3 \text{ d.p.})$$

3. Obtain the binomial expansions as far as the term suggested and the values of x for which the series converges for

(a) $(1 + 4x)^{1/2}$ $O(x^2)$, (i.e. retain terms up to the order of x^2)

(b) $(4 - 3x)^{2/3}$ $O(x^3)$

(c) $(2 - 3x^2)^{4/5}$ $O(x^6)$

Solution

(a) $(1 + 4x)^{1/2} = 1 + \frac{1}{2}(4x) + \frac{\frac{1}{2}\left(-\frac{1}{2}\right)}{1 \times 2}(4x)^2 + \cdots = 1 + 2x - 2x^2 + O(x^3)$

($O(x^3)$ in this context means terms of the order of x^3 and higher)

The expansion is valid when $|4x| < 1$, i.e. when $|x| < \frac{1}{4}$ or $-\frac{1}{4} < x < \frac{1}{4}$

(b) $(4 - 3x)^{2/3} = \left(4\left(1 - \frac{3}{4}x\right)\right)^{2/3} = 4^{2/3}\left(1 - \frac{3}{4}x\right)^{2/3}$

$$= 4^{2/3}\left(1 + \frac{\frac{2}{3}}{1}\left(-\frac{3}{4}x\right) + \frac{\frac{2}{3}\left(-\frac{1}{3}\right)}{1 \times 2}\left(-\frac{3}{4}x\right)^2\right.$$

$$\left. + \frac{\frac{2}{3}\left(-\frac{1}{3}\right)\left(-\frac{4}{3}\right)}{1 \times 2 \times 3}\left(-\frac{3}{4}x\right)^3 + \cdots\right)$$

$$= 4^{2/3}\left(1 - \frac{1}{2}x - \frac{1}{16}x^2 - \frac{1}{48}x^3\right) + O(x^4)$$

The expansion is valid when $\left|-\dfrac{3}{4}x\right| < 1$, i.e. when $\left|\dfrac{3}{4}x\right| < 1$ or $-\dfrac{4}{3} < x < \dfrac{4}{3}$

(c) $(2 - 3x^2)^{4/5} = 2^{4/5}\left(1 - \dfrac{3}{2}x^2\right)^{4/5}$

$$= 2^{4/5}\left(1 + \dfrac{\frac{4}{5}}{1}\left(-\dfrac{3}{2}x^2\right) + \dfrac{\frac{4}{5}\left(-\frac{1}{5}\right)}{1 \times 2}\left(-\dfrac{3}{2}x^2\right)^2\right.$$

$$\left. + \dfrac{\frac{4}{5}\left(-\frac{1}{5}\right)\left(-\frac{6}{5}\right)}{1 \times 2 \times 3}\left(-\dfrac{3}{2}x^2\right)^3 + \cdots\right)$$

$$= 2^{4/5}\left(1 - \dfrac{6}{5}x^2 - \dfrac{9}{50}x^4 - \dfrac{27}{250}x^6\right) + O(x^8)$$

The expansion is valid when $\left|-\dfrac{3}{2}x^2\right| < 1$, i.e. when $|x^2| < \dfrac{2}{3}$ or $-\sqrt{\dfrac{2}{3}} < x < \sqrt{\dfrac{2}{3}}$

4. Obtain the binomial expansion in negative powers of x when x is large.

If x is large then $\dfrac{1}{x}$ is small and we may assume that $\left|\dfrac{1}{x}\right| < 1$. Hence

$$\dfrac{1}{x-1} = \dfrac{1}{x\left(1 - \dfrac{1}{x}\right)} = \dfrac{1}{x}\left(1 - \dfrac{1}{x}\right)^{-1} = \dfrac{1}{x}\left(1 + \dfrac{1}{x} + \dfrac{1}{x^2} + \cdots\right)$$

$$= \dfrac{1}{x} + \dfrac{1}{x^2} + \dfrac{1}{x^3} + \cdots, \qquad\qquad |x| > 1$$

∎

Application of the binomial theorem

An important application of the binomial theorem, particularly in the context of engineering and science, arises when a system is subjected to a **perturbation**, i.e. a small change in one or more of the parameters of the system.

Example

The pressure p and volume V of a fixed mass of gas which expands adiabatically, i.e. without the input of external energy, are related by the formula $pV^\gamma = C$ where C is a constant and $\gamma = 1.4$ for a gas such as oxygen.

(a) Calculate the percentage increase in pressure which results from a percentage decrease of 0.8 in the volume.

(b) Calculate the percentage increase in pressure which results from a percentage decrease of 0.5 in the pressure.

Solution

After the change, let the new pressure and volume be $p + \delta p$ and $V + \delta V$ respectively. Note that $(p + \delta p)(V + \delta V)^{\gamma} = C = pV^{\gamma}$, so dividing by pV^{γ} we obtain

$$\left(1 + \frac{\delta p}{p}\right)\left(1 + \frac{\delta V}{V}\right)^{\gamma} = 1$$

hence

$$\left(1 + \frac{\delta p}{p}\right) = \left(1 + \frac{\delta V}{V}\right)^{-\gamma}$$

$$= 1 - \gamma \frac{\delta V}{V} + \frac{(-\gamma)(-\gamma - 1)}{2}\left(\frac{\delta V}{V}\right)^{2} - \cdots$$

We may assume that $\left(\dfrac{\delta V}{V}\right)^{2}$ and higher powers are small enough to be neglected, so that

$$\frac{\delta p}{p} \approx -\gamma \frac{\delta V}{V}$$

(a) We are given that $\dfrac{\delta V}{V} = -0.008$, therefore

$$\frac{\delta p}{p} \approx -1.4 \times (-0.008) = 0.0112$$

Hence the percentage increase in pressure is 1.12% (2 d.p.).

(b) If $\dfrac{\delta p}{p} = -0.005$ then

$$\frac{\delta V}{V} \approx -\frac{(-0.005)}{1.4} = 0.0036 \quad (4 \text{ d.p.})$$

i.e. a percentage increase in volume of 0.36% (2 d.p.). ∎

Exercise 3.2

1   Expand the polynomials $(1 + x)^{n}$ for $n = 5, 6$ and 7. For some powers of your choice, verify that the coefficient of x^{r} is $^{n}C_{r}$, e.g. the coefficient of x^{3} in the expansion of $(1 + x)^{6}$ is $^{6}C_{3} = 20$.

2 (a)   Write down the coefficients of

(i) x^4 in $(1+x)^{11}$ (ii) x^{78} in $(1+x)^{79}$

(b)   Find the value of K if the coefficient of x^5 in the expansion of $(K+x)^6$ is 18.

3* Use the identity $r\binom{n}{r} = n\binom{n-1}{r-1}$ where $r \geq 1$ to deduce that

$$\sum_{r=0}^{n}(r+1)\binom{n}{r} = (n+2) \times 2^{n-1}$$

4 Use the property $\binom{n}{r-1} + \binom{n}{r} = \binom{n+1}{r}$ to generate Pascal's triangle as far as the eleventh row, i.e. for the coefficients in the expansion of $(1+x)^{10}$.

5 (a)   Use Pascal's triangle to expand

(i) $(x+y)^4$ (ii) $(x+y)^6$

(b)   Expand

(i) $(2x-3y)^4$ (ii) $(3x+4y)^6$ (iii) $(x-y)^5$

6* By multiplying together $(1+x)^2$ and $(1+x)^5$ on the one hand and $(1+x)^3$ and $(1+x)^4$ on the other, show that

$$\sum_{r=0}^{5}\binom{5}{r} = 2\sum_{r=0}^{4}\binom{4}{r}$$

7 What is the coefficient of x in the expansion of $(1+x)(1+2x)(1+3x)\ldots(1+nx)$? (Note that we need an x from one bracket and the number 1 from each of the others.) What is the coefficient of x^n in the expansion?

8 Find the expression as far as the term in x^3 of

(a) $(1+x)^{1/3}$ (b) $(1+x)^{-1/3}$

(c) $\dfrac{1}{1+x+x^2} \equiv (1+x(1+x))^{-1}$

9 (a)   As far as the terms stated, obtain the expansions of

(i) $\dfrac{1}{2+x}$ $O(x^4)$ (ii) $\dfrac{1}{(1-2x)^{1/3}}$ $O(x^4)$

(iii) $\dfrac{1}{1+3\sqrt{x}}$ $O(x^{2.5})$ (iv) $(1+x)^p$ $O(x^2)$

(v) $\dfrac{(1+x)^{1/3}}{1+x+x^2}$ $O(x^3)$

(*Hint*: for part (v) multiply together the expansions in Question 8 parts (a) and (b)).

(b)   Using suitable series expansions obtain approximations to 4 s.f. of

(i) $(1.01)^{1/3}$ (ii) $(4.04)^{1/2}$ (iii) $(0.99)^{-1/2}$ (iv) $(9.18)^{-1/2}$

Check the results on your calculator.

10   (a)   Expand $\left(x-\dfrac{1}{x}\right)^{-1}$ as far as $O(x^4)$. Remove the factor $\dfrac{1}{x}$ first.

(b)   Note that $(x+y)^p \equiv y^p\left(1+\dfrac{x}{y}\right)^p$. Expand it as far as the term in $\left(\dfrac{x}{y}\right)^3$. What could we do when $x > y$? What would happen if p were a positive integer?

(c)   If terms $O(x^{-4})$ can be neglected when $|x|$ is large, find expansions for

(i) $\dfrac{1}{1+x}$ (ii) $\dfrac{1}{x-2}$ (iii) $\dfrac{1}{1+x+x^2}$ (iv) $\dfrac{1}{x^2-5x+6}$

Note that

$$\frac{1}{1+x} \equiv \frac{1}{x\left(1+\dfrac{1}{x}\right)}$$

11   Ignoring terms $O(x^4)$, expand the following and state in each case the convergence criterion:

(a) $(3-2x)^{3/4}$ (b) $(5+6x^2)^{-4/5}$

(c) $(8-10x-3x^2)^{-1}$

Note that

$$\frac{1}{(8-10x-3x^2)} = \frac{3}{14(2-3x)} + \frac{1}{14(4+x)}$$

This provides an alternative approach to (c).

12   (a)   A spherical balloon of radius r has volume $\dfrac{4}{3}\pi r^3$. If the balloon is inflated so that the radius increases by a small amount of δr where $\dfrac{\delta r}{r} = 0.01$, find the approximate increase in volume.

(b)   The periodic time T of a simple pendulum is given by

$$T = 2\pi\sqrt{\frac{l}{g}}$$

where l is the length of the pendulum in metres and g, the acceleration due to gravity, is $9.81\,\mathrm{m\,s^{-2}}$. If l is increased by 0.5%, by how much approximately is T increased?

13   The pressure p, volume V and absolute temperature T of a fixed mass of an ideal gas are related by the formula $pV = RT$ where R is the **universal gas constant**.

(a)   If the temperature is held constant find the percentage change in volume if the pressure increases by

(i) 1%   (ii) 2%

(b)   If the pressure and volume increase by 3% and 4% respectively, what is the percentage change in T? (Ignore terms $O\left(\dfrac{\delta T}{T}\right)^2$, etc.)

14   The acceleration due to gravity on the Earth's surface is given by $g = \dfrac{GM}{R^2}$ where G is the gravitational constant, M is the mass of the Earth and R is its radius. At height h above the Earth's surface the acceleration decreases to

$$g' = \frac{GM}{(R+h)^2}$$

Deduce that $g' \approx g\left(1 - \dfrac{2h}{R}\right)$ which h is small compared to R.

3.3 PARTIAL FRACTIONS

Rational numbers and rational functions

A **rational number** is a ratio of integers of the form $r = \dfrac{p}{q}$. The sum, difference, product and quotient of two rational numbers are also rational numbers.

Example
$$\frac{2}{5} + \frac{3}{8} = \frac{16}{40} + \frac{15}{40} = \frac{31}{40} \qquad \frac{2}{5} - \frac{3}{8} = \frac{16}{40} - \frac{15}{40} = \frac{1}{40}$$

$$\frac{2}{5} \times \frac{3}{8} = \frac{6}{40} \qquad\qquad\qquad \frac{2}{5} \div \frac{3}{8} = \frac{16}{15}$$
■

A **rational function** is a ratio of polynomials. The simplest non-trivial example is where the numerator is a constant and the denominator is a linear polynomial, e.g. $\dfrac{5}{x-2}$.

A **proper** rational function is one in which the degree of the polynomial in the numerator is less than the degree of the polynomial in the denominator. Rational functions can be combined by the four arithmetic operations to give further rational functions. The functions can be expressed as single rational functions.

Example

Express $\dfrac{5}{x-2} + \dfrac{4}{x-3}$ as a single rational function

$$\frac{5}{x-2} + \frac{4}{x-3} \equiv \frac{5(x-3)}{(x-2)(x-3)} + \frac{4(x-2)}{(x-2)(x-3)}$$

$$\equiv \frac{5(x-3) + 4(x-2)}{(x-2)(x-3)}$$

$$\equiv \frac{5x - 15 + 4x - 8}{(x-2)(x-3)}$$

$$\equiv \frac{9x - 23}{(x-2)(x-3)}$$ ∎

We must be careful to exclude from the domain of a rational function values of x which lead to a zero in the denominator of the function.

Partial fractions: basic ideas

Very often we need to reverse the process carried out in the previous example. The two rational functions on the left-hand side are called the **partial fractions** of the function on the right-hand side. One use of partial fractions is to make the integration of rational functions more straightforward. The denominators of the partial fractions are factors of the denominator of the given function.

Partial fractions of a given rational function are obtained in two stages:

(i) Recognise which type of partial fractions to use.

(ii) Obtain the coefficients which form part or all of the numerators of the fractions.

The following examples contain the more common types of partial fraction calculation. Later we set out a systematic strategy.

Examples

1. Two linear factors. Split $\dfrac{9x - 23}{x^2 - 5x + 6}$ into its partial fractions. The denominator factorises into the product $(x-2)(x-3)$, so we may write

$$\frac{9x-23}{x^2-5x+6} \equiv \frac{9x-23}{(x-2)(x-3)} \equiv \frac{A}{(x-2)} + \frac{B}{(x-3)}$$

We need to find A and B. If we multiply both sides of the identity by $(x-2)(x-3)$ we obtain the new identity

$$9x-23 \equiv A(x-3)+B(x-2) \quad \text{or} \quad A(x-3)+B(x-2) \equiv 9x-23$$

We can substitute any value of x and the resulting equation will be valid; we choose values of x which simplify the left-hand side.

Putting $x=2$ we obtain the equation $\quad A(-1)=18-23=-5 \quad$ so $\quad A=5$

Putting $x=3$ we obtain the equation $\quad B(1)=27-23=4 \quad$ so $\quad B=4$

$$\text{Hence} \quad \frac{9x-23}{x^2-5x+6} \equiv \frac{5}{(x-2)} + \frac{4}{(x-3)}$$

2. Three linear factors. Find the partial fractions of $\dfrac{4x^2-15x+15}{(x-1)(x-2)(x-3)}$

$$\frac{4x^2-15x+15}{(x-1)(x-2)(x-3)} \equiv \frac{A}{(x-1)} + \frac{B}{(x-2)} + \frac{C}{(x-3)}$$

$$\equiv \frac{A(x-2)(x-3)+B(x-1)(x-3)+C(x-1)(x-2)}{(x-1)(x-2)(x-3)}$$

Multiply the identity by $(x-1)(x-2)(x-3)$ to obtain the new identity

$$A(x-2)(x-3)+B(x-1)(x-3)+C(x-1)(x-2) \equiv 4x^2-15x+15$$

Putting $x=1$ we obtain the equation $\quad A(-1)(-2)=4-15+15=4 \quad$ so $\quad A=2$

Putting $x=2$ we obtain the equation $\quad B(1)(-1)=16-30+15=1 \quad$ so $\quad B=-1$

Putting $x=3$ we obtain the equation $\quad C(2)(1)=36-45+15=6 \quad$ so $\quad C=3$

$$\text{Hence} \quad \frac{4x^2-15x+15}{(x-1)(x-2)(x-3)} \equiv \frac{2}{(x-1)} - \frac{1}{(x-2)} + \frac{3}{(x-3)}$$

3. Repeated linear factor.   Find the partial fractions of $\dfrac{5x^2 - 17x + 15}{(x-1)(x-2)^2}$

$$\frac{5x^2 - 17x + 15}{(x-1)(x-2)^2} \equiv \frac{A}{(x-1)} + \frac{B}{(x-2)} + \frac{C}{(x-2)^2}$$

$$\equiv \frac{A(x-2)^2 + B(x-1)(x-2) + C(x-1)}{(x-1)(x-2)^2}$$

Multiply by $(x-1)(x-2)^2$ to obtain

$$A(x-2)^2 + B(x-1)(x-2) + C(x-1) \equiv 5x^2 - 17x + 15$$

Putting $x = 1$ we obtain the equation $A(-1)^2 = 5 - 17 + 15 = 3$ so $A = 3$

Putting $x = 2$ we obtain the equation $C(1) = 20 - 34 + 15 = 1$ so $C = 1$

To find B we could try another value of x, e.g. $x = 0$. It is usually easier to compare coefficients of a particular power of x. In this case we compare the coefficients of x^2 on both sides. This gives

$$A + B = 5 \quad \text{so that} \quad B = 2$$

Hence

$$\frac{5x^2 - 17x + 15}{(x-1)(x-2)^2} \equiv \frac{3}{(x-1)} + \frac{2}{(x-2)} + \frac{1}{(x-2)^2}$$

(Note that when the denominator contains a repeated linear factor you need to have partial fractions which have denominators of all powers of the repeated factor up to the one which appears in the denominator of the given function. Hence if this function contains $(x-a)^3$ in its denominator, we need partial fractions which contain respectively $(x-a)$, $(x-a)^2$ and $(x-a)^3$ in their denominators.)

4. Irreducible quadratic factor.   When a quadratic factor in the denominator does not factorise then we need to include a partial fraction of the form

$$\frac{Bx + C}{\text{quadratic factor}}$$

Find the partial fractions of $\dfrac{3x^2 + 4x + 13}{(x-1)(x^2 + 4x + 5)}$.

$$\frac{3x^2 + 4x + 13}{(x-1)(x^2 + 4x + 5)} \equiv \frac{A}{(x-1)} + \frac{Bx + C}{x^2 + 4x + 5}$$

$$\equiv \frac{A(x^2 + 4x + 5) + (Bx + C)(x-1)}{(x-1)(x^2 + 4x + 5)}$$

Multiply by $(x - 1)(x^2 + 4x + 5)$ to obtain

$$A(x^2 + 4x + 5) + (Bx + C)(x - 1) \equiv 3x^2 + 4x + 13$$

Putting $x = 1$ gives $A(10) = 3 + 4 + 13 = 20$ so $A = 2$

Putting $x = 0$ gives $A - C = 13$ so $C = -3$

Comparing coefficients of x^2 gives $A + B = 3$, therefore $B = 1$. Hence

$$\frac{3x^2 + 4x + 13}{(x - 1)(x^2 + 4x + 5)} \equiv \frac{2}{x - 1} + \frac{x - 3}{x^2 + 4x + 5}$$ ∎

Dealing with improper rational functions

An improper rational function is one in which the polynomial in the numerator is of degree equal to or greater than that of the polynomial in the denominator. Before we can apply the process of finding partial fractions we must carry out a process of long division to leave a proper rational function as the 'remainder'.

Example Write the function $\dfrac{x^3 + 4x^2 - 7x}{(x - 3)(x - 4)}$ as the sum of a linear polynomial and a proper rational

function.

First we multiply out the denominator to obtain $x^2 - 7x + 12$. The term with the highest power in the numerator is x^3 and the term with the highest power in the denominator is x^2. The ratio of these terms is $x^3 \div x^2 = x$.

Multiply the denominator by x and subtract the result from the numerator:

$$x(x^2 - 7x + 12) \equiv x^3 - 7x^2 + 12x$$

then $x^3 + 4x^2 - 7x - (x^3 - 7x^2 + 12x) \equiv 11x^2 - 19x$

So far we have deduced that

$$\frac{x^3 + 4x^2 - 7x}{(x - 3)(x - 4)} \equiv \frac{x^3 + 4x^2 - 7x}{x^2 - 7x + 12} \equiv x + \frac{11x^2 - 19x}{x^2 - 7x + 12}$$

The rational function on the right-hand side is still improper, so we must repeat the steps above.

The term with the highest power in the numerator is $11x^2$ and the term with the highest power in the denominator is x^2. The ratio of these terms is $11x^2 \div x^2 = 11$.

Multiply the denominator by 11 and subtract the result from the numerator:

first, $11(x^2 - 7x + 12) \equiv 11x^2 - 77x + 132$

then $11x^2 - 19x - (11x^2 - 77x + 132) \equiv 58x - 132$

Finally

$$\frac{x^3 + 4x^2 - 7x}{(x-3)(x-4)} \equiv x + 11 + \frac{58x - 132}{(x-3)(x-4)}$$ ∎

Exercise 3.3

1 Find integer constants which satisfy the equations

(a) $\dfrac{1}{30} = \dfrac{A}{5} - \dfrac{B}{6}$

(b) $\dfrac{23}{12} = \dfrac{A}{3} + \dfrac{B}{4} + \dfrac{C}{6}$

Show that your choice for the first equation is not unique by finding a second pair of constants which satisfy the equation.

2 Express $2x + 1 + \dfrac{2x-1}{x^2+x+1} - \dfrac{1}{x}$ as a single rational function. Factorise the numerator as far as possible.

3 Show that the sum, difference, product and ratio of the rational functions $\dfrac{2}{x-1}$ and $\dfrac{x^2+2x-1}{x-1}$ are themselves rational functions.

4 Express the following improper rational functions as a polynomial plus a proper rational function:

(a) $\dfrac{x^3 - 14x^2 + 5x - 9}{x^2 - 3x + 4}$

(b) $\dfrac{3x^2 - 19x + 23}{x^2 + 1}$

(c) $\dfrac{x^4 - 2x^3 + x + 5}{x^2 - 2x + 4}$

(d) $\dfrac{x^2 - 5x + 7}{3x^2 - 2x + 1}$

(e)* $\dfrac{x^5 - 4x^3 + 2x^2 - 1}{x^3 - 2x^2 + x + 4}$

(f)* $\dfrac{0.61x^3 - 2.32x^2 + 7.31x - 9.16}{6.31x^2 - 2.56x + 1.14}$

Parts (e) and (f) are suitable for computer algebra; in (f) retain 4 s.f.

5 Express the following proper rational functions as the sum of partial fractions:

(a) $\dfrac{2x}{(x+1)(x-1)}$

(b) $\dfrac{2x+5}{(x+3)(x+2)}$

(c) $\dfrac{x+1}{(x+4)(x+3)}$

(d) $\dfrac{2x+9}{(x+3)(x-3)}$

(e) $\dfrac{3x+2}{(x+2)(x+3)}$

6 Express the following as the sum of partial fractions:

(a) $\dfrac{x+7}{(x+2)(x+3)^2}$

(b) $\dfrac{2x^2-6x+5}{(x-1)(x-2)(x-3)}$

(c) $\dfrac{x-3}{(x^2+3)(x+1)}$

(d) $\dfrac{3x^2-x+6}{(x^2+4)(x-2)}$

(e) $\dfrac{3x^2+8x-1}{(x^2+x+3)(x-2)}$

(f) $\dfrac{1}{(x^2-4x+4)(x-1)}$

(g)* $\dfrac{x^5+2x^3+1}{(x-1)(x-2)^2(x-3)^3}$

(h)* $\dfrac{2x^2+8x+7}{(x+5)(x-6)}$

7* Express the following rational functions in the form shown, determining the unknown constants:

(a) $\dfrac{3x+5}{2x^2-5x+3} \equiv \dfrac{A}{x-1} + \dfrac{B}{2x-3}$

(b) $\dfrac{2x^2+3x+1}{x^3+3x^2-4} \equiv \dfrac{A}{x-1} + \dfrac{B}{x+2} + \dfrac{C}{(x+2)^2}$

(c) $\dfrac{1}{x^4-x^3-x-1} \equiv \dfrac{Ax+B}{x^2+1} + \dfrac{Cx+D}{x^2-x-1}$

(d) $\dfrac{3x^4+2x^2+1}{x^3-1} = Ax+B+ \dfrac{C}{x-1} + \dfrac{Dx+E}{x^2+x+1}$

8* Express the following functions in a suitable form involving partial fractions:

(a) $\dfrac{2}{3x^2+26x+35}$

(b) $\dfrac{5x+6}{2x^2-3x-20}$

(c) $\dfrac{x^2+x+1}{x^3-5x^2+4}$

(d) $\dfrac{1}{x^3+1}$

(e) $\dfrac{x^2-2x+1}{x^3-6x^2+11x-6}$

(f) $\dfrac{1}{x^5-1}$

(g) $\dfrac{x^4}{x^3+3x^2-4}$

9* Find the roots α and β of the quadratic equation $x^2 + x - 1 = 0$. Express the following functions as the sum of partial fractions in the form shown:

$$\frac{1}{x^2 + x - 1} \equiv \frac{A}{x - \alpha} + \frac{B}{x - \beta}$$

10* Repeat Question 9 for the functions

(a) $\quad \dfrac{1}{x^2 - 3x + 1}$
(b) $\quad \dfrac{x + 1}{x^2 + 4x - 2}$
(c) $\quad \dfrac{3x + 1}{x^2 - x - 7}$

11* If x is large enough for $O(x^{-3})$ terms to be neglected, show that

$$\frac{x^4 + 6x^3 - 15x^2 + 57x + 200}{x^5 + 3x^4 + 2x^3 + 19x^2 + 100x + 1000} \approx \frac{5}{x + 1} - \frac{4}{x + 2}$$

12* Using an appropriate binomial series expand $R(x) \equiv \dfrac{3x^2 + 5x + 4}{3x^3 - 15x^2 + 2x + 1}$ as far as $O(x^5)$ assuming that x is small. Hence obtain an approximate value for

$$\int_0^{0.1} \frac{3x^2 + 5x + 4}{3x^3 - 15x^2 + 2x + 1} dx$$

by integrating your series term by term and retaining 6 s.f.

13 Obtain partial fractions for the following:

(a) $\quad \dfrac{1}{x^4 + 5x^2 + 4}$
(b) $\quad \dfrac{x + 1}{x^4 + 3x^2 + 2}$

(c) $\quad \dfrac{1}{a^2 - x^2}$
(d) $\quad \dfrac{x^2}{a^4 - x^4}$

14* (a) Determine a and b which satisfy the identity

$$\exp\left(\frac{a}{x + 1}\right) \times \exp\left(\frac{b}{x + 2}\right) \equiv \exp\left(\frac{1}{x^2 + 3x + 2}\right)$$

(b) Find the natural logarithm of

$$\frac{\exp\left(\dfrac{1}{x + 1}\right) \times \exp\left(\dfrac{1}{x}\right)}{\exp\left(\dfrac{1}{x^2}\right)}$$

3.4 GRAPHS OF RATIONAL FUNCTIONS

Figure 3.3 shows sketch graphs of the three functions $f(x) = \dfrac{1}{x}$, $f(x) = \dfrac{1}{x^2}$ and $f(x) = \dfrac{2x-5}{x-3}$. Each curve has a **vertical asymptote** and a **horizontal asymptote** and the first and third functions take both positive and negative values.

Since $\dfrac{2x-5}{x-3} \equiv 2 + \dfrac{1}{x-3}$ we deduce that the graph of the function $f(x) = \dfrac{2x-5}{x-3}$ is that of the function $f(x) = \dfrac{1}{x}$ moved to the right by 3 units and moved upwards by 2 units, in other words it is centred at the point $(3, 2)$.

In general, the curve $y = \dfrac{ax+b}{cx+d}$ is a rectangular hyperbola centred at $\left(-\dfrac{d}{c}, \dfrac{a}{c}\right)$, but if $c = 0$ or if $ad = bc$ then it is a straight line. This was discussed in *Foundation Mathematics*, Chapter 8.

The graphs of other rational functions may vary considerably from those in Figure 3.3. The key points to consider when attempting to sketch them are

(i)   changes of sign in the numerator and denominator

(ii) horizontal and vertical asymptotes

(iii) behaviour for large $|x|$.

Example

Compare and contrast the graphs

(a) $y = \dfrac{x-1}{x^2 + 4x + 3}$ (b) $y = \dfrac{x-1}{x^2 + 4x + 5}$

In both cases there is a change of sign at $x = 1$, where the common numerator is zero. Further, if $|x|$ is large, we can examine the behaviour of the function by dividing by the largest power of x in either of the numerator or denominator. Hence in case (a)

$$y = \frac{x-1}{x^2 + 4x + 3} = \frac{x\left(1 - \dfrac{1}{x}\right)}{x^2\left(1 + \dfrac{4}{x} + \dfrac{3}{x^2}\right)} = \frac{\left(1 - \dfrac{1}{x}\right)}{x\left(1 + \dfrac{4}{x} + \dfrac{3}{x^2}\right)}$$

so that for large $|x|$, $y \approx \dfrac{1}{x}$. A similar conclusion is reached for case (b). So far the functions show similar behaviour, but the denominators, although almost the same, do have an important difference in that:

$$x^2 + 4x + 3 \equiv (x+1)(x+3) \text{ and is zero when } x = -1 \text{ or } x = -3, \text{ whereas}$$

$$x^2 + 4x + 5 \equiv (x+2)^2 + 1 \geq 1 \text{ for all } x$$

We now examine each curve in turn.

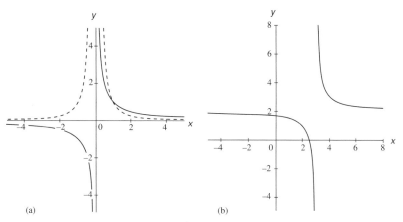

Figure 3.3 Graphs of (a) $f(x) = \dfrac{1}{x}$ and $f(x) = \dfrac{1}{x^2}$, and (b) $f(x) = \dfrac{2x - 5}{x - 3}$

(a) $y = \dfrac{x - 1}{(x + 1)(x + 3)}$ changes sign at $x = 1$ (numerator) and at $x = -1, x = -3$ (denominator). We now investigate the sign of $f(x)$.

Region	Sign of $f(x)$
$x < -3$	$\dfrac{-}{(-)(-)} = -$
$-3 < x < -1$	$\dfrac{-}{(-)(+)} = +$
$-1 < x < 1$	$\dfrac{-}{(+)(+)} = -$
$x > 1$	$\dfrac{+}{(+)(+)} = +$

The graph of the function is shown in Figure 3.4. By writing

$$y = \frac{x - 1}{(x + 1)(x + 3)} = \frac{2}{x + 3} - \frac{1}{x + 1}$$ we note that the graph is the combination of the

graphs of two rectangular hyperbolas; see Chapter 7.

(b) $y = \dfrac{x - 1}{x^2 + 4x + 5}$ changes sign only at $x = 1$, since the denominator is always positive.

The graph is shown in Figure 3.5 and you should notice that the values of y on the graph are bounded above and below. In Chapter 5 we shall see how to determine these bounding values using differentiation.   ∎

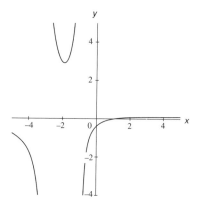

Figure 3.4 Graph of $y = \dfrac{x-1}{(x+1)(x+3)}$

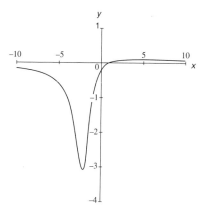

Figure 3.5 Graph of $y = \dfrac{x-1}{x^2+4x+5}$

A proper rational function $f(x)$ has a numerator of lower degree than its denominator so that for large $|x|, f(x) \to 0$. An improper rational function can be written as the sum of a polynomial function and a proper rational function so that for large $|x|$, its behaviour approximates the behaviour of the polynomial.

Example
Consider

$$y = \frac{(x-1)(x-2)(x-3)}{x^2} \equiv \frac{x^3 - 6x^2 + 11x - 6}{x^2} \equiv x - 6 + \frac{11x - 6}{x^2}$$

For large $|x|, y \approx x - 6$ and the graph is shown in Figure 3.6. ∎

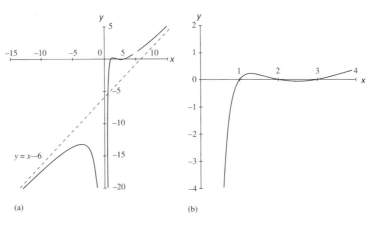

(a)

(b)

Figure 3.6 Graph of $y = \dfrac{(x-1)(x-2)(x-3)}{x^2}$

The principle of the sign change of a rational function can be used to solve inequalities involving rational functions.

Example Determine the values of x for which the inequality $\dfrac{x}{x+1} > \dfrac{x}{x+2}$ is satisfied. The given inequality is equivalent to the inequality

$$\frac{x}{x+1} - \frac{x}{x+2} > 0, \text{ i.e. } \frac{x}{(x+2)(x+1)} > 0$$

The sign changes of the rational function, $f(x)$ say, on the left-hand side of this inequality occur when $x = -2, -1, 0$. Then we have

Region	Sign of $f(x)$
$x < -2$	$\dfrac{-}{(-)(-)} = -$
$-2 < x < -1$	$\dfrac{-}{(-)(+)} = +$
$-1 < x < 0$	$\dfrac{-}{(+)(+)} = -$
$x > 0$	$\dfrac{+}{(+)(+)} = +$

so that the inequality is satisfied when either $-2 < x < -1$ or when $x > 0$.

Systematic curve sketching

Certain features of a curve can be deduced readily from a brief examination of its equation:

(i)   symmetry in either coordinate axis

(ii) points where the curve crosses the axes

(iii) regions where the graph does not exist

(iv) the behaviour of $y = f(x)$ as $x \to \pm\infty$

(v)   linear asymptotes.

Examples

1. Sketch the graph of $y = \dfrac{x^2 - 3x + 1}{x - 1}$

   (i)   There are no symmetries about the axes

   (ii) The y-axis is crossed when $x = 0$, i.e. at $y = -1$. The x-axis is crossed when $y = 0$, i.e. where $x^2 - 3x + 1 = 0$. Therefore $x = \dfrac{3 \pm \sqrt{5}}{2} \approx 0.15, 2.62$. The crossing points are $(0.15, 0)$, $(2.62, 0)$ and $(0, 1)$.

   (iii) There are three sign change and we see that

Region	Sign of $f(x)$
$x < 0.15$	$\dfrac{(-)(-)}{-} = -$
$0.15 < x < 1$	$\dfrac{(+)(-)}{-} = +$
$1 < x < 2.62$	$\dfrac{(+)(-)}{+} = -$
$x > 2.62$	$\dfrac{(+)(+)}{(+)} = +$

   (iv), (v) There is a vertical asymptote at $x = 1$. Dividing the denominator in the numerator, we obtain

   $$y = x - 2 - \frac{1}{x - 1}$$

   Hence as $x \to \pm\infty$, $y \approx x - 2$. The graph of the curve is shown in Figure 3.7.

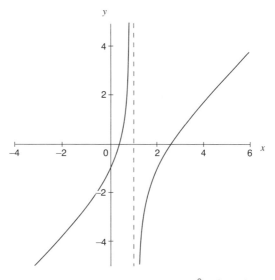

Figure 3.7 Graph of $y = \dfrac{x^2 - 3x + 1}{x - 1}$

2. Sketch the graph of $y = f(x) = \sqrt{x^2 - 1}$.

   (i) $f(-x) = f(x)$ so that $f(x)$ is an even function whose graph is symmetrical about the y-axis

   (ii) Note that $f(x) \geq 0$ and that $f(-1) = f(1) = 0$. Putting $x = 0$ leads to an invalid result, so the y-axis is not met by the curve.

   (iii) Since $f(x) \geq 0$, the third and fourth quadrants are excluded. Since we can only take the square root of a non-negative number then $|x| < 1$

   (iv) The behaviour of $f(x)$ as $x \to \pm\infty$ is found as follows

$$f(x) = (x^2 - 1)^{1/2} = |x| \left(1 - \frac{1}{x^2}\right)^{1/2}$$

$$= |x| \left(1 - \frac{1}{2x^2} + O(x)^{-4}\right) \qquad \text{(using the binomial expansion)}$$

$$= |x| - \frac{1}{2|x|} + O(|x|^{-3})$$

   Therefore $f(x)$ approaches $|x|$

   (v) There are no vertical or horizontal linear asymptotes, but $y = -x$ and $y = x$ are asymptotes if $x < 0$ or $x \geq 0$. These are shown dashed

The graph is sketched in Figure 3.8.

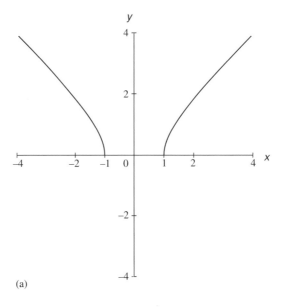

(a)

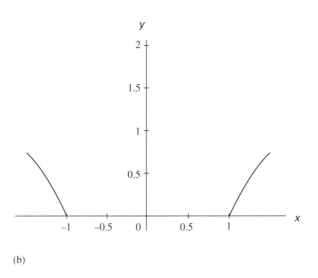

(b)

Figure 3.8 Graph of $y = f(x) = \sqrt{x^2 - 1}$: (a) $-4 \leq x \leq 4$ and (b) enlarged section for $-1.2 \leq x \leq 1.2$

3. Sketch the curve given by $x^2 = y^2(1 - y)$.

(i)   The curve is symmetrical about the y-axis since x appears only as x^2

(ii) The curve crosses the axes when $x = 0$ and when $y = 0$ and $y = 1$, i.e. at the points $(0, 0)$ and $(0, 1)$

(iii) Both sides of the equation must be non-negative so that $y \leq 1$

(iv) If $|x|$ is large then the equation $x^2 = y^2 - y^3$ can be approximated by $x^2 \approx -y^3$. It follows that y must be negative for large values of $|x|$.

The curve does *not* represent a function of x since its equation contains both x^2 and y^2. Replacing x by $-x$ does not alter the equation and this is another indicator of the symmetry of the curve about the y-axis. The equation can be rewritten as $x = \pm y\sqrt{1 - y}$, and because of the symmetry we need only sketch the graph of $x = y\sqrt{1 - y}$ in the first and fourth quadrants and reflect this curve in the y-axis. The full curve is presented in Figure 3.9.

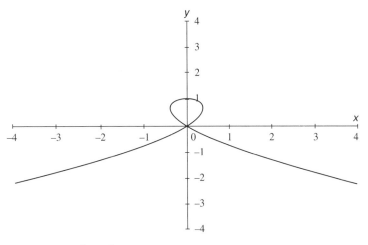

Figure 3.9 Graph of $x^2 = y^2(1 - y)$

Exercise 3.4

1   Consider the function $y = f(x) = \dfrac{x - 2}{(x + 2)(x - 1)}$

(a)   Show that it has three changes of sign and two vertical asymptotes.

(b)   Show that y is $O\left(\dfrac{1}{x}\right)$ for large $|x|$.

(c)   Sketch the graph of $y = f(x)$.

(d)   Express $f(x)$ as a sum of partial fractions and differentiate twice to prove that the curve has a maximum at $\left(4, \dfrac{1}{9}\right)$ and a minimum at $(0, 1)$.

$\left(\text{You may assume that } \dfrac{d}{dx}\left(\dfrac{c}{x + a}\right) = -\dfrac{c}{(x + a)^2}.\right)$

2 For the following rational functions determine where the sign changes occur, any vertical asymptotes and the behaviour for large $|x|$.

(a) $\dfrac{(x - 3)(x - 4)}{(x + 1)(x - 1)}$

(b) $\dfrac{x(x - 3)}{(x - 1)(x^2 + 1)}$

(c) $\dfrac{2x^2 + 1}{(x - 3)^2}$

(d) $\dfrac{x^2 + 2x - 3}{x^2 - 4x + 4}$

(e) $\dfrac{x^2 + 4x + 17}{x^2 + 2x + 7}$

(f) $\dfrac{x^2 - 7x + 12}{x^3 - 1}$

(g) $\dfrac{5x + 6}{2x^2 - 3x - 20}$

(h) $\dfrac{x + 1}{x^2 + 4x - 2}$

(i) $\dfrac{5x^4}{x^3 - 6x^2 + 11x - 6}$

(j) $\dfrac{3x - 5}{x^4 + 5x^2 + 4}$

3 Given that $y = \dfrac{x(x - 3)(x + 2)}{(x - 1)^2}$

(a)   Prove that $y = x + 1 - \dfrac{(5x + 1)}{(x - 1)^2}$

(b)   Sketch the graph. Note any sign changes, asymptotes, excluded regions and any other features of interest.

4 On the same axes sketch the graphs of

(a) $y = x^2 + 1$ and $y = \dfrac{1}{x^2 + 1}$

(b) $y = (x + 1)(x - 1)(x - 4)$ and $y = \dfrac{1}{(x + 1)(x - 1)(x - 4)}$

What do you notice about the relation between maxima, minima and zeros for the two curves in each case?

5 The rational function

$$y = \frac{x}{x^3 + 8x^2 + 21x + 18}$$

has a repeated zero for its denominator. Find the factors of the denominator. Sketch the graph of the function and indicate the vertical asymptotes.

6 Repeat Question 5 with the function

$$y = \frac{x}{x^3 + 8x^2 + 22x + 20}$$

Compare the two graphs.

7 Find a rational function which has all the following properties

(a) $f(x) \approx \dfrac{2}{x}$ when $|x|$ is large   (b) $f(-1) = f(3) = 0$

(c) $f(x)$ has opposite signs either side of $x = 1$ but $f(1)$ does not exist.

(There are many possible answers.)

8 Consider the graph of the function $f(x) = \dfrac{x^2 - 3x + 1}{x - 1}$. Where is the inequality $f(x) > 1$ satisfied? By considering the graph of $g(x) = \dfrac{x^2 - 3x + 1}{|x - 1|}$ estimate where the inequality $g(x) > 1$ is satisfied.

9 Sketch the graphs of

(a) $y = \sqrt{x} = x^{1/2}$ (b) $y = x^{1/3}$

and state the domain and range of each function.

10 Sketch the graph of $y = \sqrt{x^2 + 4}$. Describe how the function behaves

(a) when $|x|$ is small   (b) when $|x|$ is large

11 (a) Sketch the graph of

$$y = f(x) = \frac{\sqrt{x^2 - 4}}{x - 1}$$

From the graph determine when

(i) $f(x) > 1$ (ii) $|f(x)| > 1$

(b)   Verify your results by solving the appropriate equation(s).

12 (a)   Sketch on the same axes the graphs of $y = x, y = e^{-x}$ and $y = xe^{-x}$. From these sketch graphs it appears that $y = xe^{-x}$ is always to the right of $y = x$; verify that this is so.

(b)   Verify that $f(x) = xe^{-x^2}$ is an odd function. From which quadrants is its graph excluded? Sketch its graph in the first quadrant and use the oddness of the function to complete the graph.

13 By inputting very small positive values of x into a calculator decide how $f(x) = x \ln x$ behaves near $x = 0$. Sketch the graph of the function and indicate the line $y = x$ on the same axes.

14 Sketch the graph of the piecewise defined function given by

$$f(x) = \begin{cases} (x+5)(x-2)(x-1) & x < -5 \\ \dfrac{x+5}{x+1} & -5 \leq x \leq 0 \\ 5 + x \ln x & x > 0 \end{cases}$$

and estimate the values of x for which $|f(x)| < 1$.

15 Sketch the graph of $y^2 = x^3$, noting any symmetries.

16 Consider the relationship $y^2 = \dfrac{x^2(1 - x^2)}{1 + x^2}$

(a)   Establish any symmetries in the curve of the relationship.

(b)   What restrictions are there on x for the relationship to be valid?

(c)   Prove that $|y| \leq 1$. In which region of the x–y plane is the curve confined?

(d)   Sketch the curve in the first quadrant and use the symmetry properties to complete the graph.

SUMMARY

- **Factorial notation**: $n! = n \times (n-1) \times \cdots \times 2 \times 1$
- **Permutations** of m objects from a set of n objects: the answer is given by

$$^nP_m = \frac{n!}{(n-m)!} = n \times (n-1) \times \cdots \times (n-m+1)$$

- **Combinations** of m objects from a set of n objects: the number is given by

$$^nC_m = \frac{n!}{m! \times (n-m)!} = \frac{n \times (n-1) \times \cdots \times (n-m+1)}{1 \times 2 \times \cdots \times m} = {}^nC_{n-m}$$

- **Binomial expansion** for a positive integer index n is

$$(1+x)^n \equiv 1 + \frac{n}{1}x + \frac{n \times (n-1)}{1 \times 2}x^2$$
$$+ \cdots + \frac{n \times (n-1) \times \cdots \times (n-r+1)}{1 \times 2 \times \cdots \times r}x^r$$
$$+ \cdots + \frac{n}{1}x^{n-1} + x^n$$

- **Binomial expansion** for a general index p is

$$(1+x)^p = 1 + \frac{p}{1}x + \frac{p(p-1)}{1 \times 2}x^2 + \frac{p(p-1)(p-2)}{1 \times 2 \times 3}x^3$$
$$+ \cdots + \frac{p(p-1)(p-2) \times \cdots \times (p-r+1)}{1 \times 2 \times 3 \times \cdots \times r}x^r + \cdots$$

provided that $|x| < 1$

- **Partial fractions of a rational function**

denominator $(x-a)(x-b)$ fractions $\dfrac{A}{(x-a)} + \dfrac{B}{(x-b)}$

denominator $(x-a)(x-b)(x-c)$ fractions $\dfrac{A}{(x-a)} + \dfrac{B}{(x-b)} + \dfrac{C}{(x-c)}$

denominator $(x-a)^2(x-b)$ fractions $\dfrac{A}{(x-a)} + \dfrac{B}{(x-a)^2} + \dfrac{C}{(x-b)}$

denominator $(x-a)(x^2+px+q)$ fractions $\dfrac{A}{(x-a)} + \dfrac{Bx+C}{(x^2+px+q)}$

- **Poles, zeros and asymptotes**: the rational function $r(x) = \dfrac{p(x)}{q(x)}$ has a zero where $p(x) = 0$ and a pole where $q(x) = 0$.
  If $r(x) \to a$ as $x \to \infty$ or as $x \to -\infty$ then the function has a horizontal asymptote $y = a$. It has a vertical asymptote at each pole.

- **Systematic curve sketching**: points to consider are whether the curve is symmetric in either coordinate axis, where the curve crosses the axes, regions where the graph does not exist, the behaviour of $y = f(x)$ as $x \to \pm\infty$, linear asymptotes.

Answers

Exercise 3.1

1 (a) 10

(b) Note how often both squares and $n + 1$ are factors.

n	$n! + 1$	Factors	n	$n! + 1$	Factors
1	2	prime	7	5 041	71×71
2	3	prime	8	40 321	61×661
3	7	prime	9	362 881	$19 \times 71 \times 269$
4	25	5×5	10	3 628 801	11×329891
5	121	11×11	11	39 916 801	prime
6	721	7×103	12	479 001 601	$13 \times 13 \times 2834329$

2 (a) $5! = 120$ (b) $4! = 24$

(The arrangements shown are considered to be the same, so that once A is seated there are 4! choices left).

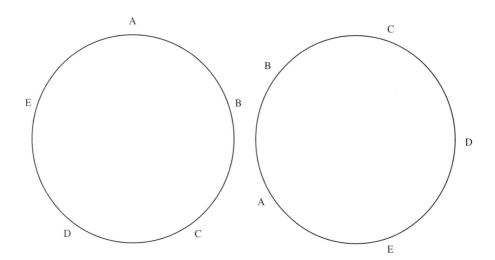

(c) $(n - 1)!$

3 (a) 30, 7920, $n(n-1)(n-2)$ (b)   388 944

4 (a) 1296
   (b) 360 (the device is really a permutation lock)

5 (a) $(23)^2 \times 9999 = 5\,289\,471$ (b) $(23)^3 \times 999 = 12\,154\,833$

6 (a) $(10)^{10} - 1 = 999\,999\,999$ (b) $10! = 3\,628\,800$
   (c) $10 \times 9^9 = 387\,420\,490$

(Option (b) is a severe restriction, whereas option (c), designed to minimise errors, is much less restrictive.)

7 102

8 (a) (i) 20   (ii) 455   (iii) 330
   (b) (ii) 84

(Selecting 6 is equivalent to eliminating 3; elimination is a kind of selection.)

9 (a) (i)   24   (ii) 12   (iii) 120

   (b) (i)   720   (ii) $\dfrac{11!}{5! \times 2! \times 2!} = 83\,160$

10 (a) $(6!)^2 = 518\,400$ (b) $4 \times 3 \times 2 \times 2 \times 2 \times 1 \times 1 = 96$

11 (a) (i) $\dfrac{n!}{3! \times (n-3)!} = \dfrac{n(n-1)(n-2)}{6}$ (ii) $\dfrac{n(n-10(-2)(n-3)}{24}$

      (iii) $\dfrac{n(n+1)}{2}$ (iv) $\dfrac{(2n)!}{(n!)^2}$

   (b)   0.0412

12 (a) $n = 6, n_1 = 3, n_2 = 1, n_3 = 2;$ $\dfrac{6!}{3! \times 1! \times 2!} = 60$

   (b) $n = 6, n_1 = 2, n_2 = 2, n_3 = 2;$ $\dfrac{6!}{3! \times (2!)^3} = 15$

(Since the rooms are identical we need to divide by a further 3!).

13 $^6C_3 \times {}^4C_2 + {}^6C_4 \times {}^4C_1 + {}^6C_5 = 186,$ $^4C_3 \times {}^6C_2 + {}^6C_1 = 66$

14 (a) (i) $^7C_2 = 21$ (ii) $^7C_5 \times {}^5C_1 = 35$ (iii) $^5C_2 = 10$

   (b) $\dfrac{10}{66} = \dfrac{5}{33}$

Exercise 3.2

1 $1 + 5x + 10x^2 + 10x^3 + 5x^4 + x^5$
$1 + 6x + 15x^2 + 20x^3 + 15x^4 + 6x^5 + x^6$
$1 + 7x + 21x^2 + 35x^3 + 35x^4 + 21x^5 + 7x^6 + x^7$

2 (a)   (i) 330   (ii) 79   (b)   3

4 The coefficients in the last row are 1, 10, 45, 120, 210, 252, 210, 120, 45, 10, 1

5 (a)   (i) $x^4 + 4x^3y + 6x^2y^2 + 4xy^3 + y^4$
        (ii) $x^6 + 6x^5y + 15x^4y^2 + 20x^3y^3 + 15x^2y^4 + 6xy^5 + y^6$
   (b)   (i) $16x^4 - 96x^3y + 216x^2y^2 - 216xy^3 + 81y^4$
        (ii) $729x^6 + 5832x^5y + 19\,440x^4y^2 + 34\,560x^3y^3 + 34\,560x^2y^4 + 18\,432xy^5$
           $+ 4096y^6$
        (iii) $x^5 - 5x^4y + 10x^3y^2 - 10x^2y^3 + 5xy^4 - y^5$

7 $1 + 2 + 3 + \cdots + n = \dfrac{n(n+1)}{2}, \quad n!$

8 (a) $1 + \dfrac{x}{3} - \dfrac{x^2}{9} + \dfrac{5x^3}{81} - \cdots$ (b) $1 - \dfrac{x}{3} + \dfrac{2x^2}{9} - \dfrac{14x^3}{81} + \cdots$
   (c) $1 - x + x^3 - \cdots$

9 (a)   (i) $\dfrac{1}{2} - \dfrac{x}{4} + \dfrac{x^2}{8} - \dfrac{x^3}{16} + \dfrac{x^4}{32} + O(x^5)$
        (ii) $\dfrac{1}{2} + \dfrac{2x}{3} + \dfrac{8x^2}{9} + \dfrac{112x^3}{81} + \dfrac{560x^4}{243} + O(x^5)$
        (iii) $1 - 3\sqrt{x} + 9x - 27x\sqrt{x} + 81x^2 - 243x^2\sqrt{x} + O(x^3)$
        (iv) $1 + px + \dfrac{p(p-1)}{2} + O(x^3)$
        (v) $1 - \dfrac{2x}{3} - \dfrac{4x^2}{9} + \dfrac{95x^3}{81} + O(x^4)$

10 (a) $-x - x^3 + O(x^5)$
   (b) $y^p \left(1 + p\dfrac{x}{y} + \dfrac{p(p-1)}{2!} \left(\dfrac{x}{y}\right)^2 + \dfrac{p(p-1)(p-2)}{3!} \left(\dfrac{x}{y}\right)^3 + \cdots \right)$

If $x > y$ we reverse the roles of x and y. If p is a positive integer then the series is finite and is valid for all values of x and y.

(c)　(i)　$\dfrac{1}{x} - \dfrac{1}{x^2} + \dfrac{1}{x^3}$　　(ii)　$\dfrac{1}{x} + \dfrac{2}{x^2} + \dfrac{4}{x^3}$

　　　(iii)　$\dfrac{1}{x^2} - \dfrac{1}{x^3}$　　(iv)　$\dfrac{1}{x^2} + \dfrac{5}{x^3}$

11　(a)　$3^{3/4}\left(1 - \dfrac{x}{2} - \dfrac{x^2}{24} - \dfrac{5x^3}{432} + O(x^4)\right): |x| < \dfrac{3}{2}$

　　(b)　$5^{-4/5}\left(1 - \dfrac{24x^2}{25} + O(x^4)\right): |x| < \sqrt{\dfrac{5}{6}}$

　　(c)　$\dfrac{1}{8}\left(1 + \dfrac{5x}{4} + \dfrac{31x^2}{16} + \dfrac{185x^3}{64} + O(x^4)\right): |x| < \min\left(\dfrac{2}{3}, 4\right)$, so $|x| < \dfrac{2}{3}$

12　(a)　0.03　　　　　　　　　　　　　(b)　0.25%

13　(a)　(i) 1% decrease　　(ii) 2% decrease
　　(b)　7%

Exercise 3.3

1　(a)　e.g. $A = B = 1, A = 6, B = 7$　　(b)　e.g. $A = 4, B = 1, C = 2$

2　$\dfrac{(x^3 + 2x^2 + 3x + 1)(2x - 1)}{x(x^2 + x + 1)}$

4　(a)　$x - 11 + \dfrac{(35 - 32x)}{x^2 - 3x + 4}$　　(b)　$3 + \dfrac{(20 - 19x)}{x^2 + 1}$

　　(c)　$x^2 - 4 + \dfrac{7(3 - x)}{x^2 - 2x + 4}$　　(d)　$\dfrac{1}{3} + \dfrac{(20 - 13x)}{3(3x^2 - 2x + 1)}$

　　(e)　$x^2 + 2x - 1 + \dfrac{1}{2(x + 1)} + \dfrac{(2 - 13x)}{2(x^2 - 3x + 4)}$ (denominator factorises)

　　(f)　$0.097x - 0.328 + \dfrac{(6.359x - 8.786)}{6.31x^2 - 2.56x + 1.14}$

5　(a)　$\dfrac{1}{x + 1} + \dfrac{1}{x - 1}$　　(b)　$\dfrac{1}{x + 3} + \dfrac{1}{x + 2}$　　(c)　$\dfrac{3}{x + 4} - \dfrac{2}{x + 3}$

　　(d)　$\dfrac{5}{2(x - 3)} - \dfrac{1}{2(x + 3)}$　　(e)　$\dfrac{7}{x + 3} - \dfrac{4}{x + 2}$

6　(a)　$\dfrac{5}{x + 2} - \dfrac{5}{x + 3} - \dfrac{4}{(x + 3)^2}$　　(b)　$\dfrac{1}{2(x - 1)} - \dfrac{1}{x - 2} + \dfrac{5}{2(x - 3)}$

(c) $\dfrac{x}{x^2+3}-\dfrac{1}{x+1}$

(d) $\dfrac{2}{x-2}+\dfrac{x+1}{x^2+4}$

(e) $\dfrac{3}{x-2}+\dfrac{5}{x^2+x+3}$

(f) $\dfrac{1}{x-1}-\dfrac{1}{x-2}+\dfrac{1}{(x-2)^2}$

(g) $-\dfrac{1}{2(x-1)}-\dfrac{202}{x-2}-\dfrac{49}{(x-2)^2}+\dfrac{407}{2(x-3)}-\dfrac{143}{(x-3)^2}+\dfrac{149}{(x-3)^2}$

(h) $2-\dfrac{17}{11(x+5)}+\dfrac{127}{11(x-6)}$

7 (a) $-\dfrac{8}{x-1}+\dfrac{19}{2x-3}$

(b) $\dfrac{2}{3(x-1)}+\dfrac{4}{3(x+2)}-\dfrac{1}{(x+2)^2}$

(c) $\dfrac{x-2}{5(x^2+1)}+\dfrac{(3-x)}{5(x^2-x-1)}$

(d) $3x+\dfrac{2}{x-1}+\dfrac{1}{x^2+x+1}$

8 (a) $\dfrac{3}{8(3x+5)}-\dfrac{1}{8(x+7)}$

(b) $\dfrac{1}{2x+5}+\dfrac{2}{x-4}$

(c) $-\dfrac{3}{7(x-1)}+\dfrac{5(1+2x)}{7(x^2-4x+4)}$

(d) $\dfrac{1}{3(x+1)}-\dfrac{(2-x)}{3(x^2-x+1)}$

(e) $\dfrac{2}{x-3}-\dfrac{1}{x-2}, x\neq 1$ $(x-1)$ is a factor of top and bottom.

(f) $\dfrac{1}{5(x-1)}-\dfrac{(x^3+2x^2+3x+4)}{5(x^4+x^3+x^2+x+1)}$

(g) $x-3+\dfrac{1}{9(x-1)}+\dfrac{80}{9(x+2)}-\dfrac{16}{3(x+2)^2}$

9 $\alpha=-\dfrac{1}{2}(\sqrt{5}+1),\ \beta=\dfrac{1}{2}(\sqrt{5}-1);\ A=-\dfrac{1}{\sqrt{5}},\ B=\dfrac{1}{\sqrt{5}}$

10 (a) $\dfrac{1}{\sqrt{5}}\left(\dfrac{1}{x-\frac{1}{2}(3+\sqrt{5})}-\dfrac{1}{x-\frac{1}{2}(3-\sqrt{5})}\right)$

(b) $\dfrac{1-3/\sqrt{2}}{2(x+\sqrt{2}-2)}+\dfrac{1-3/\sqrt{2}}{2(x-\sqrt{2}-2)}$

(c) $\dfrac{3-5/\sqrt{29}}{2(x+\frac{1}{2}(\sqrt{29}-1))}+\dfrac{3+5/\sqrt{29}}{2(x-\frac{1}{2}(\sqrt{29}-1))}$

12 $4-3x+69x^2-195x^3+1434x^4-6000x^5+O(x^6);\ \displaystyle\int_0^{0.1}R(x)\,dx\approx 0.4050$

13 (a) $\dfrac{1}{3}\left(\dfrac{1}{x^2+1}-\dfrac{1}{x^2+4}\right)$

(b) $\dfrac{x+1}{x^2+1}-\dfrac{x+1}{x^2+2}$

(c) $\dfrac{1}{2a}\left(\dfrac{1}{x+a}-\dfrac{1}{x-a}\right)$

(d) $\dfrac{1}{4a}\left(\dfrac{1}{x+a}-\dfrac{1}{x-a}\right)-\dfrac{1}{2(x^2+a^2)}$

14   (a) $a = 1, b = -1$ (b) $\dfrac{2x^2 - 1}{x^2(x + 1)}$

Exercise 3.4

1   (a) $x = -2, x = 1$ are vertical asymptotes

Region	Sign of $f(x)$
$x < -2$	$\dfrac{-}{(-)(-)} = -$
$-2 < x < 1$	$\dfrac{-}{(+)(-)} = +$
$1 < x < 2$	$\dfrac{-}{(+)(+)} = -$
$x > 2$	$\dfrac{+}{(+)(+)} = +$

(c)

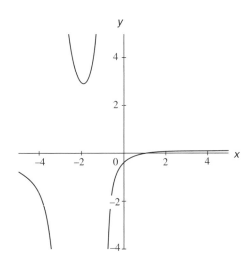

2   In the following s.c. stands for sign changes, v.a. stands for vertical asymptotes.

(a)   s.c. at $x = -4, -1, 1, 3$ v.a. at $x = -1, x = 1$ $f(x) \approx 1, |x|$ large
(b)   s.c. at $x = 0, 1, 3$ v.a. at $x = 1$ $f(x) \approx 1/x, |x|$ large
(c)   no s.c.;   v.a. at $x = 3$ $f(x) \approx 2, |x|$ large

(d) s.c. at $x = -3, 1$ v.a. at $x = 2$ $f(x) \approx 1$, $|x|$ large

(e) no s.c. no v.a. $f(x) > 0$ for all x $f(x) \approx 1$, $|x|$ large

(f) s.c. at $x = 1, 3, 4$ v.a. at $x = 1$ $f(x) \approx 1/x$, $|x|$ large

(g) s.c. at $x = -5/2, -5/6, 4$ v.a. at $x = -5/2, 4$
$f(x) \approx 5/2x$, $|x|$ large

(h) s.c. at $x = -\sqrt{6} - 2, -1, \sqrt{6} - 2$ v.a. at $x = \pm\sqrt{6} - 2$
$f(x) \approx 1/x$, $|x|$ large

(i) s.c. at $x = 0, 1, 2, 3$ v.a. at $x = 1, 2, 3$
$f(x) \approx 5x$, $|x|$ large

(j) s.c. at $x = 5/3$ no v.a. $f(x) \approx 3/x^3$, $|x|$ large

3 (b)

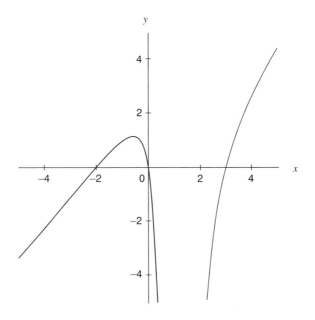

4   (a)   Since $y > 0$ for all x then $1/y > 0$; the local minimum at $(0, 1)$ becomes a local maximum; even function symmetry (left–right) is preserved.

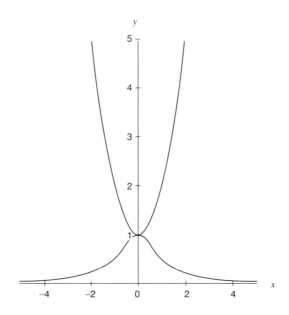

(b)   Sign changes are preserved, local maxima and minima are inverted and zeros become vertical asymptotes.

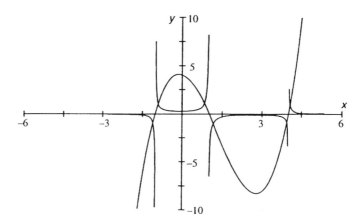

5,6   The two curves are very similar; where the curves in Question 6 differs from the curve in Question 5 it is shown dotted; it does not have a vertical asymptote at $x = -3$.

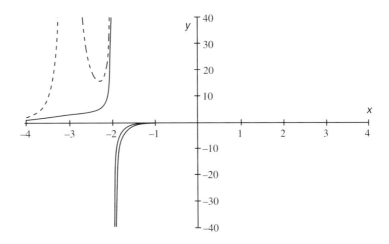

7   A simple choice is $\dfrac{2(x + 1)(x - 3)}{(x - 1)(x^2 + 1)}$

8 $2 - \sqrt{2} < x < 1, x > 2 + \sqrt{2}$ or $0.586 < x < 1, x > 3.414;\ x < 0, x > 2 + \sqrt{2}$

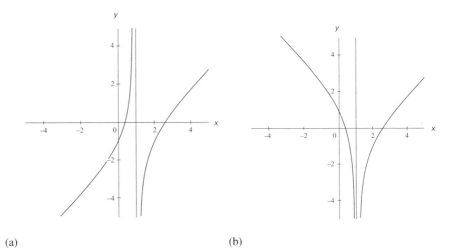

(a)   (b)

9 (a)   Domain: non-negative real numbers   Range: non-negative real numbers

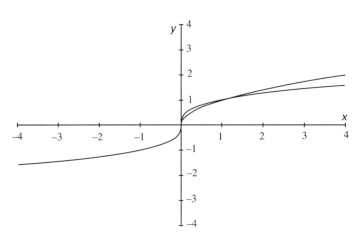

(b)   Domain: all real numbers   Range all real numbers.

10 (a) $y = 2 + \dfrac{x^2}{4} + O(x^4)$ (b) $y = x + \dfrac{2}{x} + O(x^{-3})$

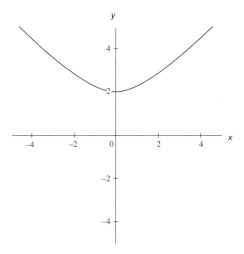

11 (a), (b) $x > 2.5$; the domain is $|x| \geq 2$

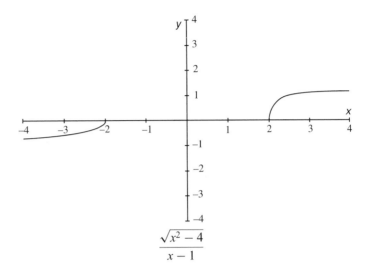

$$\frac{\sqrt{x^2 - 4}}{x - 1}$$

12   If $x > 0$, $xe^{-x} < x$ because $e^{-x} < 1$. Hence $y = xe^{-x}$ lies below and to the right of $y = x$. If $x = 0$ the curves meet. If $x < 0$, $e^{-x} > 1$ then $xe^{-x} < x$, being more negative.

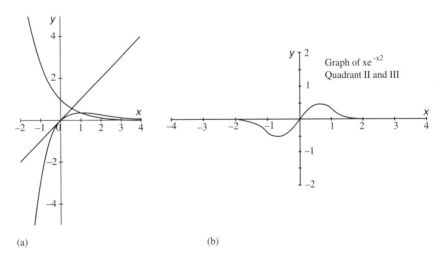

Graph of xe^{-x^2}
Quadrant II and III

(a)   (b)

13   It appears that the function is tending to zero despite the fact that the logarithm is tending to negative infinity.

x	0.1	0.01	0.001	0.0001
$x \ln x$	-0.230	-0.046	-0.007	-0.001

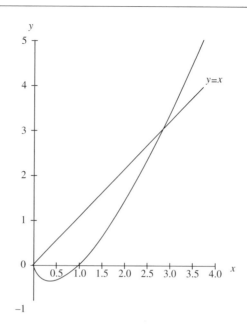

14 $|f(x)| < 1$ for $-5.05 < x < -3$ (approx.) Note that $f(x)$ is continuous but $\lim_{x\to 0-} f(x) = -4$ but $\lim_{x\to 0+} f'(x) =$ negative infinity. At $x = -5$, the discontinuity in $f'(x)$ is more obvious.

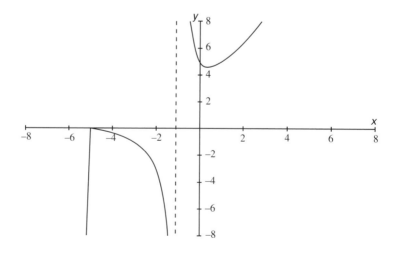

15

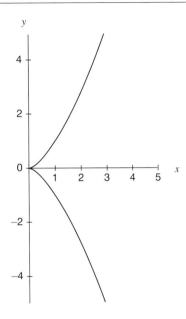

16 (a) Symmetry about both axes   (b) $|x| \leq 1$ so that $y^2 > 0$

(c)   If $|x| \leq 1$ then $\sqrt{1-x^2} < \sqrt{1+x^2}$ so that $\left| \dfrac{x\sqrt{1-x^2}}{\sqrt{1+x^2}} \right| \leq |x| \leq 1$, i.e. $|y| \leq 1$. The

graph is confined to the square $|x| \leq 1$, $|y| \leq 1$.

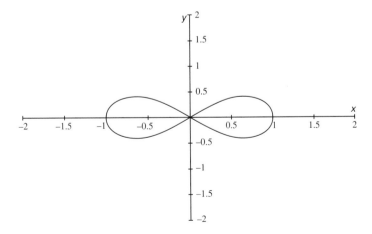

4 VECTORS

Introduction

Vectors can be used to solve many problems in geometry and mechanics. They allow results about distances and angles to be derived simultaneously. A single vector equation can describe motion in two or three dimensions. This equation can be resolved into the appropriate components, one for each spatial dimension.

Objectives

After working through this chapter you should be able to

- Represent a vector in graphical form
- Represent graphically the sum and difference of two vectors
- Represent graphically a scalar multiple of a given vector
- Identify the Cartesian coordinates of a vector
- Carry out the operations of addition, subtraction and scalar multiplication in component form
- Find the magnitude of a vector and obtain the unit vector in the same direction
- Define the scalar product of two vectors and calculate its value
- Use the scalar product to find the angle between two straight lines
- Use the scalar product to calculate the work done in moving an object against a force
- Define and calculate the vector product of two vectors
- Use the vector product to calculate the torque exerted by a force
- Obtain the equation of a plane in three dimensions and a straight line in three dimensions.

4.1 IDEAS, DEFINITIONS AND GRAPHS

Physical quantities can be divided into two categories: those which can be specified completely by magnitude only and those which require both magnitude and direction for complete specification. The former are known as **scalars**; examples are temperature, mass, energy, work and electrostatic potential. The latter are known as **vectors**; examples are force, velocity, momentum and acceleration. Every vector has a direction associated with a quantity. Hence we may talk about a speed of $3\,\mathrm{m\,s}^{-1}$ but a velocity of $3\,\mathrm{m\,s}^{-1}$ north-east.

All vector quantities combine in precisely the same manner and obey the laws of vector algebra. We classify vectors under four headings. A **free vector** has magnitude and direction but no particular position associated with it. A **displacement vector** moves every point in space through the same distance and in the same direction; in particular, the displacement vector of 2 units in a given direction may be represented graphically by a line segment of 2 units drawn in that direction from any point in space. A **line vector** is located along a particular straight line in space; if a force acts on a rigid body then the force can be moved along its line of action without changing its effect on the body. If vectors are specified relative to a fixed origin then they are called **position vectors**; we can specify the position of a point A relative to a fixed origin O by the directed line segment $\overrightarrow{OA}$. Unless we state the contrary we shall develop the vector algebra of free vectors.

Vectors may be represented graphically by directed line segments where the length of the segment in suitable units is the magnitude of the vector and the direction of the line segment is the direction of the vector. Figure 4.1 shows four representations of a vector of magnitude 2 units in the direction north-east.

The first three directed line segments are denoted $\overrightarrow{AB}$, $\overrightarrow{CD}$ or $\overrightarrow{EF}$. The fourth segment is denoted $\boldsymbol{a}$. Often, vectors are denoted by bold lower case letters, for example $\boldsymbol{a}, \boldsymbol{b}, \boldsymbol{c}$. In

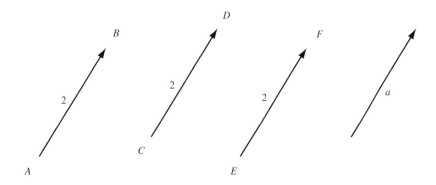

Figure 4.1 Representing a vector by a line segment

written work we underline the letter, e.g. $\underline{a}$, $\underline{b}$, $\underline{c}$. The magnitude of the vector a is denoted a or $|a|$. Two vectors are considered **equal** if they have the same magnitude and direction.

Vector addition, subtraction and multiplication by a scalar

The addition of two vectors is defined by the **triangle law**. In Figure 4.2 $\vec{AB}$ and $\vec{BC}$ represent the vectors a and b. The sum of a and b, written $a + b$, is then represented by $\vec{AC}$. Therefore $\vec{AB} + \vec{BC} = \vec{AC}$. The sum of two vectors is often called their **resultant**.

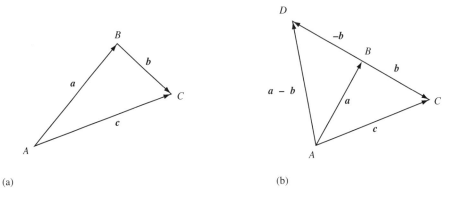

(a) (b)

Figure 4.2 Using the triangle law for (a) vector addition and (b) vector subtraction

Furthermore, the vector $-b$ is defined to be equal in magnitude to b but opposite in direction. In Figure 4.2(a) it is represented by the line segment $\vec{CB}$.

We can define the subtraction of two vectors by

$$a - b = a + (-b)$$

In Figure 4.2(b) the vector $-b$ is represented by $\vec{BD}$ since $BD = BC$ and the directed line segments $\vec{BC}$ and $\vec{BD}$ are in opposite directions.

The vector ka is a vector in the same direction as the vector a if $k > 0$ and in the opposite direction if $k < 0$; its magnitude is $|k|$ times that of a, i.e. $|ka| = |k||a|$. As an example, the vector $2a$ is in the same direction as a and has twice its magnitude; the vector $-\dfrac{1}{2}a$ is in the opposite direction to a and its magnitude is half that of a.

Laws of vector algebra

We state the following laws which are obeyed by vectors.

1. $a + b = b + a$ \qquad Commutative law of addition

2. $a + (b + c) = (a + b) + c$ Associative law of addition

3. $ka = ak$ Commutative law of scalar multiplication

4. $k(la) = (kl)a$ Associative law of scalar multiplication

5. $k(a + b) = ka + kb$ Distributive law

6. There is a **zero vector**, $\mathbf{0}$ such that for every vector a, $a + 0 = 0 + a = a$

7. For every vector a there is a vector $-a$ such that $a + (-a) = (-a) + a = 0$

8. $a - b = a + (-b)$

We shall make constant use of these laws in the rest of this chapter.

Examples

1. With reference to Figure 4.3 prove that $\overrightarrow{BC} = -3 \overrightarrow{DA}$. Note that we are establishing two results. Firstly that BC and AD are parallel (since BC is a multiple of AD, and secondly that the length of BC is three times the length of AD (in the opposite direction to DA).

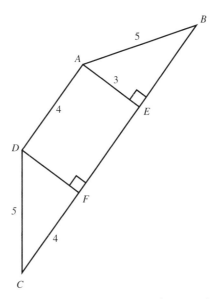

Figure 4.3 Proving that $\overrightarrow{BC} = -3\overrightarrow{DA}$

Triangles AEB and CDF are both 3, 4, 5 triangles, so $DF = 3$ and therefore $AEFD$ is a rectangle with sides 3 and 4. Also, EF is part of BC, so DA is parallel to BC. Now $BE = EF = FC = 4$, so $BC = 12$ and is in the opposite sense to DA, therefore

$$\overrightarrow{BC} = -3 \overrightarrow{DA}$$

2. Show diagrammatically that $|\lambda a + \mu b| \leq \lambda |a| + \mu |b|$. Draw a and b non-parallel and take $\lambda > 0$ and $\mu < 0$, for example (Figure 4.4).

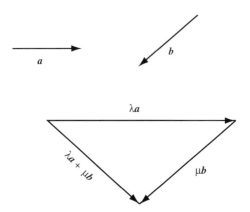

Figure 4.4 Demonstrating that $|\lambda \boldsymbol{a} + \mu \boldsymbol{b}| \leq \lambda |\boldsymbol{a}| + \mu |\boldsymbol{b}|$

No matter which option we choose, $\lambda a + \mu b$ is always the third side of a triangle whose length cannot exceed the sum of the lengths of the other two sides. Only when a and b are parallel does equality occur; then λa, μb and $\lambda a + \mu b$ can be represented as part of the same straight line.

3. $ABCD$ is a trapezium in which AB is parallel to CD. Show that

$$\overrightarrow{BA} + \overrightarrow{DA} + \overrightarrow{DC} + \overrightarrow{BC} = 2(\overrightarrow{DC} + \overrightarrow{BA})$$

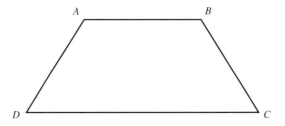

Figure 4.5 Relationship between sides of a trapezium

Referring to Figure 4.5, we have

$$\vec{DA} = \vec{DC} + \vec{CA} \quad \text{and} \quad \vec{CA} = \vec{CB} + \vec{BA}$$

therefore

$$\vec{DA} = \vec{DC} + \vec{CB} + \vec{BA}$$

Hence

$$\vec{DA} + \vec{BC} = \vec{DC} + \vec{CB} + \vec{BC} + \vec{BA} = \vec{DC} + \vec{BA}, \text{ since } \vec{CB} = -\vec{BC}$$

therefore

$$\vec{BA} + \vec{DA} + \vec{DC} + \vec{BC} = 2(\vec{DC} + \vec{BA})$$

4. In triangle ABC (Figure 4.6), M and N are the midpoints of sides BC and AC respectively. G is a point on AM such that $AG : GM = 2 : 1$. Use vector methods to show that B, G and N are collinear.

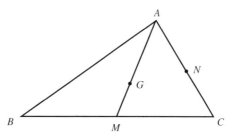

Figure 4.6   Proving that B, G and N are collinear

Let $\vec{AB} = \boldsymbol{b}$ and $\vec{AC} = \boldsymbol{c}$

In triangle ABC, $\vec{AB} + \vec{BC} = \vec{AC}$ i.e. $\boldsymbol{b} + \vec{BC} = \boldsymbol{c}$

Then $\vec{BC} = \boldsymbol{c} - \boldsymbol{b}$ and $\vec{BM} = \dfrac{1}{2}\vec{BC} = \dfrac{1}{2}(\boldsymbol{c} - \boldsymbol{b})$

Next, $\vec{AM} = \vec{AB} + \vec{BM} = \dfrac{1}{2}(\boldsymbol{b} + \boldsymbol{c})$

Therefore $\vec{AG} = \dfrac{2}{3}\vec{AM} = \dfrac{1}{3}(\boldsymbol{b} + \boldsymbol{c})$

Now $\vec{BG} = \vec{BA} + \vec{AG} = \dfrac{1}{3}(\boldsymbol{c} - 2\boldsymbol{b})$

and $\vec{BN} = \vec{BA} + \vec{AN} = -\boldsymbol{b} + \dfrac{1}{2}\boldsymbol{c} = \dfrac{1}{2}(\boldsymbol{c} - 2\boldsymbol{b})$

Therefore BN is parallel to BG, so B, G and N are collinear, since $\overrightarrow{BN}$ is a multiple of $\overrightarrow{BG}$.

5. Relative to a given origin, let A, B, C have position vectors a, b, c respectively (Figure 4.7). Let P, Q, R be the midpoints of AB, BC, CA respectively and have position vectors p, q, r. Show that $a + b + c = p + q + r$

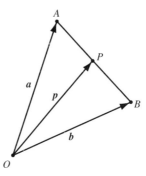

Figure 4.7   Finding the vector representation of P, the midpoint of side AB

Referring to Figure 4.7, $\overrightarrow{OB} - \overrightarrow{OA} = \overrightarrow{AB} = b - a$ and $\overrightarrow{AP} = \frac{1}{2}\overrightarrow{AB} = \frac{1}{2}(b - a)$

Then $\overrightarrow{OP} = \overrightarrow{OA} + \overrightarrow{AP} = a + \frac{1}{2}(b - a)$ so that $p = \frac{1}{2}(a + b)$

Similarly, $q = \frac{1}{2}(b + c)$ and $r = \frac{1}{2}(c + a)$

Hence $p + q + r = \frac{1}{2}(a + b) + \frac{1}{2}(b + c) + \frac{1}{2}(c + a) = a + b + c$ ◼

Exercise 4.1

1   (a)   Four points A, B, C, D have position vectors $a, b, 2a + 3b, a - 2b$ respectively. Express $\overrightarrow{AC}$, $\overrightarrow{DB}$, $\overrightarrow{BC}$, $\overrightarrow{CA}$ in terms of a and b.
    (b)   If E has position vector $3a - 2b$ show that A, B and E are collinear.

2   If the diagonals of a parallelogram are represented by the vectors a and b, show that the sides are represented by the vectors $\frac{1}{2}(a + b)$ and $\frac{1}{2}(a - b)$

3   Find the position vectors of
    (a)   the midpoint of P_1P_2, given that $\overrightarrow{OP_1} = r_1$ and $\overrightarrow{OP_2} = r_2$
    (b)   the point which divides P_1P_2 in the ratio $m : n$

4 The vectors r_1, r_2, r_3 and r_4 are the position vectors of the vertices of the quadrilateral *ABCD* from an origin *O*.

(a) Write down the vectors which correspond to the sides $\overrightarrow{AB}$ and $\overrightarrow{CD}$ and the diagonal $\overrightarrow{CA}$.

(b) If $r_1 - r_2 + r_3 - r_4 = 0$ prove that *ABCD* is a parallogram.

5 Using the notation of Question 4, write down the position vectors of the midpoints P, Q, R and S of the quadrilateral *ABCD*. Prove that *PQRS* is a parallelogram and illustrate the result by drawing any quadrilateral with internal angles all less than $180°$ and constructing the quadrilateral *PQRS*.

6 (a) The vectors $\overrightarrow{OA} = a$, $\overrightarrow{OB} = b$ and $\overrightarrow{OC} = c$ are the position vectors of the triangle *ABC* from an origin *O*. Following on from Question 4, prove that if the centroid *G* of the triangle has position vector $\overrightarrow{OG}$ then $\overrightarrow{OG} = \dfrac{1}{3}(a + b + c)$

(b) If A, B, C, D are four non-coplanar points in space with position vectors $\overrightarrow{OA} = a$, $\overrightarrow{OB} = b$, $\overrightarrow{OC} = c$ and $\overrightarrow{OD} = d$ give a geometric interpretation of the vector $\overrightarrow{OH} = \dfrac{1}{4}(a + b + c + d)$

7 If *a* and *b* are the vectors $\overrightarrow{AB}$ and $\overrightarrow{BC}$ representing two adjacent sides of a regular hexagon *ABCDEF*, write down the vectors which represent the other sides taken in order.

8 (a) The vertices A, B and C of triangle *ABC* have position vectors *a*, *b* and *c* respectively. Write down expressions for the vectors $\overrightarrow{AB}$, $\overrightarrow{BC}$ and $\overrightarrow{CA}$ and deduce that $\overrightarrow{AB} + \overrightarrow{BC} + \overrightarrow{CA} = 0$. Did you expect this result?

(b) If *G* is the centroid of triangle *ABC*, prove that $\overrightarrow{GA} + \overrightarrow{GB} + \overrightarrow{GC} = 0$.

9 R, P, Q are the midpoints of the sides AB, BC, CA of the triangle *ABC*.

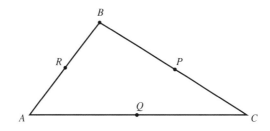

(a)   If $\overrightarrow{OA} = a$, $\overrightarrow{OB} = b$, $\overrightarrow{OC} = c$ when referred to an origin O, determine in terms of a, b and c the vectors $\overrightarrow{OP} = p$, $\overrightarrow{OQ} = q$, $\overrightarrow{OR} = r$.

(b)   Prove that $\overrightarrow{RP} = \dfrac{1}{2} \overrightarrow{AC}$.

(c)   Deduce that triangle PQR is similar to triangle ABC.

(d)   Further prove that the triangles ARQ, BPR, CQP and PQR are congruent. Determine the area of each as a function of the area of the triangle ABC.

(e)   Where is the centroid G of the triangle PQR? What is the vector $\overrightarrow{OG}$?

10 $ABCD$ is a tetrahedron in three-dimensional space, where P, Q, R and S are the midpoints of the edges AB, BC, CD and DA respectively.

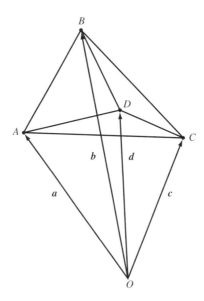

(a)   Prove that $\overrightarrow{PQ}$ and $\overrightarrow{RS}$ are equal and parallel.

(b)   Prove that P, Q, R and S are not only coplanar but form a parallelogram in their plane.

(c)   Prove that the diagonals of $PQRS$ meet at the centroid of $ABCD$. (Note the answer to Question 6(b).)

4.2 UNIT VECTORS AND CARTESIAN COORDINATES

A vector defines a direction in a two-dimensional plane or in a three-dimensional space. In each direction we can define a vector of unit magnitude, known as a **unit vector** $\hat{a}$.

Given a vector a, a unit vector in the direction of a is defined as

$$\hat{a} = \frac{a}{|a|}$$

Hence $a = |a|\hat{a}$.

In the two-dimensional Cartesian plane we define i and j to be the unit vectors in the x and y directions respectively. Any two-dimensional vector can be uniquely represented as a linear combination of i and j, i.e. in the form $\alpha i + \beta j$, where α and β are scalars. For example, referring to Figure 4.8, we have

$$\overrightarrow{OA} = 9i + j \qquad \overrightarrow{OB} = 6i + 6j \qquad \overrightarrow{OC} = 3i - 2j \qquad \overrightarrow{OD} = -2i + 2j$$

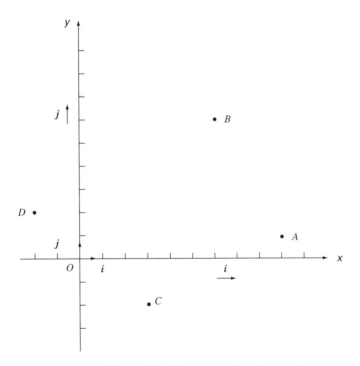

Figure 4.8 Components of a vector

A general two-dimensional vector $a = a_1 i + a_2 j$ has magnitude $|a| = a = (a_1^2 + a_2^2)^{1/2}$ with projections on the coordinate axes at angles with cosines $\dfrac{a_1}{a}$ and $\dfrac{a_2}{a}$ respectively (Figure 4.9). We call a_1 and a_2 the **components** of a and we call the ratios $\dfrac{a_1}{a} = \cos \alpha$ and $\dfrac{a_2}{a} = \cos \beta$ the **direction cosines of** a. It is straightforward to show that $\cos^2 \alpha + \cos^2 \beta = 1$.

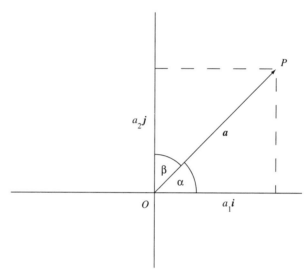

Figure 4.9 Components of a general two-dimensional vector

Note that we can write $\overrightarrow{OP} = (a_1, a_2) = a$. If we multiply a by a factor λ to obtain $\lambda a = (\lambda a_1, \lambda a_2)$ then it is clear that the i and j components are in the same ratio for both vectors. Consequently, we call the ratio $a_1 : a_2$ the **direction ratio** of a, a quantity which remains fixed.

Example

Find the magnitude, direction cosines and direction ratio for the vectors $\overrightarrow{OC}$ and $\overrightarrow{BA}$ in Figure 4.8.

Now $\overrightarrow{OC} = 3i - 2j$ so that $|\overrightarrow{OC}| = (3^2 + (-2)^2)^{1/2} = \sqrt{13}$. The direction cosines of $\overrightarrow{OC}$ are therefore $\dfrac{3}{\sqrt{13}}$ and $\dfrac{-2}{\sqrt{13}}$. The direction ratio $\overrightarrow{OC}$ is $3 : -2$. The vector $\overrightarrow{BA} = \overrightarrow{OA} - \overrightarrow{OB} = 9i + j - 6i - 6j = 3i - 5j$, so its magnitude is

$$|\overrightarrow{BA}| = (3^2 + (-5)^2)^{1/2} = \sqrt{34}$$

Its direction cosines are $\dfrac{3}{\sqrt{34}}$ and $\dfrac{-5}{\sqrt{34}}$ and its direction ratio is $3 : -5$. ∎

Three dimensions

In three dimensions we denote the unit vectors along the x, y and z axes by i, j and k respectively (Figure 4.10(b)). Given any vector a its components in the directions i, j and k are a_1, a_2 and a_3 respectively (Figure 4.10(a)).

We represent a by a line segment emanating from the origin in the appropriate direction with length $|a|$. We take a_1 to be the length of projection of this segment on the x-axis, etc., taking account of the direction.

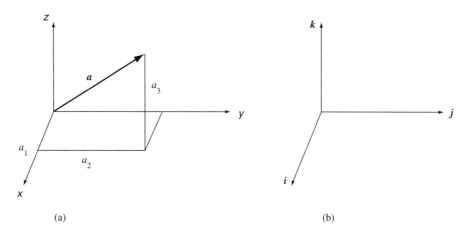

(a)   (b)

Figure 4.10 Components of a vector in three dimensions: (a) vector a has components a_1, a_2 and a_3; (b) the x, y, z axes can be replaced by the i, j, k directions

Now $a = a_1 i + a_2 j + a_3 k$ and by Pythagoras' theorem, applied twice, we have

$$a = |a| = \sqrt{a_1^2 + a_2^2 + a_3^2}$$

Generally, we write $\overrightarrow{OP} = r$ as the position vector of the point $P(x, y, z)$ with reference to the origin, so $r = xi + yj + zk$ and $r = |r| = \sqrt{x^2 + y^2 + z^2}$. The notation r is used to represent a variable vector.

Similar definitions exist for two- and three-dimensional cases regarding the definitions of direction cosines. As before, let the unit vector $\hat{a} = \dfrac{a}{a}$ and if the direction cosines in the x, y and z directions are l, m and n respectively, then

(a) $l = \dfrac{a_1}{a}, m = \dfrac{a_2}{a}, n = \dfrac{a_3}{a}$

(b) $l^2 + m^2 + n^2 = 1$

(c) If α, β and γ denote the angles between a and the coordinate axes then

$$\cos \alpha = \frac{a_1}{a} \qquad \cos \beta = \frac{a_2}{a} \qquad \cos \gamma = \frac{a_3}{a}$$

The ratios $a_1 : a_2 : a_3$ constitute the **direction ratios** of the respective components. Note that the vectors $(1, -2, 3)$ and $(-3, 6, -9)$ have the same direction ratios.

Example

Calculate the magnitude, direction cosines and direction ratios for the vector $a = 2i + 3j + 6k$.

The magnitude of a is $a = |a| = \sqrt{2^2 + 3^2 + 6^2} = 7$.

The direction cosines of a are $\dfrac{2}{7}, \dfrac{3}{7}$ and $\dfrac{6}{7}$. The direction ratios of a are $2 : 3 : 6$. ∎

Moments

An important application of vector arithmetic is the determination of moments in mechanics (Figure 4.11).

Masses $m_1, m_2, \ldots, m_6$ are shown and the position vector of the mass m_i relative to the origin O is r_i.

The total moment of the masses about O is

$$m_1 r_1 + m_2 r_2 + \cdots + m_6 r_6$$

In general, the total moment for n masses m_i with position vectors r_i is

$$\sum_{i=1}^{n} m_i r_i$$

The position vector $\bar{r}$ of the centre of mass relative to the origin is given by the ratio

$$\frac{\text{sum of the moments}}{\text{sum of the masses}}$$

i.e. $\bar{r} = \dfrac{\displaystyle\sum_{i=1}^{n} m_1 r_i}{\displaystyle\sum_{i=1}^{n} m_i} = \dfrac{\displaystyle\sum_{i=1}^{n} m_i r_i}{M}$

where M is the total mass.

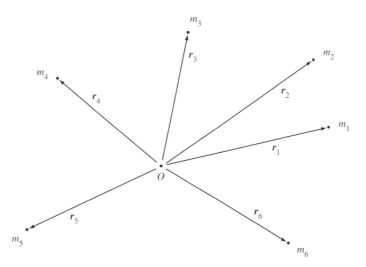

Figure 4.11 Moments in three dimensions

In other words, the particles of mass m_i at positions r_i can be replaced by a single particle of mass M at position $\bar{r}$ in order to generate the same moment about the origin.

Exercise 4.2

1 If the vectors a, b, c have components $(3, 2), (-2, 1), (6, 11)$ respectively, find the components of

(a) $2a + b - c$ (b) $-2b + 3c$ (c) $\dfrac{1}{3}(a + b + c)$

In case (c) verify by a suitable sketch that the point with position vector $\dfrac{1}{3}(a + b + c)$ lies inside the triangle whose vertices have position vectors a, b, c.

2 For the vectors (a) to (c) in Question 1 find the magnitudes, direction cosines and the unit vectors in the same direction.

3 Express the vector c in Question 1 in the form $\lambda a + \mu b$. Can we express any vector $d = d_1 i + d_2 j$ in the form $\lambda a + \mu b$? Give formulae for the components in terms of λ and μ. Note that the i and j components are distinct and must compare identically.

4   The vectors a, b, c have components $(2, 1, -1), (-3, 0, 1), (0, 1, -2)$ respectively. Find the components of

(a) $3a + 2b - 7c$ (b) $4a + 5b$

(c) $3b + 7c$ (d) $a + b - c$

5   For each of the vectors a, b, c in Question 4 find the magnitude, direction cosines and a unit vector in the same direction.

6   Draw on a piece of paper two vectors a and b emerging from a common point in different directions. Then draw a third vector c of your choice. Construct a parallogram with sides parallel to a and b, having c as one of its diagonals. This shows that we can write $c = \lambda a + \mu b$ where λ and μ are scalars.

7   (a)   Show that the vector $\dfrac{1}{ab}(ba + ab)$ bisects the angle between the vectors a and b.

(b)   Find the direction cosines of the bisector of the angle between the vectors $3i - 6j + 2k$ and $i - 2j - 2k$

8   The points A, B and C have coordinates $(1, 2, 3), (-2, 1, 2)$ and $(-1, 0, 5)$ relative to a fixed origin.

(a)   Find the components of the position vectors $b = \overrightarrow{AB}$ and $c = \overrightarrow{AC}$.

(b)   Which of $(2, 1, 2), (0, -4, 8)$ and $(0, 0, 0)$ lie in the plane ABC?

9   Relative to the origin, $\overrightarrow{OP_1} = r_1 = x_1 i + y_1 j + z_1 k$ and $\overrightarrow{OP_2} = r_2 = x_2 i + y_2 j + z_2 k$. Given that $OP_1^2 = x_1^2 + y_1^2 + z_1^2 = r_1^2$ find expressions for

(a) $(OP_2)^2$ (b) $(P_1 P_2)^2$

Use the cosine formula to show that $\cos P_1 \hat{O} P_2 = \dfrac{x_1 x_2 + y_1 y_2 + z_1 z_2}{r_1 r_2}$

10   The velocity of a boat relative to the water in a river is represented by the vector $3i + 4j$ and that of the water relative to the river bed by the vector $i - 2j$. What is the velocity of the boat relative to the river bed if the magnitude of each unit vector is $1 \, \text{km hr}^{-1}$?

11   Find the combined moment of the following masses, referred to O, in metres east and north respectively: $2 \, \text{kg}$ at $(1, 0)$, $3 \, \text{kg}$ at $(2, 3)$, $4 \, \text{kg}$ at $(0, -1)$, $1 \, \text{kg}$ at $(-6, 0)$ and $3 \, \text{kg}$ at $(-2, 2)$.

12   Repeat Question 11 this time taking moments about the point $(-2, 1)$.

13 Find the centre of mass of the following: $3m$ at $(1, 1, 0)$, $2m$ at $(-1, 0, 1)$, m at $(-4, 2, 1)$, $3m$ at $(-1, -1, 6)$, $5m$ at $(0, 0, -1)$, $4m$ at $(1, -1, 1)$ and $2m$ at $(-2, -1, 3)$.

14 An aeroplane aims to fly from airport A to airport B, a distance of 700 km, in one hour. B is $10°$W of N from A. The meteorological information predicts mean wind speeds as follows: $100 \, \text{km h}^{-1}$ due W for the first 20 minutes, zero wind for the next 20 minutes and $90 \, \text{km h}^{-1}$ in a direction $80°$E of N for the last 20 minutes. Choose flying velocities for each stage of the journey so that the aeroplane is travelling on the direct line from A to B in the middle 20 minutes.

4.3 THE SCALAR PRODUCT OF TWO VECTORS

There is no intuitive feel for what the product of two vectors might be. In this section and in Section 4.5 we define two kinds of product, each of which has many applications. If a and b are the magnitudes of the vectors a and b, and θ is the acute angle between their directions then we define the **scalar product** $a.b$ of the vectors as

$$a.b = ab \cos \theta = |a||b| \cos \theta$$

In geometrical terms $a.b$ is the product of the length of one of the vectors and the projection of the other vector onto it (Figure 4.12).

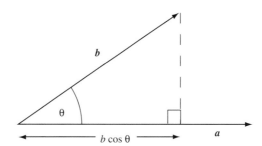

Figure 4.12   Concept of a scalar product

Properties of the scalar product

(a) $a.b = b.a$ commutative law

(b) $a.(b + c) = a.b + a.c$ distributive law

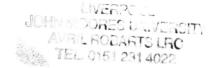

(c) $a.(kb) = ka.b$

(d) If a and b are perpendicular then $a.b = 0$ ($\theta = 90°$ so that $\cos\theta = 0$)

(e) If $a = 0$ or $b = 0$ then $a.b = 0$

Note the results $i.j = j.k = k.i = 0,\quad j.i = k.j = i.k = 0,\quad i.i = j.j = k.k = 1.$

Cartesian form

Let $a = a_1 i + a_2 j + a_3 k$, then

$$a.b = (a_1 i + a_2 j + a_3 k).(b_1 i + b_2 j + b_3 k)$$
$$= a_1 i.b_1 i + a_1 i.b_2 j + a_1 i.b_3 k + a_2 j.b_1 i + a_2 j.b_2 j + a_2 j.b_3 k$$
$$+ a_3 k.b_1 i + a_3 k.b_2 j + a_3 k.b_3 k$$

Now $a_1 i.b_2 j = a_1 b_2 i.j = 0$ and $a_1 i.b_1 i = a_1 b_1 i.i = a_1 b_1$, etc., so

$$a.b = a_1 b_1 + a_2 b_2 + a_3 b_3$$

Angle between two vectors in three dimensions

Since $a.b = ab\cos\theta = a_1 b_1 + a_2 b_2 + a_3 b_3$ then

$$\cos\theta = \frac{a_1 b_1 + a_2 b_2 + a_3 b_3}{(a_1^2 + a_2^2 + a_3^2)^{1/2}(b_1^2 + b_2^2 + b_3^2)^{1/2}}$$

Note that the result was proved in Question 9 of Exercise 4.2 using the cosine rule.

Application: Work done by a force

The work done by a force F in moving an object along the displacement r is $W = F.r$ (Figure 4.13). If F and r are parallel then the work done is a maximum, whereas if F and r are perpendicular then the work done is zero.

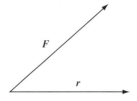

Figure 4.13 Work done by a force

Examples

1. Find the angle between the two vectors with components $(2, 2, -1)$ and $(1, 4, 1)$. Let $a = 2i + 2j - k$, $b = i + 4j + k$

   First $a = \sqrt{2^2 + 2^2 + (-1)^2} = \sqrt{9} = 3$ and $b = \sqrt{1^2 + 4^2 + 1^2} = \sqrt{18} = 3\sqrt{2}$.
   Then $a.b = 2 \times 1 + 2 \times 4 + (-1) \times 1 = 2 + 8 - 1 = 9$.

   Finally $\cos\theta = \dfrac{a.b}{|a||b|} = \dfrac{9}{3 \times 3\sqrt{2}}$, so $\theta = \cos^{-1}\left(\dfrac{1}{\sqrt{2}}\right) = \dfrac{\pi}{4}$

2. Find the work done in moving an object along the displacement $r = 3i + 2j - 5k$ by an applied force $F = 2i - j - k$. Repeat for a force $F = 3i + 2j + k$ and a displacement $r = i - j - k$.

   The work done is $F.r = (2i - j - k).(3i + 2j - 5k) = 6 - 2 + 5 = 9$.

   If the force is $F = 3i + 2j + k$ and the displacement is $r = i - j - k$ then the work done is $F.r = 3 - 2 - 1 = 0$. The zero result indicates that the direction of the force is at right angles to the displacement. ∎

Geometrical application: Pythagoras' theorem

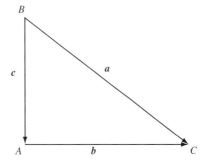

Figure 4.14 Pythagoras' theorem

The triangle ABC has a right angle at A (Figure 4.14). Note that $b.c = c.b = 0$. Now $\overrightarrow{BC} = \overrightarrow{BA} + \overrightarrow{AC}$, i.e. $a = c + b = b + c$. Taking the scalar product of this equation with a gives

$$a.a = a.(b + c) = (b + c).(b + c) = b.b + c.c$$

i.e. $a^2 = b^2 + c^2$, which is Pythagoras' theorem.

Several theorems in plane geometry can be proved quite easily using vector methods. Here is an example.

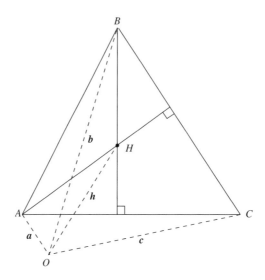

Figure 4.15 Deriving a vector equation relating the position vectors of the vertices to the position vector of H, the intersection of all three altitudes

Example

The vertices of the triangle ABC have position vectors $\boldsymbol{a}$, $\boldsymbol{b}$ and $\boldsymbol{c}$. H is the point of intersection of the altitudes of the triangle and $\overrightarrow{OH} = \boldsymbol{h}$ (Figure 4.15). $\overrightarrow{AH}$ is perpendicular to $\overrightarrow{BC}$ and $\overrightarrow{BH}$ is perpendicular to $\overrightarrow{CA}$, so

$$(\boldsymbol{h} - \boldsymbol{a}).(\boldsymbol{b} - \boldsymbol{c}) = 0$$
$$(\boldsymbol{h} - \boldsymbol{b}).(\boldsymbol{c} - \boldsymbol{a}) = 0$$

Multiplying out the equations and adding the results, we obtain the equation

$$(\boldsymbol{h} - \boldsymbol{c}).(\boldsymbol{a} - \boldsymbol{b}) = 0$$

The derivation is left to you as an exercise.

This last equation shows that $\overrightarrow{CH}$ is perpendicular to $\overrightarrow{AB}$. We therefore conclude that the altitudes of a triangle are concurrent, i.e. they meet at one point, namely H. ■

This example shows that far-reaching results can be obtained using relatively straightforward algebraic vector operations coupled with symmetry. In this case it was the permutation $\boldsymbol{a} \rightarrow \boldsymbol{b} \rightarrow \boldsymbol{c} \rightarrow \boldsymbol{a}$ which came to our aid.

Exercise 4.3

 1 Determine the angles between the following pairs of vectors:

(a) $2i - j$ and $3i + 2j$ (b) $6i$ and $2i - j$

(c) $-\pi i + \sqrt{2}j$ and $\sqrt{2}i + \pi j$

2 Find the magnitudes of the vectors $4i + 2j - k$ and $2i - 8j + 4k$ and the angle between them.

3 Find the angle between the vectors $3i - 4j$ and $2i + j - 2k$

4 For the triangle ABC with vertices $(3, -1, 4)$, $(2, -2, 1)$ and $(5, 1, 3)$ find the lengths of its sides, the cosines of its angles and its area.

5 Let A, B, C, D be the points $(2, 3, 4)$, $(1, 2, 3)$, $(1, 0, 2)$, $(2, 3, -2)$ respectively. Show that AB is perpendicular to CD.

6 There are two unit vectors perpendicular to the vector at $ai + bj$. One of them is k, write down the other one.

7 The vector c is perpendicular to both a and b.

(a)   Write down the scalar product equations which express these statements.

(b)   Find the unit vectors perpendicular to both $2i - j + k$ and $3i + 4j - k$

8 Use the laws of scalar products to simplify the following expressions:

(a) $(a + b).(a - b)$ (b) $(a + b).(a - 3b)$

(c) $(a + b).c - (a - b).c$

9* Prove the following results for the vectors a, b and c:

(a)   If $(b.c)a = (a.b)b$ then a is parallel to b.

(b)   If $b + c$ is perpendicular to $b - c$ then $|b| = |c|$.

(c)   If $|a + c| = |a - c|$ then a is perpendicular to c.

In case (c) use the property $(a + c).(a + c) = |a + c|^2$

10   Earlier on for triangle ABC in which the vectors a, b, c and h were defined, we found that $(h - a).(b - c) = 0$ and $(h - b).(c - a) = 0$. By expanding out the left-hand sides and adding the resulting equations, prove that $(h - c).(a - b) = 0$.

11   For the triangle ABC of Question 10 write down the displacement vectors of the midpoints of AB and AC. K is the point where the perpendicular bisectors of AB and AC meet. It follows that both $\left(k - \frac{1}{2}(a + b)\right).(a - b) = 0$ and $\left(k - \frac{1}{2}(a + c)\right).(c - a) = 0$, where $k = \overrightarrow{OK}$. Draw a sketch to illustrate the situation. Add these two equations and interpret the result.

12   An unknown vector x satisfies the equation $px + (x.b)a = c$ where p is a non-zero scalar and a, b and c are given vectors. By writing $x = \frac{1}{p}c - \lambda a$ determine the value of λ. Assume that $p + a.b \neq 0$.

13   The forces $F_1 = 2i - j + k, F_2 = -i + 3k, F_3 = 2i - 3j - 4k, F_4 = -i - 11j - k$ all act at the same point on an object. What is the work done by the resultant force (i.e. the sum of the given forces) in moving the object along the displacement $-9i - j + 6k$? Determine a direction along which the work done is zero.

4.4 STRAIGHT LINES IN TWO AND THREE DIMENSIONS

Straight lines in two dimensions

In two dimensions we derive the equation of a straight line as follows. Figure 4.16 shows that the line lies in the direction of the vector b. The position vector of a fixed point A on the line is a, referred to the origin. The position vector of a general point P on the line is r. Three such examples are shown in the figure, with dashed lines.

In triangle OAP, $\overrightarrow{OP} = \overrightarrow{OA} + \overrightarrow{AP}$. Now $\overrightarrow{AP}$ is a vector in the direction of b and can be represented by tb where t is a scalar. Hence the vector equation of the line is

$$r = a + tb$$

Example   Find the vector equation of the straight line which passes through the point $(-1, 1)$ in the direction, $(2, 1)$, i.e. through the point whose position vector is $-i + j$ in the direction $2i + j$. Deduce the Cartesian equation of the line.

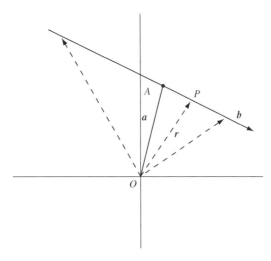

Figure 4.16   Straight line in two dimensions

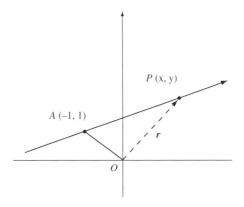

Figure 4.17   Vector equation of a line through a given point in a given direction

Referring to Figure 4.17, the vector equation of the line is

$$r = -i + j + t(2i + j) = (-1 + 2t)i + (1 + t)j$$

But $r = xi + yj$ is the position vector of a general point, and comparing the i and j components separately, we obtain

$$x = -1 + 2t, \qquad y = 1 + t$$

Eliminating t gives the equation $x + 1 = 2(y - 1)$, which can be rearranged to give

$$y = \frac{1}{2}(x + 3)$$ ∎

Notice that in formulating the vector equation of the line it is the ratio of b_1 and b_2 which is important, not their actual values.

Line through two given points

We know that a unique straight line passes through two given points. If these points have position vectors r_1 and r_2 then the straight line is in the direction of $r_2 - r_1$ and therefore its equation is

$$r = r_1 + t(r_2 - r_1)$$

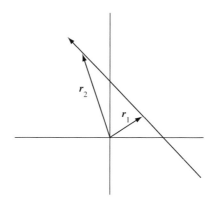

Figure 4.18 Straight line passing through two given points

Note that the precise form of the vector equation of the line is not unique because any two fixed points on the line can be chosen as r_1 and r_2, as in Figure 4.18, and it is only the direction ratios of its line of action which are important.

Note that any two lines in two dimensions that are not parallel must intersect.

Example

Find the point of intersection of the lines $r = i + j + t(-2i + j)$ and

$$r = -2i - j + s(i - 3j).$$

The equations can be rewritten as $r = (1 - 2t)i + (1 + t)j$ and $r = (-2 + s)i - (1 + 3s)j$. The coefficients of i and j must match, i.e. $1 - 2t = -2 + s$ and $1 + t = -1 - 3s$. Solving these equations gives $t = \frac{11}{5}, s = -\frac{7}{5}$, so the point of intersection is $\left(-\frac{17}{5}, \frac{16}{5}\right)$, which has position vector $-\frac{17}{5}i + \frac{16}{5}j$ ∎

Straight lines in three dimensions

In three dimensions exactly similar arguments can be followed to derive the vector equation of a straight line. The line can be defined uniquely using (a) a fixed point on the line and a given direction, (b) two distinct fixed points on the line.

In case (a) we have the situation shown in Figure 4.19(a). The line is in the direction d and the fixed point has position vector r_1. Now $\overrightarrow{OP} = \overrightarrow{OA} + \overrightarrow{AP}$, i.e. $r = r_1 + td$. Again the dashed arrows denote choices for $\overrightarrow{OP}$.

In case (b) we have the situation shown in Figure 4.19(b), where the fixed points P_1 and P_2 have position vectors r_1 and r_2. We have $\overrightarrow{OP} = \overrightarrow{OP_1} + \overrightarrow{P_1 P_2}$, i.e. $r = r_1 + (r_2 - r_1)$.

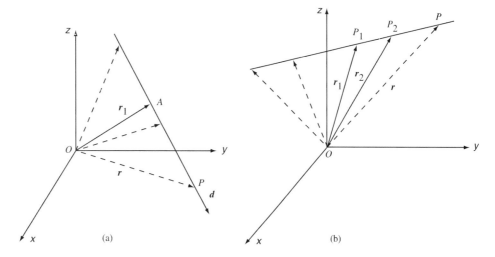

Figure 4.19 Lines in three dimensons: (a) through a given point in a given direction and (b) through two given points

The difference from the two-dimensional case becomes evident when we produce the Cartesian equations. We use the form $r = r_1 + td$ where $r = xi + yj + zk$ corresponds to the point (x, y, z), $r_1 = x_1 i + y_1 j + z_1 k$ or (x_1, y_1, z_1) and $d = (d_1, d_2, d_3)$.

Comparing the coefficients of i, j and k in turn gives

$$x - x_1 = td_1 \qquad y - y_1 = td_2 \qquad z - z_1 = td_3$$

Eliminating t we obtain we arrive at the equation

$$\frac{x - x_1}{d_1} = \frac{y - y_1}{d_2} = \frac{z - z_1}{d_3}$$

where each fraction is equal to t.

For the other form, using $r_2 = x_2 i + y_2 j + z_2 k$ or (x_2, y_2, z_2), we obtain the equations

$$\frac{x - x_1}{x_2 - x_1} = \frac{y - y_1}{y_2 - y_1} = \frac{z - z_1}{z_2 - z_1}$$

Note that two equalities are necessary because a point in three dimensions needs three coordinates to locate it. If it lies on a line then it has lost two of those degrees of freedom, needing only one piece of information to determine its position, namely the value of t.

Examples

1. Find the equation of the straight line passing through the point $(1, -2, 1)$ in the direction of $(3, 2, 1)$.
   The vector equation is

   $$r = i - 2j + k + t(3i + 2j + k)$$

   and the Cartesian equations, obtained by comparing coefficients, are

   $$\frac{x - 1}{3} = \frac{y + 2}{2} = \frac{z - 1}{1}$$

2. Find the equation of the straight line joining the points $(1, -2, 1)$ and $(-1, 0, 2)$. What is the angle between the direction of this line and the direction of the line in the first example?
   The equation of the line is

   $$\frac{x - 1}{-1 - 1} = \frac{y + 2}{0 - 2} = \frac{z - 1}{2 - (-1)}$$

   i.e. $\quad \dfrac{x - 1}{-2} = \dfrac{y + 2}{-2} = \dfrac{z - 1}{3}$

   The direction of this line is $-2i - 2j + 3k$, found from the denominators, so the angle between the two lines is

   $$\cos^{-1}\left(\frac{(3i + 2j + k).(-2i - 2j + 3k)}{(3^2 + 2^2 + 1^2)^{1/2}((-2)^2 + (-2)^2 + 3^2)^{1/2}}\right) = \cos^{-1}\left(\frac{-7}{\sqrt{14}\sqrt{17}}\right)$$
   $$= 116.98°$$

   or, if we want the acute angle, $63.02°$. ∎

Skew lines and transversals

Lines in two dimensions which are not parallel always meet, but in three dimensions it is relatively rare for two lines to meet. Lines which are not parallel but which do not meet are **skew**. At their closest point of approach a unique **transversal** can be drawn connecting the closest points. This is a straight line perpendicular to the two skew lines.

Let $b = b_1 i + b_2 j + b_3 k$ and $d = d_1 i + d_2 j + d_3 k$. A unit vector $\hat{n} = n_1 i + n_2 j + n_3 k$ perpendicular to both b and d must satisfy the scalar product equations $\hat{n}.b = 0$ and $\hat{n}.d = 0$, i.e. $n_1 b_1 + n_2 b_2 + n_3 b_3 = 0$ and $n_1 d_1 + n_2 d_2 + n_3 d_3 = 0$.

Example

Find a vector perpendicular to both $i - j + k$ and $2i + j - k$.

Let the vector be $a = a_1 i + a_2 j + a_3 k$. Then $a_1 - a_2 + a_3 = 0$ and $2a_1 + a_2 - a_3 = 0$. Adding these equations gives $a_1 = 0$ and either of the original two equations then yields $a_3 = a_2$. If we take $a_2 = 1$ then a perpendicular vector is $j + k$.

Consider two skew lines $r = a + tb$, $|b| = 1$ and $r = c + sd$, $|d| = 1$ where $\overrightarrow{OP_0} = a$ and $\overrightarrow{OQ_0} = c$ with P_1 and Q_1 being the endpoints of the transversal (Figure 4.20).

First, $\overrightarrow{OP_1} = \overrightarrow{OP_0} + \overrightarrow{P_0 P_1} = a + tb$ and $\overrightarrow{OQ_1} = \overrightarrow{OQ_0} + \overrightarrow{Q_0 Q_1} = c + sd$. Let the transversal be $\overrightarrow{P_1 Q_1} = p\hat{n}$ so that its length is p.

But $\overrightarrow{OQ_1} - \overrightarrow{OP_1} = \overrightarrow{P_1 Q_1}$, i.e. $c + sd - (a + tb) = p\hat{n}$, where $\hat{n}$ represents a unit vector in the direction of the transversal. Forming the scalar product of this last equation with $\hat{n}$ and noting that $\hat{n}.b = \hat{n}.d = 0$, we obtain $c.\hat{n} - a.\hat{n} = p$, i.e. $p = (c - a).\hat{n}$. It is usual to take $p = |(c - a).\hat{n}|$ since it is a magnitude.

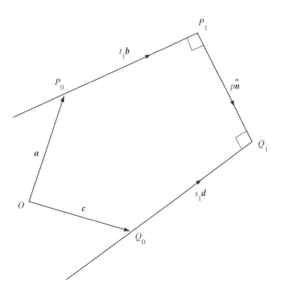

Figure 4.20 Transversal to two skew lines

Example

What is the shortest distance apart of the skew lines $r = i + j + t(i - j + k)$ and $r = j + 2k + s(2i + j - k)$?

Note first that the transversal is perpendicular to both $i - j + k$ and $2i + j - k$, and from the previous example, it is in the direction $j + k$.

A unit vector in that direction is $\dfrac{1}{\sqrt{2}}(j + k)$, or $\dfrac{1}{\sqrt{2}}(0, 1, 1)$ using the component form.

Then the shortest distance between the lines is

$$p = |(c - a).\hat{n}| = \left|(-i + 2k).\frac{1}{\sqrt{2}}(j + k)\right| = \sqrt{2}$$

The next step would be to find the endpoints of the transversal and this could be done by forming the scalar product of the equation with b and d respectively. This would provide two simultaneous equations in s and t. Although this can be done, the algebra may be awkward.

Example

Find which of the following pairs of straight lines intersect, which are parallel and which are skew. Find in each case the point of intersection or the shortest distance apart, if possible.

(a) $r = i + 2j + 3k + \lambda(3i - 2j + k)$ $r = i - j + 2k + \mu(-6i + 4j - 2k)$

(b) $r = -i + 3j - 2k + \lambda(-3i + 2j + k)$ $r = 7j - 7k + \mu(i - 3j + 2k)$

(c) $r = i + j + k + \lambda(i - j + k)$ $r = i - j - k + \mu(2i - j - k)$

Solution

(a) Note that $-6i + 4j - 2k = -2(3i - 2j + k)$, so the lines are parallel. Because the lines are in the same direction, it is not possible to find the common perpendicular using the method developed earlier.

(b) If the lines meet, then by writing the equations of the lines as $r = -(1 + 3\lambda)i + (3 + 2\lambda)j - (2 - \lambda)k$ and $r = \mu i + (7 - 3\mu)j - (7 - 2\mu)k$ and noting that the coefficients of i, j and k must agree separately, we obtain the scalar equations

$$x = -1 - 3\lambda = \mu$$
$$y = 3 + 2\lambda = 7 - 3\mu$$
$$z = -2 + \lambda = -7 + 2\mu$$

Solving simultaneously the first two of these equations gives $\lambda = -1, \mu = 2$. These values also satisfy the third equation, so the lines meet at the point $(2, 1, -3)$.

(c) Carrying out the same steps as in (b) gives the three equations

$$x = 1 + \lambda = 1 + 2\mu$$
$$y = 1 - \lambda = -1 - \mu$$
$$z = 1 + \lambda = -1 - \mu$$

By solving the first two equations and substituting the solutions into the third, we find that it is not satisfied. The equations are **not consistent**. The lines are therefore skew. To find their shortest distance apart, assume that a normal vector to both lines is $a\mathbf{i} + b\mathbf{j} + c\mathbf{k}$.

Take its scalar product with the direction vector of each line in turn to obtain the equations $a - b + c = 0$ and $2a - b - c = 0$. Adding these equations gives $3a = 2b$, and the second equation gives $c = 2a - b = \dfrac{1}{2}a$

Hence $a : b : c = 2 : 3 : 1$ which suggests a vector $2\mathbf{i} + 3\mathbf{j} + \mathbf{k}$ as being in the direction of the perpendicular. A unit vector in this direction is $\dfrac{1}{\sqrt{14}}(2\mathbf{i} + 3\mathbf{j} + \mathbf{k})$. Then the shortest distance apart is

$$p = \left| (-2\mathbf{j} - 2\mathbf{k}) . \frac{1}{\sqrt{14}}(2\mathbf{i} + 3\mathbf{j} + \mathbf{k}) \right| = \frac{8}{\sqrt{14}} \qquad \blacksquare$$

Exercise 4.4

1 Prove that $\mathbf{r} = \mathbf{j} + t(\mathbf{i} + 3\mathbf{j})$ and $\mathbf{r} = -2\mathbf{i} - 5\mathbf{j} + s(-2\mathbf{i} - 6\mathbf{j})$ are vector equations of the same straight line.

2 Find the vector equation of the straight line which passes through the points $(4, 1)$ and $(-2, 3)$.

3 Find the point of intersection of the straight lines $\mathbf{r} = 4\mathbf{i} + 3\mathbf{j} + t(\mathbf{i} + \mathbf{j})$ and $\mathbf{r} = 2\mathbf{i} - \mathbf{j} + s(\mathbf{i} - \mathbf{j})$. Obtain the Cartesian equations of the lines.

4 Find the vector form of the equations of the straight lines in two dimensions given by the equations

(a) $3x + 2y = 4$ (b) $x - 4y + 3 = 0$

 5 Take the scalar product of the direction vectors of the straight lines in Question 4 to find the angle between the lines.

6 Verify that the equation $r = -ci + dj + t(di + cj)$ represents a straight line in the form $r = a + tb$ where a and b are perpendicular. What is the shortest distance of this line from the origin?

7 Write down the Cartesian forms of the equations

(a) $r = -2i + 3j - k + \lambda(i - j + k)$

(b) $r = i - 2j - 5k + \lambda(-2i + j - 2k)$

(c) $r = -3i + j - 2k + \lambda(i + j - 3k)$

and write down the vector form of the equations

(d) $\dfrac{x-3}{-2} = \dfrac{y+1}{6} = \dfrac{z-5}{7}$ (e) $\dfrac{x+5}{4} = \dfrac{y-6}{3} = \dfrac{3z-11}{5}$

(f) $\dfrac{x-\sqrt{2}}{e} = \dfrac{y+\pi}{\sqrt{3}} = \dfrac{z+16}{\sqrt{2}}$

Which of these lines passes through the point (3, 12, 7)?

8 Write down the vector equations of the following straight lines:

(a)   the line passing through the points $(-1, 0, 2)$ and $(6, 1, 5)$

(b)   the line passing through the point $(-1, 1, 1)$ in the direction $(2, 1, 3)$

What is the angle between these lines?

9 The horizontal plane is $z = 0$, i.e. it consists of all points whose z-coordinate is zero. At which points do the lines in Question 8 pass through this plane?

10 A piece of space debris is observed by an orbital space station. At the first sighting it is 100 km from the station and is approaching obliquely. The direction of first sight is fixed as the x-axis. One minute later the object is observed at position $(30, 40, 50)$, where the component distances are given in kilometres. Assuming that the object is moving in a straight line at constant speed, find that speed and the direction in which the object is moving.

11 A straight line passes through the point $P_0(2, 1, 3)$ in the direction $(1, 1, 1)$. Write down its vector equation. The point $P_2(5, 4, 2)$ is not on the line. By writing down $\overrightarrow{PP_2}$, where P is a general point on the line and forming the scalar product of $\overrightarrow{PP_2}$ with the vector $(1, 1, 1)$, find which point P_3 on the line is closest to P_2, the distance of P_2 from the line and the vector $\overrightarrow{P_2P_3}$.

12 Which of the following pairs of straight lines intersect, which are parallel and which are skew? Find the point of intersection or (in the case of skew lines) the shortest distance apart and the direction of the transversal.

(a) $r = k + \lambda(2i + j + k)$ $r = 2i + j - k + \mu(-i - j + k)$

(b) $r = 3i - 4j + 2k + \lambda(-2i + 4j - 8k)$ $r = 9i - 16j + 5k + \mu(i - 2j + 4k)$

(c) $r = 4i + j + 2k + \lambda(3i + k)$ $r = -i - k + \mu(2i + j + 2k)$

13 A straight line is drawn outward from the origin to the point $(2, 4, 2)$. At which point does it meet the straight line joining the points $(2, 1, 1)$ and $(0, 3, 1)$?

4.5 THE VECTOR PRODUCT AND THE PLANE

Two vectors can be multiplied together to form a scalar, as we have seen in Section 4.3; they can also be multiplied together to form a vector.

The **vector product**, or cross product, of two vectors a and b is written $a \times b$ and defined to be the vector with magnitude $ab \sin \theta$, where θ is the angle between a and b, and with a direction as indicated in Figure 4.21; the vector product is denoted c.

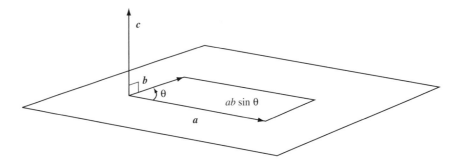

Figure 4.21   Defining the vector product

The vector $\hat{n}$ is a unit vector perpendicular to both a and b, and we write

$$c = a \times b = ab \sin \theta \, \hat{n}$$

In effect the vector a is rotated into the direction of the vector b through the angle θ and the **right-hand rule** gives the direction of $a \times b$, which is perpendicular to the plane containing the vectors a and b.

Note that $ab \sin \theta$ is the area of the parallelogram with adjacent sides a and b. Hence $a \times b$ is a vector perpendicular to the plane containing a and b and with a magnitude equal to the area of the parallelogram formed by a and b. This property is sometimes called **vector area**.

Note also that θ is measured in the sense $a \rightarrow b$. Were we to form the vector product the other way we would get $b \times a$, which has magnitude $ba \sin \theta$ but in the direction opposite to $a \times b$. Hence, referring to Figure 4.22, we find

$$b \times a = -a \times b$$

We say that the vector product is **anticommutative**. This completely contradicts our previous experience of product, which has been commutative.

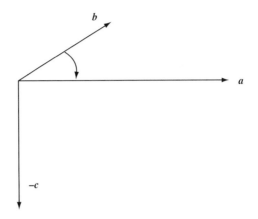

Figure 4.22 The vector product $\boldsymbol{b} \times \boldsymbol{a}$

Properties of the vector product

(a) $\boldsymbol{b} \times \boldsymbol{a} = -\boldsymbol{a} \times \boldsymbol{b}$ Anticommutative

(b) $\boldsymbol{a} \times (\boldsymbol{b} + \boldsymbol{c}) = \boldsymbol{a} \times \boldsymbol{b} + \boldsymbol{a} \times \boldsymbol{c}$ Distributive

(c) $k\boldsymbol{a} \times l\boldsymbol{b} = (kl)\boldsymbol{a} \times \boldsymbol{b}$ k and l scalars

(d) If $\boldsymbol{a} = \boldsymbol{0}$ then $\boldsymbol{a} \times \boldsymbol{b} = \boldsymbol{0}$

(e) If $\boldsymbol{a}$ is parallel to $\boldsymbol{b}$ then $\boldsymbol{a} \times \boldsymbol{b} = \boldsymbol{0}$

(f) $\boldsymbol{a} \times \boldsymbol{a} = \boldsymbol{0}$

Note the important results

$$i \times i = j \times j = k \times k = 0$$
$$i \times j = k, \quad j \times i = -k, \quad j \times k = i, \quad k \times j = -i, \quad k \times i = j, \quad i \times k = -j$$

Vector product in Cartesian form

If $\boldsymbol{a} = a_1 \boldsymbol{i} + a_2 \boldsymbol{j} + a_3 \boldsymbol{k}$ and $\boldsymbol{b} = b_1 \boldsymbol{i} + b_2 \boldsymbol{j} + b_3 \boldsymbol{k}$ then, using the results above, we obtain

$$\boldsymbol{a} \times \boldsymbol{b} = (a_2 b_3 - a_3 b_2)\boldsymbol{i} + (a_3 b_1 - a_1 b_3)\boldsymbol{j} + (a_1 b_2 - a_2 b_1)\boldsymbol{k}$$

Examples

1. Find the product of $\boldsymbol{a} = 2\boldsymbol{i} - 3\boldsymbol{j} + 4\boldsymbol{k}$ and $\boldsymbol{b} = \boldsymbol{i} - \boldsymbol{j} + 2\boldsymbol{k}$, and check that it is perpendicular to both $\boldsymbol{a}$ and $\boldsymbol{b}$.

First $a \times b = (-6 + 4)i + (4 - 4)j + (-2 + 3)k = -2i + k$. Also

$$(2i - 3j + 4k).(-2i + k) = -4 + 4 = 0$$

and $(i - j + 2k).(-2i + k) = -2 + 2 = 0$. This proves that the vector product is perpendicular to both a and b.

2. Find the shortest distance apart of the skew lines $r = i + j + t(i - j + k)$ and $r = i + 2k + s(2i + j - k)$

The normal to both lines is in the direction

$$(i - j + k) \times (2i + j - k) = (1 - 1)i + (2 + 1)j + (2 + 1)k = 3j + 3k$$

A unit normal is therefore

$$\frac{1}{\sqrt{2}}(j + k)$$

so the required distance is

$$p = |((i + j) - (j + 2k)).\hat{n}| = \sqrt{2}$$ ∎

The vector product has been defined in this apparently strange way because of its application to three-dimensional mechanics; here are two examples.

Torque as a vector product

The moment about of a force F acting at the point shown in Figure 4.23 is considered as a vector G whose direction $\hat{n}$ is perpendicular to the plane containing F and r. The sense of

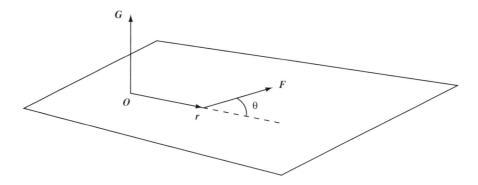

Figure 4.23 Torque acts along the axis of rotation

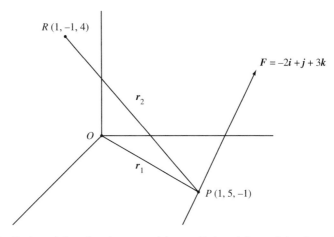

Figure 4.24 Determining the torque of force **F** about the origin O and the point R

the vector is determined by the right-hand rule and the magnitude of G is $G = Fr \sin \theta$. Hence we define

$$G = r \times F$$

G is usually called **torque** and its direction is the **axis of rotation**.

Example   Determine the torque about (a) the origin (b) the point $R(1, -1, 4)$ of a force $F = -2i + j + 3k$ acting at the point $P(1, 5, -1)$. Refer to Figure 4.24.

(a) $r \times F = (i + 5j - k) \times (-2i + j + 3k) = 16i - j + 11k$

(b) $r = (i + 5j - k) - (i - j + 4k) = 6j - 5k$ so that
$r \times F = (6j - 5k) \times (-2i + j + 3k) = 23i + 10j + 12k$ ∎

Force on a conductor

The vector point is also at the heart of the theory of electromagnetic induction. The force at a point on a conductor carrying electric current in a uniform magnetic field can be expressed as a vector product (Figure 4.25).

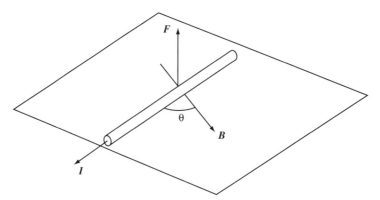

Figure 4.25 Force on a conductor

A straight-line conductor carries a constant current I in a uniform magnetic field of flux density B. A point on the conductor will experience a force proportional to I and the component of B perpendicular to the conductor. Therefore $F = IB \sin \theta$.

Then F is the vector product of I and B, so $F = I \times B = IB \sin \theta \hat{n}$ where $\hat{n}$ is the unit normal in the direction perpendicular to the plane containing I and B.

Vector description of the plane

Let Π be a plane containing the point $P_1(x_1, y_1, z_1)$ with position vector r_1 and let the unit normal to the plane be $\hat{n}$ (Figure 4.26).

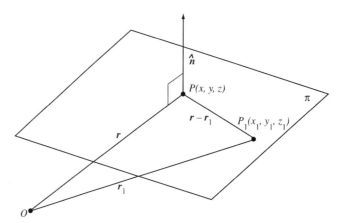

Figure 4.26 Deriving the equation of a plane

The point $P(x, y, z)$ with position vector r is a general point in the plane, and the line segment $\overrightarrow{PP_1} = r - r_1$ lies in the plane. This is perpendicular to $\hat{n}$. Hence $(r - r_1).\hat{n} = 0$, i.e. $r.\hat{n} = r_1.\hat{n}$

If we write $\hat{n} = li + mj + nk$, $r = xi + yj + zk$ and $r_1 = x_1i + y_1j + z_1k$ then $lx + my + nz = p$ where $p = lx_1 + my_1 + nz_1$. Therefore the Cartesian equation of a plane has the form

$$ax + by + cz = d$$

where the constants (a, b, c) are the direction ratios of the normal to the plane.

The two-dimensional equivalent of this equation is $ax + by = d$, which is the equation of a straight line (in two dimensions).

Examples

1. Find the equation of the plane which contains the straight lines

$$r = i + 3j - 2k + \lambda(-3i + 2j + k) \text{ and } r = 7i - 7j + \mu(i - 3j + 2k)$$

It was shown in Section 4.4 that these two lines meet at the point $(2, 1, 3)$. The normal to the plane which contains the lines is perpendicular to both lines and therefore parallel to their vector product $(-3i + 2j + k) \times (i - 3j + 2k) = 7i + 7j + 7k$

The direction ratios of this vector are $1 : 1 : 1$, so the equation of the plane is of the form $x + y + z = k$. Since $(2, 1, 3)$ lies in the plane then $2 + 1 + 3 = k$, so the equation of the plane is $x + y + z = 6$.

2. (a) Show that the plane $x + y + 2z = 4$ contains the line $x - 1 = y - 1 = 1 - z$

   (b) Show that the plane is parallel to the line $\dfrac{x - 2}{1} = \dfrac{y - 3}{-3} = \dfrac{z - 4}{1}$

   (c) Show that the line $x = -y = z$ is neither parallel to the plane nor lies in the plane. Where does it meet the plane?

Solution

(a) The given line has direction ratios $(1, 1, -1)$, and the scalar product $(1, 1, -1).(1, 1, 2) = 0$, so the line is in a direction parallel to the plane. Also, the line passes through the point $(1, 1, 1)$; this point also lies in the plane since $1 + 1 + 2 = 4$. Hence the line is contained in the plane.

(b) The scalar product $(1, -3, 1).(1, 1, 2) = 0$, so the line is in a direction parallel to the plane. But the point $(2, 3, 4)$ through which the line passes does not lie in the plane since the coordinates of the point do not satisfy the equation of the plane.

(c) The scalar product $(1, -1, 1).(1, 1, 2) = 2 \neq 0$. The line therefore intersects the plane. Substituting $x = -y = z$ in the equation of the plane gives $x - x + 2x = 4$, so $x = 2$. The point of intersection is $(2, -2, 2)$.

3. Find the equation of the common line of intersection of the planes $2x + 2y - z = 5$ and $6x + y + z = 22$

   The line must be perpendicular to the normals of both planes, i.e. it is in the direction of $(2, 2, -1) \times (6, 1, 1) = (3, -8, -10)$.

   With the direction established, we need to determine a point on the line. Since the line is not parallel to any of the coordinate axes, choose $x = 0$ without loss of generality and put this value into the equations of the planes; then solve simultaneously the reduced equations.

   Hence we solve $2y - z = 5, y + z = 22$ to obtain $y = 9, z = 13$. This shows that the point $(0, 9, 13)$ lies on the line whose equation is therefore $\dfrac{x}{3} = \dfrac{y - 9}{-8} = \dfrac{z - 13}{-10}$

4. Find the shortest distance from the plane $ax + by + cz + d = 0$ to the origin.

   Referring to Figure 4.27, $\overrightarrow{OP}$ is perpendicular the plane and therefore is parallel to $\hat{n}$, so $x : y : z = a : b : c$.

   The distance OP is $(x^2 + y^2 + z^2)^{1/2}$. Now for some constant $k, x = ka, y = kb, z = kc$, so that

$$(x^2 + y^2 + z^2)^{1/2} = k(a^2 + b^2 + c^2)^{1/2}$$

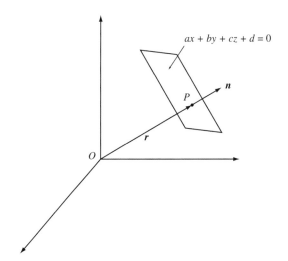

Figure 4.27   Finding the shortest distance from the origin to plane $ax + by + cz + d = 0$

Also, from the equation of the plane, $k(a^2 + b^2 + c^2) + d = 0$. Substituting for k we find that the required distance is

$$\frac{|d|}{(a^2 + b^2 + c^2)^{1/2}}$$ ■

It can be proved that the distance of the point (x_1, y_1, z_1) from the plane $ax + by + cz + d = 0$ is

$$\frac{|ax_1 + by_1 + cz_1 + d|}{(a^2 + b^2 + c^2)^{1/2}}$$

Exercise 4.5

1   Find a unit vector perpendicular to the vectors $2i - j + k$ and $3i + 4j - k$.

2   Using the definition of vector product, find the area of the triangle with vertices $A(3, -1, 4), B(2, -2, 1)$ and $C(5, 1, 3)$.

3   The points A and B have position vectors $a = 2i + 2j - k$ and $b = i - j + 2k$. Calculate the torque of the force $F = -2i + 2j - 3k$ about A if its line of action passes through B.

4   If a particle at the point P, whose position vector r referred to a point is O, rotates with angular velocity ω about an axis through O, the velocity of the particle is given by $v = \omega \times r$. A rigid body is rotating with angular velocity $5 \, \text{rad s}^{-1}$ in the direction $2i + j$ through the point $A(0, 1, -4)$. Find the velocity of the point $B(1, 4, 1)$ on the body.

5   A rigid body is rotating about an axis through the origin. Instantaneously the point $A(2, 0, 1)$ on the body has a velocity $2i + 3j - 4k$, whereas the point $B(0, 3, -1)$ has velocity $-8i + j + 3k$. Find the angular speed of rotation of the body and the direction of the axis of rotation. (Note that the angular velocity vector must be perpendicular to the velocity vector.)

6   Find the magnitude and direction of the force on a straight cable carrying a current of 3 A in the direction $(-1, 2, -2)$ in a magnetic field of 3×10^{-3}, Webers per square metre acting in the direction $(2, 1, 2)$.

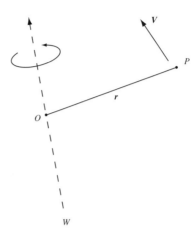

7 Find the equation of the plane whose normal is in the direction $\left(\dfrac{6}{5}, \dfrac{2}{3}, \dfrac{1}{15}\right)$ and which contains the point $(1, 1, 1)$.

8 Find the equation of the plane which contains both the point $(5, 6, 7)$ and the line

$$\frac{x-3}{2} = \frac{y-3}{3} = \frac{z-3}{4}$$

9 Find the equation of the plane which contains the three points $(1, 0, 1), (1, 1, 1)$ and $(2, 1, -1)$. Note that the general equation of the plane can be divided by any number within reason.

10 A plane passes through the points $(0, 1, 2)$ and $(1, -1, 0)$. It is parallel to the direction $(1, -1, 1)$. Find the equation of the plane and the position of the foot of the perpendicular to it from the point $(3, 0, 3)$.

11 Find the equation of the line joining the points $(2, 1, 1)$ and $(1, 3, -2)$. Find the point where this line meets the plane $x + 3y + 2z = 6$.

12 Show that the straight lines

$$\frac{x+1}{-3} = \frac{y-3}{2} = \frac{z+2}{1} \quad \text{and} \quad \frac{x}{1} = \frac{y-7}{-3} = \frac{z+7}{2}$$

intersect. Find the coordinates of the point of intersection and the equation of the plane containing the lines.

13   Prove that the distance of the point (x_1, y_1, z_1) from the plane $ax + by + cz + d = 0$ is

$$\frac{|ax_1 + by_1 + cz_1 + d|}{(a^2 + b^2 + c^2)^{1/2}}$$

(Follow the example in the text, where the distance from the origin was found, by transforming to the origin using the new coordinates $X = x - x_1$, $Y = y - y_1$, $Z = z - z_1$.)

Deduce that the distance apart of the parallel planes $ax + by + cz + d_1 = 0$ and $ax + by + cz + d_2 = 0$ is

$$\frac{|d_2 - d_1|}{(a^2 + b^2 + c^2)^{1/2}}$$

14   Find the distance from the plane $x + y + 2z = 4$ to the parallel line

$$\frac{x - 2}{1} = \frac{y - 3}{-3} = \frac{z - 4}{1}$$

15   Find the equation of the line common to the planes $2x - y - z + 2 = 0$ and $x + y + z - 5 = 0$. Find the angle between the planes.

16   (a)   If $(p, q, 0)$ are the direction ratios of a horizontal line contained by the plane $ax + by + cz + d = 0$, prove that $ap + bq = 0$.

(b)   Verify that the vector $(2, -1, 0)$ represents the direction ratios of a horizontal line contained by the plane $x + 2y + 3z + 4 = 0$.

(c)   The line of steepest slope within an inclined plane is perpendicular both to the normal to the plane and to any horizontal line contained in the plane. Find the direction ratios of the line of steepest slope in the plane $x + 2y + 3z + 4 = 0$ and the angle which it makes with the horizontal.

SUMMARY

- **A vector** requires both magnitude and direction for its complete specification
- **Resultant of two vectors**: $a + b$ is defined by the parallelogram law
- **Scalar multiple**: λa is a vector in the same direction as a if $\lambda > 0$, in the opposite direction if $\lambda < 0$, and of magnitude $|\lambda|$
- **Zero vector**: 0 is such that $a + 0 = a$ for any vector a
- **Free vectors**: used in geometry perhaps where two opposite sides of a parallelogram can be represented by the same vector
- **Unit vector**: magnitude 1, a single unit vector exists in each coordinate direction
- **Cartesian components**: if $a = a_1 i + a_2 j + a_3 k$, the components are (a_1, a_2, a_3) and $a = |a| = (a_1^2 + a_2^2 + a_3^2)^{1/2}$; the direction cosines are $\dfrac{a_1}{a}, \dfrac{a_2}{a}, \dfrac{a_3}{a}$ (for two dimensions put $a_3 = 0$)
- **Scalar product**: two non-zero vectors are perpendicular if $a.b = 0$, where

$$a.b = |a||b|\cos\theta = a_1 b_1 + a_2 b_2 + a_3 b_3$$

- **Angle between two vectors**

$$\cos\theta = \frac{a.b}{|a||b|} = \frac{a_1 b_1 + a_2 b_2 + a_3 b_3}{(a_1^2 + a_2^2 + a_3^2)^{1/2}(b_1^2 + b_2^2 + b_3^2)^{1/2}}$$

- **Straight line**: through point a in the direction of b is $r = a + tb$; through points r_1 and r_2 is $r = r_1 + t(r_2 - r_1)$; these results apply in two and three dimensions
- **Vector product**: two non-zero vectors are parallel if $a \times b = 0$, where

$$a \times b = ab\sin\theta\,\hat{n} = (a_2 b_3 - a_3 b_2)i + (a_3 b_1 - a_1 b_3)j$$
$$+ (a_1 b_2 - a_2 b_1)k = -b \times a$$

- **Equation of a plane**: $ax + by + cz = d$; normal to the plane is in the direction $ai + bj + ck$.

Answers

Exercise 4.1

1 (a) $a + 3b, 3b - a, 2(a + b), -(a + 3b)$

3 (a) $\dfrac{1}{2}(r_1 + r_2)$ (b) $\dfrac{1}{m + n}(nr_1 + mr_2)$

4 (a) $r_2 - r_1, r_4 - r_3, r_3 - r_1$

6 (b) H is the centroid of the tetrahedron $ABCD$.

7 $a, b, b - a, -a, -b, a - b$

8 (a) Vector sum is zero, returning to A.

9 (a) $\dfrac{1}{2}a, \dfrac{1}{2}b, \dfrac{1}{2}c$

(d) All are one-quarter the size of the triangle ABC.

(e) $p + q + r = a + b + c$ so that $\dfrac{1}{3}(p + q + r) = \dfrac{1}{3}(a + b + c)$, same centroid as that of the triangle ABC.

Exercise 4.2

1 (a) $(-2, -6)$ (b) $(22, 31)$ (c) $\dfrac{1}{3}(7, 14)$

2 (a) $2\sqrt{10}; -\dfrac{1}{\sqrt{10}}, \dfrac{3}{\sqrt{10}}; \dfrac{1}{\sqrt{10}}(-1, 3)$

(b) $\sqrt{1445}; \dfrac{22}{\sqrt{1445}}, \dfrac{31}{\sqrt{1445}}; \dfrac{1}{\sqrt{1445}}(22, 31)$

(c) $\dfrac{7\sqrt{5}}{3}; \dfrac{1}{\sqrt{5}}, \dfrac{2}{\sqrt{5}}; \dfrac{1}{\sqrt{5}}(1, 2)$

3 $(6, 11) = 4(3, 2) + 3(-2, 1)$ or $c = 4a + 3b$; $d_1 i + d_2 j \equiv k_1(3i + 2j) + k_2(-2i + j)$; $d_1 = k_1 - 2k_2, d_2 = 2k_1 + k_2$; $k_1 = \dfrac{1}{5}(d_1 + 2d_2), k_2 = \dfrac{1}{5}(-2d_1 + d_2)$

4 (a) $(0, 4, -13)$ (b) $(-7, 4, 1)$
(c) $(-9, 7, -11)$ (d) $(-1, 0, 2)$

5 (a) $\sqrt{6}; \dfrac{2}{\sqrt{6}}, \dfrac{1}{\sqrt{6}}, \dfrac{-1}{\sqrt{6}}; \dfrac{1}{\sqrt{6}}(2, 1, -1)$

    (b) $\sqrt{10}; \dfrac{-3}{\sqrt{10}}, 0, \dfrac{1}{\sqrt{10}}; \dfrac{1}{\sqrt{10}}(-3, 0, 1)$

    (c) $\sqrt{5}; 0, \dfrac{1}{\sqrt{5}}, \dfrac{-2}{\sqrt{5}}; \dfrac{1}{\sqrt{5}}(0, 1, -12)$

7 (b) $\dfrac{1}{\sqrt{21}}(2, -4, -1)$

8 (a) $\boldsymbol{b} = (-3, -1, -1), \boldsymbol{c} = (-2, -2, 2)$

    (b) $(0, -4, 8) = -2\boldsymbol{b} + 3\boldsymbol{c}$

10 $2\sqrt{5}\,\text{km}\,\text{h}^{-1}$ in direction $2\boldsymbol{i} + \boldsymbol{j}$ i.e. $60°$ E of N

11 $(-4, 11)\,\text{kg}\,\text{m}$

12 $(22, -2)\,\text{kg}\,\text{m}$

13 $(-0.3, -0.2, 1.3)$

14 $\overrightarrow{AP} = \overrightarrow{PQ} = \overrightarrow{QB}$; $\overrightarrow{XA}$: $17°$ W of N at $724\,\text{km}\,\text{h}^{-1}$; $\overrightarrow{PQ}$: $10°$ W of N at $700\,\text{km}\,\text{h}^{-1}$; $\overrightarrow{QY}$: $9°$ E of N at $740\,\text{km}\,\text{h}^{-1}$

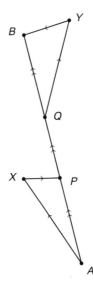

Exercise 4.3

1 (a) 60.26° (b) 36.57° (c) 90°

2 $\sqrt{21}, 2\sqrt{21}, \cos^{-1}\left(-\dfrac{2}{21}\right)$

3 $\cos^{-1}\left(\dfrac{2}{15}\right) = 82.3°$

4 $AB = \sqrt{11}, \; AC = 3, \; BC = \sqrt{22}$
   $\cos A = -\dfrac{1}{3\sqrt{11}}, \quad \cos B = \dfrac{12}{11\sqrt{2}}, \quad \cos C = \dfrac{10}{3\sqrt{22}}, \quad \text{area} = \dfrac{7\sqrt{2}}{2}$

6 $-\mathbf{k}$

7 (a) $\mathbf{c}.\mathbf{a} = \mathbf{c}.\mathbf{b} = 0$ (b) $\dfrac{\pm(-3\mathbf{i} + 5\mathbf{j} + 11\mathbf{k})}{\sqrt{155}}$

8 (a) $a^2 - b^2$ (b) $a^2 - 2\mathbf{a}.\mathbf{b} + b^2$ (c) $2\mathbf{b}.\mathbf{c}$

11 K is drawn exterior to the triangle. This would be so for an obtuse-angled triangle.

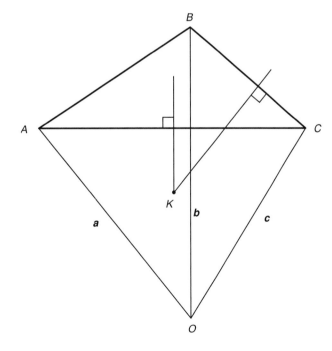

12 $\lambda = \dfrac{b.c}{p(p + a.b)}$

13 $\displaystyle\sum_{i=1}^{4} F_i = 2i - 15j - k = F$, say

$F.r = -18 + 15 - 6 = -9$, in the direction $-9i - j + 6k$. For the work to be zero, F would have to be perpendicular to r. Let $r = ai + bj + ck$. As an example, $i + 2k : a = 1, b = 0, c = 2$.

Exercise 4.4

2 $r = 4i + j + t(3i - j)$

3 Intersection at $(1, 0)$. Cartesian equations $x - y = 1, x + y = 1$

4. (a) $r = 2j + t(2i - 3j)$ (b) $r = -3i + t(4i + j)$

5. $\cos^{-1}\left(\dfrac{5}{\sqrt{13}\sqrt{17}}\right) = 70.35°$

6 $(c^2 + d^2)^{1/2}$

7 (a) $\dfrac{x+2}{1} = \dfrac{y-3}{-1} = \dfrac{z+1}{1}$ (b) $\dfrac{x-1}{-2} = \dfrac{y+2}{1} = \dfrac{z+5}{-2}$

     (c) $\dfrac{x+3}{1} = \dfrac{y-1}{1} = \dfrac{z+2}{-3}$ (d) $r = 3i - j + 5k + \lambda(-2i + 6j + 7k)$

     (e) $r = -5i + 6j + \dfrac{11}{3}k + \lambda\left(4i + 3j + \dfrac{5}{3}k\right)$

     (f) $r = \sqrt{2}i - \pi j - 16k + \lambda(ei + \sqrt{3}j + \sqrt{2}k)$

     Line (e) passes through the given point.

8 (a) $r = -i + 2k + \lambda(7i + j + 3k)$

     (b) $r = -i + j + k + \lambda(2i + j + 3k)$

        angle is $\cos^{-1}\left(\dfrac{24}{\sqrt{59}\sqrt{14}}\right) = 33.38°$

9 (a) $\lambda = -\dfrac{2}{3}, \left(-\dfrac{17}{3}, \dfrac{4}{3}, 0\right)$ (b) $\lambda = -\dfrac{1}{3}, \left(-\dfrac{5}{3}, \dfrac{2}{3}, 0\right)$

10 $1.581 \, \text{km s}^{-1}$. Track has direction ratios $(-7, 4, 5)$

11 $r = 2i + j + 3k + \lambda(i + j + k)$ or $(2 + \lambda, 1 + \lambda, 3 + \lambda)$; $\overrightarrow{PP_2} = (3 - \lambda, 3 - \lambda, -1, -\lambda)$
  $\lambda = \dfrac{5}{3}, P_3 = \left(\dfrac{11}{3}, \dfrac{8}{3}, \dfrac{14}{3}\right)$, $\overrightarrow{P_2P_3}$ parallel to $(1, 1, -2)$

12   (a)   lines skew, $\hat{n}$ parallel to $2i - 3j - k$, distance apart is $\dfrac{3}{\sqrt{14}}$
     (b)   parallel
     (c)   intersect at $(1,1,1)$

13 $(1, 2, 1)$

Exercise 4.5

1 $\pm \dfrac{1}{\sqrt{155}}(-3i + 5j + 11k)$

2 $\dfrac{7}{2}\sqrt{2}$

3 $G = 3i - 9j - 8k$

4 $v = 5\sqrt{5}(i - 2j + k)$

5 $|\omega| = 3$ in the direction of $i + 2j + 2k$

6 $\sqrt{65} \times 10^{-3}$ in the direction $(6, -2, 5)$

7 $18x + 10y + z = 29$

8 $x - 2y + z = 0$

9 $2x + z = 3$

10 $4x + 3y - z = 1$, $\left(\dfrac{23}{13}, \dfrac{12}{13}, \dfrac{43}{13}\right)$

11 $\dfrac{x - 2}{-1} = \dfrac{y - 1}{2} = \dfrac{z - 1}{-3}$, $(1, 3, -2)$

12 $(2, 1, -3)$, $x + y + z = 0$

14 $\dfrac{9}{\sqrt{6}}$

15 $x = 1, \dfrac{y}{-1} = z - 4$; direction $(0, -1, 1)$; planes are perpendicular

16 (c) $(-3, -6, 5)$; angle with the horizontal is $\sin^{-1}\left(\dfrac{5}{\sqrt{70}}\right) = 36.70°$

5 DIFFERENTIATION

INTRODUCTION

Accurate calculation of rates of change is a key requirement whenever variation is involved. Differentiation can be used to calculate rates of change as well as to calculate maximum and minimum values of a function. Many functions require special differentiation techniques; this chapter derives some of the common sorts and develops methods to handle composite functions. It is useful to obtain a simple polynomial approxiation to a function near points of interest in order to study its behaviour there. Sometimes we need to obtain the approximate rate of change of a quantity when the only information we have is a table of values; then we must use numerical methods.

Objectives

After working through this chapter you should be able to

- Use and apply the product rule for differentiation
- Use and apply the quotient rule for differentiation
- Use and apply the function of a function rule for differentiation
- Use and apply the technique of logarithmic differentiation
- Use and apply the technique of implicit differentiation
- Obtain the derivative of inverse functions
- Obtain the Maclaurin series of a function
- Find the limiting behaviour of a rational function for both large and small values of its argument
- Obtain the Taylor series of a function
- Obtain numerical approximations to derivatives of a function specified by a table of values.

5.1 COMPOSITE DIFFERENTIATION

The product rule

Suppose we wish to differentiate the product uv where u and v are functions of x. Let $y = uv$ and let an increment δx in x produce corresponding increments δu in u, δv in v and δy in y. Then

$$y + \delta y = (u + \delta u)(v + \delta v)$$

so that

$$\begin{aligned}
\delta y &= (u + \delta u)(v + \delta v) - uv \\
&= uv + v\delta u + u\delta v + (\delta u)(\delta v) - uv \\
&= v\delta u + u\delta v + (\delta u)(\delta v)
\end{aligned}$$

Then $\dfrac{\delta u}{\delta x} = v\dfrac{\delta u}{\delta x} + u\dfrac{\delta v}{\delta x} + \delta u\dfrac{\delta v}{\delta x}$

When $\delta x \to v,$ $\delta u \to 0,$ and $\delta v \to 0$

and

$$\frac{\delta u}{\delta x} \to \frac{du}{dx}, \qquad \frac{\delta v}{\delta x} \to \frac{dv}{dx}, \qquad \frac{\delta y}{\delta x} \to \frac{dy}{dx}, \qquad \text{and} \qquad \delta u\frac{\delta v}{\delta x} \to 0$$

This last result follows because $\delta u\delta v$ is an order of magnitude smaller than either δu or δv.

Hence we obtain the **product rule**

$$\frac{dy}{dx} = v\frac{du}{dx} + u\frac{dv}{dx} \tag{5.1}$$

Examples

1. Differentiate the functions

   (a) $y = (x^3 + 1)\sqrt{x}$ (b) $y = xe^x$

Solution

(a) Put $u = x^3 + 1$ and $v = \sqrt{x}$

Then $\dfrac{dy}{dx} = 3x^2\sqrt{x} + (x^3 + 1) \times \dfrac{1}{2\sqrt{x}}$ (via the product rule)

$$= 3x^{5/2} + \frac{1}{2}x^{5/2} + \frac{1}{2}x^{-1/2}$$

$$= \frac{7}{2}x^{5/2} + \frac{1}{2}x^{-1/2}$$

You could obtain the same result by multiplying out y first of all.

(b) Here, $u = x$ and $v = e^x$, so

$$\frac{dy}{dx} = 1 \times e^x + xe^x = (1 + x)e^x$$

2. (a) Obtain the equations of the tangent and the normal to the curve $y = x \ln x$ at the point $(1, 0)$.

   (b) Draw a sketch of $y = x \ln x$ and determine the coordinates of the minimum points on the curve.

Solution

(a) $\dfrac{dy}{dx} = \ln x + x \times \dfrac{1}{x} = 1 + \ln x$. At $(1, 0)$, the gradient $\dfrac{dy}{dx} = 1 + \ln 1 = 1 + 0 = 1$, so the tangent has the equation

$$y - 0 = 1 \times (x - 1), \text{ i.e. } y = x - 1.$$

The normal has slope $m = -1$, so its equation is

$$y - 0 = -1 \times (x - 1), \text{ i.e. } x + y = 1, \text{ or } y = -x + 1$$

(b) Refer to Figure 5.1. Note that $x \ln x$ is defined only when $x > 0$. The turning point on the curve occurs where $\dfrac{dy}{dx} = 1 + \ln x = 0$. Hence $\ln x = -1$, so $x = e^{-1} = \dfrac{1}{e}$

The point is $x = \dfrac{1}{e}, y = -\dfrac{1}{e}$. Furthermore $\dfrac{d^2y}{dx^2} = \dfrac{1}{x} > 0$ for all $x > 0$, hence the point is a **local minimum**.

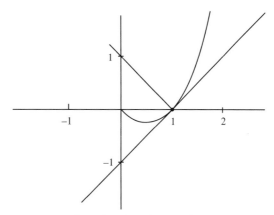

Figure 5.1 The curve $y = x \ln x$ showing tangent and normal at the point $(1, 0)$

■

The quotient rule

Let $y = \dfrac{u}{v}$ where u and v are two functions of x. Let an increment δx in x produce corresponding increments δu in u, δv in v and δy in y. Then

$$y + \delta y = \frac{u + \delta u}{v + \delta v}$$

so that
$$\delta y = \frac{u + \delta u}{v + \delta v} - \frac{u}{v}$$

$$= \frac{v(u + \delta u) - u(v + \delta v)}{v(v + \delta v)}$$

$$= \frac{uv + v\delta u - uv - u\delta v}{v(v + \delta v)}$$

$$= \frac{v\delta u - u\delta v}{v(v + \delta v)}$$

Then
$$\frac{\delta y}{\delta x} = \frac{v\dfrac{\delta u}{\delta x} - u\dfrac{\delta v}{\delta x}}{v(v + \delta v)}$$

$$\delta x \to 0,$$

$$\frac{\delta u}{\delta x} \to \frac{du}{dx}, \qquad \frac{\delta v}{\delta x} \to \frac{dv}{dx}, \qquad v(v + \delta v) \to v^2 \quad \text{and} \quad \frac{\delta y}{\delta x} \to \frac{dy}{dx}$$

Hence we obtain the **quotient rule**.

$$\frac{dy}{dx} = \frac{v\dfrac{du}{dx} - u\dfrac{dv}{dx}}{v^2} \tag{5.2}$$

Examples

1. Differentiate the functions

(a) $\dfrac{e^x}{x+1}$

(b) $\dfrac{1}{5+4x^2}$

(c) $\tan x$

(d) $\operatorname{sech} x$

Solution

(a) Let $y = \dfrac{e^x}{x+1}$, then

$$\frac{dy}{dx} = \frac{(x+1)e^x - e^x \times 1}{(x+1)^2} = \frac{xe^x}{(x+1)^2}$$

(b) Let $y = \dfrac{1}{5+4x^2}$, then

$$\frac{dy}{dx} = \frac{(5+4x^2) \times 0 - 1 \times 8x}{(5+4x^2)^2} = \frac{-8x}{(5+4x^2)^2}$$

(c) Let $y = \tan x = \dfrac{\sin x}{\cos x}$, then

$$\frac{dy}{dx} = \frac{\cos x \cos x - \sin x(-\sin x)}{\cos^2 x}$$
$$= \frac{\cos^2 x + \sin^2 x}{\cos^2 x}$$
$$= \frac{1}{\cos^2 x}$$
$$= \sec^2 x$$

(d) Let $y = \operatorname{sech} x = \dfrac{1}{\cosh x}$, then

$$\frac{dy}{dx} = \frac{\cosh x \times 0 - \sinh x \times 1}{\cosh^2 x} = -\frac{\sinh x}{\cosh^2 x}$$

$$= -\operatorname{sech} x \ \tanh x$$

2. The displacement x (metres) of a particle at time t (seconds) is given by

$$x(t) = \frac{1 - 3t}{1 + 4t + t^2} \qquad t \geq 0$$

(a) Determine the velocity of the particle in metres per second ($\mathrm{m\,s}^{-1}$).

(b) Establish when the particle
    (i) passes through $x = 0$ (ii) is stationary.

(c) Describe the life-cycle displacement of the particle.

(d) What is the acceleration, in $\mathrm{m\,s}^{-2}$ at any time $t \geq 0$?

Solution

(a) The velocity $\dfrac{dx}{dt}$, often written $\dot{x}$, is given by

$$\frac{dx}{dt} = \frac{(1 + 4t + t^2)(-3) - (1 - 3t)(4 + 2t)}{(1 + 4t + t^2)^2}$$

$$= \frac{3t^2 - 2t - 7}{(1 + 4t + t^2)^2}$$

(b) (i) $x(t) = 0$ when $t = \dfrac{1}{3}$ only

(ii) $\dfrac{dx}{dt} = 0$ when $3t^2 - 2t - 7 = 0$, i.e. when $t = -1.230$ or 1.897 (3 d.p.) but only

$t = 1.897$ is valid

(c) When $t = 0, x = 1$ and $\dfrac{dx}{dt} < 0$. The particle moves to the left until $t = 1.897$, when

$x = -0.385$ m.

If t is large, $x(t) \sim -\dfrac{3}{t}$, so $\dot{x}(t) \sim -\dfrac{3}{t^2}$. This means the particle ultimately comes

to rest at the origin, approaching it from the left (Figure 5.2).

Figure 5.2   The particle ultimately comes to rest at the origin

(d) The acceleration $\dfrac{d^2x}{dt^2}$ or $\ddot{x}$, is given by

$$\frac{d^2x}{dt^2} = \frac{(1+4t+4t^2)(6t-2) - (3t^2-2t-7)\dfrac{d}{dt}(1+4t+t^2)^2}{(1+4t+t^2)^4}$$

This reduces eventually to

$$\frac{d^2x}{dt^2} = \frac{6(9+7t+t^2-t^3)}{(1+4t+t^2)^3}$$

This detail is left for you to verify.   ■

Function of a function

We met the concept in *Foundation Mathematics*, Chapter 10. For example, given $f(x) = \sqrt{x}$ and $g(x) = x^2 + 1$, then

$$f(g(x)) = \sqrt{x^2 + 1}$$

To differentiate a general function of a function, $f(g(x))$, let an increment δx in x produce increments δg in g and δf in f; then

$$\frac{\delta f}{\delta x} = \frac{\delta f}{\delta g} \times \frac{\delta g}{\delta x}$$

When $\delta x \to 0$, $\delta g \to 0$, $\delta f \to 0$, but $\dfrac{\delta f}{\delta g} \to \dfrac{df}{dg}$ and $\dfrac{\delta g}{\delta x} \to \dfrac{dg}{dx}$, so we obtain the **chain rule**

$$\frac{df}{dx} = \frac{df}{dg} \times \frac{dg}{dx} \tag{5.3}$$

If we write $u = g(x)$ then an alternative notation is

$$\frac{df}{dx} = \frac{df}{du} \times \frac{du}{dx} \tag{5.4}$$

Therefore if $y = \sqrt{x^2 + 1}$ then $y = u^{1/2}$ where $u = x^2 + 1$. Hence

$$\frac{dy}{dx} = \frac{dy}{du} \times \frac{du}{dx} = \frac{1}{2} u^{-1/2} \times 2x$$

$$= \frac{1}{2} \times \frac{1}{\sqrt{x^2 + 1}} \times 2x = \frac{x}{\sqrt{x^2 + 1}}$$

Examples   1. Differentiate

(a) $(x^3 + 1)^4$ (b) $\dfrac{1}{5 + 4x^2}$ (c) e^{2x}

(d) $\cos(x^2)$ (e) $\sinh(\sqrt{x})$ (f) $\ln(\sqrt{x^2 + 1})$

Solution

(a) Put $y = u^4$ where $u = x^3 + 1$, then

$$\frac{dy}{du} = 4u^3 \quad \text{and} \quad \frac{du}{dx} = 3x^2$$

$$\frac{dy}{dx} = \frac{dy}{du} \times \frac{du}{dx}$$

$$= 4u^3 \times 3x^2$$

$$= 12x^2(x^3 + 1)^3$$

(b) Here $y = \dfrac{1}{u}$ where $u = 5 + 4x^2$, so

$$\frac{dy}{du} = -\frac{1}{u^2} \quad \text{and} \quad \frac{du}{dx} = 8x.$$

$$\frac{dy}{dx} = -\frac{8x}{(5 + 4x^2)^2}$$

This is simpler than differentiating the function as a quotient.

(c) $\dfrac{d}{dx}(e^{2x}) = e^{2x} \times (2) = 2e^{2x}$

(d) $\dfrac{d}{dx}(\cos(x^2)) = -\sin(x^2) \times (2x) = -2x\sin(x^2)$

(e) $\dfrac{d}{dx}(\sinh\sqrt{x})) = \cosh(\sqrt{x}) \times \dfrac{1}{2\sqrt{x}} = \dfrac{\cosh\sqrt{x}}{2\sqrt{x}}$

(f) Apply the chain rule three times

$$\frac{d}{dx}\left(\ln\left(\sqrt{x^2+1}\right)\right) = \frac{1}{\sqrt{x^2+1}} \times \frac{d}{dx}\left(\sqrt{x^2+1}\right)$$

$$= \frac{1}{\sqrt{x^2+1}} \times \frac{1}{2\sqrt{x^2+1}} \times \frac{d}{dx}(x^2+1)$$

$$= \frac{x}{x^2+1} \qquad \text{(eventually)}$$

Observe that $\ln(\sqrt{x^2+1}) = \dfrac{1}{2}\ln(x^2+1)$. By taking $\dfrac{1}{2}\ln(x^2+1)$ we need only two applications of the chain rule, instead of three.

Differentiation may combine both function of a function and the product and quotient values.

2. Differentiate the following with respect to the named variable:

(a) $(3x+1)^4(x+1)^2$ (x) (b) $\sin 2nt \cos(3nt + \varepsilon)$ $(t;\, n,\, \varepsilon \text{ constant})$

(c) $\cos pt \exp(-p^2 t)$ $(t;\, p \text{ constant})$ (d) $\dfrac{\sinh(x^2-1)}{\cosh(x^2+1)}$ (x)

Solution

(a) $\qquad \dfrac{d}{dx}(3x+1)^4 = 4(3x+1)^3(3) = 12(3x+1)^3$

$\qquad \dfrac{d}{dx}(x+1)^2 = (x+1)(2) = 2(x+1)$

Using the product rule

$$\frac{d}{dx}\{(3x+1)^4(x+1)^2\}$$

$$= (x+1)^2 \times 12(3x+1)^3 + (3x+1)^4 \times 2(x+1)$$

$$= 2(x+1) \times (3x+1)^3\{6(x+1) + (3x+1)\}$$

$$= 2(x+1)(3x+1)^3(9x+7)$$

(b)
$$\frac{d}{dt}\sin 2nt = 2n\cos 2nt$$

$$\frac{d}{dt}\cos(3nt + \varepsilon) = -3n\sin(3nt + \varepsilon)$$

Hence

$$\frac{d}{dt}\{\sin 2nt \cos(3nt + \varepsilon)\}$$
$$= 2n\cos 2nt \cos(3nt + \varepsilon) - 3n\sin 2nt \times \sin(3nt + \varepsilon)$$

(c)
$$\frac{d}{dt}\cos pt = -p\sin pt$$

$$\frac{d}{dt}\exp(-p^2 t) = -p^2 \exp(-p^2 t)$$

Hence

$$\frac{d}{dt}\{\cos pt \exp(-p^2 t)\} = -p(\sin pt + p\cos pt)\exp(-p^2 t)$$

(d)
$$\frac{d}{dx}\sinh(x^2 - 1) = 2x\cosh(x^2 - 1)$$

$$\frac{d}{dx}\cosh(x^2 + 1) = 2x\sinh(x^2 + 1)$$

$$\frac{d}{dx}\left\{\frac{\sinh(x^2 - 1)}{\cosh(x^2 + 1)}\right\}$$
$$= \frac{\cosh(x^2 + 1) \times 2x\cosh(x^2 - 1) - 2x\sinh(x^2 + 1)\sinh(x^2 - 1)}{\cosh^2(x^2 + 1)}$$

$$= \frac{2x\cosh 2}{\cosh^2(x^2 + 1)}$$

using the result that $\cosh A \cosh B - \sinh A \sinh B \equiv \cosh(A - B)$.

3. The deflection θ, in radians, at time t (seconds) of a galvanometer needle is given by

$$\theta = e^{-0.5t}\sin\frac{5\pi t}{12}$$

At what time does the deflection reach its first maximum?

We need to find when $\dfrac{d\theta}{dt} = 0$.

Now $\quad \dfrac{d\theta}{dt} = -0.5e^{-0.5t}\sin\dfrac{5\pi t}{12} + e^{-0.5t}\dfrac{5\pi}{12}\cos\dfrac{5\pi t}{12}$

Hence $\quad \dfrac{d\theta}{dt} = 0$ when

$$e^{-0.5t}\sin\dfrac{5\pi t}{12} = \dfrac{5\pi}{12}e^{-0.5t}\cos\dfrac{5\pi t}{12}$$

i.e. $\qquad \tan\dfrac{5\pi t}{12} = \dfrac{5\pi}{6}$

then $\qquad \dfrac{5\pi t}{12} = \tan^{-1}\dfrac{5\pi}{6} = 1.206$

i.e. $\qquad t = 0.921$ s

Note that $\dfrac{d\theta}{dt} > 0$ for $t = 0.9\,$s and $\dfrac{d\theta}{dt} < 0$ for $t = 1.0\,$s, so the deflection is a (local) maximum.

4. If $y = \dfrac{\sin x}{x}$ prove that y satisfies the **differential equation**

$$\dfrac{d^2y}{dx^2} + \dfrac{2}{x}\dfrac{dy}{dx} + y = 0$$

We could use the brute force method of differentiating $\dfrac{\sin x}{x}$ twice followed by substituting into the left-hand side of the differential equation, but we can avoid this by writing

$$xy = \sin x$$

and differentiating twice. Note that

$$\frac{d^2}{dx^2}(\sin x) = \frac{d}{dx}(\cos x) = -\sin x = -xy$$

and $\quad \frac{d}{dx}(xy) = x\frac{dy}{dx} + y \qquad$ (product rule)

also $\quad \frac{d^2}{dx^2}(xy) = \frac{d}{dx}\left(x\frac{dy}{dx}\right) + \frac{dy}{dx} = x\frac{d^2y}{dx^2} + 2\frac{dy}{dx} \qquad$ (product rule)

Hence we obtain

$$\frac{d^2y}{dx^2} + \frac{2}{x}\frac{dy}{dx} + y = 0 \qquad \blacksquare$$

Logarithmic differentiation

If y is a function of x then $\dfrac{d}{dx}\ln y = \dfrac{1}{y} \times \dfrac{dy}{dx}$ using the chain rule. This is of particular use when differentiating functions with variable exponents.

Example Differentiate the functions

(a) a^x, $\qquad a > 0$ $\qquad\qquad\qquad\qquad$ (b) x^x, $\qquad x > 0$

(c) y^x, $\qquad y$ is a function of x

Solution

(a) First, $\ln(a^x) = x \ln a$, so that

$$\frac{d}{dx}(\ln a^x) = \ln a = \frac{1}{a^x}\frac{d}{dx}(a^x) \qquad \text{(chain rule)}$$

then $\quad \dfrac{d}{dx}(a^x) = \ln a \times a^x$

Note that when $a = e$, $\ln e = 1$ and $\dfrac{d}{dx}(e^x) = e^x$

(b) $\qquad \ln(x^x) = x \ln x$

and $\quad \dfrac{d}{dx}(x \ln x) = x \times \dfrac{1}{x} + \ln x = 1 + \ln x$

so $\quad \dfrac{d}{dx}(\ln(x^x)) = x \times \dfrac{1}{x} + \ln x = 1 + \ln x$

Hence $\quad \dfrac{d}{dx}(x^x) = (1 + \ln x)x^x$

(c) $\qquad \ln(y^x) = x \ln y$

$$\dfrac{d}{dx}(x \ln y) = \ln y + \dfrac{x\,dy}{y\,dx}$$

but $\quad \dfrac{d}{dx}(\ln(y^x)) = \dfrac{1}{y^x} \times \dfrac{d}{dx}(y^x)$

so $\quad \left(\dfrac{d}{dx}(y^x)\right) = y^x\left(\ln y + \dfrac{x\,dy}{y\,dx}\right)$ ∎

We have assumed throughout this section that

$$\dfrac{d}{dx}(x^r) = rx^{r-1}$$

where $r = \dfrac{p}{q}$ is a rational constant and p, q are integers. So far we only know for sure that the result is true when $r = n$, a positive integer.

If $r = -n$, a negative integer, then via the quotient rule

$$\dfrac{d}{dx}\left(\dfrac{1}{x^n}\right) = \dfrac{(x^n \times 0) - (1 \times n \times x^{n-1})}{x^{2n}} = \dfrac{-n}{x^{n+1}} = -nx^{-n-1}$$

so the general result is true for a negative integer.

Now set $y = x^{p/q}$, and assume $q > 0$, (p can be of either sign). Then

$$y^q = x^p$$

and hence

$$\frac{d}{dx}(y^q) = qy^{q-1} \times \frac{dy}{dx} \qquad \text{and} \qquad \frac{d}{dx}(x^p) = px^{p-1}$$

Therefore

$$\frac{dy}{dx} = \frac{px^{p-1}}{qy^{q-1}} = \frac{p}{q}\frac{y}{x} \times \frac{x^p}{y^q} = \frac{py}{qx} \qquad \text{since} \quad y^q = x^p$$

Hence

$$\frac{dy}{dx} = \frac{p}{q}x^{p/q-1} = rx^{r-1} \qquad \left(y = x^{p/q}, r = \frac{p}{q}\right)$$

as we required. This shows the general result is true for a rational index r.

Exercise 5.1

1   Differentiate the following functions with respect to x using the product and quotient rules as appropriate. Validate the results by differentiating directly, having multiplied or divided out.

(a) $x^2(x + 3)$

(b) $\sqrt{x}\left(2 - \dfrac{1}{x}\right)$

(c) $\dfrac{(5 - x)(2 + 3x)}{x^{1/3}}$

(d) $(x + 1)(x + 2)(x + 3)$

(e) $(a + bx)(c + dx)$

(f) $\left(9 - \dfrac{1}{x^2}\right)^2 (2 - 7x^3)$

2   Differentiate the following functions with respect to x:

(a) $e^x \ln x$

(b) $x \tan x$

(c) $\dfrac{x}{x^2 + 1}$

(d) $\dfrac{2x - 1}{x^3 - 1}$

(e) $\dfrac{2x^2}{1 + \sin x}$

(f) $\dfrac{\cos x}{x}$

(g) $\dfrac{\ln x}{3x + 1}$

(h) $\dfrac{e^x - e^{-x}}{e^x + e^{-x}} = \tanh x$

(i) $\dfrac{5}{(\ln x)^2}$

(j) $\dfrac{x^2 + 1}{\tan x}$

3 Prove that the curve $y = xe^{-x}$ has only one stationary point. Determine its location and nature.

4 (a) Consider the rational function

$$y = \frac{2x - 3}{x + 4}$$

Prove that it cannot possess any turning points.

(b) Differentiate the general bilinear rational function

$$y = \frac{ax + b}{cx + d} \qquad ad \neq bc \qquad (a, b, c, d \text{ constant})$$

and prove in the general case that there are no turning points.

5 Prove that the rational function

(a) $\dfrac{x(x - 2)}{x - 1}$ has no turning points.

(b) Now prove that the rational function

$$\frac{x^2 + 4x + 17}{x - 1}$$

has two turning points, a maximum at $(-3.69, -3.38)$ and a minimum at $(5.69, 15.38)$, coordinates quoted to 2 d.p.

6 By using function of a function principles, differentiate the following:

(a) $\ln(x + 1)$ (b) $\sin 2x$

(c) e^{3x} (d) $\tan 2x$

(e) $\cos(4x - 1)$ (f) $\ln(x^2 + 1)$

(g) $\ln x^2$ (h) e^{x^2}

(i) $\sin(2x^2 - 3x + 1)$

(j) $(ax + b)^n$, n integer; a, b constant

(k) e^{kx} (l) $\tan(cx + d)$ with c, d constant

(m) $\exp\left(-\dfrac{1}{2}\dfrac{(x - \mu)^2}{\sigma^2}\right)$ with μ, σ constant

(n) $\ln(x^3 + x + 1)$ (o) $\ln(x^2 + 1)^{1/3}$

Why is the answer to (o) exactly one-third of the answer to (f)?

7 Differentiate the following functions with respect to x:

(a) $\sin^2 2x$ (b) $\ln(\sin 3x)$

(c) $\tan \sqrt{x}$ (d) $(x + 2)^5 (x - 1)^3$

(e) $\ln\{(x - 1)(x + 1)(x + 2)\}$ (f) $\dfrac{\cos x + \sin x}{\cos x - \sin x}$

(g) $\dfrac{1}{\ln x}$ (h) $\dfrac{e^{ax+b}}{\sin(ax + b)}$ (a, b constant)

(i) $\cos^2 \sqrt{x}$ (j) $\ln(\cos^2 \sqrt{x})$

(k) $(ax + b)^n (cx + d)^m$, all parameters except x are constant. Regroup the factors.

(l) 2^{2x+1}

Evaluate the derivative when $x = 1$

8 If $y = f(x) = \dfrac{u(x)}{v(x)}$ and $\dfrac{dy}{dx} = uv' + u'v$, prove that

$$\frac{d^2 y}{dx^2} = uv'' + 2u'v' + u''v \qquad \left(u' = \frac{du}{dx}, u'' = \frac{du}{dx^2}, \text{etc.}\right)$$

Use this result to write down the second derivatives of

(a) $x \ln x$ (b) $x \sin x$

(c) $x^3 e^x$ (d) $\ln kx$

Express $\dfrac{d^3y}{dx^3}$, the third derivative, in terms of u, v and their derivatives. Can you see a pattern emerging?

9 Use logarithmic differentiation to determine the ratio $f'(x)/f(x)$ where $y = f(x) = (x + a)^m(x + b)^n(x + c)^p$; a, b, etc., constant. For the case $a = 0, b = 1$, $c = 2, m = 1, n = p = -1$ determine $f'(x)$ and verify the property above.

10 It can be proved that

$$\frac{d}{dx}(xe^x) = P_1(x)e^x$$

$$\frac{d^2}{dx^2}(x^2e^x) = P_2(x)e^x$$

$$\frac{d^3}{dx^3}(x^3e^x) = P_3(x)e^x$$

where $P_n(x)$ are polynomials of degree n ($n = 1, 2, 3$), called **Legendre polynomials**. Determine $P_n(x)$ for $n = 1, 2, 3$.

11 Given $y = \dfrac{\sin x}{(1 + 4\sin x)^{1/2}}$

   (a)   Identify the domain of $y = f(x)$, noting that the denominator is not real for certain values of x.

   (b)   Hence prove that stationary points occur when $\cos x = 0$ only.

   (c)   Prove that y has local maxima at $\left((4n + 1)\dfrac{\pi}{2}, \dfrac{1}{\sqrt{5}}\right)$, n integer.

   (d)   Sketch the graph of $y = f(x)$.

12 Use logarithmic differentiation to find the derivatives of the following functions:

   (a) x^{-x} (b) $2^{-5x}\tan(1 + x^2)$ (c) $\dfrac{x^x - e^x}{x^x + e^x}$

13 Define the domain of $f(x) = x^{2x-3}$. Sketch the graph of $f(x)$ for $0 \le x \le 3$. Determine approximately the coordinates of the minimum point (3 s.f.).

14 Differentiate the following functions:

   (a) $\left(\dfrac{1 + x}{1 - x}\right)^{1/2}$ (b) $\dfrac{(1 - 3x)^{1/2}}{(1 - x)^3}$

using first standard then logarithmic differentiation.

15 (a)   Prove that

(i) $\dfrac{d}{dx}(xy) = y + x\dfrac{dy}{dx}$ where y is a function of x

(ii)   Hence find $\dfrac{d^2}{dx^2}(xy)$

(b) $y = \dfrac{A\cos kx + B\sin kx}{x}$ where A, B and k are arbitrary constants, prove that

$$\frac{d^2y}{dx^2} + \frac{2}{x}\frac{dy}{dx} + k^2y = 0$$

16 The displacement of a particle, $x(t)$, is governed by the law

$$x(t) = \frac{t(1-t)}{1+t^2} \quad (t \geq 0)$$

(a)   Determine the velocity of the particle and establish when

(i) it passes through $x = 0$
(ii) it is stationary

(b)   Describe the life-cycle displacement of the particle.

(c)   What is the acceleration?

17 The displacement of an object undergoing simple harmonic motion satisfies the law

$$x(t) = A\sin(nt + \phi)$$

where A, n and ϕ are constants.

(a)   Determine
(i) $\dot{x}(t)$ (ii) $\ddot{x}(t)$

(b)   Deduce that $x(t)$ satisfies the differential equation

$$\ddot{x} + n^2x = 0$$

18 In damped harmonic motion the law of displacement is

$$x(t) = Ae^{-kt}\sin(nt + \phi)$$

where A, k, n and ϕ are constants. Show that $x(t)$ satisfies the differential equation

$$\ddot{x} + 2k\dot{x} + (k^2 + n^2)x = 0$$

19   A mountain biker needs to ride across country from A to B to complete the final leg of a tournament. A lies in open flat grassland, where an average speed of $40 \, \text{km hr}^{-1}$ can be maintained, and B lies at the end of the straight road CB, where an average speed of $50 \, \text{km hr}^{-1}$ is possible. Given the distances shown, the cyclist follows the path ADB to minimise the overall time, where $CD = x \, \text{km}$.

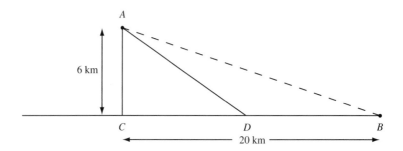

(a)   Show that the time T for the total journey is given by

$$T = \frac{(36 + x^2)^{1/2}}{40} + \frac{20 - x}{50}$$

(b)   Prove that this is a minimum when $x = 8 \, \text{km}$.

20   A sphere of given radius a is completely enclosed by a right circular cone.

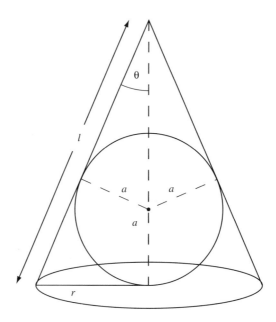

(a)   If θ is the semi-angle of the cone, r the base radius and l the slant height, prove that

(i) $r = l \sin \theta$
(ii) $l - r = a \cos \theta$

(b)   Given that the total surface area of the cone is $\pi r(r + l)$, deduce that this is a minimum when $\theta = \sin^{-1}\left(\dfrac{1}{3}\right)$

5.2 IMPLICIT AND INVERSE DIFFERENTIATION

Implicit differentiation

An explicit relationship between y and x takes the form $y = f(x)$ where f is some function, e.g. $y = x^2 + 2$. A relationship such as $x^2 + y^3 = 1$ *could* be rearranged to $y = (1 - x^2)^{1/3}$ but as it stands it is an **implicit relationship** between x and y. In some cases we cannot obtain an explicit form, e.g. $x^2 + y^2 = \cos y$. In the case of an implicit relationship, $\dfrac{dy}{dx}$ can be obtained in terms of both variables. What we do is to *differentiate* the expression *as it stands*.

Examples

1. If $x^2 + y^2 = \cos y$, then differentiating produces

$$2x + 2y\frac{dy}{dx} = -\sin y \frac{dy}{dx}$$

so
$$\frac{dy}{dx} = \frac{-2x}{2y + \sin y}$$

Note that $\dfrac{d}{dx}(y^2) = 2y\dfrac{dy}{dx}$, by the chain rule.

A non-separable relationship such as $x^2 + y^2 = \cos y$ is known as an implicit relationship and the process of finding $\dfrac{dy}{dx}$ is called **implicit differentiation**. We know that it is totally possible to sketch curves defined in terms of implicit functions. Now we see that tangents and normals at a given point can also be found.

2. An ellipse has equation $(x - 3)^2 + 4(y - 2)^2 = 25$. Show that the ellipse passes through the origin and find the equation of the tangent and normal to the ellipse at this point.

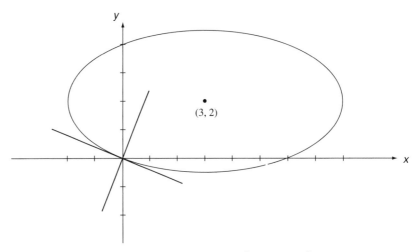

Figure 5.3 The ellipse $(x - 3)^2 + 4(y - 2)^2 = 25$

The equation reduces to $x^2 + 4y^2 = 6x + 16y$, which differentiates to

$$2x + 8y\frac{dy}{dx} = 6 + 16\frac{dy}{dx}$$

hence $\quad \dfrac{dy}{dx} = \dfrac{6 - 2x}{8y - 16} = \dfrac{3 - x}{4(y - 2)}$

At $(0, 0)$, $\dfrac{dy}{dx} = -\dfrac{3}{8}$. The tangent is therefore $y = -\dfrac{3x}{8}$ and the normal $y = \dfrac{8x}{3}$

The curve is sketched in Figure 5.3 and the tangent and normal are marked on the graph. ∎

Inverse differentiation

Consider Figure 5.4. At the point $P, y = f(x)$ differentiates to

$$\frac{dy}{dx} = \tan \psi = f'(x)$$

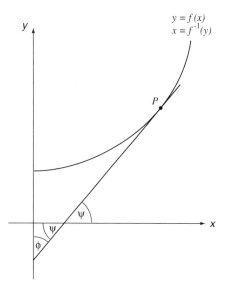

Figure 5.4 Inverse differentiation

But if we write $x = f^{-1}(y)$ instead, then

$$\tan \phi = \frac{d}{dy}(f^{-1}(y))$$

Now $\psi + \phi = \frac{\pi}{2}$, so that

$$\tan \phi = \frac{1}{\tan \psi}$$

Therefore

$$\frac{dx}{dy} = \left(\frac{1}{\dfrac{dy}{dx}} \right) \tag{5.5}$$

Expressed in words:

> The rate of change of x with respect to y is the reciprocal of the rate of change of y with respect to x.

This result is perhaps not unexpected, but when we go to second and higher order derivatives, matters are less straightforward.

Examples

1. Express $\dfrac{d^2x}{dy^2}$ in terms of $\dfrac{dy}{dx}$ and $\dfrac{d^2y}{dx^2}$

Since $\quad \dfrac{dx}{dy} = \dfrac{1}{\left(\dfrac{dy}{dx}\right)}$

then $\quad \dfrac{d^2x}{dy^2} = \dfrac{d}{dy}\left(\dfrac{dx}{dy}\right) \qquad$ (by definition)

$$= \dfrac{d}{dy}\left(\dfrac{1}{\left(\dfrac{dy}{dx}\right)}\right)$$

$$= \dfrac{d}{dx}\left(\dfrac{1}{\left(\dfrac{dy}{dx}\right)}\right) \times \dfrac{dx}{dy} \qquad \text{(chain rule)}$$

If we write $\dfrac{dy}{dx} = p$, then

$$\dfrac{d}{dx}\left(\dfrac{1}{p}\right) = -\dfrac{1}{p^2} \times \dfrac{dp}{dx} \qquad \text{(function of a function)}$$

and $\quad \dfrac{dp}{dx} = \dfrac{d}{dx}\left(\dfrac{dy}{dx}\right) = \dfrac{d^2y}{dx^2},$

so

$$\dfrac{d^2x}{dy^2} = \dfrac{d}{dx}\left(\dfrac{1}{p}\right) \times \dfrac{1}{p}$$

$$= -\dfrac{1}{p^2} \times \dfrac{dp}{dx} \times \dfrac{1}{p} = -\dfrac{1}{p^3} \times \dfrac{dp}{dx}$$

$$= -\dfrac{\dfrac{d^2y}{dx^2}}{\left(\dfrac{dy}{dx}\right)^3}$$

We can differentiate inverse functions by removing the inverse property and using implicit differentiation.

2. Determine derivatives of the functions

    (a) $\sin^{-1} x$ (b) $\tan^{-1} x$ (c) $\cosh^{-1} x$ (d) $\ln x$

Solution

(a) Set $y = \sin^{-1} x$, then $x = \sin y$. Differentiation with respect to x gives

$$1 = \cos x \frac{dy}{dx} \quad \text{so} \quad \frac{dy}{dx} = \frac{1}{\cos y}$$

but

$$\cos y = \sqrt{1 - \sin^2 y} = \sqrt{1 - x^2} \quad \text{so} \quad \frac{dy}{dx} = \frac{1}{\sqrt{1 - x^2}}$$

(b) Set $y = \tan^{-1} x$, then $x = \tan y$. Differentiation gives

$$1 = \sec^2 y \times \frac{dy}{dx} \quad \text{so} \quad \frac{dy}{dx} = \frac{1}{\sec^2 y}$$

$$= \frac{1}{1 + \tan^2 y} = \frac{1}{1 + x^2}$$

(c) Set $y = \cosh^{-1} x$, then $x = \cosh y$. Differentiation gives

$$1 = \sinh y \frac{dy}{dx} \quad \text{so} \quad \frac{dy}{dx} = \frac{1}{\sinh y}$$

$$= \frac{1}{(\cosh^2 y - 1)^{1/2}} = \frac{1}{\sqrt{x^2 - 1}}$$

(d) Set $y = \ln x$, then $x = e^y$, Differentiation gives

$$1 = e^y \frac{dy}{dx} \quad \text{so} \quad \frac{dy}{dx} = \frac{1}{e^y} = \frac{1}{x}$$

■

Exercise 5.2

1 Find $\dfrac{dy}{dx}$ in terms of x and y in each of the following cases, using implicit differentiation.

(a) $x^2 y^2 = y$

(b) $x^2 + xy + y^3 = 1$

(c) $x^3 + y^3 = 3(x + y)$

(d) $\sqrt{x} + \sqrt{y} = 3$

(e) $x^p y^q = k$ p, q, k constant

(f) $x^2 = e^y$

(g) $\cos y = \ln x$ (h) $x^2 \cos y = y^2 \cos x$

(i) $\tan\left(\dfrac{x}{y}\right) = x$ (j) $\sin\left(\dfrac{x}{y}\right) = \dfrac{1}{y}$

2   Determine the gradient of the tangent to the curve

$$xy^3 - 2x^2y^2 + x^4 - 1 = 0$$

at the point $(1, 2)$.

3   If $y = \dfrac{\cos ax}{1+x}$, prove that

$$\frac{d^2y}{dx^2} + \frac{2}{1+x}\frac{dy}{dx} + a^2y = 0$$

4   Determine the equations of (a) tangent and (b) the normal to the curve

$$x^2 - 3xy + 2y^2 = 1 + y$$

at the point $(1, 0)$.

5   Given that $y = \left\{x - (1 + x^2)^{1/2}\right\}^{5/2}$, prove that

(a) $(1 + x^2)^{1/2}y' = -\dfrac{5}{2}y$

(b) $4(1 + x^2)y'' + 4xy' = 25y$

6   If $y = x\sin x$, determine

(a) $\dfrac{dy}{dx}$ (b) $\dfrac{d^2y}{dx^2}$ (c) $\dfrac{d^2x}{dy^2}$

7   The pressure p and volume V of an adiabatically expanding gas obey the law $pV^\gamma = c$, where c is a constant. Prove that the **volumetric elasticity**, $-V\dfrac{dp}{dV}$, varies in direct proportion to the pressure.

8   Use logarithmic differentiation to determine $\dfrac{dy}{dx}$ in the following cases:

(a) $x^y = 1$ (b) $x^y = e^{2x}$

(c)* $x^x = y^y$ (d)* $5^{x/y} = 2^{x^2}$

9   Differentiate the following functions with respect to x:

(a) $\tan^{-1} x$ (b) $x \tan^{-1} x$

(c) $\sin^{-1}\left\{ \dfrac{1 - x}{1 + x} \right\}$ (d) $\cos^{-1}(\tan \sqrt{x})$

(e) $\sinh^{-1} x$ (f) $\cosh^{-1} x$

(g) $\tanh^{-1} x$ (h)* $(1 + x^2)^{1/2} \sinh^{-1} x$

(i)* $\cosh^{-1} \sqrt{x^2 - 1}$

10   If $y = \sin^{-1}(\cos x)$, prove that $\dfrac{dy}{dx} = -1$

11   If $y = (\tan^{-1} x)^2$, prove that

$$\frac{d}{dx}\left\{ (1 + x^2)\frac{dy}{dx} \right\} = \frac{2}{1 + x^2}$$

12   The triangle ABC is isosceles with $AB = AC = a$, BD is the median bisecting AC.

(a)   Prove that

(i) $BD^2 = \dfrac{a^2}{4}(5 - 4 \cos \theta)$

(ii) $\sin \phi = \dfrac{\sin \theta}{(5 - 4 \cos \theta)^{1/2}}$

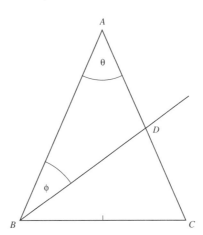

(b)   Deduce that the maximum value of ϕ is $\pi/6$, which occurs when the triangle is equilateral.

13* The **curvature** κ of a curve $y = f(x)$ at any point $P(x, y)$ upon it is defined to be the rate of change in direction of the inclination of the tangent line per unit length of arc.

(a)   Prove that the definitions

$$\kappa = -\frac{\dfrac{d^2y}{dx^2}}{\left[1 + \left(\dfrac{dy}{dx}\right)^2\right]^{3/2}} = \frac{-\dfrac{d^2x}{dy^2}}{\left[1 + \left(\dfrac{dx}{dy}\right)^2\right]^{3/2}}$$

are equivalent.

(b)   For $y = \ln x$ show that

$$\kappa = \frac{-x}{(1 + x^2)^{3/2}}$$

Where is this a maximum?

14* The **circle of curvature** or **osculating circle** of a curve at a point P is defined to be the circle of the same curvature lying on the concave side of the curve and tangential to it at P, as illustrated.

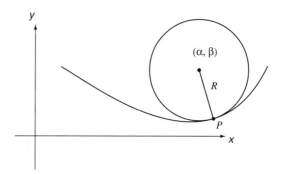

The **radius of curvature**, $\rho = \dfrac{1}{|\kappa|}$, where κ is the curvature, and the centre of the circle has coordinates

$$\alpha = x - \frac{\dfrac{dy}{dx}\left[1 + \left(\dfrac{dy}{dx}\right)^2\right]}{\dfrac{d^2y}{dx^2}} \qquad \beta = y + \frac{\left[1 + \left(\dfrac{dy}{dx}\right)^2\right]}{\dfrac{d^2y}{dx^2}}$$

Determine the equation of the circle of curvature for the hyperbola $2xy + x + y = 4$ at the point $(1, 1)$. Sketch the hyperbola and its circle of curvature.

15* In road building, adjacent sections are matched together by using piecewise smooth curves or **splines**. If the road is flat this is modelled by function curves in which adjacent functions plus their first and second derivatives are equal at the endpoints.

(a)   Prove that the curvature of each function must be equal at an endpoint

(b)   A **spline-fit** curve satisfies

$$x = y^3 + ay^2 + by + c \qquad 0 < x < 1$$
$$y = \ln x \qquad x \geq 1$$

and is such that the function and its first and second derivatives are equal at $(1, 0)$. Determine a, b, c.

16* The equation of a conic is of the form

$$ax^2 + 2hxy + by^2 + 2gx + 2fy + c = 0$$

where a, h, b, g, f, c are constant. Show that

(a) $\dfrac{dy}{dx} = -\dfrac{(ax + hy + g)}{(hx + by + f)}$

(b) $\dfrac{d^2y}{dx^2} = \dfrac{abc + 2fgh - af^2 - bg^2 - ch^2}{(hx + by + f)^3}$

17 In Question 16 examine what happens when

(a) $ab = h^2 \qquad g = f = c = 0$

(b) $abc + 2fgh = af^2 + bg^2 + ch^2$

5.3 MACLAURIN SERIES

Let us assume that $f(x)$ can be represented by the series

$$f(x) = a_0 + a_1 x + a_2 x^2 + a_3 x^3 + \cdots + a_n x^n + \cdots$$

To find the coefficients $a_0, a_1, a_2, \ldots, a_n \ldots$ we will assume that we can differentiate the series term by term to obtain

$$f'(x) = a_1 + 2a_2 x + 3a_3 x^2 + \cdots + na_n x^{n-1} + \cdots$$

$$f''(x) = 2a_2 + 6a_3 x + \cdots + n(n-1)a_n x^{n-2} + \cdots$$

$$f''(x) = 3!a_3 + \cdots + n(n-1)(n-2)a_n x^{n-3} + \cdots$$

If we differentiate an arbitrary number of times n, we obtain

$$f^{(n)}(x) = n! \times a_n + \text{powers of } x$$

Putting $x = 0$ in each case we find that

$$f(0) = a_0 \qquad \text{hence} \qquad a_0 = f(0)$$

$$f'(0) = a_1 \qquad \text{hence} \qquad a_1 = f'(0)$$

$$f''(0) = 2a_2 \qquad \text{hence} \qquad a_2 = \frac{f''(0)}{2!}$$

$$f'''(0) = 3! \times a_3 \quad \text{hence} \quad a_3 = \frac{f'''(0)}{3!}$$

$$\vdots \qquad\qquad\qquad \vdots$$

$$f^{(n)}(0) = n! \times a_n \quad \text{hence} \quad a_n = \frac{f^{(n)}(0)}{n!}$$

This gives **Maclaurin's expansion**

$$f(x) = f(0) + xf'(0) + \frac{x^2}{2!} f''(0) + \frac{x^3}{3!} f'''(0) + \cdots + \frac{x^n}{n!} f^{(n)}(0) + \cdots$$

provided that $f(x)$ and its derivatives exist at the origin, and the series converges. The **radius of convergence** R gives the range of x, $|x| \leq R$, for which the series represents $f(x)$. Such a series is called a **power series** as it contains the powers of x

Examples   1. Find the Maclaurin expansions of

(a) e^x (b) $\sin x$ (c) $\cos x$

(d) $\ln(1 + x)$ (e) $(1 + x)^p$, p is real

Solution

(a) e^x is the only function of x which is invariant under the operation of differentiation. This means that

$$\frac{d}{dx}(e^x) = e^x$$

So $f(0) = e^0 = 1$ and $f'(0) = f''(0) = f'''(0) = 1$, etc. The Maclaurin series for e^x must therefore be

$$e^x = 1 + x + \frac{x^2}{2!} + \frac{x^3}{3!} + \cdots + \frac{x^n}{n!} + \cdots = \sum_{r=0}^{\infty} \frac{x^r}{r!} \quad (0! = 1) \tag{5.6}$$

It happens that this series exists for any real number x i.e. $R = \infty$

(b) For $\sin x$, note that

$$\frac{d}{dx}(\sin x) = \cos x \qquad \frac{d^2}{dx^2}(\sin x) = -\sin x$$

$$\frac{d^3}{dx^3}(\sin x) = -\cos x \qquad \frac{d^4}{dx^4}(\sin x) = \sin x, \text{ etc.}$$

The cycle repeats after every four differentiations. But $\sin 0 = 0$, $\cos 0 = 1$, so

$$f(0) = 0, \qquad f'(0) = 1, \qquad f'''(0) = 0, \text{ etc.}$$

Hence

$$\sin x = x - \frac{x^3}{3!} + \frac{x^5}{5!} - \cdots + (-1)^{n-1}\frac{x^{2n-1}}{(2n-1)!} + \cdots = \sum_{r=0}^{\infty} \frac{(-1)^{r-1}x^{2r-1}}{(2r-1)!}$$

$$\tag{5.7}$$

(c) For $\cos x$ a totally parallel argument to (b) follows to prove that

$$\cos x = 1 - \frac{x^2}{2!} + \frac{x^4}{4!} + \cdots + (-1)^{n-1} \frac{x^{2n}}{(2n)!} + \cdots = \sum_{r=0}^{\infty} \frac{(-1)^{r-1} x^{2r}}{(2r)!} \qquad (5.8)$$

The series for $\sin x$ and $\cos x$ also converge for all values of x.

(d)
$$\frac{d}{dx} \ln(1+x) = -\frac{1}{1+x} \qquad f(0) = 1$$

$$\frac{d^2}{dx^2} \ln(1+x) = -\frac{1}{(1+x)^2} \qquad f'(0) = -1$$

$$\frac{d^3}{dx^3} \ln(1+x) = \frac{(-1) \times (-2)}{(1+x)^3} \qquad f''(0) = 2$$

Taking this forward we can prove that

$$\frac{d^n}{dx^n} \ln(1+x) = \frac{(-1) \times (-2) \times \cdots \times (-(n-1))}{(1+x)^n}$$

so that

$$f^n(0) = (-1)^{n-1} \times (n-1)!$$

and $\dfrac{f^{(n)}(0)}{n!} = \dfrac{(-1)^{n-1}}{n}$

Then

$$\ln(1+x) = x - \frac{x^2}{2} + \frac{x^3}{3} - \frac{x^4}{4} + \cdots + (-1)^{n-1} \frac{x^n}{n} + \cdots = \sum_{r=1}^{\infty} \frac{(-1)^{r-1} x^r}{r}$$

$$(5.9)$$

This time the range of values for which the series converges is $|x| < 1$. It is left as an exercise for you to investigate why the function $\ln x$ does not possess a Maclaurin expansion.

(e) $\qquad (1+x)^p, \quad p \text{ real} \qquad f(0) = 1$

$$\frac{d}{dx}(1+x)^p = p(1+x)^{p-1} \qquad f'(0) = p$$

$$\frac{d^2}{dx^2}(1+x)^p = p(p-1)(1+x)^{p-2} \qquad f''(0) = p(p-1)$$

$$\frac{d^3}{dx^3}(1+x)^p = p(p-1)(p-2)(1+x)^{p-3} \qquad f'''(0) = p(p-1)(p-2)$$

The Maclaurin expansion is therefore

$$(1+x)^p = 1 + px + \frac{p(p-1)}{2!} + \frac{p(p-1)(p-2)}{3!}x^3 + \cdots \tag{5.10}$$

This series is identical to the binomial expansion. This is no coincidence; a power series for a function is **unique**, no matter how it is found.

Series can be used to find approximations for composite functions if x is sufficiently small.

2. Obtain the Maclaurin series for

$$y = \frac{\cos x}{1+x} \quad \text{to} \quad O(x^4)$$

Use the binomial expansion for $(1+x)^{-1}$

$$y = (1+x)^{-1} \cos x$$

$$= (1 - x + x^2 - x^3 + x^4 + O(x^5))\left(1 - \frac{x^2}{2} + \frac{x^4}{24} + O(x^6)\right)$$

$$= 1 - x + \frac{x^2}{2} - \frac{x^3}{2} + \frac{13x^4}{24} + O(x^5)$$

after multiplying out and combining corresponding powers. ∎

Limits

Note that

$$\frac{\sin x}{x} = \frac{x - \frac{x^3}{6} + O(x^5)}{x} = 1 - \frac{x^2}{6} + O(x^4)$$

for small x. Even though $\frac{\sin x}{x}$ appears to be of the form $\frac{0}{0}$ when x is zero, it is perfectly well behaved and approaches 1 as $x \to 0$. We write $\lim\limits_{x \to 0} \frac{\sin x}{x} = 1$

We will now determine other limits, but you should note that the form $\dfrac{0}{0}$ *may* or *may not* possess a **finite limit**, and subsequent analysis is always required.

Example

For the following functions determine the limits as $x \to 0$, where they exist.

(a) $\dfrac{1 - (1+x)^{1/2}}{1 - (1+x)^{1/3}}$

(b) $\dfrac{\sin x - x}{\tan x - x}$

(c) $\dfrac{x - \ln(1+x)}{2x^2}$

Solution

(a) Use the binomial expansion for $(1+x)^{1/2}$ and $(1+x)^{1/3}$:

$$\frac{1 - (1+x)^{1/2}}{1 - (1+x)^{1/3}} = \frac{1 - \left(1 + \dfrac{x}{2} + O(x^2)\right)}{1 - \left(1 + \dfrac{x}{3} + O(x^2)\right)} = \frac{-\dfrac{x}{2} + O(x^2)}{-\dfrac{x}{3} + O(x^2)} = \frac{\dfrac{1}{2} + O(x)}{\dfrac{1}{3} + O(x)}$$

cancelling top and bottom by x. This is permissible provided that x is not actually zero. As $x \to 0$ the limit is $\dfrac{3}{2}$. You could verify this on your calculator by taking very small values of x, e.g. 0.01, 0.001.

Note that we write $+O(x^2)$ rather than $-O(x^2)$ since all we indicate by the notation is that terms of $O(x^2)$ are not included.

(b) The Maclaurin expansion for $\tan x$ is

$$\tan x = x + \frac{x^3}{3} + O(x^5)$$

This is left to the exercises. Using the power series for $\tan x$ and $\sin x$, we obtain

$$\frac{\sin x - x}{\tan x - x} = \frac{\left(x - \dfrac{x^3}{6} + O(x^5)\right) - x}{\left(x + \dfrac{x^3}{3} + O(x^5)\right) - x} = \frac{x^3}{x^3}\left[\frac{-\dfrac{1}{2} + O(x^2)}{1 + O(x^2)}\right] \to -\frac{1}{2} \quad \text{as } x \to 0$$

(c) Putting terms up to $O(x^2)$ in the expansion for $\ln(1+x)$, we obtain

$$\frac{x - \ln(1+x)}{2x^2} = \frac{x - \left(x - \dfrac{x^2}{2} + O(x^3)\right)}{2x^2} \to \frac{1}{4} \quad \text{as } x \to 0$$

cancelling by x^2.

∎

In general, we may write

$$\frac{f(x)}{g(x)} \equiv \frac{f(0) + xf'(0) + \frac{x^2}{2!}f''(0) + \cdots + \frac{x^r}{r!}f^{(r)}(0) + \cdots}{g(0) + xg'(0) + \frac{x^2}{2!}g''(0) + \cdots + \frac{x^r}{r!}g^{(r)}(0) + \cdots}$$

Then $$\lim_{x \to 0} \frac{f(x)}{g(x)} = \frac{f(0)}{g(0)}$$

unless $f(0) = g(0) = 0$

If $f(0) = g(0) = 0$, then cancelling x,

$$\lim_{x \to 0} \frac{f(x)}{g(x)} = \frac{f'(0)}{g'(0)}, \quad \text{unless } f'(0) = g'(0) = 0$$

And if $f'(0) = g'(0) = 0$, then cancelling $\frac{x^2}{2}$,

$$\lim_{x \to 0} \frac{f(x)}{g(x)} = \frac{f''(0)}{g''(0)}$$

unless $f''(0) = g''(0) = 0$, etc.

In other words, if a function $\sim \frac{0}{0}$ as $x \to 0$ then differentiate both numerator and denominator as many times as you need until a derivative ratio other than $\frac{0}{0}$ is found. Remember that 0, a finite non-zero value and ∞ may be the limit. This process is a special case of **L'Hôpital's rule**.

Example   Repeat the previous example using L'Hôpital's rule.

(a) The function becomes $\frac{0}{0}$ when $x = 0$. Differentiate top and bottom to obtain

$$\lim_{x \to 0} \frac{1 - (1+x)^{1/2}}{1 - (1+x)^{1/3}} = \lim_{x \to 0} \frac{\frac{1}{2}(1+x)^{-1/2}}{\frac{1}{3}(1+x)^{4/3}} = \frac{3}{2}$$

(b) $$\lim_{x \to 0} \frac{\sin x - x}{\tan x - x} = \lim_{x \to 0} \frac{\cos x - 1}{\sec^2 x - 1} = \frac{0}{0}$$

Differentiating a second time gives

$$\lim_{x \to 0} \frac{-\sin x}{2 \sec x \tan x} = -\lim_{x \to 0} \frac{1}{2 \sec^2 x} = -\frac{1}{2}$$

(c) $$\lim_{x \to 0} \frac{x - \ln(1 + x)}{2x^2} = \lim_{x \to 0} \frac{1 - \dfrac{1}{1 + x}}{4x} = \lim_{x \to 0} \frac{\dfrac{1 + x - 1}{1 + x}}{4x}$$

$$= \lim_{x \to 0} \frac{1}{4(1 + x)} \to \frac{1}{4} \qquad \blacksquare$$

Exercise 5.3

1 Use Maclaurin's expansion to express each of the following functions as a power series as far as the term in x^5.

(a) $\cos x$ (b) $\sinh x$ (c) $\cosh x$

(d) $\tan x$ (e) $\ln(3 + 2x)$ (f) $\tan^{-1} x$

(g) $\sinh^{-1} x$ (h) $\tanh^{-1} x \left(= \dfrac{1}{2} \ln\left(\dfrac{x + 1}{x - 1}\right) \right)$

2 Verify the Maclaurin expansion for $\tan x$ by taking the ratio of the series for $\sin x$ and $\cos x$.

3* Determine Maclaurin series as far as $O(x^3)$ for

(a) $e^{-x}(1 + 2x)^{-2}$ (b) $(a + x)^p$

                        (i) p real
                         (ii) p a positive integer less than 3

4 Why does $\ln x$ not possess a Maclaurin expansion?

5* (a) Prove that $y = \dfrac{\sin x}{1 - x^2}$ satisfies the differential equation

$$(1 - x^2)\frac{d^2 y}{dx^2} - 4x\frac{dy}{dx} + (1 + x^2)y = 0$$

Hence obtain $y(0)$, $y'(0)$ and $y''(0)$.

(b)   Twice differentiate the differential equation and determine values for $y^{(3)}(0)$ and $y^{(4)}(0)$, i.e. the values of $\dfrac{d^3y}{dx^3}$ and $\dfrac{d^4y}{dx^4}$ when $x = 0$.

(c)   Write out the Maclaurin expansion for y up to $O(x^5)$.

6* (a)   Prove that $\exp(\sin^{-1}x)$ satisfies the differential equation

$$(1 - x^2)\frac{d^2y}{dx^2} - x\frac{dy}{dx} - y = 0$$

(b)   Differentiate this equation a further three times and determine $y^{(n)}(0)$ for $n = 1\,(1)\,5$.

(c)   Hence write out the Maclaurin series as far as $O(x^5)$.

7   (a)   Obtain the Maclaurin series for $(1 + x)^{1/2}$ and verify it is identical to the binomial series.

(b)   Using the Maclaurin (binomial) series, determine limits as $x \to 0$ of

(i) $\dfrac{1 - \sqrt{1+x}}{1 - \sqrt{1-x}}$
(ii) $\dfrac{1 + \dfrac{x}{2} - \sqrt{1+x}}{1 + \dfrac{x}{3} - (1+x)^{1/3}}$

8   Use Maclaurin series to determine limits as $x \to 0$ of the following:

(a) $\dfrac{\sin x}{x}$

(b) $\dfrac{1 - \cos x}{2x^2}$

(c) $\dfrac{e^x - \left(1 + x + \dfrac{x^2}{2} + \dfrac{x^3}{6} + \dfrac{x^4}{24}\right)}{x^5}$

(d) $\dfrac{\sin x + \left(x - \dfrac{x^3}{6}\right)}{\tan x - \left(x - \dfrac{x^3}{3}\right)}$

Check (b) and (c) by differentiating numerator and denominator and setting $x = 0$.

5.4 NUMERICAL DIFFERENTIATON

The Taylor expansion

The Maclaurin expansion or Maclaurin series

$$f(x) = f(0) + \frac{x}{1!}f'(0) + \frac{x^2}{2!}f''(0) + \cdots + \frac{x^n}{n!}f^{(n)}(0) + \cdots$$

enables us to compute $f(x)$ in a **neighbourhood** of $x = 0$ when $f(x)$ and all of its derivatives are known at $x = 0$, and are finite, i.e.

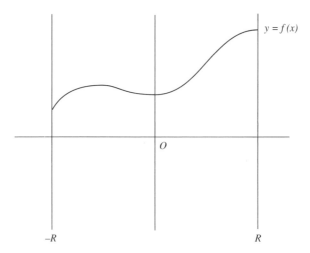

Figure 5.5   The graph of $y = f(x)$ within $|x| < R$

The expansion is valid for $|x| < R$, the radius of convergence, i.e. $-R < x < R$.

It may happen that f and all of its derivatives are defined at some other arbitrary point x. If so, it can similarly be proved that

$$f(x + h) = f(x) + \frac{h}{1!} f'(x) + \frac{h^2}{2!} f''(x) + \cdots + \frac{h^n}{n!} f^{(n)}(x) + \cdots$$

provided that h is small. In other words, we can compute f within a neighbourhood, i.e. a small window, centred upon x.

This expansion (Figure 5.6) is called the **Taylor expansion** or the **Taylor series**.

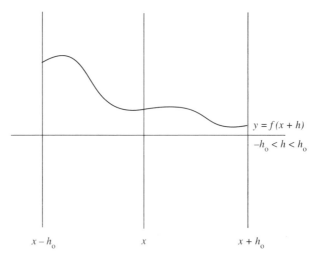

Figure 5.6   Deriving the Taylor expansion

Approximations and increments

Taking the first two terms of the Taylor expansion, we obtain

$$f(x + h) \approx f(x) + hf'(x) \quad \text{i.e.} \quad f(x) + h\frac{df}{dx}$$

Hence, if x is changed by a small amount or **increment** h, then $f(x)$ is changed approximately by an amount $hf'(x)$.

Example

Find an approximation to $\sin 31°$ without resorting to a calculator.

$$31° = \left(\frac{\pi}{6} + \frac{\pi}{180}\right)^c \quad \left(x = \frac{\pi}{6}, h = \frac{\pi}{180}\right)$$

If $f(x) = \sin x$, then $\dfrac{df}{dx} = \cos x$, so that

$$\sin 31° \approx \sin 30° + \cos 30° \times \frac{\pi}{180}$$

i.e.
$$\sin 31° \simeq 0.5 + \frac{\sqrt{3}}{2} \times \frac{\pi}{180}$$
$$\simeq 0.5 + 0.015\,11$$
$$= 0.515\,11$$

To 5 d.p. $\sin 31° = 0.515\,04$, so the approximation is good. ■

As an alternative notation we write $y = f(x)$ and replace h by δx, so $y + \delta y = f(x + \delta x)$. Then

$$\delta y = f(x + h) - f(x) \simeq \delta x \frac{dy}{dx}$$

Expressed in words:

increment in $y \approx$ increment in $x \times$ local value of the derivative

or increment ratio $\approx$ local value of derivative

Be aware that many books use the notation Δx rather than δx in the context of the practical determination of increments or fractional changes, i.e.

$$\Delta y \approx \Delta x \times \frac{dy}{dx}$$

We will therefore use both styles of notation.

Example

If α is the positive root of the cubic equation $px^3 - 3x = 1$, find the approximate change in α when p is increased from 4.0 (for which $\alpha = 1$) to 4.1.

Since $p\alpha^3 - 3\alpha - 1 = 0$, then

$$p = f(\alpha) = \frac{3\alpha + 1}{\alpha^3} = \frac{3}{\alpha^2} + \frac{1}{\alpha^3}$$

$$f'(\alpha) = -\frac{6}{\alpha^3} - \frac{3}{\alpha^4}$$

Now $\Delta p = f'(\alpha)\Delta\alpha$ and $\Delta\alpha = \dfrac{\Delta p}{f'(\alpha)}$

If $\Delta p = 0.1$ when $\alpha = 1$, and $f'(1) = -9$, then $\Delta\alpha \approx -0.011$ and $\alpha \approx 0.989$. In other words, the positive root of the equation $4.1x^3 - 3x - 1 = 0$ is estimated to be 0.989. (To 4 d.p. the root is 0.9891.) ∎

The **fractional change** in a variable x is defined to be $\dfrac{\delta x}{x}$ when δx is the increment and x the local value.

Example

A mass of air is compressed by 1% adiabatically. The pressure p and volume V are connected by the adiabatic law $pV^\gamma = k$, where k is constant. What is the corresponding percentage decrease in volume, given $\gamma = 1.4$?

Differentiating with respect to V, we obtain the equation

$$\frac{dp}{dV} V^\gamma + \gamma p V^{\gamma - 1} = 0 \tag{1}$$

If δp and δV are the increments of p and V, then

$$\delta p \approx \frac{dp}{dV} \delta V$$

Multiplying equation (1) by δV and substituting, we obtain

$$\delta p V^\gamma + \gamma p V^{\gamma - 1} \delta V = 0$$

The fractional increase in pressure $\approx \dfrac{\delta p}{p}$, so dividing by p we have

$$\left(\frac{\delta p}{p} + \gamma \frac{\delta V}{V} \right) V^\gamma = 0$$

Since $V \neq 0$

$$\frac{\delta V}{V} \approx -\frac{1}{\gamma}\frac{\delta p}{p}$$

$$\approx -\frac{1}{1.4} \times 0.01$$

Hence the approximate change in V is a 0.714% decrease. ∎

Logarithmic derivatives

A good way of estimating fractional change is to determine incremental change in the logarithm, i.e.

$$\Delta(\ln x) \approx \Delta x \times \frac{d}{dx}(\ln x) = \frac{\Delta x}{x}$$

Example

A simple pendulum has a periodic time of 2 s. What happens if its length is increased by 1%?

The periodic time T is given by $T = 2\pi\sqrt{\frac{l}{g}}$, where l is the length and g is the acceleration due to gravity, a constant.

Taking logarithms we obtain

$$\ln T = \ln 2\pi + \frac{1}{2}\ln l - \frac{1}{2}\ln g$$

The fractional change in a constant must be zero, i.e. $\Delta g = 0$, so

$$\frac{\Delta T}{T} = \frac{1}{2}\frac{\Delta l}{l}$$

hence $\quad \dfrac{\Delta T}{T} = 0.0005, \qquad \Delta T = 0.01 \text{ s}$

The periodic time is increased by 0.01 s, change of 0.005%. ∎

Numerical rates of change

The derivative of a function represents the rate of change of one variable with respect to another, and this value in itself may be of more significance than the relative incremental sizes of the variables. The following examples illustrate the point.

Examples

1. A balloon of maximum diameter 50 cm is being inflated at $800\,\text{cm}^3\,\text{s}^{-1}$. How long does it take to inflate it fully? At what rate is its diameter changing when it has a radius of 25 cm

$$\text{Volume } V = \frac{4\pi}{3}(25)^3 = 65\,450\,\text{cm}^3; \text{ inflation time} \approx 81.81\,\text{s}.$$

Given $\qquad V = \frac{4}{3}\pi r^3$

then $\qquad \dfrac{dV}{dt} = \dfrac{dV}{dr} \times \dfrac{dr}{dt} = 4\pi r^2 \dfrac{dr}{dt}$

Therefore

$$800 = 4\pi\,(12.5)^2 \times \frac{dr}{dt}$$

and so $\quad \dfrac{dr}{dt} = 0.407\ \text{cm s}^{-1}$

As the diameter $D = 2r$, then $\dfrac{dD}{dt} = 2\dfrac{dr}{dt} = 0.814\ \text{cm s}^{-1}$.

2. An electrical capacitor is initially charged to a value Q_0. After t seconds the residual charge Q (following leakage as electric current through a resistor) is given by

$$Q = Q_0 \exp(-t/CR)$$

$\exp(-t/CR)$ denotes e to the power $-t/CR$.

   If the resistor $R = 10\ M\Omega$ (10 megohm $= 10 \times 10^6$ ohms), the capacitor $C = 5\ \mu F$ (5 microfarads $= 5 \times 10^{-6}$ farads) and t is measured in seconds, calculate the proportionate residual charge after a one-minute time interval. Also calculate the current leakage at that time, in amps, if $Q_0 = 0.2\,C$ (1 amp $= 1$ coulomb s^{-1}).

Note that $\quad CR = 50 \quad$ and $\quad t = 60$

Therefore $\quad \dfrac{Q}{Q_0} = \exp\left(-\dfrac{6}{5}\right) = 0.3012$

Now the current

$$I = \frac{dQ}{dt}$$

So $\qquad I = -RCQ_0 \exp\left(-\frac{t}{CR}\right)$

$$= 50 \times 0.2 \times 0.3012 = 3.012 \text{ amps} \qquad \blacksquare$$

Exercise 5.4

1 Determine the Taylor expansion for $f(x) = \frac{1}{x}$ in the form

$$\frac{1}{x} = \frac{1}{a} + hf'(a) + \frac{h^2}{2!}f''(a) + \frac{h^3}{3!}f'''(a) + \dots$$

where $x = a + h$.

2 By substituting $h = x - 11$ or otherwise, prove that

(a) $\qquad \frac{1}{12 - x} = 1 + (x - 11) + (x - 11)^2 + (x - 11)^3 + \dots$

(b) $\qquad \frac{1}{(12 - x)^2} = 1 + 2(x - 11) + 3(x - 11)^2 + 4(x - 11)^3 + \dots$

when h is small enough, i.e. x close to 11.

(c) For what values of x do the series converge?

3 The general form of L'Hôpital's rule states that

$$\lim_{x \to a} \frac{f(x)}{g(x)} = \frac{f(a)}{g(a)}$$

unless $f(a) = g(a) = 0$.

If $f(a) = g(a) = 0$ then

$$\lim_{x \to a} \frac{f(x)}{g(x)} = \frac{f'(a)}{g'(a)}$$

unless $f'(a) = g'(a) = 0$

If $f'(a) = g'(a) = 0$ then

$$\lim_{x \to a} \frac{f(x)}{g(x)} = \frac{f''(a)}{g''(a)}$$

unless $f''(a) = g''(a) = 0$, etc.
Determine the following limits:

(a) $\displaystyle\lim_{x \to a} \frac{\exp(x - a) - 1}{x - a}$

(b) $\displaystyle\lim_{\omega \to n} \frac{\omega E(\cos \omega t - \cos nt)}{L(n^2 - \omega^2)}$

(All parameters except ω may be treated as constants.)

(c) $\displaystyle\lim_{x \to 1} \frac{x - 1}{\sin \pi x - 1}$

4 Use increments to determine the following to 4 s.f.:

(a) $\cos 61°$ (b) $\cos 59°$

(c) $\sqrt{101}$ (d) $(65)^{1/3}$

(e) $\ln(e^5 + 1)$ (f) $\ln(101) - \ln(100)$

5 The cubic equation

$$5x^3 - px - 1 = 0$$

has a positive root $x = 1$ when $p = 4$. If p is increased by 0.5% what is the approximate percentage change in the root?

6 We can determine approximate fractional changes of two or more variables using logarithmic differentiation, e.g. change in the volume of cylinder given small changes in the linear dimensions:

i.e. $V = \pi r^2 l$

        $\ln V = \ln \pi + \ln r^2 + \ln l$

So $\dfrac{\Delta V}{V} = \Delta(\ln \pi) + \Delta(\ln r^2) + \Delta(\ln l)$

          $= 2\dfrac{\Delta r}{r} + \dfrac{\Delta l}{l}, \qquad \text{since } \Delta(\ln \pi) = 0$

(a)   Find the approximate percentage change in the volume of a cylinder with:
       (i) a 2% increase in diameter (i.e. radius) and a 1% increase in length
       (ii) a 3% decrease in diameter and a 2% increase in length

(b)   Find the approximate percentage change in the volume of a cone with:
       (i) a 2% increase in length and a 2% increase in diameter
       (ii) a 2% increase in length and a 2% decrease in diameter

(c)   What is the approximate percentage change in the volume of a cuboid, sides
       a, b, c, with a 1% increase in the length of each side?

(d)   Find the percentage change in the volume of a sphere whose surface area is
       increased by 2%.

7 A compound pendulum has periodic time

$$T = 2\pi\sqrt{\frac{l^2 + 3h^2}{3gh}}$$

l is the length of the pendulum and h the distance from the pivot to the centre of gravity.

Find the percentage change in T when $h = \dfrac{2l}{3}$ and l is increased by 2%; h does not change.

8 The efficiency of an ideal gas engine working with a compression ratio r may be expressed as

$$E = 1 - \left(\frac{1}{r}\right)^{0.4}$$

If $r = 5$ find the percentage decrease in efficiency corresponding to a 5% decrease in r.

9* A vessel in the shape of an inverted cone of height 13 cm and diameter 10 cm is being filled with water at a rate of $10\,\text{cm}^3\,\text{s}^{-1}$.

(a)   How long does it take to fill it up?

(b)   When is the water at a depth of 10 cm?

(c)   At what rate is its depth increasing when the depth is 10 cm?

Note that the depth h is directly proportional to the radius r.

10 A hanging chain takes the shape of a catenary, i.e.

$$y = c \cosh\left(\frac{x}{c}\right)$$

If $c = 2$ determine the slope of the chain when $x = 3$.

11 The current I in an inductively coiled electric circuit builds up to a peak value according to the equation

$$I = \frac{E}{R}(1 - e^{-Rt/L})$$

If $E = 40$ V, $R = 20\,\Omega$ and $L = 2$ H, write down the time when $I = \frac{E}{2R}$, i.e. half its maximum value. Determine its rate of change at that time.

12 The differential equation model of limited population growth, given a maximum sustainable population N takes the form

$$\frac{dP}{dt} = r(N - P)P$$

i.e. the growth in population is proportional to both the population present and the difference between the actual population and its maximum value.

The differential equation integrates to

$$P(t) = \frac{NP_0}{P_0 + (N - P_0)e^{-rN(t-t_0)}}$$

at time t given a start population P_0 when $t = t_0$

(a) Show that $\frac{dP}{dt}$ is greatest when $P = \frac{N}{2}$

(b) What is the ultimate population as $t \to \infty$?

(c) Sketch versus t the graphs of

(i) $P(t)$ (ii) $\frac{dP}{dt}$

(d) Show that $P(t)$ increases to the value $N/2$ after a time

$$\tau = t_0 - \left(\frac{1}{rN}\right)\ln\left(\frac{P_0}{N - P_0}\right)$$

SUMMARY

- **Product rule**: if $y = uv$ then

$$\frac{dy}{dx} = v\frac{du}{dx} + u\frac{dv}{dx}$$

- **Quotient rule**: if $y = \dfrac{u}{v}$ then

$$\frac{dy}{dx} = \frac{v\dfrac{du}{dx} - u\dfrac{dv}{dx}}{v^2}$$

- **Chain rule**: if y is a function of u and u is a function of x, then

$$\frac{dy}{dx} = \frac{dy}{du} \times \frac{du}{dx}$$

- **Logarithmic differentiation**: if y is a function of x then

$$\frac{d}{dx}(\ln y) = \frac{1}{y} \times \frac{dy}{dx}$$

- **Leibniz' theorem**

$$\frac{d^n}{dx^n}(uv) = \sum_{r=0}^{n} \binom{n}{r} u^{(r)} v^{(n-r)}$$

- **Implicit differentiation**: take $x^2 + y^2 = a^2$ as an example; implicit differentiation gives

$$2x + 2y\frac{dy}{dx} = 0$$

- **Inverse differentiation**

$$\frac{dx}{dy} = \frac{1}{\dfrac{dy}{dx}}$$

- **Maclaurin's series** for a function is

$$f(x) = f(0) + xf'(0) + \frac{x^2}{2!} f''(0) + \cdots + \frac{x^n}{n!} f^{(n)}(0) + \ldots$$

- **Radius of convergence** for a Maclaurin series in x: the radius of convergence R is such that the series converges when $|x| < R$

- **Taylor expansion**: defined for a function $f(x)$ as

$$f(x + h) = f(x) + hf'(x) + \frac{h^2}{2!} f''(x) + \cdots + \frac{h^n}{n!} f^{(n)}(x) + \ldots$$

- **L'Hôpital's rule**: if $f(0) = g(0) = 0$ then

$$\lim_{x \to 0} \frac{f(x)}{g(x)} = \lim_{x \to 0} \frac{f'(x)}{g'(x)}$$

- **Approximate change**: the approximate change in $f(x)$ when x changes by δx is $\left(\dfrac{df}{dx}\right)\delta x$.

Answers

Exercise 5.1

1 (a) $3x(x+2)$

(b) $\dfrac{2x+1}{2x^{3/2}}$

(c) $-\dfrac{(10-26x+15x^2)}{3x^{4/3}}$

(d) $3x^2+12x+11$

(e) $ad+bc+2bdx$

(f) $\dfrac{-(9x^2-1)(189x^5+7x^3-8)}{x^5}$

2 (a) $e^x\left(\ln x+\dfrac{1}{x}\right)$

(b) $\tan x+x\sec^2 x$

(c) $\dfrac{1-x^2}{(1+x^2)^2}$

(d) $\dfrac{-(4x^3-3x^2+2)}{(x^3-1)^2}$

(e) $\dfrac{4x(1+\sin x)-2x^2\cos x}{(1+\sin x)^2}$

(f) $-\dfrac{x\sin x+\cos x}{x^2}$

(g) $\dfrac{1}{x(3x+1)}-\dfrac{3\ln x}{(3x+1)^2}$

(h) $\dfrac{4}{(e^x+e^{-x})^2}=\dfrac{1}{\cosh^2 x}$

(i) $\dfrac{-10}{x(\ln x)^3}$

(j) $-\dfrac{(x^2+1)}{\tan^2 x}+\dfrac{2x}{\tan x}-x^2-1$

3 Maximum at $\left(1,\dfrac{1}{e}\right)$

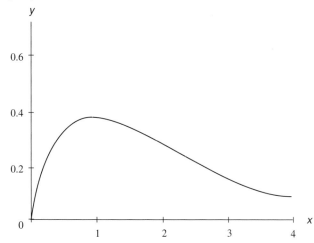

4 (b) $\dfrac{dy}{dx}=\dfrac{ad-bc}{(cx+d)^2}$; sign always the same, hence never zero

5 (a) $\dfrac{dy}{dx}\geq 1$

6 (a) $\dfrac{1}{x+1}$ (b) $2\cos 2x$

(c) $3e^{3x}$ (d) $2\sec^2 2x$

(e) $-4\sin(4x-1)$ (f) $\dfrac{2x}{x^2+1}$

(g) $\dfrac{2}{x}$ (h) $2xe^{x^2}$

(i) $(4x-3)\cos(2x^2-3x+1)$ (j) $an(ax+b)^{n-1}$

(k) ke^{kx} (l) $c\sec^2(cx+d)$

(m) $-\dfrac{(x-\mu)}{\sigma^2}\exp\left(-\dfrac{1}{2}\dfrac{(x-\mu)^2}{\sigma^2}\right)$ (n) $\dfrac{x(3x+1)}{x^3+x+1}$

(o) $\dfrac{2x}{3(x^2+1)}, \quad \ln(x^2+1)^{1/3}=\dfrac{1}{3}\ln(x^2+1)$

7 (a) $4\sin 2x\cos 2x$ (b) $\dfrac{3}{\tan 3x}$

(c) $\dfrac{\sec^2(\sqrt{x})}{2\sqrt{x}}$ (d) $(8x+1)(x-1)^2(x+2)^4$

(e) $\dfrac{3x^2+4x-1}{(x-1)(x+1)(x+2)}$ (f) $\dfrac{2}{(\cos x-\sin x)^2}$

(g) $-\dfrac{1}{x(\ln x)^2}$ (h) $ae^{ax+b}\left(\dfrac{\sin(ax+b)-\cos(ax+b)}{\sin^2(ax+b)}\right)$

(i) $-\dfrac{\sin\sqrt{x}\cos\sqrt{x}}{\sqrt{x}}$ (j) $-\dfrac{\tan\sqrt{x}}{\sqrt{x}}$

(k) $(ax+b)^{n-1}(cx+d)^{m-1}\{acx(m+n)+adn+bcm\}$

(l) $\ln 2(2^{2x+2}),\ 16\ \ln 2$

8 (a) $\dfrac{1}{x}$ (b) $2\cos x-x\sin x$ (c) $x(x^2+6x+6)e^x$

(d) $-\dfrac{1}{x^2}$

$\dfrac{d^3y}{dx^3}=uv'''+3u'v''+3u''v'+u'''v$

Coefficients are those of Pascal's triangle. In general,

$\dfrac{d^n}{dx^n}(uv)=\sum_{r=0}^{n}\binom{n}{r}u^{(r)}v^{(n-r)}$

This result is known as **Leibniz's theorem**.

9    $\dfrac{m}{x+a} + \dfrac{n}{x+b} + \dfrac{p}{x+c}$ $\dfrac{2 - x^2}{x(x+1)(x+2)}$

10    $(x+1)$, $(x^2 + 4x + 2)$, $(x^3 + 9x^2 + 18x + 6)$

11   (a)   Domain excludes $\sin x < -\dfrac{1}{4}$

      (d)

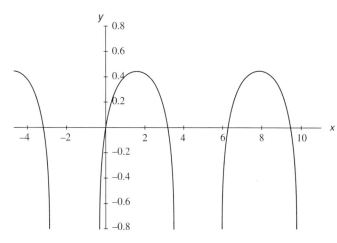

12   (a) $-x^{-x}(1 + \ln x)$ (b) $-2^{-5x}\{5 \ln 2 \tan(1 + x^2) - 2x \sec^2(1 + x^2)\}$

      (c) $\dfrac{2(ex)^x \ln x}{(x^x + e^x)^2}$

13   Domain $x > \dfrac{3}{2}$

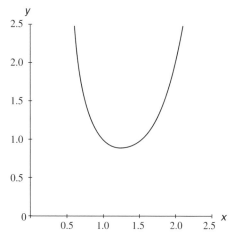

Minimum $\approx (1.24, 0.91)$

14 (a) $\dfrac{1}{(1-x)^{3/2}(1+x)^{1/2}}$ (b) $\dfrac{3(1-5x)}{2(1-x)^4(1-3x)^{1/2}}$

15 (a) (ii) $\dfrac{d^2}{dx^2}(xy) = 2\dfrac{dy}{dx} + x\dfrac{d^2y}{dx^2}$

16 (a) $\dot{x} = -\dfrac{(t^2 + 2t - 1)}{(t^2 + 1)^2}$

    (i) $t = 0, 1\,\text{s}$
    (ii) $t = \sqrt{2} - 1 = 0.414\,\text{s}$

(b) Particle moves to the right to $x = \dfrac{1}{2}(\sqrt{2} - 1) \approx 0.207\,\text{m}$ when $t = 0.414\,\text{s}$ before reversing and decelerating to rest at $x = -1$.

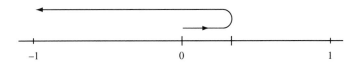

(c) $\ddot{x} = \dfrac{2(t^3 + 3t^2 - 3t - 1)}{(t^2 + 1)^3}$

17 (a) (i) $nA\cos(nt + \phi)$
    (ii) $-n^2 A\sin(nt + \phi)$

Exercise 5.2

1 (a) $\dfrac{2xy^2}{1 - 2x^2 y}$ (b) $-\dfrac{(2x + y)}{(x + 3y^2)}$

(c) $\dfrac{x^2 - 1}{1 - y^2}$ (d) $-\sqrt{\dfrac{y}{x}}$

(e) $-\dfrac{py}{qx}$ (f) $2xe^{-y}$

(g) $\dfrac{-1}{x\sin y}$ (h) $\dfrac{y^2\sin x + 2x\cos y}{x^2\sin y + 2y\cos x}$

(i) $-\dfrac{y}{x}\left\{(y - 1)\cos^2\left(\dfrac{x}{y}\right) - \sin^2\left(\dfrac{x}{y}\right)\right\}$

(j) $\dfrac{y\cos\left(\dfrac{x}{y}\right)}{x\cos\left(\dfrac{x}{y}\right) - 1}$

2 1

4 (a) $2y - x + 1 = 0$ (b) $y + 2x - 2 = 0$

6 (a) $\sin x + x \cos x$ (b) $2 \cos x + x \sin x$

(c) $\dfrac{x \sin x - 2 \cos x}{(\sin x + x \cos x)^3}$

7 $-V\dfrac{dP}{dV} = \gamma p$

8 (a) $-\dfrac{y}{x \ln x}$ $(y = 0)$ (b) $\dfrac{2e^{2x} - yx^{y-1}}{x^y \ln x} = \dfrac{2 - \dfrac{y}{x}}{\ln x}$

(c) $\dfrac{x^x(\ln x + 1)}{y^y(\ln y + 1)}$ (d) $\dfrac{y}{x} - \dfrac{2 \ln 2 y^2 2^{x^2} 5^{-x/y}}{\ln 5}$

9 (a) $\dfrac{1}{1 + x^2}$ (b) $\tan^{-1} x + \dfrac{x}{1 + x^2}$

(c) $-\dfrac{1}{\sqrt{x}(x + 1)}$ (d) $-\dfrac{1}{2\sqrt{x} \cos \sqrt{x}(2 \cos^2 \sqrt{x} - 1)^{1/2}}$

(e) $\dfrac{1}{\sqrt{x^2 + 1}}$ (f) $\dfrac{1}{\sqrt{x^2 - 1}}$

(g) $-\dfrac{1}{x^2 - 1}$ (h) $1 + \dfrac{x \sinh^{-1} x}{\sqrt{1 + x^2}}$ (i) $\dfrac{x}{\{(x^2 - 2)(x^2 - 1)\}^{1/2}}$

13 $\left(\dfrac{1}{\sqrt{2}}, -\dfrac{1}{2}\ln 2\right)$

14 $\left(x - \dfrac{5}{2}\right)^2 + \left(y - \dfrac{5}{2}\right)^2 = \dfrac{9}{2}$

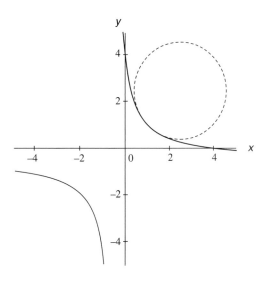

15 (b) $a = \dfrac{1}{2}, b = 1, c = 1$

17 (a) $\dfrac{dy}{dx} = K$, constant, $\Rightarrow y = Kx$, integrating, a straight line

Take $a = 1, (x - Ky)^2 \equiv x^2 + 2hxy + by^2$

so $h = -K, b = K^2$

(b) $\dfrac{d^2y}{dx^2} = 0$, so $y = mx + d$, integrating twice

The conic equation is just $(y - mx - d)^2 = 0$, a repeated straight line

Exercise 5.3

1 (a) $1 - \dfrac{x^2}{2!} + \dfrac{x^4}{4!}$

(b) $1 + x + \dfrac{x^2}{2!} + \dfrac{x^3}{3!} + \dfrac{x^4}{4!} + \dfrac{x^5}{5!}$

(c) $1 + \dfrac{x^2}{2!} + \dfrac{x^4}{4!}$

(d) $x + \dfrac{x^3}{3} + \dfrac{2x^5}{15}$

(e) $\ln 3 + \dfrac{2x}{3} - \dfrac{2x^2}{9} + \dfrac{8x^3}{81} - \dfrac{4x^4}{81} + \dfrac{32x^5}{1215}$

(f) $x - \dfrac{x^3}{3} + \dfrac{x^5}{5}$

(h) $x + \dfrac{x^3}{3} + \dfrac{x^5}{5}$

3 (a) $1 - 5x + \dfrac{33x^2}{2} - \dfrac{277}{6}x^3 + \ldots$

(b)   (i) $a^p + pa^{p-1}x + \dfrac{p(p-1)}{2!}a^{p-2}x^2 + \dfrac{p(p-1)(p-2)}{3!}a^{p-3}x^3 + \ldots$ (p real)

(ii) $p = 0$, the series is 1

$p = 1$, the series is $a + x$

$p = 2$, the series is $a^2 + 2ax + x^2$

4   The function $\ln x$ does not possess a Maclaurin expansion because $\ln x$ and all its derivatives are infinite when $x = 0$.

5 (a) $y(0) = 0, y'(0) = 1, y''(0) = 0$

(b) $y^{(3)}(0) = 5, y^{(4)}(0) = 0, y^{(5)}(0) = 101$

(c) $y(x) = x + \dfrac{5x^3}{6} + \dfrac{101x^5}{120} + O(x^7)$

6 (c) $y(x) = 1 + x + \dfrac{x^2}{2} + \dfrac{x^3}{3} + \dfrac{5x^4}{24} + \dfrac{x^5}{6} + O(x^6)$

7 (a) $1 + \dfrac{x}{2} - \dfrac{x^2}{8} + \dfrac{x^3}{16} - \dfrac{5x^4}{128} + \dfrac{7x^5}{256}$ 　　　(b) 　(i) -1

(ii) $\dfrac{9}{8}$

8 (a) 1 　　　　　　　　　　　　　　(b) $\dfrac{1}{4}$

(c) $\dfrac{1}{120}$ 　　　　　　　　　　(d) $\dfrac{1}{16}$

Exercise 5.4

1 $\dfrac{1}{x} = \dfrac{1}{a+h} = \dfrac{1}{a} - \dfrac{h}{a^2} + \dfrac{h}{a^3} - \cdots$

2 (c) $10 < x < 12$

3 (a) 1 　　　　(b) $\dfrac{Et \sin nt}{2L}$ 　　　　(c) $-\dfrac{1}{\pi}$

4 (a) 0.4848 　　　　　　　　　(b) 0.5150

(c) 10.05 　　　　　　　　　　(d) 4.021

(e) $5 + \dfrac{1}{e^5} = 5.007$ 　　　　(f) 0.0100

5 0.18% increase

6 (a) (i) 5% increase 　　(ii) 4% decrease

(b) (i) 6% increase 　　(ii) 2% decrease

(c) 3% increase

(d) 3% increase

7 0.85% increase

8   2.29%

9   (a)   34.03 s   (b)   15.49 s   (c)   0.2151 cm s^{-1}

10   sinh 1.5 = 2.129

11 $t = \dfrac{L}{R} \ln 2 = 0.0693$ ($L = 2, R = 20$), 1 amp s^{-1}

12   (b) N

   (c)   (i)

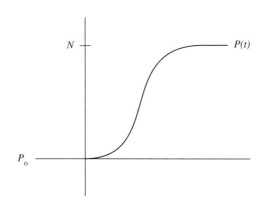

   (ii)

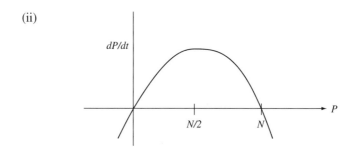

6 INTEGRATION

Introduction

Integration is a versatile tool in mathematics. Interpretable as the reverse of differentiation, it can be applied to finding plane areas, volumes of irregular solids, the location of centres of gravity and the mean value of a function over a given interval. Very few integrations can be carried out by using a table of simple integrals, so many techniques have been developed to extend the range of functions which can be integrated. These techniques often rely on transforming the integral to one or more simpler integrals. If there is no obvious analytical approach, the value of a definite integral may be obtained by a numerical method.

OBJECTIVES

After working through this chapter you should be able to

- Understand and apply the method of substitution to find both definite and indefinite integrals

- Recognise that the integrals $\int_0^a f(a - x)dx$ and $\int_0^a f(x)dx$ are equal

- Recognise that the integrals $\int_a^b f(a + b - x)dx$ and $\int_a^b f(x)dx$ are equal

- Use the method of substitution when the integrand involves powers of trigonometric functions

- Use the t substitution for rational trigonometric integrands

- Find integrals where the integrand is of the form $\dfrac{f'(x)}{f(x)}$

- Use the method of integration by parts

- Understand the derivation of reduction formulae

- Find integrals when the integrand is a rational function by using partial fractions

- Use and apply the trapezium rule and Simpson's rule

- Estimate the errors involved in obtaining a trapezium approximation and a Simpson approximation.

6.1 INTEGRATION BY SUBSTITUTION

In Chapter 12 of *Foundation Mathematics* the integral was introduced as the limit of a sum and we then saw how this process could be linked to the reversal of differentiation, i.e. if for some function $F(x)$, $\dfrac{dF}{dx} = f(x)$ then $\displaystyle\int f(x)dx = F(x) + C$

In this chapter we examine some further methods of integration. We start with a very powerful method of finding integrals analytically. The **method of substitution** is, in principle, straightforward. The independent variable is changed from say x to a new variable, say u, by means of a simple connecting formula. The resulting new integral is hopefully simple to evaluate.

If the original integral is indefinite then the substitution process produces a function of the new variable u. Applying the substitution formula will give a function of the original variable x, which is the solution to the original problem. If the original integral is definite then the substitution method will produce a new definite integral which has the same numerical value as the original one.

Figure 6.1 shows the approximation of the area under a curve as the sum of the areas of vertical strips. As the width of the strips is allowed to decrease to zero, in the limit, the area is given **exactly** by a definite integral:

$$\int f(x)dx = \lim_{\Delta x \to 0} \sum f(x_k).\Delta x$$

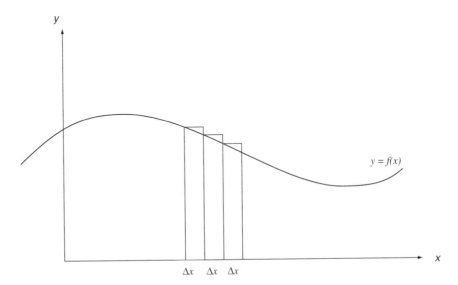

Figure 6.1 Area under a curve

If we change from x to u then

$$\Delta x \approx \frac{dx}{du}\Delta u$$

Hence

$$f(x).\Delta x \approx f(x(u)).\frac{dx}{du}.\Delta u = g(u).\Delta u \text{ (say)}$$

so that $\int f(x)dx = \int g(u)du$

where $g(u) = f(x).\dfrac{dx}{du} = f(x) \div \dfrac{du}{dx}$

Figure 6.2 shows how a typical strip appears in the two systems.
We now give some examples of the method of substitution in action.

Example   Integrate the following functions using the substitution given:

(a) $(x+3)^4$, $u = x+3$ (b) $\dfrac{1}{(3x-2)^2}$, $u = 3x-2$

(c) $(5x-3)^{1/3}$, $u^3 = 5x-3$ (d) e^{ax}, $u = ax$

(e) $\dfrac{1}{(a^2-x^2)^{1/2}}$, $x = a\sin\theta$ given than $|x| < a$

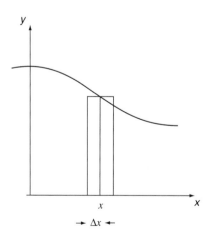
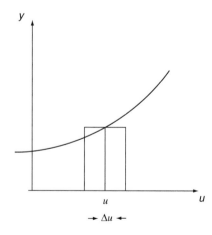

Figure 6.2 Comparing areas

Solution

(a) Differentiate the given substitution formula with respect to x to obtain $\dfrac{du}{dx} = 1$, from which it follows that $\dfrac{dx}{du} = 1$. Then

$$\int (x+3)^4 \, dx = \int u^4 \frac{du}{dx} \, dx = \int u^4 \, du = \frac{u^5}{5} + C = \frac{(x+3)^5}{5} + C$$

(Had we expanded and integrated the result term by term we would have obtained a slightly different result, in effect $\dfrac{(x+3)^5}{5} - \dfrac{81}{5} + C'$. Since the discrepancy is a number, it is allowed for by the arbitrary constants C and C'.)

(b) Differentiate $u = 3x - 2$ to obtain $\dfrac{du}{dx} = 3$ so that $\dfrac{dx}{du} = \dfrac{1}{3}$. In effect this means we can replace dx by $\dfrac{1}{3} du$ in the integral. We write $dx = \dfrac{1}{3} du$; this is not a valid mathematical equation, merely a statement of what we are going to do. Then

$$\int \frac{1}{(3x-2)^2} \, dx = \int \frac{1}{u^2} \times \frac{1}{3} \, du = -\frac{1}{3u} + C = -\frac{1}{3(3x-2)} + C$$

(c) Differentiating $u^3 = 5x - 3$ gives $3u^2 \dfrac{du}{dx} = 5$, i.e. '$dx = \dfrac{3u^2}{5} du$'. Note that $u = (5x-3)^{1/3}$. Then

$$\int \frac{1}{(5x-3)^{1/3}} \, dx = \int \frac{1}{u} \times \frac{3u^2}{5} \, du = \int \frac{3u}{5} \, du = \frac{3u^2}{10} + C = \frac{3(5x-3)^{2/3}}{10} + C$$

(d) If $u = ax$ then $\dfrac{du}{dx} = a$, so '$dx = \dfrac{1}{a} du$'. Then

$$\int e^{ax} \, dx = \int e^u \times \frac{1}{a} \, du = \frac{1}{a} e^u + C = \frac{1}{a} e^{ax} + C$$

(e) If $x = a \sin \theta$ then $\dfrac{dx}{d\theta} = a \cos \theta$, so '$dx = a \cos \theta \, d\theta$'. Furthermore,

$$(a^2 - x^2)^{1/2} = (a^2 - a^2 \sin^2 \theta)^{1/2} = (a^2 \cos^2 \theta)^{1/2} = a \cos \theta. \text{ Then}$$

$$\int \frac{1}{(a^2 - x^2)^{1/2}} \, dx = \int \frac{a \cos \theta}{a \cos \theta} \, d\theta = \int 1 \times d\theta = \theta + C = \sin^{-1}\left(\frac{x}{a}\right) + C \qquad \blacksquare$$

This example gave the substitution to be used. You will learn by experience which substitution to choose in a given problem. A summary at the end of this section gives some guidelines to the choice of substitution. Sometimes a problem may require more than one substitution for its solution, often combined with the use of a trigonometric identity. However, no matter how clever we try to be, there are some integrals which defy all attempts to evaluate them in terms of our standard functions: an example is $\int e^{x^2} dx$.

It is easy to lose one's way in a long chain of substitutions and identities, and it is wise if time permits to check your 'answer' by differentiating it to verify that you obtain the original function.

The definite integral

In applying the method of substitution to definite integrals, we have one extra task: to obtain new limits of integration. The definite integral has a fixed value and the transformed integral must have the same value. In this case, having obtained the value, there is no need to refer back by using the substitution formula. (It would be possible to ignore the limits temporarily, use the method of substitution on the corresponding indefinite integral and apply the original limits to the resulting function; only rarely is this the better course of action.) Figure 6.3 shows the general idea.

When $x = a, u = c$ and when $x = b, u = d$. Then

$$\int_a^b f(x)dx = \int_c^d g(u)du$$

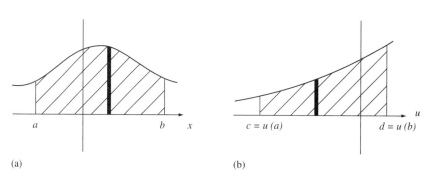

(a) (b)

Figure 6.3 Definite integration by substitution: the shaded areas are equal

Example

Evaluate the following definite integrals using the substitution shown.

(a) $\int_0^1 \frac{1}{2x+1}dx$, $\quad u = 2x + 1$ (b) $\int_5^{12} \frac{1}{\sqrt{4+x}}dx$, $\quad u^2 = 4 + x$

(c) $\displaystyle\int_0^{\pi/2} \sin^3\theta\cos d\theta,$ $u = \sin\theta$ (d) $\displaystyle\int_{\sqrt{5}}^{2\sqrt{6}} \frac{3s}{(s^2+3)^{1/3}}\,ds,$ $u^3 = s^2 + 3$

(e) $\displaystyle\int_0^{\ln 2} (e^x + 1)^2 e^x\,dx,$ $u = e^x + 1$

Solution

(a) If $u = 2x + 1$ then '$du = 2\,dx$' so that '$dx = \dfrac{1}{2}du$'

When $x = 0$, $u = 1$ and when $x = 1$, $u = 3$

Hence $\displaystyle\int_0^1 \frac{1}{2x+1}\,dx = \frac{1}{2}\int_1^3 \frac{1}{u}\,du = \frac{1}{2}[\ln u]_1^3 = \frac{1}{2}\ln 3$

(b) If $u^2 = 4 + x$ then '$2u\,du = dx$'

When $x = 5$, $u^2 = 9$ so that $u = 3$ and when $x = 12$, $u^2 = 16$ so that $u = 4$

Hence $\displaystyle\int_5^{12} \frac{1}{\sqrt{4+x}}\,dx = \int_3^4 \frac{2u}{u}\,du = [2u]_3^4 = 2$

(c) If $u = \sin\theta$ then '$du = \cos\theta\,d\theta$'

When $\theta = 0$, $u = \sin 0 = 0$ and when $\theta = \dfrac{\pi}{2}$, $u = \sin\dfrac{\pi}{2} = 1$

Hence $\displaystyle\int_0^{\pi/2} \sin^3\theta\cos\theta\,d\theta = \int_0^1 u^3\,du = \left[\frac{1}{4}u^4\right] = \frac{1}{4}$

(d) If $u^3 = s^2 + 3$ then '$3u^2\,du = 2s\,ds$'

When $s = \sqrt{5}$, $u^3 = 8$ so that $u = 2$ and when $s = 2\sqrt{6}$, $u^3 = 27$ so that $u = 3$

Hence $\displaystyle\int_{\sqrt{5}}^{2\sqrt{6}} \frac{3s}{(s^2+3)^{1/3}}\,ds = \int_2^3 \frac{3}{2}\frac{3u^2}{u}\,du = \left[\frac{9}{4}u^2\right]_2^2 = \frac{45}{4}$

(e) If $u = e^x + 1$ then '$du = e^x\,dx$'

When $x = 0$, $u = 2$ and when $x = \ln 2$, $u = 3$

Hence $\displaystyle\int_0^{\ln 2} (e^x + 1)^2 e^x\,dx = \int_2^3 u^2\,du = \left[\frac{1}{3}u^3\right]_2^3 = \frac{19}{3}$ ∎

Properties of the definite integral

If the function $F(x)$ is an indefinite integral of the function $f(x)$ then we know that

$F'(x) = f(x)$ and $F(x) = \displaystyle\int f(x)dx$

Furthermore, $\displaystyle\int_a^b f(x)dx = F(b) - F(a)$ represents the area enclosed by the lines $x = a$ and $x = b$, the curve $y = f(x)$ and the x-axis. So if a and b are interchanged, we reverse the sign of the definite integral, i.e.

$$\int_b^a f(x)dx = F(a) - F(b) = -\int_a^b f(x)dx$$

If we consider a reflection in the y-axis of the curve $y = f(x)$ between $x = 0$ and $x = a$ then by setting $u = a - x$ we obtain

$$\int_0^a f(a - x)dx = -\int_a^0 f(u)du = -\int_a^0 f(x)dx$$

Remember that we can rename the variable of integration if we so wish. But

$$\int_a^0 f(x)dx = -\int_0^a f(x)dx$$

so $$\int_0^a f(a - x)dx = \int_0^a f(x)dx$$

Figure 6.4 illustrates the principle of reflection.

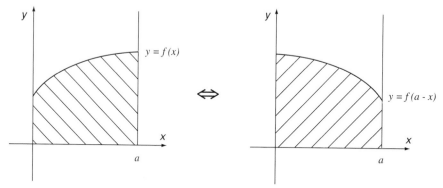

Figure 6.4 The principle of reflection confirms '0' as lower $\int$ limit

In general, over the interval $[a, b]$

$$\int_a^b f(x)dx = \int_a^b f(a + b - x)dx$$

You can prove this by using the substitution $u = a + b - x$

Examples   1. Evaluate

(a) $\displaystyle\int_{1}^{6} \frac{1}{\sqrt{10-x}}\,dx$ (b) $\displaystyle\int_{0}^{\pi/2} \sin^3 x\,dx$

Solution

(a) If $u^2 = 10 - x$ then '$2u\,du = -dx$'. When $x = 1$, $u = 3$ and when $x = 6$, $u = 2$. On substituting we obtain

$$\int_{1}^{6} \frac{1}{\sqrt{10-x}}\,dx = \int_{3}^{2} -\frac{2u}{u}\,du = \int_{2}^{3} 2\,du = [2u]_{2}^{3} = 2$$

(b) Put $u = \cos x$. Then '$du = -\sin x\,dx$'. Also, $\sin^2 x = 1 - u^2$. When $x = 0$, $u = 1$ and when $x = \dfrac{\pi}{2}$, $u = 0$, hence

$$\int_{0}^{\pi/2} \sin^3 x\,dx = \int_{1}^{0} (1 - u^2) \times (-1)du = \int_{0}^{1} (1 - u^2)du = \left[u - \frac{1}{3}u^2\right]_{0}^{1} = \frac{2}{3}$$

2. Without evaluating the integrals directly, prove that

$$\int_{0}^{\pi/2} \sin^2 x\,dx = \int_{0}^{\pi/2} \cos^2 x\,dx = \frac{\pi}{4}$$

Since $\cos x = \sin\left(\dfrac{\pi}{2} - x\right)$ then $\cos^2 x = \sin^2\left(\dfrac{\pi}{2} - x\right)$. But $\displaystyle\int_{0}^{a} f(a - x)dx = \int_{0}^{a} f(x)dx$,

so $$\int_{0}^{\pi/2} f\left(\frac{\pi}{2} - x\right)dx = \int_{0}^{\pi/2} f(x)dx$$

and $$\int_{0}^{\pi/2} \cos^2 x\,dx = \int_{0}^{\pi/2} \sin^2\left(\frac{\pi}{2} - x\right)dx = \int_{0}^{\pi/2} \sin^2 x\,dx$$

Adding the integrals gives

$$\int_{0}^{\pi/2} (\cos^2 x + \sin^2 x)dx = \int_{0}^{\pi/2} 1\,dx = \frac{\pi}{2}$$

Therefore

$$\int_{0}^{\pi/2} \sin^2 x\,dx = \int_{0}^{\pi/2} \cos^2 x\,dx = \frac{\pi}{4} \qquad\blacksquare$$

Special substitutions

Trigonometric power integrals

These integrals have the form $\int \sin^n x \, dx$ or $\int \cos^n x \, dx$ where n is a positive integer. There are two cases to consider:

(i) n odd   For $\sin^n x$ put $u = \cos x$ and use the identity $\sin^2 x \equiv 1 - \cos^2 x$.
     For $\cos^n x$ put $u = \sin x$ and use the identity $\cos^2 x \equiv 1 - \sin^2 x$.

(ii) n even   We use the double-angle formulae.

An example of the use of the double-angle formulae:

$$\sin^2 x \equiv \frac{1}{2}(1 - \cos 2x) \quad \text{so} \quad \int \sin^2 x \, dx = \frac{1}{2}\left(x - \frac{1}{2}\sin 2x\right) + C$$

With $\sin^4 x \equiv \frac{1}{4}(1 - \cos 2x)^2$, replace $\cos^2 2x$ by $\frac{1}{2}(1 + \cos 4x)$, etc.

We return to these integrals in the next section.

The t substitution

This is used when the integrand is a fraction whose denominator is a trigonometric function. Assume, for the moment, that the integrand is a function of θ.

The idea is to put $t = \tan\dfrac{\theta}{2}$ so that '$dt = \dfrac{1}{2}\sec^2\theta \, d\theta$' i.e. '$dt = \dfrac{1}{2}(1 + \tan^2\theta)d\theta$', hence

'$d\theta = \dfrac{2}{1 + t^2}dt$'; furthermore $\sin\theta = \dfrac{2t}{1 + t^2}$ and $\cos\theta = \dfrac{1 - t^2}{1 + t^2}$

Example

$$\int \frac{1}{1 + \sin\theta}d\theta = \int \frac{2}{1 + t^2} \times \frac{1}{\left(1 + \dfrac{2t}{1 + t^2}\right)}dt = \int \frac{2}{1 + t^2} \times \frac{1}{\left(\dfrac{1 + t^2 + 2t}{1 + t^2}\right)}dt$$

$$= \int \frac{2}{(1 + t)^2}dt = -\frac{2}{1 + t} + C = -\frac{2}{1 + \tan\dfrac{\theta}{2}} + C \qquad \blacksquare$$

The 'derivative' substitution

An integral of the form $\int (f(x))^p f'(x)dx$ where p is a rational number can be determined by putting $u = f(x)$ so that '$du = f'(x)dx$'; provided $p \neq -1$, this means that

$$\int (f(x))^p f'(x)dx = \int u^p \, du = \frac{u^{p+1}}{p + 1} + C = \frac{(f(x))^{p+1}}{p + 1} + C$$

As an example

$$\int \frac{x}{(x^2+4)^{3/2}}\,dx = \frac{1}{2}\int \frac{2x}{(x^2+4)^{3/2}}\,dx = \frac{1}{2}\frac{(x^2+4)^{-1/2}}{-\frac{1}{2}}+C = -(x^2+4)^{-1/2}+C$$

Note the special case when $p=-1$.

Then $\int \frac{f'(x)}{f(x)}\,dx = \ln|f(x)|+C = \ln|Kf(x)|$, where K is another arbitrary constant.

Integrands of this type occur often.
    As an example

$$\int \frac{2x+1}{x^2+x+5}\,dx = \ln|x^2+x+5|+C$$

Summary

1. Choose a simple substitution which removes the awkward factor, e.g. $u^3 = x^2+4$ in the case $(x^2+4)^{1/3}$.
2. The choice of substitution may not be unique: many books give a wide range of substitutions.
3. If one substitution does not work then consider an alternative but remember that not all integrals can be evaluated by calculus, let alone by the method of substitution.

Exercise 6.1

1   Using the information at the very end of the question, integrate the following functions using the substitution given.

(a) $(x+5)^5$, $u = x+5$

(b) $\dfrac{1}{(x+4)^3}$, $u = x+4$

(c) $\dfrac{1}{\sqrt{9-2x}}$, $u^2 = 9-2x$

(d) e^{2x}, $u = 2x$

(e) e^{ax}, $u = ax$

(f) 5^x, $e^{ax} = 5^x$

(g) $\dfrac{1}{(1-x^2)^{1/2}}$, $x = \sin\theta$

(h) $\dfrac{x}{(1-x^2)^{1/2}}$, $u^2 = 1-x^2$

(i) $\dfrac{1}{x^2+9}$, $x = 3\tan\theta$

(j) $\dfrac{1}{(4x^2-1)^{1/2}}$, $x = \dfrac{1}{2}\cosh\theta$

(k) $\dfrac{1}{(x^2+25)^{1/2}}$, $u = 5\sinh\theta$

(l) $(ax+b)^p$ $u = ax + b$

In (c) $x < 4.5$, in (f) identify the value of a and proceed as in (e); in (g) and (h) $|x| < 1$; in (j) $|x| > 0.25$ and in (l) p is any real number.

2   Evaluate the following definite integrals using the substitution given.

(a) $\int_0^1 e^{4x+1}\, dx,$ $u = 4x + 1$ (b) $\int_0^1 \frac{1}{1+x^2}\, dx,$ $u = \tan\theta$

(c) $\int_0^1 \frac{x^3}{1+x^4}\, dx,$ $u = 1 + x^4$ (d) $\int_0^1 x(4+x^2)^{1/2}\, dx,$ $u = 4 + x^2$

(e) $\int_1^2 \frac{1}{(x^2-1)^{1/2}}\, dx,$ $x = \cosh\theta$ (f) $\int_0^{\pi/2} \cos^4\theta \sin\theta\, d\theta,$ $u = \cos\theta$

(g) $\int_2^3 \frac{1}{x^2 - 4x + 13},$ $u = x - 2$ (h) $\int_2^4 \frac{1}{(x^2 - 4x + 9)^{1/2}},$ $x - 2 = \cosh\theta$

3   (a)   Given that $\sin^2 x \equiv \frac{1}{2}(1 - \cos 2x)$ determine $\int \sin^2 x\, dx.$

(b)   Evaluate $\int_0^a (a^2 - y^2)^{1/2}\, dy$

(c)   Replace $\sin^2 x$ by $1 - \cos^2 x$ and determine $\int \sin^3 x\, dx$

4   Looking at the following as being of the form $\int (f(x))^p f'(x)dx$, use a suitable substitution in order to determine the integral in each case.

(a) $\int x(1 + x^2)^{1/3}\, dx$ (b) $\int \cos x(1 + \sin^2 x)dx$

(c) $\int 7xe^{x^2}\, dx$ (d) $\int x^3(1 + x^4)^{1/2}\, dx$

(e) $\int (2x + 5)(x^2 + 5x)^4\, dx$ (f) $\int \frac{\sin x}{(\cos x)^{1/2}}\, dx$

(g) $\int_0^1 \frac{x+4}{(x^2 + 4)^{1/2}}\, dx$ (h) $\int_e^{e^2} \frac{1}{x(\ln x)^2}\, dx$

5   Noting that $\int \frac{f'(x)}{f(x)}\, dx = \ln|f(x)| + C$, determine the following integrals:

(a) $\int \frac{x^2 + 1}{x^3 + 3x}\, dx$ (b) $\int \frac{1}{x^2 - 1}\, dx$ (c) $\int \frac{1}{e^{2x} - 3e^x}\, dx$

(d) $\int \frac{\sin x \cos x}{1 - \cos x}\, dx$ (e) $\int_0^{1/2} \frac{x^2}{1 - x^3}\, dx$ (f) $\int_0^{\pi/4} \frac{\tan x}{1 + \cos^2 x}\, dx$

(*Hints*: in (b) where $|x| > 1$ use partial fractions; in (c) where $x > \ln 3$ let $u = e^x$ and in (d) let $\cos x = u$.)

6   Determine the following integrals:

(a) $\displaystyle\int \cos^2 x \, dx$

(b) $\displaystyle\int \cos^3 x \, dx$

(c) $\displaystyle\int \sin^4 x \, dx$

(d) $\displaystyle\int \cos^5 x \, dx$

(e) $\displaystyle\int \cos^2 x \sin^2 x \, dx$

(f) $\displaystyle\int_0^{\pi/4} \tan^3 x \, dx$

(g) $\displaystyle\int \frac{\cos^3 x}{1 - \sin x} \, dx$

(h) $\displaystyle\int_0^{\pi/2} \sin^3 x \cos^3 x \, dx$

7   Using suitable trigonometric identities, determine the following integrals, which involve multiple angles:

(a) $\displaystyle\int \sin x \sin 2x \, dx$

(b) $\displaystyle\int \cos x \cos 3x \, dx$

(c) $\displaystyle\int \sin 3x \sin 2x \, dx$

(d) $\displaystyle\int \cos^4 2x \sin^3 2x \, dx$

(e) $\displaystyle\int_0^{\pi/2} \sin 5x \sin 2x \, dx$

8   Use the identities

$$\sin(m \pm n)x \equiv \sin mx \cos nx \pm \cos mx \sin nx$$

and $$\cos(m \pm n)x \equiv \cos mx \cos nx \mp \sin mx \sin nx$$

to prove that for m and n integers

(a) $\displaystyle\int_0^{2\pi} \sin mx \cos nx \, dx = 0$ for all integer values of m and n

(b) $\displaystyle\int_0^{2\pi} \sin mx \sin nx \, dx = \begin{cases} 0 & m \neq n \\ \pi & m = n \neq 0 \\ 0 & m = n = 0 \end{cases}$

9   If $t = \tan \dfrac{x}{2}$ prove that '$dx = \dfrac{2}{1 + t^2} dt$', that $\sin x = \dfrac{2t}{1 + t^2}$ and that $\cos x = \dfrac{1 - t^2}{1 + t^2}$.

Hence determine the integrals

(a) $\displaystyle\int \frac{1}{3 - 2\cos x} \, dx$

(b) $\displaystyle\int_0^{\pi/2} \frac{1}{2 + \cos x} \, dx$

10   Use the substitution $t = \tan\dfrac{x}{2}$ to determine indefinite integrals of the following:

(a) $\dfrac{1}{1 + \sin x + \cos x}$

(b) $\sec x$

(c) $\dfrac{1}{3 + 4\sin x}$

(d) $\operatorname{cosec} x$

Evaluate the definite integrals

(e) $\displaystyle\int_0^{\pi/2} \dfrac{1}{5 + \cos x}\,dx$

(f) $\displaystyle\int_{\pi}^{2\pi} \dfrac{1}{1 + \sin x}\,dx$

11   (a)   The function $\operatorname{sech} x \equiv \dfrac{1}{\cosh x}$; prove that $\operatorname{sech} x \equiv \dfrac{\cosh x}{1 + \sinh^2 x}$, hence find its indefinite integral.

(b)   Prove that $\tanh^2 x + \operatorname{sech}^2 x \equiv 1$ and thus find $\displaystyle\int \tanh^2 x\,dx$

(c)*   Write $u = \dfrac{x - 1}{4}$ and use a suitable substitution involving hyperbolic functions to

find $\displaystyle\int \dfrac{1}{\sqrt{x^2 - 2x + 17}}\,dx$. Give your answer in terms of a logarithm.

(d)   By setting $x = 3\cosh u$ determine $\displaystyle\int \sqrt{x^2 - 9}\,dx$

(e)   Evaluate the following:

(i) $\displaystyle\int_0^1 \dfrac{1}{\sqrt{x^2 + 1}}\,dx$ (ii) $\displaystyle\int_0^1 \dfrac{x}{\sqrt{x^2 + 1}}\,dx$

(iii) $\displaystyle\int_0^1 \dfrac{x^2}{\sqrt{x^2 + 1}}\,dx$

In (iii) note that $x^2 \equiv (x^2 + 1) - 1$

12   Evaluate the following definite integrals:

(a) $\displaystyle\int_0^{\pi/4} \tan^4 x\,dx$

(b) $\displaystyle\int_2^3 x\sqrt{x - 2}\,dx$

(c) $\displaystyle\int_0^{\pi/4} \dfrac{1 - x}{x(2 - x)}\,dx$

(d) $\displaystyle\int_a^b e^{-x}\,dx$

(e) $\displaystyle\int_1^X \dfrac{1}{x^p}\,dx$

(f) $\displaystyle\int_0^{\pi/4} xe^{x^2}\,dx$

6.2 INTEGRATION BY PARTS

When the integrand is a product of two functions and one of the functions can be readily differentiated, a method worth considering is the method of **integration by parts**. It is derived from the product rule for differentiation:

$$\frac{d}{dx}(uv) = u\frac{dv}{dx} + v\frac{du}{dx}$$

where u and v are functions of x.

If we integrate both sides with respect to x, we obtain

$$\int \frac{d}{dx}(uv)dx = \int u\frac{dv}{dx}dx + \int v\frac{du}{dx}dx$$

i.e. $$uv = \int u\frac{dv}{dx}dx + \int v\frac{du}{dx}dx = \int u\frac{dv}{dx}dx + \int \frac{du}{dx}v\,dx$$

so $$\int u\frac{dv}{dx}dx = uv - \int \frac{du}{dx}v\,dx$$

In words we say 'the integral of a product is the first term times the integral of the second minus the integral of the derivative of the first term times the integral of the second.'

Notice that in the first instance the first term is copied or differentiated while the second term is integrated and the result copied.

Example

Find $\int xe^x\,dx$

Put $u = x$ and $\frac{dv}{dx} = e^x$, so $\frac{du}{dx} = 1$ and $v = e^x$

Then $\int xe^x\,dx = xe^x - \int 1 \times e^x\,dx = xe^x - e^x + C = x(e^x - 1) + C$ ■

Notice that the arbitrary constant appears at the end of the working and can be withheld during the earlier integrations. As always, the result can be checked by differentiation.

In the process we have to carry out two integrations: the second term in the product must be integrated at the first stage and the hope is that the second integration is straightforward, as in the example. We can represent the method in the following notation:

$$\int p(x)q(x)\,dx = p(x)Q(x) - \int \frac{dp}{dx}Q(x)\,dx$$

where $\quad Q(x) = \int q(x)\,dx$

Note that the ordering of terms in the product may be important. If one of the terms cannot immediately be integrated, it must be made the first term in the integrand. Even when both terms can be integrated directly, one ordering may lead to an easier working than the other.

Examples

Find the indefinite integrals of the functions

(a) $x \sin x$ 　　　　　　 (b) $x^2 e^x$ 　　　　　　 (c) $x \ln x$

Solution

(a) We can integrate either x or $\sin x$. However, integration of x produces $\frac{1}{2}x^2$ and this only makes the problem more complicated, whereas it differentiates to 1, a simplification. On the other hand, either integration or differentiation of sine leads to the cosine, which is no worse that the original. Hence we leave the product in the order given.

Applying the rule gives

$$\int x \sin x \, dx = x \times (-\cos x) - \int 1 \times (-\cos x)\,dx$$

$$= -x \cos x + \int \cos x \, dx$$

$$= -x \cos x + \sin x + C$$

(b) Here, the integration of e^x merely repeats the function, whereas differentiation of x^2 leads to a simplification. Again, we keep to the order given. Then

$$I = \int x^2 e^x \, dx = x^2 e^x - \int 2xe^x \, dx = x^2 e^x - 2\int xe^x \, dx$$

Although we have reduced the power of x by 1, we are still not at the stage where we can determine the integral on sight. To make progress we dump unnecessary baggage and concentrate on that integral. We write $J = \int xe^x \, dx$ then apply the method of parts again:

$$J = \int xe^x \, dx = xe^x - \int 1 \times e^x \, dx = xe^x - e^x$$

Now we return to collect the earlier result:

$$I = x^2 e^x - 2(xe^x - e^x) + C = (x^2 - 2x + 2)e^x + C$$

Notice that we put in the arbitrary constant only at the last stage.

(c) We cannot yet integrate $\ln x$ but differentiation produces $\dfrac{1}{x}$. This counteracts the increase in the index when integrating $x = x^1$. We start by re-ordering the product:

$$\int x \, \ln x \, dx = \int \ln x \times x \, dx$$

$$= \ln x \times \frac{1}{2} x^2 - \int \left(\frac{1}{x} \times \frac{1}{2} x^2 \right) dx$$

$$= \frac{1}{2} x^2 \ln x - \frac{1}{2} \int x \, dx$$

$$= \frac{1}{2} x^2 \ln x - \frac{1}{4} x^2 + C \qquad \blacksquare$$

So far, each application of the 'parts' formula has produced a simplification in the integral to be determined. The following example seems to lead us round in a circle; its moral is, Do not give up too easily.

Example   Find the indefinite integral $I = \int e^x \sin x \, dx$.

We note first that two components of the product give similar results whether differentiated or integrated; this is not a good sign. Applying the method of parts:

$$I = \int e^x \sin x \, dx$$

$$= e^x(-\cos x) - \int e^x(-\cos x) dx$$

$$= -e^x \cos x + \int e^x \cos x \, dx$$

Writing $J = \int e^x \cos x \, dx$ we apply the method to this integral and obtain

$$J = \int e^x \cos x \, dx = e^x \sin x - \int e^x \sin x \, dx = e^x \sin x - I$$

Putting this result back into earlier result gives

$$I = -e^x \cos x + e^x \sin x - I$$

Rearranging, we obtain

$$2I = -e^x \cos x + e^x \sin x$$

hence

$$I = \frac{1}{2}(-e^x \cos x + e^x \sin x) + C = \frac{1}{2}e^x(\sin x - \cos x) + C,$$

adding the arbitrary constant at the end. ∎

Definite integrals

In many cases the application of the method of parts to a definite integral means that those parts of the integral already evaluated can be removed at each step, as the next example shows.

Example

Find the indefinite integrals

(a) $\displaystyle\int_0^{\pi} e^x \cos x \, dx$

(b) $\displaystyle\int_0^1 (ax + b)e^{-x} \, dx$

Solution

(a) Applying the method of integration by parts, we obtain

$$I = \int_0^{\pi} e^x \cos x \, dx = [e^x \sin x]_0^{\pi} - \int_0^{\pi} e^x \sin x \, dx$$

$$= 0 - 0 - \int_0^{\pi} e^x \sin x \, dx$$

$$= \int_0^{\pi} e^x \times (-\sin x)dx$$

$$= [e^x \cos x]_0^{\pi} - \int_0^{\pi} e^x \cos x \, dx$$

$$= -e^{\pi} - 1 - I$$

so that

$$I = -\frac{1}{2}(e^{\pi} + 1) = -12.070$$

(b)
$$\int_0^1 (ax + b)e^{-x}\,dx = [-(ax + b)e^{-x}]_0^1 - \int_0^1 a \times (-e^{-x})\,dx$$

$$= -\frac{1}{e}(a + b) + b + a[-e^{-x}]_0^1$$

$$= -\frac{1}{e}(a + b) + b + a\left(1 - \frac{1}{e}\right)$$

$$= a\left(1 - \frac{2}{e}\right) + b\left(1 - \frac{1}{e}\right) \qquad \blacksquare$$

Special tricks

Some integrals, which appear not to involve a product, can be determined by the method of parts by introducing an artificial product. This is especially useful when the original integrand can be readily differentiated.

Example   Find the integrals

(a) $\displaystyle\int \ln x\,dx$ (b) $\displaystyle\int_{1/2}^1 \sin^{-1} x\,dx$

Solution

(a) Write $\ln x$ as $\ln x \times 1$, then

$$\int \ln x\,dx = \int \ln x \times 1\,dx = \ln x \times x - \int \frac{1}{x} \times x\,dx = x \ln x - x + C$$

(b) Write $\sin^{-1} x$ as $\sin^{-1} x \times 1$, then

$$\int_{1/2}^1 \sin^{-1} x\,dx = \int_{1/2}^1 \sin^{-1} x \times 1\,dx = [\sin^{-1} x \times x]_{1/2}^1 - \int_{1/2}^1 \frac{x}{\sqrt{1 - x^2}}\,dx$$

$$= \frac{\pi}{2} - \frac{1}{2} \times \frac{\pi}{6} + [\sqrt{1 - x^2}]_{1/2}^1 = \frac{5\pi}{12} - \frac{\sqrt{3}}{2} = 0.4430 \qquad \blacksquare$$

Reduction formulae

An important application of the method of parts is in the integration of functions such as $\int \sin^n x \, dx$ where n is a positive integer.

First we write $\sin^n x = \sin^{n-1} x \times \sin x$. Then

$$
\begin{aligned}
I_n = \int \sin^n x \, dx &= \int \sin^{n-1} x \times \sin x \, dx \\
&= \sin^{n-1} x \times (-\cos x) - \int (n-1) \cos x \times \sin^{n-2} x \times (-\cos x) dx \\
&= -\cos x \times \sin^{n-1} x + (n-1) \int \sin^{n-2} x \times (1 - \sin^2 x) dx \\
&= -\cos x \times \sin^{n-1} x + (n-1)(I_{n-2} - I_n)
\end{aligned}
$$

where $I_{n-2} = \int \sin^{n-2} x \, dx$

Rearranging to make I_n the subject gives

$$
nI_n = -\cos x \times \sin^{n-1} x + (n-1)I_{n-2}
$$

i.e. $\quad I_n = -\dfrac{\cos x \times \sin^{n-1} x}{n} + \dfrac{(n-1)}{n} I_{n-2}$

This formula, which relates I_n to I_{n-2} is an example of a **reduction formula**.

Example

Use the reduction formula to evaluate $I_n = \displaystyle\int_0^{\pi/2} \sin^n x \, dx$ in the cases (a) n is even and (b) n is odd. Then find numerical values for I_7 and I_6.

$$
\begin{aligned}
I_n &= \left[-\dfrac{\cos x \times \sin^{n-1} x}{n} \right]_0^{\pi/2} + \dfrac{(n-1)}{n} I_{n-2} \\
&= 0 - 0 + \dfrac{(n-1)}{n} I_{n-2} = \dfrac{(n-1)}{n} I_{n-2}
\end{aligned}
$$

Hence replacing n by $n-2$, $I_{n-2} = \dfrac{(n-3)}{(n-2)} I_{n-4}$, so that $I_n = \dfrac{(n-1)(n-3)}{n(n-2)} I_{n-4}$, etc.

(a) When n is odd we continue the process to obtain

$$
I_n = \dfrac{(n-1)}{n} \dfrac{(n-3)}{(n-2)} \times \cdots \times \dfrac{2}{3} I_1
$$

Since $I_1 = \displaystyle\int_0^{\pi/2} \sin x \, dx = [-\cos x]_0^{\pi/2} = 1$ then

$$I_n = \frac{(n-1)\,(n-3)}{n\quad(n-2)} \times \cdots \times \frac{2}{3} \times 1$$

(b) When n is even we continue the process to obtain

$$I_n = \frac{(n-1)\,(n-3)}{n\quad(n-2)} \times \cdots \times \frac{1}{2} I_0$$

Since $I_0 = \displaystyle\int_0^{\pi/2} 1 \, dx = \frac{\pi}{2}$ then

$$I_n = \frac{(n-1)\,(n-3)}{n\quad(n-2)} \times \cdots \times \frac{1}{2} \times \frac{\pi}{2}$$

The required numerical values are

$$I_7 = \frac{6}{7} \times \frac{4}{5} \times \frac{2}{3} \times 1 = \frac{16}{35}$$

$$I_6 = \frac{5}{6} \times \frac{3}{4} \times \frac{1}{2} \times \frac{\pi}{2} = \frac{5\pi}{32}$$

∎

Exercise 6.2

1 Integrate by parts the following functions:

(a) $x \cos x$ (b) $x^2 e^x$ (c) $x(x+1)e^x$

(d) $x^2 \sin x$ (e) $x \cos 2x$ (f) $(ax + b)\cos kx$

(g) $(ax + b)^2 e^{kx}$ (h) $\sin^2 x = \sin x \times \sin x$

2 Find the following definite integrals:

(a) $\displaystyle\int_0^1 x^2 e^{-x} \, dx$ (b) $\displaystyle\int_0^{\pi/2} x \sin 2x \, dx$ (c) $\displaystyle\int_{-1}^1 x^2(2+x)^{1/2} \, dx$

3 (a)   Use integration by parts to show that

$$\int e^{ax} \cos bx \, dx = \frac{e^{ax}(b \sin bx + a \cos bx)}{a^2 + b^2} + C$$

(b)   Assuming a similar form for $\int e^{ax} \sin bx \, dx$, find the constants A and B for which

$$\int e^{ax} \sin bx \, dx = e^{ax}(A \sin bx + B \cos bx) + C.$$

(c)   Deduce the values of

(i) $\int_0^{\pi/2} e^x \sin x \, dx$ \qquad (ii) $\int_0^{\pi/4} e^{-2x} \cos 3x \, dx$

4 By writing the integrand as a product of itself and 1, determine the following integrals:

(a) $\int \tan^{-1} x \, dx$ \qquad (b) $\int \sinh^{-1} x \, dx$ \qquad (c) $\int \tanh^{-1} x \, dx$

5 Find the integrals

(a) $\int x^2 \ln x \, dx$ \qquad (b) $\int x \sin^{-1} x \, dx$ \qquad (c) $\int_1^e x^n \ln x \, dx, \quad n > 0$

6 (a) The gradient at any point (x, y) on a curve which passes through the point $(e, 0)$ is $\ln x$. Find the equation of the curve.

(b) The tangent to the curve at the point P whose x-coordinate is e^2 meets the x-axis at T. Find (i) the gradient of the tangent, (ii) the equation of the line through P and T, (iii) the coordinates of T.

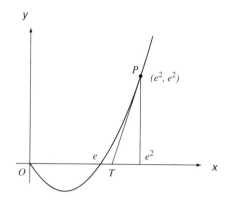

7 If $I_{m,n} = \int_0^{\pi/2} \sin^m x \cos^n x \, dx$ use integration by parts to show that

$$(m + n)I_{m,n} = (n - 1)I_{m,n-2} = (m - 1)I_{m-2,n}$$

Deduce the following results:

$$\int_0^{\pi/2} \sin^6 x \cos^4 x \, dx = \frac{3 \times 1}{10 \times 8} \int_0^{\pi/2} \sin^6 x \, dx = \frac{3\pi}{512}$$

$$\int_0^{\pi/2} \sin^6 x \cos^4 x \, dx = \frac{3 \times 1}{10 \times 8} \int_0^{\pi/2} \sin^6 x \, dx = \frac{3\pi}{512}$$

8 A particle travelling under damped harmonic motion along a straight line has a speed given by $v = t^3 e^{-at}$. Integrate to find the distance travelled (a) in the first $\dfrac{1}{a}$ seconds; (b) ultimately, i.e. by the time it comes to rest.

9 The power absorbed in a purely inductive alternating current circut is given by $P = IV$ where $I = I_0 \sin \omega t$ and $V = V_0 \cos \omega t$. Integrate the power function to determine how much power is absorbed in one-quarter of a cycle, i.e. between $t = 0$ and $t = \dfrac{\pi}{2\omega}$. How much power is absorbed in a complete cycle?

10 The total magnetic field induced at a point P by a cable of radius a carrying a current of I amps is given by

$$\delta H = \frac{2\pi a^2 In}{(a^2 + x^2)^{3/2}} \delta x$$

where the wire is wrapped round at n turns per metre.

The total flux is found by integrating from $x_1 = a_1 \cot \theta_1$ to $x_2 = a_2 \cot \theta_2$. Prove that $H = 2\pi a In(\cos \theta_2 - \cos \theta_1)$. If is an infinitely long cable, i.e. $\theta_2 = 0$ and $\theta_1 = \pi$, what is the value of H?

6.3 FURTHER METHODS OF INTEGRATION

Rational functions

The integrals of rational functions often involve logarithms, inverse functions and other rational functions. The use of partial fractions to break down a given rational function into components which are readily integrated is a key stage.

Example

Find the indefinite integral of each of the following functions:

(a) $\dfrac{3x+1}{x-5}$

(b) $\dfrac{2}{x^2-4x+3}$

(c) $\dfrac{x^2}{(x-1)(x+2)}$

(d) $\dfrac{x+1}{(x-1)^2}$

(e) $\dfrac{x+1}{x^2+2x+7}$

(f) $\dfrac{3x-1}{(x+1)(x-2)^2}$

Solution

(a) First we split the given rational function into a quotient and a remainder which is a proper rational function:

$$\frac{3x+1}{x-5} \equiv 3 + \frac{16}{x-5}$$

Then we integrate to obtain $\displaystyle\int \frac{3x+1}{x-5}\, dx = 3x + 16\ln|x-5| + C$

(b) We obtain the partial fractions of the given function:

$$\frac{2}{x^2-4x+3} \equiv \frac{1}{x-3} - \frac{1}{x-1}$$

Then integration produces

$$\int \frac{2}{x^2-4x+3}\, dx = \ln|x-3| - \ln|x-1| + C = \ln\left|K\left(\frac{x-3}{x-1}\right)\right|$$

where we have written the constant of integration as $\ln K$.

(c) This time we must divide out by the denominator then split the proper rational function remainder into its partial fractions:

$$\frac{x^2}{(x-1)(x+2)} \equiv 1 + \frac{1}{3}\left(\frac{1}{x-1} - \frac{4}{x+2}\right)$$

Upon integrating we obtain

$$\int \frac{x^2}{(x-1)(x+2)} \, dx = x + \frac{1}{3}\ln|x-1| - \frac{4}{3}\ln|x+2| + C$$

(d) First $\dfrac{x+1}{(x-1)^2} \equiv \dfrac{1}{(x-1)} + \dfrac{2}{(x-1)^2}$

   Then $\displaystyle\int \frac{x+1}{(x-1)^2} \, dx = \ln|x-1| - \frac{2}{(x-1)} + C$

(e) The denominator does not factorise but the numerator is exactly half the derivative of the denominator, i.e.

$$\frac{1}{2}\frac{d}{dx}(x^2 + 2x + 7) = \frac{1}{2}(2x+2) = x+1$$

Noting that $x^2 + 2x + 7$ is always positive, integration produces

$$\int \frac{x+1}{x^2+2x+7} \, dx = \frac{1}{2}\int \frac{2(x+1)}{x^2+2x+7} \, dx = \frac{1}{2}\ln(x^2+2x+7) + C$$

(f) First $\dfrac{3x-1}{(x+1)(x-2)^2} \equiv \dfrac{4}{9}\left(\dfrac{1}{x-2} - \dfrac{1}{x+1}\right) + \dfrac{5}{3(x-2)^2}$

   Then $\displaystyle\int \frac{3x-1}{(x+1)(x-2)^2} \, dx = \frac{4}{9}\ln\left|\frac{x-2}{x+1}\right| - \frac{5}{3(x-2)} + C$ ∎

Although the denominator in case (e) did not factorise, we were lucky in that the numerator was a multiple of the denominator and this allowed a relatively simple integration. In general we shall not be so fortunate. To make progress in these cases we need to take on board the following standard results:

$$\int \frac{1}{x^2-a^2} \, dx = \frac{1}{2a}\ln\left|\frac{x-a}{x+a}\right| + C, \qquad \int \frac{1}{x^2+a^2} \, dx = \frac{1}{a}\tan^{-1}\left(\frac{x}{a}\right) + C$$

$$\int \frac{x}{x^2 \pm a^2} \, dx = \frac{1}{2}\ln|x^2 \pm a^2| + C$$

If x is replaced by $x + b$, we obtain similar results, e.g.

$$\int \frac{1}{(x+b)^2 - a^2} \, dx = \frac{1}{2a}\ln\left|\frac{x+b-a}{x+b+a}\right| + C$$

In each case we inspect the denominator of the given rational function for factors; if there are none, we express the denominator as the sum or difference of squared terms: one involves the variable of integration and the other is a pure number. Then we use one of the standard forms above.

Example

Integrate the following functions.

(a) $\dfrac{x+3}{x^2-16}$

(b) $\dfrac{2x+5}{x^2+5}$

(c) $\dfrac{x}{x^2+2x+7}$

(d) $\dfrac{3x+1}{9x^2+14x-5}$

(e) $\dfrac{1}{x^2-4x-11}$

(f) $\dfrac{6x+4}{3x^2+4x+20}$

Solution

(a) First, $\dfrac{x+3}{x^2-16} \equiv \dfrac{1}{8}\left(\dfrac{7}{x-4}+\dfrac{1}{x+4}\right)$

Then $\displaystyle\int \dfrac{x+3}{x^2-16}\,dx = \dfrac{1}{8}(7\ln|x-4| + \ln|x+4|) + C$

(b) The partial fractions are

$$\dfrac{2x+5}{x^2+5} \equiv \dfrac{2x}{x^2+5} + \dfrac{5}{x^2+5}$$

Then we integrate to obtain

$$\int \dfrac{2x+5}{x^2+5}\,dx \equiv \ln(x^2+5) + \sqrt{5}\tan^{-1}\left(\dfrac{x}{\sqrt{5}}\right) + C$$

(c) Note that

$$\dfrac{x}{x^2+2x+7} \equiv \dfrac{x+1}{x^2+2x+7} - \dfrac{1}{(x+1)^2+6}$$

since $x^2+2x+7 \equiv (x+1)^2+6$. Hence

$$\int \dfrac{x}{x^2+2x+7}\,dx = \dfrac{1}{2}\ln(x^2+2x+7) - \dfrac{1}{\sqrt{6}}\tan^{-1}\left(\dfrac{x+1}{\sqrt{6}}\right) + C$$

where we have copied the first term from part (e) of the first example in this section and used the standard from with x replaced by $x+1$ and with $a^2=6$.

(d) First, $\dfrac{3x+1}{9x^2+14x-5} \equiv \dfrac{1}{x-1} - \dfrac{6}{9x-5}$

Then $\displaystyle\int \dfrac{3x+1}{9x^2+14x-5}\,dx = \ln|x-1| - \dfrac{2}{3}\ln|9x-5| + C$

(e) Completing the square in the denominator gives

$$\frac{1}{x^2 - 4x - 11} \equiv \frac{1}{(x-2)^2 - 15}$$

Then $\displaystyle\int \frac{1}{x^2 - 4x - 11} \, dx = \frac{1}{2\sqrt{15}} \ln \left| \frac{x - \sqrt{15} - 2}{x + \sqrt{15} - 2} \right| + C$

(f) First, $\displaystyle\frac{6x - 4}{3x^2 + 4x + 20} \equiv \frac{6x + 4}{3x^2 + 4x + 20} - \frac{8}{3\left(\left(x + \dfrac{2}{3}\right)^2 + \dfrac{56}{3}\right)}$

Then

$$\int \frac{6x - 4}{3x^2 + 4x + 20} \, dx = \ln(3x^2 + 4x + 20) - \frac{4}{\sqrt{14}} \tan^{-1}\left(\frac{3x + 2}{2\sqrt{14}}\right) + C$$

Note that $3x^2 + 4x + 20$ is always positive. We used the substitution $u = x + \dfrac{2}{3}$ and put $a^2 = \dfrac{56}{9}$ in the standard result to obtain the last term. ∎

Definite integrals

Note that with definite integrals the interval over which the integration is taken must not include points where the integrand cannot be defined.

Example

Evaluate where possible the integrals

(a) $\displaystyle\int_0^1 \frac{3x^2 + 4x + 1}{x^3 + 2x^2 + x + 1} \, dx$

(b) $\displaystyle\int_{-4}^0 \frac{1}{x^2 + 6x + 9} \, dx$

(c) $\displaystyle\int_{-4}^0 \frac{1}{x^2 + 6x - 9} \, dx$

(d) $\displaystyle\int_1^2 \frac{2x + 1}{3x^2 + 6x + 28} \, dx$

Solution

(a) The numerator is the exact derivative of the denominator, so

$$\int_0^1 \frac{3x^2 + 4x + 1}{x^3 + 2x^2 + x + 1} \, dx = [\ln(x^3 + 2x^2 + x + 1)]_0^1 = \ln 5$$

(b) Note that $x^2 + 6x + 9 \equiv (x + 3)^2$. The indefinite integral is $-\dfrac{1}{x + 3} + C$, but the definite integral does not exist because $x = -3$ is in the interval of integration.

(c) Begin with $x^2 + 6x - 9 \equiv (x+3)^2 - 18$. Then put $u = x + 3$ to obtain

$$\int_{-4}^{0} \frac{1}{x^2 + 6x - 9} \, dx = \int_{-1}^{3} \frac{1}{u^2 - 18} \, du = \frac{1}{6\sqrt{2}} \left[\ln\left(\frac{u - 3\sqrt{2}}{u + 3\sqrt{2}} \right) \right]_{-1}^{3}$$

$$= \frac{1}{6\sqrt{2}} \ln\left(\frac{17}{81 - 56\sqrt{2}} \right) = -0.264$$

(d) Putting the integrand in partial fractions gives

$$\frac{2x + 1}{3x^2 + 6x + 28} \equiv \frac{\frac{1}{3}(6x + 6)}{2x^2 + 6x + 28} - \frac{1}{3(x+1)^2 + 25}$$

Integrating gives

$$\int_{1}^{2} \frac{2x + 1}{3x^2 + 6x + 28} \, dx$$

$$= \left[\frac{1}{3} \ln(3x^2 + 6x + 28) \right]_{1}^{2} - \left[\frac{1}{3} \times \frac{\sqrt{3}}{5} \tan^{-1}\left(\frac{\sqrt{3}(x+1)}{5} \right) \right]_{1}^{2}$$

$$= \frac{1}{3}\left(2\ln 2 - \ln\frac{37}{13} \right) - \frac{\sqrt{3}}{15} \tan^{-1}\left(\frac{5\sqrt{3}}{43} \right) = 0.0905 \qquad \blacksquare$$

Square root surds

Of the vast range of integrals that can be carried out using ingenuity, patience and some cunning, we now look at some of the more common types which involve square roots.

Example   Find the following integrals:

(a) $\displaystyle\int \sqrt{x^2 + a^2} \, dx$ (b) $\displaystyle\int_{1/2}^{3/2} \sqrt{4x^2 - 4x + 5} \, dx$ (c) $\displaystyle\int \frac{5 - 4x}{\sqrt{3x - x^2 - 2}} \, dx$

Solution

(a) Put $x = \sinh u$, then '$dx = \cosh u \, du$'.

Integrating gives

$$\int \sqrt{x^2 + a^2}\, dx = \int a \cosh u \times a \cosh u\, du = \frac{a^2}{2} \int (1 + \cosh 2u)\, du$$

$$= \frac{a^2}{2}\left(u + \frac{1}{2}\sinh 2u\right) + C = \frac{a^2}{2}(u + \sinh u \times \cosh u) + C$$

and since

$$\cosh u = \left(1 + \frac{x^2}{a^2}\right)^{1/2} = \frac{(x^2 + a^2)^{1/2}}{a},$$

we write the integral as

$$\frac{a^2}{2}\sinh^{-1}\left(\frac{x}{a}\right) + \frac{x}{2}(x^2 + a^2)^{1/2} + C$$

(b) Note that $4x^2 - 4x + 5 \equiv (2x - 1)^2 + 4$. Put $2u = 2x - 1$ so that '$du = dx$' and $4x^2 - 4x + 5 \equiv 4u^2 + 4$.

When $x = \frac{1}{2}$, $u = 0$ and when $x = \frac{3}{2}$, $u = 1$. Then using part (a) with $a = 1$ and $x = u$, we obtain

$$\int_{1/2}^{3/2} \sqrt{4x^2 - 4x + 5}\, dx = \int_0^1 2\sqrt{u^2 + 1}\, du = [\sinh^{-1} u + u(u^2 + 1)^{1/2}]_0^1$$

$$= \ln(\sqrt{2} + 1) + \sqrt{2} = 2.2956$$

where we first have used the log equivalent of $\sinh^{-1}$.

(c) We first write the numerator as a multiple of the derivative of the denominator plus a numerical remainder, i.e. $5 - 4x \equiv 2(3 - 2x) - 1$.

Then $\int \frac{5 - 4x}{\sqrt{3x - x^2 - 2}}\, dx = \int \frac{2(3 - 2x)}{\sqrt{3x - x^2 - 2}}\, dx - \int \frac{1}{\sqrt{3x - x^2 - 2}}\, dx$

The first term is in a standard form and integrates to $4\sqrt{3x - x^2 - 2}$. To deal with the second term, note that $3x - x^2 \equiv \frac{1}{4} - \left(x - \frac{3}{2}\right)^2$; put $u = x - \frac{3}{2}$ so that '$du = dx$' and

$$\int \frac{1}{\sqrt{3x - x^2 - 2}}\, dx = \int \frac{1}{\left(\frac{1}{4} - u^2\right)^{1/2}}\, du$$

Now put $u = \dfrac{1}{2}\sin v$ so that '$du = \dfrac{1}{2}\cos v$'. The integral becomes

$$\int \frac{\frac{1}{2}\cos v}{\frac{1}{2}\cos v}\,dv = \int 1\,dv = v = \sin^{-1}(2u) = \sin^{-1}(2x - 3)$$

The complete result is $4\sqrt{3x^2 - x - 2} - \sin^{-1}(2x - 3) + C$ ■

In these days of computer algebra you are advised to verify the algebraic results we have obtained. The value of following the algebraic argument is that you get an insight into the subtleties of analytic methods of integration.

Exercise 6.3

1 Determine the following integrals, expressing any logarithms in modulus form. Divide out as necessary and resolve into partial fractions.

(a) $\dfrac{3x}{2x + 3}$

(b) $\dfrac{2x + 1}{x + 3}$

(c) $\dfrac{x^3}{2 - x}$

(d) $\dfrac{5}{x^2 + x - 6}$

(e) $\dfrac{3x}{(x + 1)(x - 2)}$

(f) $\dfrac{x^3 + 2}{x^2 - 1}$

(g) $\dfrac{4x + 3}{(x - 2)^2}$

(h) $\dfrac{7x + 5}{(x - 3)(x - 2)(x + 2)}$

2 Find the value of each of the following definite integrals:

(a) $\displaystyle\int_0^1 \frac{6x^2 + x - 1}{4x^3 + x^2 - 2x + 3}\,dx$

(b) $\displaystyle\int_0^1 \frac{1}{(x + 1)(x + 2)}\,dx$

(c) $\displaystyle\int_0^3 \frac{2x + 3}{(x - 5)(2x + 1)}\,dx$

(d) $\displaystyle\int_0^1 \frac{3x^2 + 2x + 1}{(x + 3)(x - 4)(x - 5)}\,dx$

(e) $\displaystyle\int_0^1 \frac{x^3 + 1}{2x^2 + x - 15}\,dx$

(f) $\displaystyle\int_2^3 \frac{3x - 5}{2x^2 + x - 15}\,dx$

(g) $\displaystyle\int_0^1 \frac{1}{3x^2 + 2x - 21}\,dx$

3 Find the following indefinite integrals: query

(a) $\displaystyle\int \frac{1}{x^2 + x - 7}\,dx$

(b) $\displaystyle\int \frac{2x - 1}{2x^2 + x - 8}\,dx$

(c) $\displaystyle\int \frac{3x + 1}{3x^2 + 11x + 4}\,dx$

Now find the definite integrals

(d) $\displaystyle\int_0^1 \frac{1}{x^2 + 3x - 5}\,dx$

(e) $\displaystyle\int_{-2}^{-1} \frac{2x + 3}{x^2 + 2x - 5}\,dx$

4 Integrate the following functions:

(a) $\dfrac{x^2 + 3x - 4}{x^2 - 2x - 8}$

(b) $\dfrac{2x - 5}{x^2 + x + 1}$

(c) $\dfrac{2x + 1}{x^2 + 2x + 5}$

(d) $\dfrac{2x^2}{(x^2 + 1)^2}$

(e) $\dfrac{x^3 + x^2 + x + 2}{x^4 + 3x^2 + 2}$

(f) $\dfrac{2x^2 + 3}{(x^2 + 1)^2}$

(g) $\dfrac{\sin x}{\cos x(1 + \cos^2 x)}$

(h) $\dfrac{(2 + \tan^2 \theta)\sec^2 \theta}{1 + \tan^3 \theta}$

Find the values of the definite integrals

(i) $\displaystyle\int_0^1 \frac{2x^3 + x^2 + 4}{(x^2 + 4)^2}\,dx$

(j) $\displaystyle\int_0^1 \frac{x^3 + x^2 - 5x + 15}{(x^2 + 5)(x^2 + 2x + 3)}\,dx$

5 Establish the following standard results:

(a) $\displaystyle\int \frac{1}{x^2 + a^2}\,dx = \frac{1}{a}\tan^{-1}\left(\frac{x}{a}\right) + C$ (use $x = a\tan u$)

(b) $\displaystyle\int \frac{a}{x\sqrt{x^2 - a^2}}\,dx = \sec^{-1}\left(\frac{x}{a}\right) + C$

(c) $\displaystyle\int \frac{1}{\sqrt{x^2 + a^2}}\,dx = \sinh^{-1}\left(\frac{x}{a}\right) + C$ or $\ln\left|\dfrac{x + \sqrt{x^2 + a^2}}{a}\right| + C$

(d) $\displaystyle\int \frac{1}{\sqrt{x^2 - a^2}}\,dx = \cosh^{-1}\left(\frac{x}{a}\right) + C$ or $\ln\left|\dfrac{x + \sqrt{a^2 - x^2}}{a}\right| + C$

(e) $\displaystyle\int \frac{1}{a^2 - x^2}\,dx = \frac{1}{a}\tanh^{-1}\left(\frac{x}{a}\right) + C$ or $\dfrac{1}{2a}\ln\left|\dfrac{a + x}{a - x}\right| + C$

6 Use substitutions of the form $x + a = b\sin\theta$, $x + a = b\sinh u$ or $x + a = b\cosh u$ to determine the indefinite integrals of the following functions:

(a) $\dfrac{a + x}{(a^2 - x^2)^{3/2}}$

(b) $\dfrac{x - 1}{(x^2 + 2x + 3)^{1/2}}$

(c) $\dfrac{1}{(x - 1)(2x^2 - 8x - 1)^{1/2}}$ $\left(\text{first use } x - 1 = \dfrac{1}{z}\right)$

(d) $\dfrac{1}{(x + a)(x^2 - a^2)^{1/2}}$

(e) $\dfrac{9x + 2}{(3x^2 + 4x - 4)^{1/2}}$

7 Find the indefinite integral of $\dfrac{1}{a + b\sin\theta}$ when $a + b > 0$. Consider separately the cases $a > b$ and $a < b$.

8 If $f(x) < g(x)$ over the interval $[a, b]$ then $\displaystyle\int_a^b f(x)dx < \int_a^b g(x)dx$. Prove that over the interval $1 < u < 1 + x$

$$\int_1^{1+x} \frac{1}{1+x}du < \int_1^{1+x} \frac{1}{u}du < \int_1^{1+x} du$$

Carry out the integrations to prove that $\dfrac{x}{1+x} < \ln(1+x) < x$ and deduce that

$$\frac{x}{1-x} > \ln\frac{1}{(1-x)} > x \qquad (0 < x < 1)$$

9 A is the maximum on the graph of $y = ax^{-x}$. Determine the magnitude of the area shaded in the figure

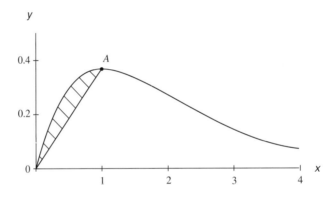

10 In a particular chemical reaction the amount x of a substance remaining at time t is given by the equation

$$\int \frac{1}{(3-x)(6-x)}dx = Kt$$

If $x(0) = 0$ and $x(10) = 1$ show that $K = \dfrac{1}{30}\ln\left(\dfrac{5}{4}\right)$ and determine $x(30)$.

11 The current i in an inductive d.c. circuit satisfies the equation

$$\int \frac{L}{E - Ri}di = t$$

where t is time, L, R and E are the (constant) inductance, resistance and applied voltage respectively.

(a) Determine $i(t)$ given that $i(0) = 0$.

(b) If $L = 0.2$ henries, $R = 40$ ohms and $E = 40$ volts, find $i(0.01)$.

6.4 NUMERICAL INTEGRATION

In this section we introduce two of the simpler methods of obtaining approximate values for definite integrals. We use the idea of a definite integral representing the area under a curve to explain them.

Trapezium rule

In Figure 6.5 B and C are two points on the curve $y = f(x)$ with x-coordinates x_0 and x_1 respectively where $x_1 - x_0 = h$.

The area under the curve between the ordinates $x = x_0$ and $x = x_1$ is approximately equal to the area of the trapezium $ABCD$; the approximation improves as x_1 is brought closer to x_0. The area of the trapezium is $\frac{1}{2} AD \times (AB + CD)$.

If we write f_0 for $f(x_0)$ and f_1 for $f(x_1)$ then the area under the curve is given approximately by $\frac{1}{2} h \, (f_0 + f_1)$.

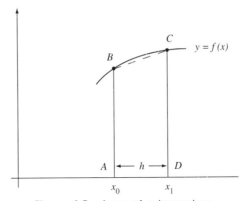

Figure 6.5 Area of a trapezium

In Figure 6.6 the area under the curve $y = f(x)$ between $x = a$ is split into n strips of equal width h. If $a = x_0$ and $b = x_n$ then $x_1 - x_0 = x_2 - x_1 = \cdots = x_n - x_{n-1} = h$, so that

$$h = \frac{b - a}{n}$$

The area under the curve is given approximately by the sum of the areas of the trapeziums. If $f_r = f(x_r)$ for $r = 0, 1, 2, \ldots, n$ then the combined area of the trapeziums is

$$\frac{h}{2}(f_0 + f_1) + \frac{h}{2}(f_1 + f_2) + \cdots + \frac{h}{2}(f_{n-1} + f_n)$$

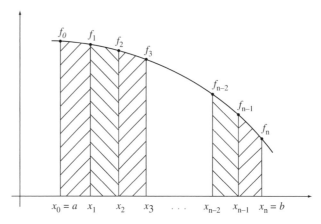

Figure 6.6 Area under several trapeziums

We therefore approximate the definite integral by the **trapezium rule** as

$$\int_a^b f(x)dx \approx \frac{h}{2}(f_0 + 2f_1 + 2f_2 + \cdots + 2f_{n-1} + f_n) \tag{6.1}$$

We call h the **step size**.

Examples
1. Use the trapezium rule with four strips then with eight strips to estimate the value of $I = \int_1^3 x^3 \, dx$. With four strips $h = 0.5$. Table 6.1 shows the preliminary calculations.

The coefficients in row 3 are the coefficients of f_r in (6.1) and the product of each coefficient with f_r is shown in row 4. The sum of the numbers in row 4 is multiplied by $\frac{1}{2}h$ to give I_T, the trapezium approximation to the integral I.

The sum of the numbers in row 4 is 82. Hence $I_T = \frac{1}{2} \times \frac{1}{2} \times 82 = 20.5$.

Table 6.1

x	1	$1\frac{1}{2}$	2	$2\frac{1}{2}$	3
x^3	1	$\frac{27}{8}$	8	$\frac{125}{8}$	27
coefficient	1	2	2	2	1
row 2 × row 3	1	$\frac{27}{4}$	16	$\frac{125}{4}$	27

Table 6.2

x	1	$1\frac{1}{4}$	$1\frac{1}{2}$	$1\frac{3}{4}$	2	$2\frac{1}{4}$	$2\frac{1}{2}$	$2\frac{3}{4}$	3
x^3	1	$\frac{125}{64}$	$\frac{27}{8}$	$\frac{343}{64}$	8	$\frac{729}{64}$	$\frac{125}{8}$	$\frac{1331}{64}$	27
coefficient	1	2	2	2	2	2	2	2	1
row 2 × row 3	1	$\frac{125}{32}$	$\frac{27}{4}$	$\frac{343}{32}$	16	$\frac{729}{32}$	$\frac{125}{4}$	$\frac{1331}{32}$	27

With eight strips the calculations are as shown in Table 6.2. The sum of the numbers in row 4 is 162. Hence $I_T = \frac{1}{2} \times \frac{1}{4} \times 161 = 20.125$. The exact value of the integral is 20. Hence with four strips the error, $e_T = I_T = 1$ is 0.5 whereas with eight strips the error is 0.125. Therefore by doubling the number of strips we have reduced the error by a factor of 4.

2. Find the approximate value of $I = \int_0^{\pi/2} \sin x \, dx$ using the trapezium rule with two

strips then with four strips, comparing your answers with the exact value.

(a) With two strips (Figure 6.7) the calculations are

$$I_T = \frac{1}{2} \times \frac{\pi}{4}\left(\sin 0 + 2\sin\frac{\pi}{4} + \sin\frac{\pi}{2}\right)$$
$$= \frac{\pi}{8}(\sqrt{2} + 1) = 0.94\,806$$

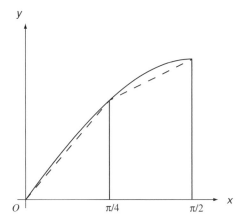

Figure 6.7   Approximating $I = \int_0^{\pi/2} \sin x \, dx$ using the trapezium rule

(b) With four strips we obtain

$$I_T = \frac{1}{2} \times \frac{\pi}{8} \left(0 + 2 \sin \frac{\pi}{8} + 2 \sin \frac{\pi}{4} + 2 \sin \frac{3\pi}{8} + 1 \right)$$
$$= 0.98\ 712$$

The actual value is $[- \cos x]_0^{\pi/2} = 1$, so the errors are as follows:

$$\varepsilon_2 = -0.05\ 194 \qquad \varepsilon_4 = -0.01\ 288$$

where ε_2 is the error using two strips and ε_4 is the error using four strips we could display the early calculations in a table, as in the first example. The error is divided by approximately 4 when the number of strips is doubled. ∎

Estimate of maximum absolute error

It can be shown that the error in a trapezium approximation to $\int_a^b f(x)dx$ is given by

$$e_T = -\frac{(b-a)}{12} h^2 \times f''(\xi)$$

where $h = \dfrac{b-a}{n}$, n being the number of strips, and where ξ is an unspecified value of x between a and b. However, we can determine an upper limit to $|e_T|$ by taking $M = \max\limits_{a \le x \le b} |f''(x)|$, i.e. M is the maximum value of the magnitude of $f''(x)$ as x is taken through all values from a to b.

We then write

$$|e_T| < \frac{(b-a)}{12} h^2 \times M$$

and this is of particular use in estimating an error when $f(x)$ is easily differentiated but when its integral cannot be found by calculus. Note that the magnitude of the error is approximately proportional to the square of the step length h. Hence if the number of strips is doubled, so the step length is halved, the magnitude of the error is approximately quartered.

Examples

1. (a) Estimate the maximum absolute error in the estimates of the integrals in the previous examples of this section.

   (b) How many strips would be needed to guarantee that the trapezium estimates of the integrals in (a) were accurate to 2 d.p., then to 4 d.p.?

Solution

(a) In the case of $I = \int_1^3 x^3 \, dx, f(x) = x^3, f''(x) = 6x, M = \max_{1 \leq x \leq 3} |6x| = 18$

Hence $|e_T| < \dfrac{(3 - 1)}{12} h^2 \times 18 = 3h^2$

With $n = 4, h = \dfrac{1}{2}, |e_T| < 0.75$; with $n = 8, h = \dfrac{1}{4}, |e_T| < 0.1875$

In the case of $I = \int_0^{\pi/2} \sin x \, dx, f(x) = \sin x, f''(x) = -\sin x$, and

$$M = \max_{0 \leq x \leq \pi/2} |-\sin x| = 1$$

Hence $|e_T| < \dfrac{\left(\dfrac{\pi}{2} - 0\right)}{12} h^2 \times 1 = \dfrac{\pi h^2}{24}$

With $n = 2, h = \dfrac{\pi}{4}, |e_T| < 0.0807$; with $n = 4, h = \dfrac{\pi}{8}, |e_T| < 0.0202$.

Notice that the actual errors were comfortably less than the predicted maximum error in each case. This is a consequence of the use of M in the error formula.

(b) For an accuracy of 2 d.p. we require $|e_T| < 0.005 = \dfrac{1}{200}$.

In the case of $I = \int_1^3 x^3 \, dx$ we require $3h^2 < \dfrac{1}{200}$ and since $h = \dfrac{3 - 1}{n} = \dfrac{2}{n}$ this

means $\dfrac{12}{n^2} < \dfrac{1}{200}$ or $n^2 > 2400$. The smallest integer which satisfies this inequality is $n = 49$. To guarantee 2 d.p. accuracy we take 49 strips; in practice we would probably take 50 strips.

For an accuracy of 4 d.p. we require $|e_T| < 0.000\,05 = \dfrac{1}{20\,000}$

This means $3h^2 < \dfrac{1}{20000}$ or $n^2 > 240\,000$. The smallest integer which satisfies this inequality is $n = 490$. To guarantee 4 d.p. accuracy we take 490 strips.

In the case of $I = \int_0^{\pi/2} \sin x \, dx$ the requirement for 2 d.p. accuracy is $\dfrac{\pi h^2}{24} < \dfrac{1}{200}$,

which leads to $n = 8$; for 4 d.p. the number of strips required is 80. Notice that the estimate of the number of strips required is likely to be pessimistic. This is a consequence of the use of M in the formula.

2. Determine $\int_1^2 e^{\sin x} \, dx$

We cannot perform the integration exactly (using calculus) but we can find an approximate value and can estimate the maximum error in that estimate. Take $h = 0.2$

so that $x_0 = 1, x_1 = 1.2, \ldots, x_5 = 2.0$. Then

$$\int_1^2 e^{\sin x}\, dx \approx \frac{0.2}{2}(f_0 + 2f_1 + \cdots + 2f_4 + f_5)$$

$$= 0.1(e^{\sin 1} + 2(e^{\sin 1.2} + \cdots + e^{\sin 1.8}) + e^{\sin 2})$$

$$\approx 2.597\ 03$$

To estimate the error we need to find $\max |f''(x)|$:

$$f'(x) = \cos x\, e^{\sin x} \qquad f''(x) = (\cos^2 x - \sin x)e^{\sin x}$$

The maximum value of $f''(x)$ occurs when $x = \dfrac{\pi}{2}$, verified by an exercise, and $f''\left(\dfrac{\pi}{2}\right) = -e$, so $|e_T| < \dfrac{(2-1)(0.2)^2(e)}{12} \approx 0.009\ 06$

Hence we may say that the value of the integral is in the interval 2.597 ± 0.009 ∎

Simpson's rule

A second method of estimating definite integrals relies on approximating a function over a double strip by a parabola rather than by two straight line segments, as in the case of the trapezium rule. This method, known as the **Simpson rule**, therefore requires an even number of strips in order to be applied. Figure 6.8(a) shows a pair of strips and the curve $y = f(x)$. First we find the area under the parabola shown in Figure 6.8(b).

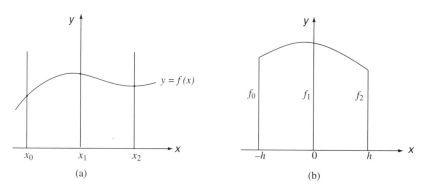

(a) (b)

Figure 6.8 Simpson's rule: (a) a pair of strips under the curve $y = f(x)$ and (b) finding the area under a parabola

Let the parabola be $y = ax^2 + bx + c$. Then since it passes through the points $(-h, f_0)$, $(0, f_1)$ and (h, f_2) we obtain the equations

$$ah^2 - bh + c = f_0 \quad \text{(i)}, \qquad c = f_1 \quad \text{(ii)}, \qquad ah^2 + bh + c = f_2 \quad \text{(iii)}$$

The exact value of the area under the parabola is given by

$$\int_{-h}^{h} (ax^2 + bx + c)dx = \left[a\frac{x^3}{3} + b\frac{x^2}{2} + cx \right]_{-h}^{h}$$

$$= \left(a\frac{h^3}{3} + b\frac{h^2}{2} + ch \right) - \left(-a\frac{h^3}{3} + b\frac{h^2}{2} + ch \right)$$

$$= 2a\frac{h^3}{3} + 2ch$$

From equation (ii) we see immediately that $c = f_1$. Adding (i) and (iii) gives $2ah^2 + 2c = f_0 + f_2$, so $2ah^2 = f_0 - 2f_1 + f_2$. The exact area is

$$\frac{h}{3}(f_0 - 2f_1 + f_2) + 2hf_1 = \frac{h}{3}(f_0 + 4f_1 + f_2)$$

Note that the actual values of x are unimportant; it is the fact they are spaced apart by h which matters.

If we now have four strips and the function values at the ordinates $x = x_0$, $x = x_1, \ldots, x = x_4$ are respectively $f_0, f_1, \ldots, f_4$ then the Simpson approximation becomes

$$\frac{h}{3}(f_0 + 4f_1 + f_2) + \frac{h}{3}(f_2 + 4f_3 + f_4) = \frac{h}{3}(f_0 + 4f_1 + 2f_2 + 4f_3 + f_4)$$

In general, Simpson's rule approximates $I = \int_{a}^{b} f(x)dx$ by

$$I_S = \frac{h}{3}(f_0 + 4f_1 + 2f_2 + \cdots + 4f_{n-1} + f_n)$$

where the number of strips, n, *must be even*.

Examples

1. Use Simpson's rule to estimate $I = \int_{0}^{\pi/2} \sin x \, dx$ using two strips then four strips. Calculate the error in each case and compare your results with those obtained by the trapezium rule. With two strips, $h = \dfrac{\pi}{4}$ and

$$I_S = \frac{\pi}{12}\left(\sin 0 + 4\sin\frac{\pi}{4} + \sin\frac{\pi}{2} \right) = 1.002\,28$$

With four strips, $h = \dfrac{\pi}{8}$ and

$$I_S = \frac{\pi}{24}\left(\sin 0 + 4\sin\frac{\pi}{8} + 2\sin\frac{\pi}{4} + 4\sin\frac{3\pi}{8} + \sin\frac{\pi}{2} \right) = 1.000\,13$$

Note how much more accurate these estimates are than the trapezium estimates with the same number of strips. In fact, the error is $O(h^4)$, rather than $O(h^2)$ as it was with the trapezium error. Hence, halving the step length reduces the error to one-sixteenth of its value.

2. Use Simpson's rule to estimate $I = \int_1^3 x^3\, dx$ using four strips. Comment on your results. With four strips $h = 0.5$. Hence the Simpson estimate is

$$I_S = \frac{1}{6}\left(1 + 4 \times \frac{27}{8} + 2 \times 8 + 4 \times \frac{125}{8} + 27\right) = 20$$

This is the exact answer. Since we are approximating a cubic function by two quadratics, this is an unexpected bonus.

Note that the amount of calculation effort in applying Simpson's rule is exactly the same as in applying the trapezium rule with the same number of strips. Since its accuracy is greater, Simpson's rule is almost always a superior method. ■

Estimates of maximum absolute error

We quote the result for the maximum magnitude of the error e_S in a Simpson approximation to $\int_a^b f(x)dx$. The error satisfies the relationship

$$|e_S| < \frac{(b-a)}{180}h^4 \times M \quad \text{where} \quad M = \max_{a \le x \le b} |f^{(\text{iv})}(x)|$$

Note how the formula is much more complicated than for the trapezium rule; since it involves a fourth derivative it can be applied only in very simple cases. Furthermore, since the fourth derivative of a cubic function is identically zero, the maximum error will be zero, confirming the result of the earlier example.

Example

Estimate the maximum error in the Simpson approximation with four strips to the integral

$$I = \int_0^{\pi/2} \sin x\, dx$$

First $M = \max_{0 \le x \le \pi/2} |\sin x| = 1$

Then $|e_S| < \dfrac{1}{180}\left(\dfrac{\pi}{2}\right)\left(\dfrac{\pi}{8}\right)^4 \times 1 = 0.000\,207\,5$

consistent with the estimate we obtained earlier. ■

Exercise 6.4

 1 Use the trapezium rule with (i) four strips and (ii) eight strips to estimate the following definite integrals:

(a) $\displaystyle\int_1^2 x^4\,dx$ 　　　　(b) $\displaystyle\int_1^2 x^5\,dx$ 　　　　(c) $\displaystyle\int_1^2 e^x\,dx$

(d) $\displaystyle\int_1^2 e^{-x}\,dx$ 　　　(e) $\displaystyle\int_1^2 \frac{1}{x}\,dx$ 　　　(f) $\displaystyle\int_0^{\pi/2} \cos x\,dx$

How do the estimates compare with the exact values?

 2 Estimate the maximum errors in the approximations of Question 1.

 3 Repeat Question 1 using Simpson's rule.

 4 Estimate the maximum errors in the approximations of Question 3.

 5 For $f(x) = e^{\sin x}$ prove that the maximum value in magnitude of $f''(x)$ over the range $[1, 2]$ occurs when $x = \pi/2$. Hence determine $\displaystyle\int_1^2 \sin x\,dx$ to 3 d.p. using the trapezium rule with $h = 0.25$. Estimate the error in the calculated values.

 6 Using $h = 0.25$ determine

$$\int_1^2 e^{\sin x}\,dx$$

using Simpson's rule. Why can we not use $h = 0.2$?

 7 Bessel functions model wave patterns with circular or spherical symmetry, e.g. the ripples on the surface of a pond formed by stone falling into it. Associated with these functions is the integral.

$$\frac{1}{\pi}\int_0^\pi \cos(\theta - \sin\theta)\,d\theta$$

Determine approximate values of the integral using

(a)　　The trapezium rule with $h = \pi/3$ 　　　　(b)　　Simpson's rule with $h = \pi/4$

Determine the error in case (a).

8 The **sinc function** which occurs in electronics and signal processing is defined by

$$\text{sinc}\,(x) = \int_0^x \frac{\sin t}{t}\, dt$$

Using Simpson's rule with $h = 0.5$, determine sinc (2); take $\dfrac{\sin 0}{0} = 1$. What is the value

of $\displaystyle\int_{-2}^2 \frac{\sin t}{t}\, dt$?

 9 The **prime number theorem** states that the number of prime numbers in an interval $[a, b]$ is approximately given

$$\int_a^b \frac{dx}{\ln x}$$

For $a = 100, b = 200$ use Simpson's rule with $h = 25$ to obtain an estimate of the number of primes that lie between 100 and 200; compare your estimate with the true value.

 10 The coordinates of points on a curve are given in the following table.

x	0	0.2	0.4	0.6	0.8	1.0	1.2	1.4
y	1	1.02	1.09	1.22	1.43	1.72	2.12	2.66

Use the trapezium rule to calculate

(a) the volume of revolution obtained by rotating the area under the curve bounded by the lines $x = 0, x = 1.4$ and the x-axis through $360°$ about the x-axis

(b) the centroid of this volume, i.e.

$$\bar{x} = \frac{\pi \displaystyle\int_0^{1.4} xy^2\, dx}{\pi \displaystyle\int_0^{1.4} y^2\, dx} \qquad (\bar{y} = 0)$$

 11 The **Normal Distribution** in probability theory gives rise to the **error function**

$$\text{erf}(x) = \frac{1}{\sqrt{2\pi}} \int_0^x e^{-z^2/2}\, dz$$

(a) Use Simpson's rule with h small enough to prove that $\text{erf}(1) = 0.3413$ to 4 d.p.

(b) Use integration by parts to prove that

$$I_2 = \int_0^1 z^2 e^{-z^2/2}\, dz = \int_0^1 e^{-z^2/2}\, dz - \frac{1}{e}$$

Deduce the value of I_2.

6.5 EXISTENCE OF INTEGRALS

There are several cases where we must look carefully at an integral to see whether it actually exists. In this section we examine some of the more straightforward instances.

Example   Discuss the existence of the following integrals. Do not attempt to evaluate them.

(a) $\displaystyle\int_a^b \frac{1}{(x-c)^2}\,dx$ (b) $\displaystyle\int_{-1}^1 \frac{1}{|x|^{1/2}(x+3)}\,dx$ (c) $\displaystyle\int \frac{1}{((x-1)(x-2)^4)^{1/3}}\,dx$

Solution

(a) If c is outside the interval $[a, b]$ then the integral exists.

(b) For very small values of $|x|$, $f(x) = \dfrac{1}{3|x|^{1/2}}$, but the indefinite integral $F(x) \approx \dfrac{2}{3}|x|^{1/2}$

and the value $x = -3$ lies outside the interval of integration, so the integral exists.

(c) Near $x = -1$, $f(x) \approx -\dfrac{1}{(x-1)^{1/3}}$ and $F(x) \approx -\dfrac{3}{2}(x-1)^{2/3}$

Near $x = 2$, $f(x) \approx -\dfrac{1}{(x-2)^{4/3}}$ and $F(x) \approx -\dfrac{3}{(x-2)^{1/3}}$

Hence the integral exists if $x = 2$ is outside the interval of integration.   ■

Infinite integrals

The definite integral of $f(x)$ is defined by $\displaystyle\int_a^b f(x)dx = F(b) - F(a)$ where $F'(x) = f(x)$.

Some integrals are defined over an infinite interval, i.e. $b \to \infty$ or $a \to -\infty$ or both, and these integrals will exist if $\lim_{b\to\infty} F(b)$ or $\lim_{a\to\infty} F(a)$ exists as necessary.

Example   Determine which of the following integrals exist:

(a) $\displaystyle\int_1^\infty x^p\,dx$ (b) $\displaystyle\int_a^\infty e^{-x}\,dx$ (c) $\displaystyle\int_2^\infty x\sqrt{x-2}\,dx$ (d) $\displaystyle\int_0^\infty xe^{x^2}\,dx$ (e) $\displaystyle\int_{-\infty}^\infty \frac{1}{1+x^2}\,dx$

Solution

(a) If X is a large positive number then $\displaystyle\int_1^X x^p\,dx = \left[\frac{x^{p+1}}{p+1}\right]_1^X = \frac{X^{p+1}-1}{p+1}$

If $p > -1$ then this fraction $\to \infty$ as $X \to \infty$.

If $p > -1$ then the fraction $\to -\dfrac{1}{p+1}$ as $X \to \infty$

If $p = -1$ then $\displaystyle\int_1^X x^p\,dx = \int_1^X \frac{1}{x}\,dx = \ln X \to \infty$ as $X \to \infty$

Hence the given integral exists when $p < -1$.

(b) $\int_a^X e^{-x}\,dx = e^{-a} - e^{-X} \to e^{-a}$ as $X \to \infty$. Hence the infinite integral exists for any $a > 0$

(c) $f(x)$ is defined when $x > 2$, and $f(x) = x^{3/2}$ for large x. Hence, for large x, $F(x) \approx \frac{2}{5}x^{5/2} \to \infty$ as $x \to \infty$. Therefore the infinite integral does not exist.

(d) $\int_0^X xe^{x^2}\,dx = \left[\frac{1}{2}e^{x^2}\right]_0^X = \frac{1}{2}(e^{X^2} - 1) \to \infty$ as $X \to \infty$. Therefore the infinite integral does not exist.

(e) If $f(x) = \dfrac{1}{1+x^2}$ then $f(-x) = f(x)$ so the integrand is an even function. Now

$$\int_0^X \frac{1}{1+x^2}\,dx = \tan^{-1}X - \tan^{-1}0 = \tan^{-1}X, \text{ since } \tan^{-1}0 = 0.$$

As $X \to \infty$, $\tan^{-1}X \to \dfrac{\pi}{2}$, so $\displaystyle\int_0^\infty \frac{1}{1+x^2}\,dx = \frac{\pi}{2}$. Therefore, because the integrand is

even, $\displaystyle\int_{-\infty}^\infty \frac{1}{1+x^2}\,dx = 2\int_0^\infty \frac{1}{1+x^2}\,dx = \pi$

Exercise 6.5

1   Which of the following integrals exist?

(a) $\displaystyle\int_{-2}^1 \sqrt{x+1}\,dx$ (b) $\displaystyle\int_{-2}^1 \sqrt{2x+11}\,dx$ (c) $\displaystyle\int_{-5}^2 \frac{1}{x(x+7)}\,dx$

(d) $\displaystyle\int_a^b \frac{e^x}{e^x+1}\,dx$ (e) $\displaystyle\int_a^b \frac{1}{x}\,dx$ (f) $\displaystyle\int_{-3}^{-2} \ln x\,dx$

(g) $\displaystyle\int_5^{10} \frac{1}{x(x-1)(x-2)}\,dx$ (h) $\displaystyle\int_a^b \frac{1}{x(x-1)(x-2)}\,dx$

(i) $\displaystyle\int_a^b \sqrt{1-4x}\,dx$ (j) $\displaystyle\int_a^b \frac{1}{\sqrt{x^2-2x+3}}\,dx$

2   Examine the existence of these infinite integrals

(a) $\displaystyle\int_{-\infty}^0 e^x\,dx$ (b) $\displaystyle\int_{-\infty}^0 e^{x^2/2}\,dx$ (c) $\displaystyle\int_0^\infty \frac{x}{1+x^2}\,dx$

3   Assuming that $\lim_{t\to\infty} t^n e^{-st} = 0$, $s > 0$ and that $\lim_{t\to\infty} f(t)e^{-st} = 0$, $s > 0$ if $|f(t)| < K$ for all t, where K is some constant, we can define the **Laplace transform** as

$$F(s) = L(f(t)) = \int_0^\infty e^{-st}f(t)\,dt$$

Find $L(1)$ and $L(t)$. Prove that $L(\sin t) = 1/(s^2+1)$. What is $L(\cos t)$?

SUMMARY

- **Integration by substitution**:

$$\int f(x)dx = \int g(u)du \qquad g(u) = f(x) \times \frac{dx}{du}$$

$$\int_a^b f(x)dx = \int_c^d g(u)du$$

when $x = a, u = c$ and when $x = b, u = d$

- **Properties of the definite integral**:

$$\int_b^a f(x)dx = -\int_a^b f(x)dx$$

$$\int_0^a f(a - x)dx = \int_0^a f(x)dx$$

$$\int_a^b f(a + b - x)dx = \int_a^b f(x)dx$$

- **Trigonometric power integrals**

n even   use double-angle formulae

n odd $\displaystyle\int \sin^n x\, dx,$ put $u = \cos x$

$\displaystyle\int \cos^n x\, dx,$ put $u = \sin x$

- **The t substitution**:

$$t = \tan\frac{x}{2} \qquad \frac{dt}{dx} = \frac{1}{2}\sec^2 x \qquad \sin x = \frac{2t}{1 + t^2} \qquad \cos x = \frac{1 - t^2}{1 + t^2}$$

- **The 'derivative' substitution:**

$$\int \frac{f'(x)}{f(x)} dx = \ln f(x)$$

- **Integration by parts**

$$\int u \frac{dv}{dx} dx = uv - \int \frac{du}{dx} v \, dx$$

- **Reduction formulae:** For example, if

$$I_n = \int_0^{\pi/2} \sin x \, dx \quad \text{then} \quad I_n = \frac{n-1}{n} I_{n-2}.$$

- **Rational functions:** use partial fractions to split the integrand.
- **Square roots:** use the substitutions $x = a \sin u$, $x = a \sinh u$, $x = a \cosh u$ for $\sqrt{a^2 - x^2}$, $\sqrt{a^2 + x^2}$, $\sqrt{x^2 - a^2}$ respectively
- **Trapezium rule**

$$\int_a^b f(x) dx \approx \frac{h}{2}(f_0 + 2f_1 + \cdots + 2f_{n-1} + f_n)$$

$$|e_S| < \frac{(b-a)h^2}{12} M \qquad M = \max_{a \le x \le b} |f''(x)|$$

- **Simpson's rule:**

$$\int_a^b f(x) dx \approx \frac{h}{3}(f_0 + 4f_1 + 2f_2 + \cdots + 2f_{n-2} + 4f_{n-1} + f_n) = I_S$$

$$|e_S| < \frac{(b-a)h^4}{180} M \qquad M = \max_{a \le x \le b} |f^{(iv)}(x)|$$

Answers

Note that in all cases an indefinite integral should include an arbitrary constant $+C$.

Exercise 6.1

1 (a) $\dfrac{(x+5)^6}{6}$

(b) $-\dfrac{1}{2(x+4)^2}$

(c) $-\sqrt{9-2x}$

(d) $\dfrac{1}{2}e^{2x}$

(e) $\dfrac{1}{a}e^{ax}$

(f) $\dfrac{5^x}{\ln 5}$

(g) $\sin^{-1}x$

(h) $-\sqrt{1-x^2}$

(i) $\dfrac{1}{3}\tan^{-1}x$

(j) $\dfrac{1}{2}\cosh^{-1}2x$

(k) $\sinh^{-1}\left(\dfrac{x}{5}\right)$

(l) $\dfrac{1}{a(p+1)}(ax+b)^{p+1}, p\neq -1$ and $\dfrac{1}{a}\ln|ax+b|, p=-1$

2 (a) $\dfrac{1}{4}e(e^4-1)$

(b) $\dfrac{\pi}{4}$

(c) $\dfrac{1}{4}\ln 2$

(d) $\dfrac{1}{3}(5\sqrt{5}-8)$

(e) $\ln(2+\sqrt{3})$

(f) $\dfrac{1}{5}$

(g) $\dfrac{1}{3}\tan^{-1}\left(\dfrac{1}{3}\right)$

(h) $\dfrac{5}{4}\ln 5+3$

3 (a) $\dfrac{1}{4}\sin 2x-\dfrac{x}{2}$

(b) $\dfrac{\pi a^2}{4}$

(c) $-\dfrac{1}{3}\cos x(2+\sin^2 x)$

4 (a) $\dfrac{3}{8}(1+x^2)^{4/3}$

(b) $\dfrac{1}{2}(1+\sin^2 x)$

(c) $\dfrac{7}{2}e^{x^2}$

(d) $\dfrac{1}{6}(1+x^4)^{3/2}$

(e) $\dfrac{1}{2}x^5(x+5)^5$

(f) 2

(g) $-4\ln\left(\dfrac{\sqrt{5}-1}{2}\right)+\sqrt{5}-2\approx 2.161$

(h) $\dfrac{1}{2}$

5 (a) $\dfrac{1}{3}\ln(x^3+3x)$

(b) $\dfrac{1}{2}\ln\left(\dfrac{x-1}{x+1}\right)$

(c) $\dfrac{1}{9}(3e^{-x}-x+\ln(e^x-3))$

(d) $\ln(1-\cos x)+\cos x$

(e) $\ln 2-\dfrac{1}{3}\ln 7$

(f) $\dfrac{1}{2}\ln\left(\dfrac{3}{2}\right)$

6 (a) $\dfrac{x}{2}+\dfrac{1}{4}\sin 2x$

(b) $\dfrac{1}{3}\sin x(2+\cos^2 x)$

(c) $\dfrac{1}{32}\sin 4x - \dfrac{1}{4}\sin 2x + \dfrac{3x}{8}$

(d) $\dfrac{1}{15}\sin x(8 + 4\cos^2 x + 3\cos^4 x)$

(e) $\dfrac{1}{8}\left(\dfrac{1}{4}\sin 4x - x\right)$

(f) $\dfrac{1}{2}(1 - \ln 2)$

(g) $\sin x + \dfrac{1}{2}\sin^2 x$

(h) $\dfrac{2}{15}$

7 (a) $\dfrac{1}{2}\sin x - \dfrac{1}{6}\sin 3x$

(b) $\dfrac{1}{4}\sin 2x + \dfrac{1}{8}\sin 4x$

(c) $\dfrac{1}{2}\sin x - \dfrac{1}{10}\sin 5x$

(d) $-\dfrac{1}{70}(\cos 2x)^5(5\sin^2 2x + 2)$

(e) $-\dfrac{2}{21}$

9 (a) $\dfrac{2}{\sqrt{5}}\tan^{-1}\left(\sqrt{5}\tan\dfrac{x}{2}\right)$

(b) $\dfrac{\pi}{3\sqrt{3}}$

10 (a) $\ln\left|\dfrac{1 + \sin x + \cos x}{1 + \cos x}\right| = \ln\left|1 + \tan\dfrac{x}{2}\right|$

(b) $\ln\left|\dfrac{1 + \sin x}{\cos x}\right|$

(c) $\dfrac{1}{\sqrt{7}}\ln\left|\dfrac{\sqrt{7}\cos x + \sin x - 1}{\sqrt{7}\cos x - \sin x - 1}\right|$

(d) $\ln\left|\dfrac{1 - \cos x}{\sin x}\right|$

(e) $\dfrac{1}{\sqrt{6}}\tan^{-1}\left(\dfrac{\sqrt{6}}{3}\right)$

(f) -2

11 (a) $\tan^{-1}(\sinh x)$ or $2\tan^{-1}(e^x)$

(b) $x + \dfrac{2}{e^{2x} + 1}$

(c) $\ln(\sqrt{x^2 - 2x + 17} + x - 1)$

(d) $\ln|\sqrt{x^2 - 9} + x|$

(e) (i) $\ln(1 + \sqrt{2})$ (ii) $\sqrt{2} - 1$

(iii) $\dfrac{1}{2}(\sqrt{2} - \ln(1 + \sqrt{2}))$

12 (a) $\dfrac{\pi}{4} - \dfrac{2}{3}$

(b) $\dfrac{26}{15}$

(c) $\dfrac{1}{2}\ln 3 - \dfrac{3}{2}\ln 2$

(d) $e^{-a} - e^{-b}$

(e) $\dfrac{1}{p - 1}(1 - x^{1-p}), \quad p > 1$

(f) $\dfrac{1}{2}(e - 1)$

Exercise 6.2

1 (a) $x \sin x + \cos x$ (b) $(x^3 - 3x^2 + 6x - 6)e^x$

     (c) $(x^2 - x + 1)e^x$ (d) $(2 - x^2) \cos x + 2x \sin x$

     (e) $\dfrac{1}{2} x \sin 2x + \dfrac{1}{4} \cos 2x$ (f) $\dfrac{a}{k^2} \cos kx + \left(\dfrac{ax + b}{k} \right) \sin kx$

     (g) $\dfrac{1}{k^3} (a^2 k^2 x^2 - 2akx(a - bk) + 2a^2 - 2abk + b^2 k^2)e^{kx}$

     (h) $\dfrac{1}{2} (x - \cos x \sin x)$

2 (a) $2 - \dfrac{5}{e}$ (b) $\dfrac{\pi}{4}$ (c) $\dfrac{2}{105} (64\sqrt{2} - 71)$

3 (b) $A = \dfrac{a}{a^2 + b^2}, \qquad B = -\dfrac{b}{a^2 + b^2}$

     (c) (i) $\dfrac{1}{2} (1 + e^{\pi/2})$ (ii) $\dfrac{4 + 5\sqrt{2}e^{-\pi/2}}{26}$

4 (a) $x \tan^{-1} x - \dfrac{1}{2} \ln(x^2 + 1)$ (b) $x \sinh^{-1} x - \sqrt{x^2 + 1}$

     (c) $x \tanh^{-1} x + \dfrac{1}{2} \ln(x^2 - 1), |x| > 1$

5 (a) $\dfrac{x^3}{3} \ln x - \dfrac{x^3}{9}$ (b) $\dfrac{1}{4} (2x^2 - 1) \sin^{-1} x + \dfrac{1}{4} x\sqrt{1 - x^2}$

     (c) $\dfrac{nx^{n+1} + 1}{(n + 1)^2}$

6 (a) $x \ln x - x$

     (b) (i) 2   (ii) $y = 2x - e^2$ (iii) $\left(\dfrac{1}{2} e^2, 0 \right)$

8 (a) $\dfrac{2}{a^4} \left(3 - \dfrac{8}{e} \right)$ (b) $\dfrac{6}{a^4}$

9 $\dfrac{I_0 V_0}{2\omega}, 0$

10 $4\pi I an$

Exercise 6.3

1 (a) $\dfrac{3x}{2} - \dfrac{9}{4}\ln|2x+3|$

(b) $2x - 5\ln|x+3|$

(c) $-\dfrac{1}{3}x(x^2 + 3x + 12) - 8\ln|x-2|$

(d) $\ln\left|\dfrac{x-2}{x+3}\right|$

(e) $\ln|(x+1)(x-2)^2|$

(f) $\ln\left|\dfrac{x-1}{x+1}\right| + \dfrac{1}{2}\ln|x^2 - 1| + \dfrac{1}{2}x^2$

(g) $4\ln|x-2| - \dfrac{11}{x-2}$

(h) $-\dfrac{9}{20}\ln|x+2| - \dfrac{19}{4}\ln|x-2| + \dfrac{26}{5}\ln|x-3|$

2 (a) $\dfrac{1}{2}\ln 2$

(b) $\ln\dfrac{4}{3}$

(c) $-\dfrac{1}{11}\left(2\ln 7 + 13\ln\dfrac{5}{2}\right) \approx -1.437$

(d) $-\dfrac{43}{4}\ln 5 - \dfrac{239}{28}\ln 3 + \dfrac{270}{7}\ln 2 \approx 0.0568$

(e) $\dfrac{1}{88}(416\ln 2 - 75\ln 3 - 133\ln 5) \approx -0.092$

(f) It does not exist, the denominator of the integrand is zero between $x = 2$ and $x = 3$.

(g) $\dfrac{1}{16}\ln\dfrac{19}{11} \approx 0.0342$

3 (a) $\dfrac{1}{\sqrt{29}}\ln\left|\dfrac{2x - \sqrt{29} + 1}{2x + \sqrt{29} + 1}\right|$ provided that $x \neq \dfrac{1}{2}(\pm\sqrt{29} - 1)$

(b) $\dfrac{1}{2}\left(\left(1 + \dfrac{3}{\sqrt{65}}\right)\ln(4x + \sqrt{65} + 1) + \left(1 - \dfrac{3}{\sqrt{65}}\right)\ln(4x - \sqrt{65} + 1)\right)$ provided that $4x \pm \sqrt{65} + 1 \neq 0$

(c) $\dfrac{1}{2}\left(\left(1 + \dfrac{9}{\sqrt{73}}\right)\ln(6x + \sqrt{73} + 11) + \left(1 - \dfrac{9}{\sqrt{73}}\right)\ln(6x - \sqrt{73} + 11)\right)$ provided that $6x \pm \sqrt{73} + 11 \neq 0$

(d) $\dfrac{1}{\sqrt{29}}\ln\left(\dfrac{1}{10}(39 - 7\sqrt{29})\right) = -0.378$

(e) $\ln 6 - \left(1 + \dfrac{1}{2\sqrt{6}}\right)\ln(1 + \sqrt{6}) - \left(1 - \dfrac{1}{2\sqrt{6}}\right)\ln(\sqrt{6} - 1) = 0.0053$

4 (a) $\ln(x + 2) + x$

(b) $\ln(x^2 + x + 1) - 4\sqrt{3}\tan^{-1}\left(\dfrac{2x + 1}{\sqrt{3}}\right)$

(c) $\ln(x^2 + 2x + 5) - \dfrac{1}{2}\tan^{-1}\left(\dfrac{x + 1}{2}\right)$

(d) $\tan^{-1}x - \dfrac{x}{x^2 + 1}$

(e) $\tan^{-1}x + \dfrac{1}{2}\ln(x^2 + 2)$

(f) $\dfrac{5}{2}\tan^{-1}x + \dfrac{5}{2(x^2 + 1)}$

(g) $\dfrac{1}{2}\ln(1 + \cos^2 x) - \ln|\cos x|$

(h) $\ln|1 + \tan\theta| + \dfrac{2}{\sqrt{3}}\tan^{-1}\left(\dfrac{2\tan\theta - 1}{\sqrt{3}}\right)$

(i) $\dfrac{1}{2}\tan^{-1}\left(\dfrac{1}{2}\right) + \ln\dfrac{5}{4} - \dfrac{1}{5} \approx 0.2550$

(j) $\sqrt{5}\left(\tan^{-1}\left(\dfrac{1}{\sqrt{5}}\right) - 2\tan^{-1}\left(\dfrac{2}{\sqrt{5}}\right)\right) + \dfrac{5}{\sqrt{2}}\left(\tan^{-1}\left(\dfrac{3}{\sqrt{2}}\right) - \tan^{-1}(\sqrt{2})\right)$
$+ \dfrac{1}{2}\ln\left(\dfrac{11}{6}\right) \approx 0.2303$

6 (a) $\dfrac{x + a}{a(a^2 - x^2)^{1/2}}$

(b) $\sqrt{x^2 + 2x + 3} - 2\ln(\sqrt{x^2 + 2x + 3} + x + 1)$

(c) $-\dfrac{1}{\sqrt{7}}\sin^{-1}\left(\dfrac{5 + 2x}{3\sqrt{2}(x - 1)}\right)$

(d) $\dfrac{1}{a}\left(\dfrac{x - a}{x + a}\right)^{1/2}$

(e) $3\sqrt{3x^2 + 4x - 4} - \dfrac{4}{\sqrt{3}}(\sqrt{9x^2 + 12x - 12} + 3x + 2)$

7 If $a > b$, $\dfrac{2}{(a^2 - b^2)^{1/2}}\tan^{-1}\left(\left(\dfrac{a - b}{a + b}\right)\tan\dfrac{\theta}{2}\right)$

If $a < b$, $\dfrac{1}{(b^2 - a^2)^{1/2}}\ln\left|\dfrac{(a + b)^{1/2} + (b - a)^{1/2}\tan\dfrac{\theta}{2}}{(a + b)^{1/2} - (b - a)^{1/2}\tan\dfrac{\theta}{2}}\right|$

9 $1 - \dfrac{5}{2e} = 0.0803$

10 $\dfrac{61}{31}$

11 (a) $i(t) = \dfrac{E}{R}(1 - e^{-Rt/L})$ (b) $1 - e^{-2} = 0.865$ amps

Exercise 6.4

1 True values

     (a)   6.2   (b)   10.5   (c)   4.6708

     (d)   0.2325   (e)   0.6931   (f)   1

Four strips

     (a)   6.3457   error $= 0.1457$ (b)   10.8896   error $= 0.3896$

     (c)   4.6951   error $= 0.0243$ (d)   0.2338   error $= 0.0013$

     (e)   0.6970   error $= 0.0039$ (f)   0.9871   error $= -0.0129$

Eight strips

     (a)   6.2365   error $= 0.0365$ (b)   10.5976   error $= 0.0976$

     (c)   4.6769   error $= 0.0061$ (d)   0.2328   error $= 0.0003$

     (e)   0.6941   error $= 0.0010$ (f)   0.8876   error $= -0.0024$

2 Four strips

     (a)   0.25   (b)   0.4167   (c)   0.0385

     (d)   0.0019   (e)   0.0043   (f)   0.0202

Eight strips

     (a)   0.0625   (b)   0.1042   (c)   0.0096

     (d)   0.0005   (e)   0.0010   (f)   0.0050.

3 Four strips

     (a)   6.20052   error $= 0.00052$ (b)   10.5039   error $= 0.0039$

     (c)   4.6709   error $= 0.0001$ (d)   0.232549   error $= 5 \times 10^{-6}$

     (e)   0.69325   error $= 0.00011$ (f)   1.000135   error $= 0.000135$

Eight strips

     (a)   6.20003   error $= 0.00003$ (b)   10.50024   error $= 0.00024$

     (c)   4.6707806   error $= 6 \times 10^{-6}$ (d)   0.2325447   error $= 3 \times 10^{-7}$

     (e)   0.693155   error $= 7 \times 10^{-6}$ (f)   1.0000083   error $= 0.8 \times 10^{-6}$

4 Four strips

(a)	0.000 52	(b)	0.0052	(c)	0.000 16
(d)	8×10^{-6}	(e)	0.000 52	(f)	0.000 21

Eight strips

(a)	0.000 033	(b)	0.000 33	(c)	1×10^{-5}
(d)	5×10^{-7}	(e)	8.2×10^{-5}	(f)	1.3×10^{-5}

5 $f'''(x) = \cos x (2 - 3 \sin x - \sin^2 x) \exp(\sin x)$

$\qquad = 0$ if $\cos x = 0$ only $\quad (1 \le x \le 2)$

$f''\left(\dfrac{\pi}{2}\right) = -\sin\left(\dfrac{\pi}{2}\right) \cdot e^{\sin \pi/2} = -e.$ Integral $= 2.597, \varepsilon_2 \approx 0.009\,06$

6 2.6048; $h = 0.2$ means $n = 5$, not possible for Simpson's rule as n is odd.

7 (a) 0.4398, $\varepsilon_2 \approx 0.3655$ $\qquad\qquad$ (b) 0.4465

8 sinc $(2) \approx 1.6055$

$\dfrac{\sin t}{t}$ is even, so $\displaystyle\int_{-2}^{2} \dfrac{\sin t}{t} dt = 2$ sinc$(2) \approx 3.2110$

9 20.066; there are 21 primes between 100 and 200

10 (a) Volume $= 10.84$ $\qquad\qquad$ (b) $(\bar{x}, \bar{y}) = (0.957, 0)$

11 (b) $I_2 = 0.4876$

Exercise 6.5

1 (a) no, $x < -1$ leads to an undefined square root

$\quad$ (b) yes

$\quad$ (c) no, $x = 0$ leads to an undefined integrand

$\quad$ (d) yes

$\quad$ (e) yes if $a > b > 0$ or if $a < b < 0$

$\quad$ (f) no, $\ln x$ requires $x > 0$

$\quad$ (g) yes

$\quad$ (h) yes if 0, 1 and 2 lie outside the interval $a \le x \le b$

(i)   yes if $a > \dfrac{1}{4}$

(j)   yes $(x^2 - 2x + 3 \geq 2 > 0)$

2 (a)   Put $u = -x$; integral exists.   (b)   Integral does not exist.

(c) $F(x) \approx \dfrac{1}{2}\ln(1 + x^2)$; integral does not exist.

3 $\dfrac{1}{s}, \dfrac{1}{s^2}, \dfrac{s}{s^2 + 1}$

7 GEOMETRY AND CALCULUS

Introduction

Conic sections have many applications in science and engineering. These curves describe such diverse phenomena as the shape of a radio telescope and the path of streamline flow around a corner. Polar coordinates allow us to describe some curves more concisely. The interaction between calculus and geometry is relevant to curvature as well as to the vector description of motion in two and three dimensions.

Objectives

After working through this chapter you should be able to

- Recognise the standard equations of the parabola, the ellipse and the hyperbola
- Identify the main features of the parabola, the ellipse and the hyperbola
- Understand the effects of axis rotation and axis translation on the equation of a two-dimensional curve
- Describe curves by polar coordinates
- Sketch the graph of a curve described in parametric form
- Understand and calculate the curvature at a point on a curve
- Differentiate a vector function of time
- Work out radial and transverse components of velocity and acceleration from the displacement of an object in vector form
- Express the length along a curve between two given points as a definite integral
- Find the area under a curve described in polar coordinates
- Describe surfaces in three dimensions.

7.1 THE CONIC SECTIONS

A curve in the x–y plane whose equation is of the form

$$ax^2 + by^2 + 2hxy + 2gx + 2fy + c = 0 \qquad (7.1)$$

where a, b, c, f, g and h are constant can be proved to be a **conic section**, i.e. the curve of intersection of a circular cone and a plane.

Types of conic

Let the axis of the cone in Figure 7.1 be vertical and let the generators be inclined at an angle β to the vertical. They meet at the vertex V. The plane cutting the cone is inclined at an angle α to the vertical. Table 7.1 summarises the cases.

We looked at the basic properties of the parabola, ellipse and hyperbola (Figure 7.2) in Chapter 9 of *Foundation Mathematics*, but it is important to know that equation (7.1) represents such curves in general, particularly if their centres are displaced from the origin and the axes of symmetry are inclined at an angle to the coordinate axes.

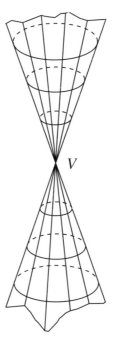

Figure 7.1 A double cone

Table 7.1

	Plane does not contain V	Plane contains V
$\alpha = \dfrac{\pi}{2}$	Circle	Point
$\beta < \alpha < \dfrac{\pi}{2}$	Ellipse	Point
$\alpha = \beta$	Parabola	Straight line (two coincident straight lines)
$\alpha < \beta$	Hyperbola (two branches)	Pair of straight lines

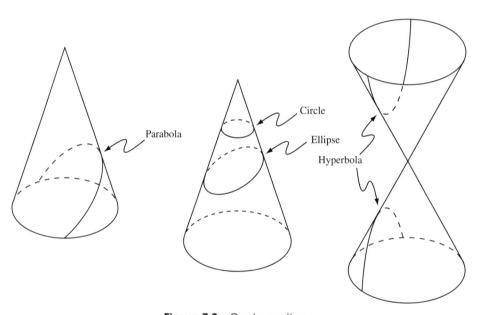

Figure 7.2   Conic sections

Figure 7.3 shows examples for the parabola, ellipse and hyperbola. In this chapter we shall see how a shift of origin or a rotation of axes relate a conic section to equation (7.1) and vice-versa, but a systematic reduction and resolution of the equation lies beyond our scope. Nonetheless, we can identify curves from the general equation in special cases.

Examples   1. Identify the conic section when the constants in the general equation (7.1) have the following values

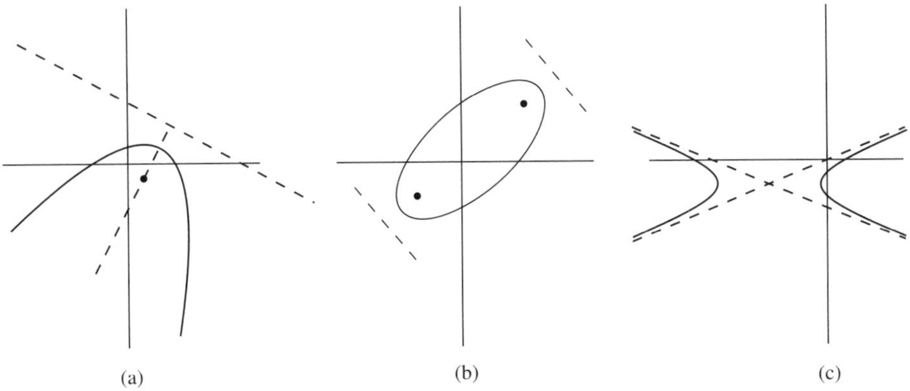

Figure 7.3 Non-standard examples of conic section: (a) parabola, (b) ellipse and (c) hyperbola

(a) $a = b = h = 0$ $\quad$ $g,\ f,\ c \neq 0$

(b) $a = b = 1$ $\quad$ $h = f = g = c = 0$

(c) $h \neq 0$ $\quad$ $a,\ b = 0$

(d) $a = 1$ $\quad$ $b = 4$ $\quad$ $h = g = f = 0$ $\quad$ $c = -1$

Solution

(a) $2gx + 2fy + c = 0$ $\quad$ a single straight line

(b) $x^2 + y^2 = 0$ $\quad$ a single point, namely the origin $(0,0)$ because $x = y = 0$

(c) $xy = 0$ $\quad$ two straight lines, namely the coordinate axes

(d) $x^2 + 4y^2 = 1$ $\quad$ an ellipse centred on the origin with semi-major and semi-minor axes of lengths 1 and $\dfrac{1}{2}$ and foci on the x-axis at $\left(\pm \sqrt{\dfrac{3}{2}}, 0 \right)$

2. (a) Identify the conic

$$2x^2 + xy - y^2 + x - 2y = 1 = 0,$$
$$\text{i.e. } (2x - y - 1)(x + y + 1) = 0$$

(b) Prove that no conic can exist and satisfy $4x^2 - 12xy + 9y^2 + 11 = 0$

Solution

(a) The factorised form leads to either

$$2x - y - 1 = 0 \quad \text{or} \quad x + y + 1 = 0$$

i.e. the conic is reduced to two straight lines.

(b) $4x^2 - 12xy + 9y^2 = (2x - 3y)^2$. The expression is non-negative, so when 11 is added it is impossible for the left-hand side of the equation to equal zero. ■

In Chapter 9 of *Foundation Mathematics* we met the principal conic sections of parabola, ellipse and hyperbola in the context of locus and examined the properties of the circle in detail. We now examine the fundamental properties of each of these curves, together with their parametric forms.

Parabola

This is the locus of a point $P(x, y)$ equidistant from a given point and a given straight line. The x-axis is taken through the vertex along the line of symmetry. The y-axis is perpendicular to the x-axis through the vertex (Figure 7.4).

Let the given line be $x = -a$ and the given point be $(a, 0)$. Then

$$PS = PM \qquad \text{and therefore } PS^2 = PM^2$$

Hence $(x - a)^2 + y^2 = (x + a)^2$. This reduces to the equation in **standard form**:

$$y^2 = 4ax$$

The parametric form is $x = at^2$, $y = 2at$ where $t > 0$ covers the upper branch in quadrant 1 and $t < 0$ the lower branch in quadrant 4; Figure 7.4(b) gives some examples.

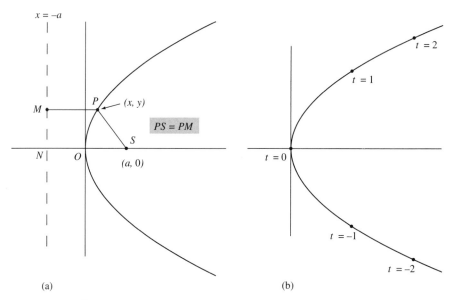

Figure 7.4 Parabola: (a) standard form and (b) parametric form

Examples

1. (a) Show that the chord joining the points $P_1(at_1^2, 2at_1)$ and $P_2(at_2^2, 2at_2)$ on the parabola $y^2 = 4ax$ has equation

$$2x - (t_1 + t_2)y + 2at_1t_2 = 0$$

(b) Hence show that the equation of the tangent at the point $P(at^2, 2at)$ is

$$x - ty + at^2 = 0$$

and that the equation of the normal at P is

$$tx + y = at^3 + 2at$$

(c) Show further that the locus of the midpoint of the chord P_1P_2, where P_1 is fixed, is itself a parabola. Draw a sketch to illustrate this.

Solution

(a) The equation of the straight line P_1P_2 is

$$\frac{y - at_1^2}{at_1^2 - at_2^2} = \frac{x - 2at_1}{2at_1 - 2at_2}$$

Cancel the common factor $a(t_1 - t_2)$ in the denominator to obtain

$$\frac{y - at_1^2}{t_1 + t_2} = \frac{x - 2at_1}{2}$$

Cross-multiplying gives the required equation.

(b) As $t_2 \to t_1(=t)$ the chord becomes a tangent at the point P, so putting $t_2 = t_1 = t$ and cancelling the factor 2 in the denominator, we obtain the required equation.

The tangent has gradient $\dfrac{1}{t}$ so the normal has gradient $-t$. Its equation is

$$(y - 2at) = -t(x - at^2)$$

i.e. $y + tx = at^3 + 2at$, as required.

(c) The midpoint of the straight line joining $(at_1^2, 2at_1)$ and $(at_1^2, 2at_2)$ has coordinates

$$\left(\tfrac{1}{2}a(t_1^2 + t_2^2), a(t_1 + t_2) \right), \quad \text{i.e. } x = \tfrac{1}{2}a(t_1^2 + t_2^2), \quad y = a(t_1 + t_2)$$

If t_1 is taken to be fixed, then

$$x - \tfrac{1}{2}at_1^2 = \tfrac{1}{2}at_2^2, \quad y - at_1 = at_2, \quad \text{so} \quad (y - at_1)^2 = 2a\left(x - \tfrac{1}{2}at_1^2 \right)$$

which is the equation of a parabola whose vertex is at $\left(\tfrac{1}{2}at_1^2, at_1 \right)$, i.e. halfway along the straight line joining the origin to P. Figure 7.5 illustrates the situation. ■

A more general equation for the parabola emerges if the vertex is displaced from the origin and/or the axis of symmetry is rotated.

Example

Determine the equations of the parabolas with

(a) focus $(0, 1)$,   directrix $y = 0$;   (b) focus $(1, 0)$,   directrix $y = x$

Draw sketches of the curves.

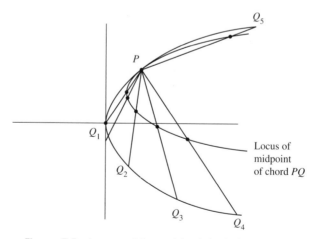

Figure 7.5 Locus of the midpoint of chord PQ

Solution

(a) Refer to Figure 7.6(a). The relationship $PS^2 = PM^2$ becomes

$$x^2 + (y-1)^2 = y, \quad \text{i.e. } y = \frac{1}{2}(x^2 + 1)$$

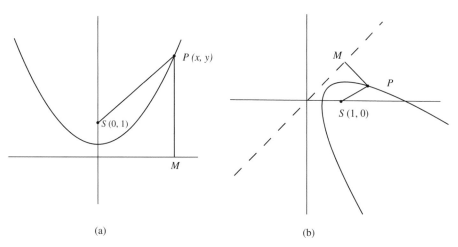

Figure 7.6 Two parabolas: (a) focus $(0,1)$, directrix $y = 0$ and (b) focus $(1,0)$, directrix $y = x$

(b) Refer to Figure 7.6(b). This time $PS^2 = PM^2$ leads to

$$(x-1)^2 + y^2 = \frac{(x-y)^2}{2}$$

so $2((x-1)^2 + y^2) = (x-y)^2$, i.e. $x^2 + 2xy + y^2 - 4x + 2 = 0$. This equation was derived from the formula giving the distance from the point $(1,0)$ to the straight line $x - y = 0$. ∎

Ellipse

The ellipse (Figure 7.7) may be defined in a way similar to a parabola with the focus–directrix property $PS = ePM$, where $e < 1$ is called the **eccentricity**.

As with the parabola it is evident that the ellipse is symmetric about the perpendicular to the directrix, SN (Figure 7.8). Also, since $PS/PM = e$, a further point Z must exist on the perpendicular and on the curve such that $ZS/ZN = e$.

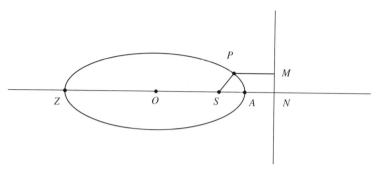

Figure 7.7 An ellipse

So far we have established that Z exists and there is symmetry about the x-axis, but nothing more. Take the origin to be the midpoint of ZA and the other distances shown $s = OS$, $a = OA$, $n = ON$. Then

$$AS = eAN \Rightarrow a - s = e(n - a)$$
$$ZS = eZN \Rightarrow a + s = e(n + a)$$

i.e. $\qquad s = ae \qquad n = \dfrac{a}{e}$

These two equations can be solved to give

$$(PS)^2 = e^2(PM)^2 \Rightarrow (x - ae)^2 + y^2 = e^2\left(\frac{a}{e} - x\right)^2$$

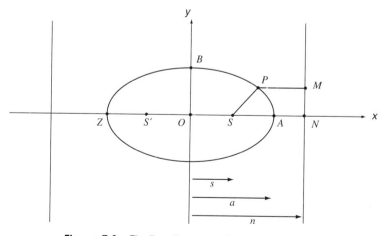

Figure 7.8 Finding the equation of an ellipse

If we write $b^2 = a^2(1 - e^2)$ then we obtain the **standard equation** of the ellipse.

$$\frac{x^2}{a^2} + \frac{y^2}{b^2} = 1 \tag{7.3}$$

Not only is there symmetry about the x-axis but also about the y-axis, and reflecting the curve in the y-axis establishes the existence of another focus and another directrix. The chosen origin is the **centre** of the ellipse.

To draw an ellipse, place two pins on a sheet of paper connected by a loose string (Figure 7.9). By keeping the string taut we can trace out the curve.

Why does this method produce an ellipse?

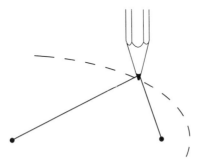

Figure 7.9 Drawing an ellipse

$$\text{Length of string} = PS + PS' = e(M'P + PM)$$
$$= eM'M = e\,\frac{2a}{e} = 2a$$

a constant, equal to the length of the string (Figure 7.10). This length is called the **major axis** of the ellipse.

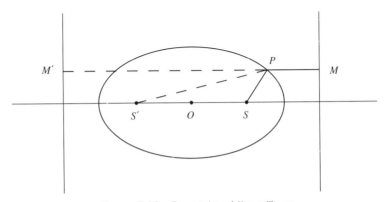

Figure 7.10 Property of the ellipse

We have the following results:

Foci	$(ae, 0)$ and $(-ae, 0)$
Directrices	$x = \pm \dfrac{a}{e}$
Semi-major axis	$OA = a$
Semi-minor axis	$OB = b$
Parametric form	$x = a\cos\theta \qquad y = b\sin\theta$

This last result follows since $\dfrac{x^2}{a^2} + \dfrac{y^2}{b^2} \equiv \cos^2\theta + \sin^2\theta = 1$

Consider Figure 7.11. $x = a\cos\theta$, $y = b\sin\theta$ are the coordinates of P on the ellipse shown, whose foci are S and S'. Also drawn is the **auxiliary circle**, a circle centred at the origin and of radius a. P' is the point on that circle which lies directly above P, i.e. has the same x-coordinate, and θ is the **polar angle** in the auxiliary circle. The ratio $PX : P'X = \dfrac{b}{a}$ at all positions of P.

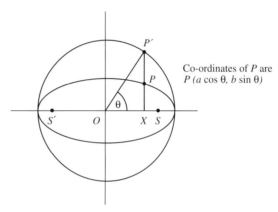

Co-ordinates of P are
$P\,(a\cos\theta, b\sin\theta)$

Figure 7.11 Auxiliary circle

This parametric form for the ellipse is very useful but it must always be remembered that θ is the polar angle $P'OX$ in the auxiliary circle; it is not the ellipse polar angle $P\hat{O}X$ subtended at P. Avoid this pitfall!

Examples

1. (a) Determine the lengths of the semi-major and semi-minor axes for the ellipse $x^2 + 2y^2 = 8$. Also determine the coordinates of the foci and equations of the directrixes. Sketch the ellipse.

   (b) Find the equations of the tangents which are parallel to the line $y = 2x$.

Solution

   (a) The equation can be rearranged to give $\dfrac{x^2}{8} + \dfrac{y^2}{4} = 1$, so $a = 2\sqrt{2}, b = 2$. The foci are thus on the x-axis at $(\pm 2, 0)$. Now $b^2 = a^2(1 - e^2)$, so $1 - e^2 = \dfrac{4}{8} = \dfrac{1}{2}$. Thus $e = \dfrac{1}{\sqrt{2}}$. The directrices are $x = \pm 4$. The ellipse is sketched in Figure 7.12.

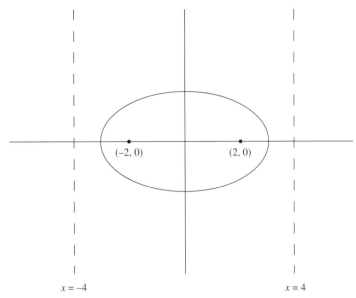

Figure 7.12 The ellipse $x^2 + 2y^2 = 8$

(b) For the tangents parallel to $y = 2x$, differentiate the equation of the ellipse to obtain the equation

$$\frac{2x}{8} + \frac{2y}{4}\frac{dy}{dx} = 0,$$

i.e. $\dfrac{dy}{dx} = -\dfrac{x}{2y},$

at the points of contact but $\dfrac{dy}{dx}$ must equal 2.

Hence $y = -\dfrac{x}{4}$ at the points which satisfy $\dfrac{x^2}{8} + \dfrac{x^2}{16} = 1$. Then $x^2 = \dfrac{16}{3}$, so $x = \pm\dfrac{4}{\sqrt{3}}$ and $y = \mp\dfrac{1}{\sqrt{3}}$. The equations of the tangents are

$$y + \frac{1}{\sqrt{3}} = 2\left(x - \frac{4}{\sqrt{3}}\right) \quad \text{and} \quad y - \frac{1}{\sqrt{3}} = 2\left(x + \frac{4}{\sqrt{3}}\right)$$

i.e. $y = 2x - 3\sqrt{3}$ and $y = 2x + 3\sqrt{3}$

2. Determine the coordinates of the points of intersection of the ellipse $\dfrac{x^2}{a^2} + \dfrac{y^2}{b^2} = 1$ and the parabola $y^2 = 4ax$ and the relationship between a and b, if the two cut at right angles. Draw an illustrative sketch and shade in the region of the plane where

$$y^2 < 4ax \qquad \frac{x^2}{a^2} + \frac{y^2}{b^2} > 1$$

The tangents to the conics have gradients satisfying the equations

$$2\frac{x}{a^2} + \frac{2y}{b^2} \times \frac{dy}{dx} = 0, \qquad \text{i.e.} \quad \frac{dy}{dx} = y' = -\frac{b^2 x}{a^2 y}$$

So $2yy' = 4ax$, i.e. $y' = \dfrac{2a}{y}$. If the two are perpendicular at the intersection points, then $\dfrac{-b^2 x}{a^2 y} \times \dfrac{2a}{y} = -1$, i.e. $ay^2 = 2b^2 x$. Now $y^2 = 4ax$, so that $4a^2 = 2b^2$ and $\dfrac{a}{b} = \dfrac{1}{\sqrt{2}}$

Figure 7.13 illustrates the situation. The foci of the ellipse lie on the y-axis.

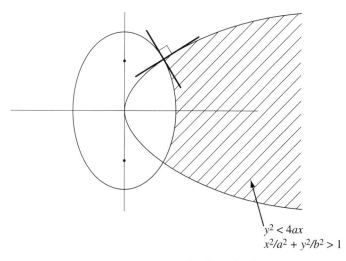

$y^2 < 4ax$
$x^2/a^2 + y^2/b^2 > 1$

Figure 7.13   Intersection of the ellipse $x^2/a^2 + y^2/b^2 = 1$ with the parabola $y^2 = 4ax$ ■

Hyperbola

The hyperbola (Figure 7.14) is also defined by the focus–directrix property, $PS = ePM$, this time with $e > 1$. With the same notation as for the ellipse

$$AS = eAN \Rightarrow s - a = e(a - n)$$
$$ZS = eZN \Rightarrow s + a = e(a + n)$$

These two equations can be solved to give

$$s = ae \qquad n = \frac{a}{e}$$

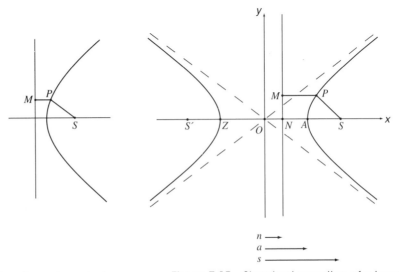

Figure 7.14 Forming a hyperbola

Figure 7.15 Standard equation of a hyperbola

Now

$$(PS^2) = e^2(PM)^2 \Rightarrow (x - ae)^2 + y^2 = e^2\left(x - \frac{a}{e}\right)^2$$

If we write $b^2 = a^2(e^2 - 1)$ then we obtain the **standard equation** of the hyperbola

$$\frac{x^2}{a^2} - \frac{y^2}{b^2} = 1 \tag{7.4}$$

The hyperbola is contained within the pair of straight lines $\dfrac{x^2}{a^2} - \dfrac{y^2}{b^2} = 0$, i.e. $y = \pm\dfrac{b}{a}x$ which are the **asymptotes**, shown as dashed lines in Figure 7.15. There is complete symmetry in the x and y axes, so a reflected branch of the hyperbola exists to the left with its own focus S', also contained within the asymptotes (Figure 7.15).

We therefore have the results:

Foci	$(\pm ae, 0)$
Directrices	$x = \pm \dfrac{a}{e}$
Parametric form	$x = a\cosh\theta \qquad y = b\sinh\theta$

This last result follows from

$$\frac{x^2}{a^2} - \frac{y^2}{b^2} = \cosh^2\theta - \sinh^2\theta \equiv 1$$

Like the parabola, the hyperbola is an open curve. It is left as an exercise for you to prove that $|SP - S'P| = 2a$, but there is an important application for this result.

Example

The principle of **sound ranging** as used in many sea and air navigational systems is based upon monitoring, the exact time taken by a sound emanating from an unknown point on the ground to reach observers at three distinct points.

The time taken for the sound to reach an observer is directly proportional to the distance travelled, so if two observers are involved, the difference in times relate directly to the differences in distances. The locus of these differences thus lies on a hyperbola whose foci are the observation points (Figure 7.16).

A template can be used to sketch the locus of P, but later in the exercises we see that a hyperbola, like an ellipse, can be drawn by a simple construction method.

When three observers are involved, A, B and C, there are three hyperbolae from A and B, B and C, and C and A. These meet in a point.

A special hyperbola is the **rectangular hyperbola**, i.e. a hyperbola whose asymptotes are at right angles. By convention (and for simplicity) we rotate the axes by $45°$ so the coordinate axes act as asymptotes. The equation becomes

$$xy = c^2 \tag{7.5}$$

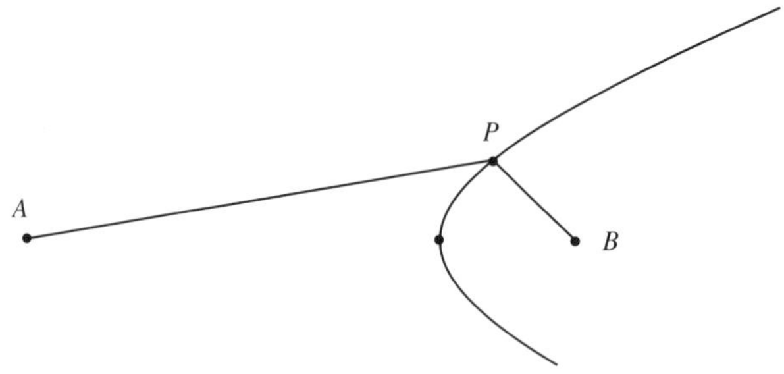

(a)

Figure 7.16 Sound ranging:

The curve is shown in Figure 7.17. The parametric form of the equation is $x = ct, y = \dfrac{c}{t}$ where $t > 0$ is in quadrant 1 and $t < 0$ is in quadrant 3.

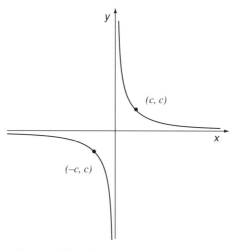

Figure 7.17 Rectangular hyperbola

Example

Determine the conditions that must be satisfied by a, b, c for the rectangular hyperbola $xy = c^2$ to intersect or to be a tangent to the ellipse $\dfrac{x^2}{a^2} + \dfrac{y^2}{b^2} = 1$. Draw an illustrative sketch.

Figure 7.18 indicates that if a is the x-coordinate of an intersection point, so is $-a$. Substituting $y = \dfrac{c^2}{x}$ in the equation of the ellipse, we obtain

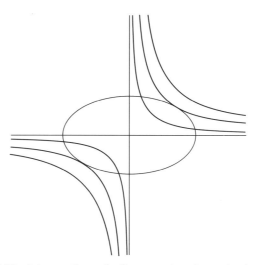

Figure 7.18 Intersection of ellipse and rectangular hyperbolas

$$\frac{x^2}{a^2} + \frac{c^4}{b^2 x^2} = 1$$

Multiplying by x and rearranging $\quad \dfrac{x^4}{a^2} - x^2 + \dfrac{c^4}{b^2} = 0$

This is a quadratic equation in x^2 with roots

$$x^2 = \frac{1 \pm \sqrt{1 - \dfrac{4c^4}{a^2 b^2}}}{\dfrac{2}{a^2}}$$

If $ab > 2c^2$ the curves intersect

If $ab = 2c^2$ the curves touch

If $ab < 2c^2$ the curves do not meet

In the case of intersection there are two positive values for x^2, i.e. four possible values for x. ■

Exercise 7.1

1 Identify the conic sections when the constants in the general conic equation have the following values:

(a) $a = b = 1$ $h = 0$ $f^2 + g^2 > c$

(b) $a \neq 0$ $b, h, g, f, c = 0$

(c) $a = 1$ $f = -\dfrac{1}{2}$ $b, h, g, c = 0$

(d) $h = \dfrac{1}{2}$ $c = -1$ $b, h, g, f = 0$

2 Determine the vertex, axis, focus and directrix of the parabolas whose equations are

(a) $y^2 = 8x$ (b) $y = x^2$ (c) $y = x^2 - 2x - 4$

3 A parabolic reflector operates on the principle that a heat or light source at the focus emits a parallel beam of energy via a mirrored reflector. Alternatively the reflector receives energy from a distinct source in parallel rays, e.g. from the sun, and collects this at the focus.

The figure depicts incoming energy and the principle depends upon the angle between PS and the normal at P being the same as the angle between the normal at P and the

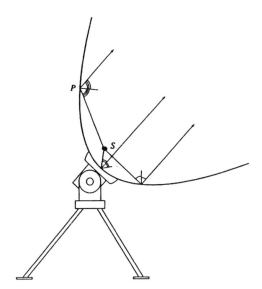

incoming ray. Prove that the two angles are the same, whatever, the position of P, using the standard form of the parabola, $y = 4ax$, with the incoming rays parallel to the x-axis.

4   Determine the equation of the parabola which is the locus of a point equidistant from the origin and the line $x + y = 2$. Where is its vertex and where does it cross the coordinate axes? Draw an illustrative sketch.

5

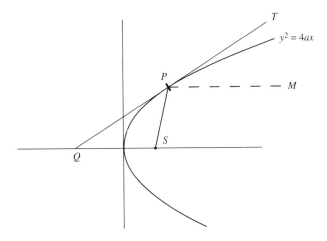

Using parameters prove that $QS = PS = a(t^2 + 1)$ where QPT is the tangent to the parabola at P. Deduce the result to Question 3. (The answer may give you a hint.)

6

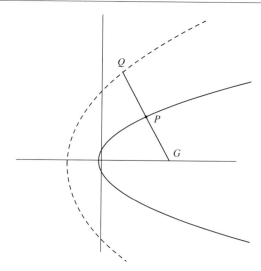

The normal to the parabola $y^2 = x$ meets the x-axis at G and $QP = PG$; P is the point $(at^2, 2at)$. Determine the coordinates of G and Q in terms of the parameter t then eliminate t to show that the locus of Q is the parabola $y^2 = 2(2x + 1)$.

7 (a)   If the straight line $y = mx + c$ is a tangent to the parabola $y^2 = 4ax$, prove that
$$c = \frac{a}{m}.$$

(b)   The **latus rectum** to $y^2 = 4ax$ is a line parallel to the y-axis passing through the focus. Determine the equation of the tangent at the endpoint of the latus rectum in the first quadrant.

(c)   Draw a sketch of $y^2 = 4ax$ showing two perpendicular tangents. Prove that they always meet on the directrix $x = -a$. (The answer gives a hint.)

8   The normal to the parabola $y^2 = 4ax$ at the point $(at^2, 2at)$ has the equation

$$tx + y = at^3 + 2at$$

A cubic polynomial in the parameter t, it may possess up to three roots t_1, t_2, t_3. This suggests that from a suitable chosen point $Q(x, y)$ three normals can be drawn. Draw a sketch to show this is possible with the concurrent point to the right of the focus and relatively close to the x-axis. Deduce that $t_1 + t_2 + t_3 = 0$.

9 Find the lengths of the axes, the eccentricity, the foci and the directrixes of the ellipses

(a) $\dfrac{x^2}{9} + \dfrac{y^2}{4} = 1$

(b) $\dfrac{x^2}{4} + \dfrac{y^2}{9} = 1$

(c) $4x^2 + 25y^2 - 8x - 50y - 46 = 0$

10 Two pegs are fixed at the points $(1,0)$ and $(0,1)$ and a string of length two units is used to trace out an ellipse. Establish that its equation is

$$3x^2 + 2xy + 3y^2 - 4x - 4y = 0$$

and draw an illustrative sketch.

11 Prove that the equations of the tangent and the normal to the ellipse $\dfrac{x^2}{a^2} + \dfrac{y^2}{b^2} = 1$ at the point (x_1, y_1) are

$$\dfrac{xx_1}{a^2} + \dfrac{yy_1}{b^2} = 1 \qquad \text{(tangent)}$$

$$\dfrac{x - x_1}{\dfrac{x_1}{a^2}} = \dfrac{y - y_1}{\dfrac{y_1}{b^2}} \qquad \text{(normal)}$$

Determine these equations for the ellipse $5x^2 + 3y^2 = 137$ at the point $(5,2)$.

12 (a) Prove that the line $y = mx + c$ intersects the ellipse $\dfrac{x^2}{a^2} + \dfrac{y^2}{b^2} = 1$ according to whether the quadratic equation $(a^2m^2 + b^2)x^2 + 2a^2mcx + a^2(c^2 - b^2) = 0$ has real roots.

(b) Deduce that the line $y = mx + \sqrt{a^2m^2 + b^2}$ must be a tangent.

(c) A perpendicular tangent has slope $-\dfrac{1}{m}$. Establish that two perpendicular tangents must lie on the **director circle**.

$$x^2 + y^2 = a^2 + b^2$$

and draw an illustrative sketch.

13 Prove that the chord PQ passing through the points $P(a\cos\theta, b\sin\theta)$ and $Q(a\sin\phi, b\cos\phi)$ on the ellipse $\dfrac{x^2}{a^2} + \dfrac{y^2}{b^2} = 1$ has the equation

$$bx\cos\left(\frac{1}{2}(\theta+\phi)\right) + ay\sin\left(\frac{1}{2}(\theta+\phi)\right) = ab\cos\left(\frac{1}{2}(\theta-\phi)\right)$$

14 Find the asymptotes, eccentricity, foci and directrixes of the following hyperbolas:

(a) $\quad \dfrac{x^2}{4} - y^2 = 1$ (b) $\quad 4x^2 - 9y^2 = 36$

15 If P is a point on a hyperbola with foci S, S' and major axis of length $2a$, use the focus–directrix property to prove that $|SP - S'P| = 2a$.

16 Prove the following properties for the hyperbola $\dfrac{x^2}{a^2} - \dfrac{y^2}{b^2} = 1$ using arguments which parallel those for the ellipse.

(a) The equation of tangent at $P(x_1, y_1)$ is

$$\frac{xx_1}{a^2} - \frac{yy_1}{b_2} = 1$$

(b) The equation $y = mx + \sqrt{a^2m^2 - b^2}$ represents a tangent with slope $m > \left|\dfrac{b}{a}\right|$

(c) The director circle has equation $x^2 + y^2 = a^2 - b^2$. What happens if the hyperbola is rectangular?

17 Find the equation of the tangent and normal to the hyperbola $x^2 - 6y^2 = 3$ at the point $(-3, -1)$.

18 The rectangular hyperbola $xy = c^2$ is represented parametrically by $x = ct, y = \dfrac{c}{t}$. Determine the equations of the following lines:

(a) the chord joining two points whose parameters are t_1 and t_2

(b) the tangent at the point t

(c) the nornmal at the point t

19 To draw a hyperbola, tie a string round a marker at P.

Ensure $l > l'$ and extend the string round two pegs F' and F.

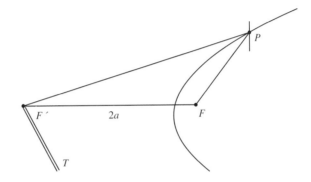

Keep both ends together at T. Extend P and pull the ends at T. Show that
$F'P - FP = 2a + l' - l$

7.2 TWO-DIMENSIONAL CURVES

Symmetry, turning points and asymptotic behaviour are highly important in drawing graphs but it may happen that transforming the equation of a curve makes the geometry simpler.

Axes

An ellipse centred at the point $P(h, k)$ rotated anticlockwise by an angle θ (Figure 7.19) will have a general conic equation in x and y of the form (7.1). However, when referred to

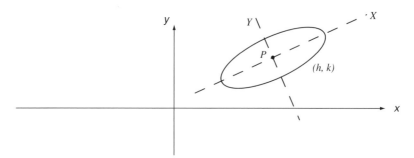

Figure 7.19 Rotated ellipse

the axes X and Y, it will have an equation in standard or **canonical form**, i.e.

$$\frac{X^2}{a^2} + \frac{Y^2}{b^2} = 1$$

This means that if we **translate** the origin of coordinates to (h, k) then **rotate** the axes by θ we will obtain the canonical form and we will be able to read off the key properties. Translation and rotation are independent actions and can be considered as part of a two-stage process that can be carried out in either order.

To translate the origin to (h, k) we apply the transformation shown in Figure 7.20:

$$X = x - h \qquad Y = y - k \tag{7.6}$$

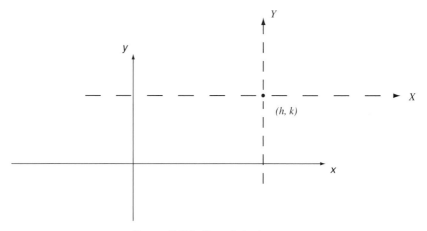

Figure 7.20 Translated axes

Rotation

From Figure 7.21 it can be deduced that

$$x = X \cos\theta - Y \sin\theta \qquad y = X \sin\theta + Y \cos\theta \tag{7.7}$$

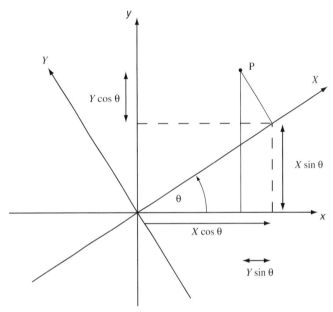

Figure 7.21   Rotated axes

Example

Transform the conic equation $2x^2 - 3xy - 2y^2 + 2x + 11y - 13 = 0$ by translating the origin to the point $(1, 2)$ then rotating the axes by an acute angle whose tangent is 3. Draw an illustrative sketch.

First take $x = X' + 1, y = Y' + 2$ to obtain the equation

$$2(X' + 1)^2 - 3(X' + 1)(Y' + 2) - 2(Y' + 2)^2 + 2(X' + 1) + 11(Y' + 2) - 13 = 0$$

i.e. $2X'^2 - 3X'Y' + 2Y'^2 = 1$

If $\tan^{-1}\theta = 3$ then $\sin\theta = \dfrac{3}{\sqrt{10}}$ and $\cos\theta = \dfrac{1}{\sqrt{10}}$, so

$$X = \frac{X' - 3Y'}{\sqrt{10}} \qquad Y = \frac{3X' + Y'}{\sqrt{10}}$$

Then $2(X - 3Y)^2 - 3(X - 3Y)(3X + Y) - 2(3x + Y)^2 = 10$

i.e. $5(Y^2 - X^2) = 2.$

Then

$$\frac{Y^2}{\left(\dfrac{2}{5}\right)} - \frac{X^2}{\left(\dfrac{2}{5}\right)} = 1$$

which represents a hyperbola (Figure 7.22). ∎

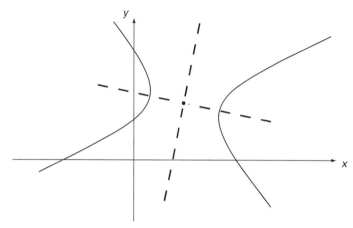

Figure 7.22 Translated and rotated conic

Parameters

We can often represent a curve mathematically in terms of a single variable t; in essence this recognizes its one-dimensional nature. The variable t is called a **parameter** and each point on the curve has a unique value of t assigned to it. Note that the choice of the form of t is not unique, but for the better-known curves there are established conventions.

Circle	$x^2 + y^2 = a^2$	$x = a \cos t$	$y = a \sin t$
Parabola	$y^2 = 4ax$	$x = at^2$	$y = 2at$
Ellipse	$\dfrac{x^2}{a^2} + \dfrac{y^2}{b^2} = 1$	$x = a \cos t$	$y = b \sin t$
Hyperbola	$\dfrac{x^2}{a^2} - \dfrac{y^2}{b^2} = 1$	$x = a \cosh t$	$y = b \sinh t$
Rectangular hyperbola	$xy = c^2$	$x = ct$	$y = c/t$
		(45° rotated)	

It is possible to define parameters for more general curves too.

Example

(a) Define parametric forms for the following:

   (i)   the circle radius 2 centred at $(2, 1)$

   (ii) the rectangular hyperbola with asymptotes $x = 1, y = 1$ and which passes through the origin.

   (iii) the ellipse $9x^2 + 4y^2 - 18x + 8y = 23$

   (iv) the circle $x^2 + y^2 = x$

(b) In the form shown, define parameters for the following curves:

   (i) $x^2 + y^2 = x$ $(y = tx)$

   (ii) $y^2(1 - x) = x^2(1 + x)$ $(y = tx)$

   (iii) $xy = (x - 1)^2$ $(y = t^2 x)$

   (iv) $\sin^{-1}\left(\dfrac{y}{x}\right) = \dfrac{1}{x}$ $(y = x \sin t)$

Solution

(a) (i) Cartesian equation:

$$(x - 2)^2 + (y - 1)^2 = 4 \qquad \text{(radius, } a = 2)$$

$$\text{Take } x = 2 + 2\cos t = 2(1 + \cos t), \qquad y = 1 + 2\sin t$$

(ii) Cartesian equation:

$$(x - 1)(y - 1) = 1 \qquad (c = 1)$$

$$\text{Take } x = 1 + t \qquad y = 1 + \tfrac{1}{t}$$

(iii) The equation becomes

$$\frac{(x - 1)^2}{4} + \frac{(y + 1)^2}{9} = 1$$

$$\text{Take } x = 1 + 2\cos t, \qquad y = -1 + 3\sin t$$

(iv) The equation becomes

$$\left(x - \frac{1}{2}\right)^2 + y^2 = \frac{1}{4}$$

$$\text{Take } x = \frac{1}{2}(1 + \cos t), \qquad y = \frac{1}{2}\sin t$$

(b) (i) $x^2(1 + t^2) = x$: so

$$x = \frac{1}{1 + t^2}, \qquad y = \frac{t}{1 + t^2}$$

(ii) $t^2(1 - x) = 1 + x$: so

$$x = \frac{t^2 - 1}{t^2 + 1}, \qquad y = \frac{t(t^2 - 1)}{t^2 + 1}$$

(iii) $t^2 = \left(\dfrac{x - 1}{x}\right)^2$: so

$$t = \frac{x - 1}{x}; \qquad x = \frac{1}{1 + t} \qquad y = \frac{t^2}{1 + t}$$

(iv) $t = \dfrac{1}{x}$: so

$$x = \frac{1}{t}, \qquad y = \frac{\sin t}{t} \qquad \blacksquare$$

It also happens that curves are defined parametrically in their own right and the Cartesian parent equation might be difficult or impossible to find.

Example

A **cycloid** is the locus traced out by a point $P(x, y)$ on the circumference of a disc rolling on a horizontal line. If the disc has radius a, the locus takes the form in Figure 7.23, when the disc has rolled through an angle θ. From the geometry note that P started at O and the amount rolled ON is equal to the length of arc PN, i.e. $ON = PN = a\theta$. Then

$$x = OM = ON - MN = ON - PL = a\theta - a\sin\theta$$
$$y = PM = SN - LN = a - a\cos\theta$$

We can eliminate θ to prove that

$$x = a\cos^{-1}\left(\frac{a - y}{a}\right) - \sqrt{2ay - y^2}$$

but this is a dreadful expression. $\qquad \blacksquare$

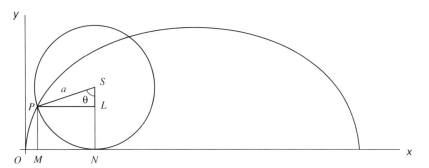

Figure 7.23   A cycloid

Parameters can play a key role in describing motion. The parameter t is the time and $r(t) = x\mathbf{i} + y\mathbf{j}$ is the **displacement** in two dimensions; $\dot{r}(t) = \dot{x}\mathbf{i} + \dot{y}\mathbf{j}$ is the **velocity**; $\dot{x} = \dfrac{dx}{dt}$ etc, and $(\dot{x}^2 + \dot{y}^2)^{1/2}$ is the **speed**.

Example

A particle is moving in two dimensions and the coordinates x, y of its displacement satisfy

$$x = (t-1)^2 \qquad y = \frac{t}{9}(t^2 - 3)$$

where t is in seconds.

(a) Trace out its path for the first four seconds.

(b) Find its speed and determine when and where it is zero.

Solution

(a) When the following values are plotted they produce the graph of Figure 7.24.

t	0	1	2	3	4
x	1	0	1	4	9
y	0	$-\dfrac{2}{9}$	$\dfrac{2}{9}$	2	$\dfrac{52}{9}$

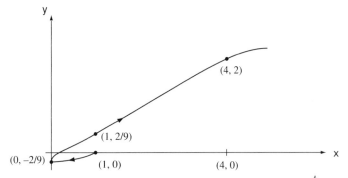

Figure 7.24   Particle path described by $x = (t - 1)^2$, $y = \dfrac{t}{9}(t^2 - 3)$

The particle moves towards the y-axis just below the origin to the point $\left(0, -\dfrac{2}{9}\right)$ at $t = 1$, then reverses round to the point $\left(1, \dfrac{2}{9}\right)$ at $t = 2$, before climbing away up to the right.

(b) $\qquad \dot{x} = 2(t - 1) \qquad \dot{y} = \dfrac{1}{3}(t - 1)(t + 1)$

The speed is $(\dot{x}^2 + \dot{y}^2)^{1/2}$, i.e.

$$|\dot{x}\boldsymbol{i} + \dot{y}\boldsymbol{j}| = |t - 1|\left\{4 + \left(\frac{t + 1}{3}\right)^2\right\}^{1/2}$$

This is zero when $t = 1$, corresponding to the turning point on the y-axis. ◼

Sketching of curves

An **implicit function** takes the form $f(x, y) = 0$, in which it is generally not possible to isolate x or y in the form $x = g(y)$ or $y = h(x)$. The locus of $f(x, y) = 0$ divides the regions of the plane where $f(x, y) > 0$ from where $f(x, y) < 0$, but because we may not be able to isolate one of the variables, or find a suitable parameter, or even a coordinate transformation, we may not be able to draw $f(x, y) = 0$ directly. However, we can glean information locally and asymptotically about $f(x, y) = 0$ and its gradient. Key points to look for are

1. Points where the curve crosses the axes

2. Asymptotes and behaviour for large x and/or y

3. Excluded regions of the x–y plane

4. Symmetries, oddness/evenness in x and y

5. Derivatives at computable points, so that gradients can be found

Example

Describe the curve $x^3 + y^3 + x - y = 0$

1. The point $(0, 0)$ is obviously on the curve.

When $x = 0$, $y^3 - y = 0$, i.e. $y = 0, 1, -1$
When $y = 0$, $x^3 + x = 0$, i.e. $x = 0$ only.
Hence $(0, -1)$, $(0, 0)$ and $(0, 1)$ are on the curve.

2. There are no vertical asymptotes; x^3 and y^3 dominate for large x and/or y. Dividing, $f(x, y)$ by the non-factorising quadratic $x^2 - xy + y^2$, we obtain

$$x + y + \frac{(x - y)}{x^2 - xy + y^2} = 0$$

so if x and y are both large, $x + y \approx 0$, i.e. $y \approx -x$

3. No regions of the plane are obviously excluded.

4. $f(-x, -y) = -f(x, y)$, so there is odd symmetry between opposite quadrants. We need only determine the curve in quadrants 1 and 2.

5. Given that

$$\frac{dy}{dx} = \frac{(1 + 3x^2)}{1 - 3y^2}$$

the value of $\frac{dy}{dx}$ is infinite when $y = \pm\frac{1}{\sqrt{3}} = \pm 0.577$

With $y = \frac{1}{\sqrt{3}}$, $x^3 + x - \frac{2}{3\sqrt{2}} = 0$, an equation which solves numerically to give $x \approx 0.344$

We also have at $(0, 0)$, $\frac{dy}{dx} = 1$, and at $(0, 1)$ and $(0, -1)$, $\frac{dy}{dx} = -\frac{1}{2}$

To obtain a plot of $f(x, y) = 0$ you will need computer graphics. Figure 7.25 shows the plot, highlighting items 1 to 5. ∎

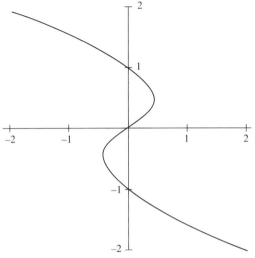

Figure 7.25   Plot of $x^3 + x + y^3 - y = 0$

Exercise 7.2

1   A conic has equation

$$17x^2 - 12xy + 8y^2 + 46x - 28y + 17 = 0$$

Transfer the origin to the point $(-1, 1)$ then rotate anticlockwise by an acute angle $\tan^{-1} 2$ to obtain the canonical form.

2   (a)   The conic with equation $4x^2 + y^2 - 4xy - 10y - 19 = 0$ is a parabola. Establish that is so by rewriting the equation in the form $(2x - y + 1)^2 = 4(x + 2y + 5)$ and verifying that the **principal axes**, i.e. the main axes of symmetry, are perpendicular. Show that the vertex lies at $\left(-\dfrac{7}{5}, -\dfrac{9}{5}\right)$

(b)   Starting with the conic equation, transfer the origin to the point $\left(-\dfrac{7}{5}, -\dfrac{9}{5}\right)$ and rotate by an angle $\tan^{-1} 2$ to prove that the conic is indeed a parabola.

(c)   Draw an illustrative sketch.

3   Define parameters for

(a)   the hyperbola $(x + y)(x - y) = 2(x + y + 1)$
(complete the square in x and y.)

(b)   the parabola $y^2 = 4a(a - x)$
(start with $y = 2at$).

(c)   the rectangular hyperbola, centred on $(-8, -2)$ with asymptotes parallel to the coordinate axes, which passes through the origin

What is the Cartesian equation for (c)?

4   Define parameters in the form suggested for the curves

(a) $x^3 + y^3 = xy$ $(y = tx)$
(b) $(x + 1)y^2 = x^2$ $(y = tx)$
(c) $y(1 - y)^2 = x^3$ $(y = t^2 x)$
(d) $2x = y^2 + 1$ $(y = \cos tx)$

5   Determine the Cartesian form, i.e. y as a function of x or vice versa, by eliminating t from the following parametric pairs of equations. Identify the curve in each case.

(a) $x = 5 \cos t,$ $y = 5 \sin t$
(b) $x = 2 + 3 \cos t,$ $y = -2 + 3 \sin t$
(c) $x = \dfrac{t^2}{2},$ $y = t$
(d) $x = -t^2,$ $y = t$
(e) $x = 2t,$ $y = t^2$
(f) $x = 1 + 2t,$ $y = 3 - t^2$
(g) $x = 2 \cos t,$ $y = 3 \sin t$
(h) $x = 1 + 2 \cos t,$ $y = \sin t$
(i) $x = 2t,$ $y = \dfrac{2}{t}$
(j) $x = -t,$ $y = \dfrac{1}{t}$

6   In the following cases, eliminate t from the pairs of parametric equations and express y as a function of x or vice versa.

(a) $x = 2t^3$ $y = 3 + 4t$

(b) $x = \dfrac{1+t}{1-t}, \qquad y = \dfrac{2-t}{2+t}$

(c) $x = t^3 - t, \qquad y = t^3 + t$

(d) $x = \ln t, \qquad y = \dfrac{t}{2t^2 - 1}$

7 Determine the equation of tangent and normal to the curve whose parameters are $x = 2t^3, y = 3 + 4t$

8 If $a, t > 0$ the parametric equation $x = at, y = \dfrac{a}{t}$ represents the branch of a rectangular hyperbola in the first quadrant. If $y = mx + c$ is a chord passing through the two points whose parameters are t_1 and t_2, determine m and c in terms of t_1, t_2

9 If the chord in Question 8 passes through the fixed point $\left(at_1, \dfrac{a}{t_1}\right)$ and an arbitrary point $\left(at, \dfrac{a}{t}\right)$, determine the locus of the midpoint of the chord.

10 A curve has the parametric equation $x = t(1 - \ln t), y = 2\dfrac{\sin t}{t}$. Sketch the curve for $0 \leq t \leq 4$.

11 Determine axis-crossing points and symmetries for the implicit function

$$f(x, y) = x^4 + y^3 + x^2 - y = 0$$

Draw the graph.

    12   Note symmetry in the line $y = x$ to draw $x + y + e^x + e^y = 0$. Are any regions of the plane excluded?

13 Draw the graph of

$$x + y + \sin(x - 2y) = 0 \quad \text{for} \quad -3 \leq x, y \leq 3$$

7.3 GEOMETRY OF MOTION

Gradient, tangent and curvature

The **gradient** of a curve described parametrically can be determined using the **chain rule**. If $(x(t), y(t))$ is a point on the curve, then

$$\frac{dy}{dx} = \frac{dy}{dt} \times \frac{dt}{dx} = \frac{\dfrac{dy}{dt}}{\dfrac{dx}{dt}} = \frac{\dot{y}}{\dot{x}}$$

Likewise we can determine higher derivatives

$$\frac{d^2y}{dx^2} = \frac{d}{dx}\left(\frac{dy}{dx}\right) = \frac{d}{dt}\left(\frac{dy}{dx}\right) \times \frac{dt}{dx} = \frac{d}{dt}\left(\frac{\dot{y}}{\dot{x}}\right) \times \frac{1}{\dot{x}} = \frac{\dot{x}\ddot{y} - \ddot{x}\dot{y}}{\dot{x}^3},$$

where $\ddot{x} = \dfrac{d^2x}{dt^2}$ and $\ddot{y} = \dfrac{d^2y}{dt^2}$

Example

Determine $\dfrac{dy}{dx}$ and $\dfrac{d^2y}{dx^2}$ for the parabola $y^2 = 4ax$ in both standard and parametric form.

First, $y^2 = 4ax$ can be written in parametric form $x = at^2, y = 2at$.

(a) $\dot{x} = 2at, \dot{y} = 2a,$ so $\dfrac{\dot{y}}{\dot{x}} = \dfrac{1}{t}$

Also $2y\dfrac{dy}{dx} = 4a,$ so $\dfrac{dy}{dx} = \dfrac{2a}{y} = \dfrac{1}{t}$

(b) $\ddot{x} = 2a, \ddot{y} = 0,$ so

$$\frac{\dot{x}\ddot{y} - \ddot{x}\dot{y}}{\dot{x}^3} = \frac{2at \times 0 - 2a \times 2a}{(2at)^3} = -\frac{1}{2at^3}$$

$$\frac{d}{dx}\left(\frac{2a}{y}\right) = -\frac{2a}{y^2}\frac{dy}{dx} = -\frac{4a^2}{y^3} = \frac{1}{2at^3}$$

The parametric form can be extended to **derivative functions**, *e.g.* curvature. The **radius of curvature** at a point on a curve is defined to be the radius of a circle whose rate of bending is the same as for the curve at that point. Figure 7.26 illustrates the concept.

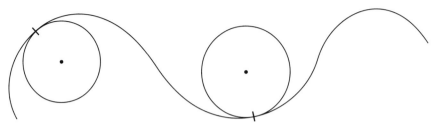

Figure 7.26 Radius of curvature

We can also illustrate curvature in terms of derivatives (Figure 7.27). The **curvature**

$$\kappa = \lim_{\delta s \to 0} \frac{\delta \psi}{\delta s} = \frac{1}{\rho}$$

where ρ is the **radius of curvature**; $\delta \psi$ and δs represent incremental elements of **tangent angle** and **arc length**. Now $\tan \psi = \dfrac{dy}{dx}$ and $\delta s^2 \approx \delta x^2 + \delta y^2$ (Figure 7.28).

Therefore $\left(\dfrac{\delta s}{\delta x} \right)^2 \cong 1 + \left(\dfrac{\delta y}{\delta x} \right)^2$; hence $\dfrac{ds}{dx} = \left(1 + \left(\dfrac{dy}{dx} \right)^2 \right)^{1/2}$

Now $\sec^2 \psi \times \dfrac{d\psi}{ds} = \dfrac{d^2 y}{dx^2} \times \dfrac{ds}{dx}$

so $\sec^2 \psi \times \dfrac{1}{\rho} = \dfrac{d^2 y}{dx^2} \times \cos \psi$

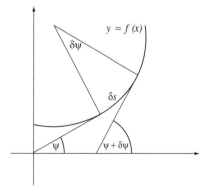

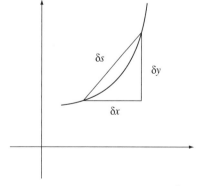

Figure 7.27 Deriving a formula for radius of curvature   **Figure 7.28** Diagram to show $\delta s^2 \equiv \delta x^2 + \delta y^2$

i.e. $\qquad \rho = \dfrac{\sec^3 \psi}{\dfrac{d^2 y}{dx^2}} = \dfrac{\left(1 + \left(\dfrac{dy}{dx}\right)^2\right)^{3/2}}{\dfrac{d^2 y}{dx^2}}$ $\qquad\qquad$ (7.8a)

It is left as an exercise for you to show that, if the curve is given in terms of a parameter t and $\dot{x} = \dfrac{dx}{dt}, \dot{y} = \dfrac{dy}{dt}$ etc. we have

$$\rho = \frac{(\dot{x}^2 + \dot{y}^2)^{3/2}}{\dot{x}\ddot{y} - \dot{y}\ddot{x}} \qquad\qquad (7.8b)$$

Example

Find the radius of curvature of the curves

(a) $x^2 - y^2 = a^2$ $\qquad\qquad$ (b) $x = at^3, \qquad y = at$

Solution

(a) Differentiating, $2x - 2yy' = 0, \quad$ i.e. $y' = \dfrac{x}{y}$

Therefore

$$y'' = \frac{y \times 1 - x \times y'}{y^2} = \frac{y^2 - x^2}{y^3} = -\frac{a^2}{y^3}$$

Hence

$$\rho = \frac{(1 + y'^2)^{3/2}}{y''} = \frac{y^3}{a^2}\left(1 + \left(\frac{x}{y}\right)^2\right)^{3/2} = \frac{(x^2 + y^2)^{3/2}}{a^2}$$

(b) $\dot{x} = 3at^2, \dot{y} = a \quad$ so $\quad \ddot{x} = 6at, \ddot{y} = 0$

$$\rho = \frac{(\dot{x}^2 + \dot{y}^3)^{3/2}}{\dot{x}\ddot{y} - \ddot{x}\dot{y}} = -\frac{a(1 + 9t^4)^{3/2}}{6t}$$ ■

Motion

A most important application in the use of parameters arises in describing motion. Consider for example the motion of a particle in a circle (Figure 7.29).

Choose the origin to be the centre of the circle, so that $r = a\cos t\, i + a\sin t\, j$ represents the position vector of a particle moving, around the circle. The velocity of the particle is given by

$$\dot{r} = \frac{dr}{dt} = -a\sin t\, i + a\cos t\, j$$

differentiating each of the components with respect to t. Putting r and $\dot{r}$ on the same diagram, we obtain Figure 7.30.

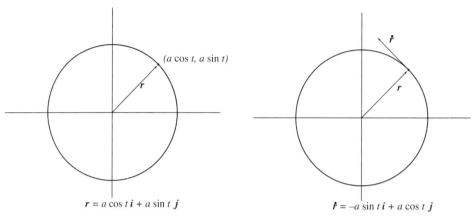

$$r = a \cos t\, \boldsymbol{i} + a \sin t\, \boldsymbol{j}$$

Figure 7.29 Motion in a circle

$$\dot{r} = -a \sin t\, \boldsymbol{i} + a \cos t\, \boldsymbol{j}$$

Figure 7.30 Displacement and velocity vectors in circular motion

Observe that $\dot{r}$ is perpendicular to $\dot{r}$, i.e. $\boldsymbol{r}.\dot{\boldsymbol{r}} = \dot{\boldsymbol{r}}.\boldsymbol{r} = 0$. This is what one would expect for motion in a circle, i.e. the velocity of the particle is directed at right angles to its position vector at all times.

The motion of a projectile subject to the constant downward force of gravity g, ignoring air resistance can be shown to satisfy

$$\ddot{x} = 0 \qquad \ddot{y} = -g$$

We will see in Chapter 11 that the coordinates x and y are given by

$$x = U \cos \alpha t \qquad y = U \sin \alpha t - \frac{1}{2} g t^2$$

where U is the initial speed of projection (Figure 7.31). You can easily verifty that $\ddot{x} = 0$ and $\ddot{y} = -g$.

The projectile is assumed to have been fired with velocity U at an angle α from a position on the ground, taken as the origin. The time t acts as a parameter which can be

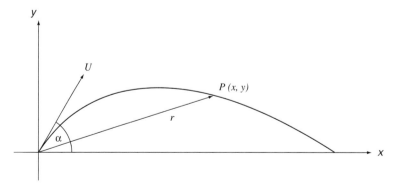

Figure 7.31 Motion of a projectile

eliminated to give the equation

$$y = x \tan \alpha - \frac{1}{2} \frac{gx^2}{U^2} \sec^2 \alpha$$

i.e. the projectile follows a path whose equation is a parabola. Were the projectile to represent a bullet fired from a gun, then U would be constant and α alone would determine the maximum height and range of an accessible target, i.e. the dotted curve in Figure 7.32.

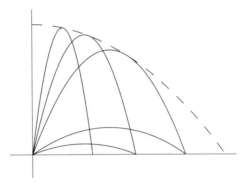

Figure 7.32   Envelope of projectile paths

This is called the **envelope** of the possible projectile paths. It is left as an exercise for you to prove it has the equation

$$y = \frac{U^2}{2g} - \frac{gx^2}{2U^2}$$

This is known as the **parabola of safety**. An object outside this parabola cannot be hit. The motion of a particle in two dimensions is represented in general by $\mathbf{r} = x\mathbf{i} + y\mathbf{j}$ and

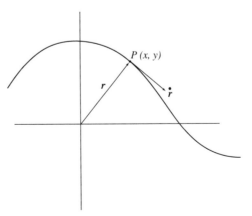

Figure 7.33   Motion of a particle in two dimensions

$\dot{r} = \dot{x}\mathbf{i} + \dot{y}\mathbf{j}$ (Figure 7.33). Vector $\mathbf{r}$ is the position vector of $P(x, y)$ and vector $\dot{\mathbf{r}}$, the velocity, is tangential to the motion curve, since $\dfrac{\dot{y}}{\dot{x}} = \dfrac{dy}{dx}$.

However, if we change to polar coordinates by setting $x = r\cos\theta, y = r\sin\theta$ and r and θ become the new variables (Figure 7.34).

Differentiating the equations $x = r\cos\theta, y = r\sin\theta$ with respect to t, we have

$$\dot{x} = \dot{r}\cos\theta - r\dot{\theta}\sin\theta, \qquad \dot{y} = \dot{r}\sin\theta + r\dot{\theta}\cos\theta$$

Differentiating again gives

$$\ddot{x} = \ddot{r}\cos\theta - 2\dot{r}\dot{\theta}\sin\theta - r\dot{\theta}^2\cos\theta - r\ddot{\theta}\sin\theta$$

$$\ddot{y} = \ddot{r}\sin\theta + 2\dot{r}\dot{\theta}\cos\theta - r\dot{\theta}^2\sin\theta + r\ddot{\theta}\sin\theta$$

for the Cartesian components of the acceleration $\ddot{\mathbf{r}} = \ddot{x}\mathbf{i} + \ddot{y}\mathbf{j}$. However, we are most interested in the resolution of those components outward in the radial (R) direction and tangential (T) to it (Figure 7.35):

R: $\ddot{x}\cos\theta + \ddot{y}\sin\theta$

T: $-\ddot{x}\sin\theta + \ddot{y}\cos\theta$

from which you can prove that in polar coordinates

R: $\ddot{r} - r\dot{\theta}^2$

T: $r\ddot{\theta} + 2\dot{r}\dot{\theta}$

A most important application in two dimensions arises when the motion of a particle is governed by a force field directed inwardly towards the origin. In this case

$$\ddot{r} - r\dot{\theta}^2 = f(r)$$

$$r\ddot{\theta} + 2\dot{r}\dot{\theta} = 0, \quad \text{i.e.} \quad \frac{d}{dt}(r^2\dot{\theta}) = 0 \quad \text{so} \quad r^2\dot{\theta} = h \qquad (h \text{ constant})$$

If $f(r)$ is known, $\dot{\theta}$ can be eliminated from the radial equation, which can be solved as a differential equation for $r(t)$. The polar equation of the orbit, i.e. r in terms of θ can then be found. Solving such differential equations is however beyond our scope.

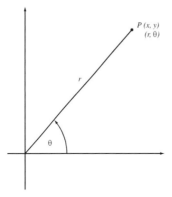

Figure 7.34 Polar coordinates

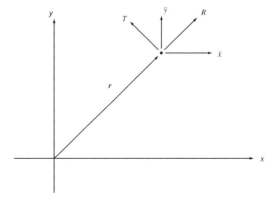

Figure 7.35 Components of acceleration

Exercise 7.3

1 Find the gradient at the point θ on the curve

$$x = a\cos^3\theta, \qquad y = a\sin^3\theta$$

2 If $x = a(\theta - \sin\theta)$, $y = a(\theta + \sin\theta)$, find $\dfrac{dy}{dx}$ and $\dfrac{d^2y}{dx^2}$ when $\theta = \dfrac{\pi}{2}$

3 If $x = 2\cos t - \cos 2t$, $y = 2\sin t - \sin 2t$, find $\dfrac{d^2y}{dx^2}$ when $t = \dfrac{\pi}{2}$

4 A particle moves in two dimensions and its displacement coordinates satisfy

$$x = a(3 + \cos t) \qquad y = a\sin t$$

Determine the Cartesian equation of its locus and prove that its speed, $(\dot{x}^2 + \dot{y}^2)^{1/2}$, is constant.

5

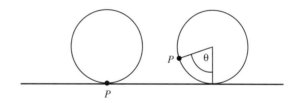

P is a point on the circumference of a rolling disc of unit radius. The locus of P can be proved to satisfy

$$x = \theta - \sin\theta \qquad y = 1 - \cos\theta \qquad (0 \le \theta < 2\pi)$$

(a) What is the speed of P when $\theta = \dfrac{2\pi}{3}$, i.e. $(\dot{x}^2 + \dot{y}^2)^{1/2}$?

(b) Determine when the speed of P is
(i) a maximum   (ii) locally zero

(c) Show that the acceleration $(\ddot{x}^2 + \ddot{y}^2)^{1/2}$ is constant. (Assume $\theta = t$)

6 Find the gradient at the origin of the curve whose parametric equation is

$$x = \theta e^{-\theta} \qquad y = \theta^2 e^{-\theta} - 2\theta$$

7 Show that the radius of the curvature is given by

$$\rho = \frac{(\dot{x}^2 + \dot{y}^2)^{3/2}}{\dot{x}\ddot{y} - \dot{y}\ddot{x}}$$

for a curve whose equation is given in terms of parametric coordinates.
 Determine ρ for the following curves:

(a) $x = t^2, y = 2t$ at the point $(1, 2)$

(b) $x = t, y = -\dfrac{8}{t}$ at the point $(2, -4)$

8 For the circle $x^2 + y^2 = a^2$ whose parametric equations are $x = a\cos t, y = a\sin t$, prove that $\rho = a$, a constant, as one would expect.

9 The displacement of a particle is given by $r = a\cos t\, i + b\sin t\, j$

(a)   Show that the particle follows the path of the ellipse

$$\frac{x^2}{a^2} + \frac{y^2}{b^2} = 1$$

(b)   Establish that $r.\dot{r} \neq 0$ in general. When is $r.\dot{r} = 0$?

(c)   Determine the values of t for which the speed is
   (i) a maximum   (ii) a minimum

10 Write down using parameters, an equation for the displacement vector $\overrightarrow{OP}$ on the curves

(a) $xy = 4$ (b) $(x - 1)^2 + y^2 = 1$

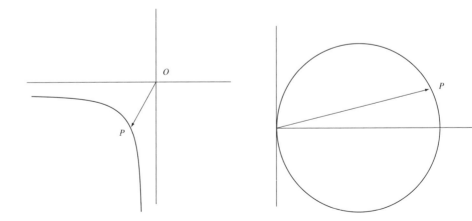

11   Determine the equation of the parabola of safety

$$y = \frac{U^2}{2g} - \frac{gx^2}{2U^2}$$

by differentiating with respect to α the trajectory equation

$$y = x \tan \alpha - \frac{1}{2} \frac{gx^2}{U^2} \sec^2 \alpha$$

and setting $\dfrac{dy}{d\alpha} = 0$. Determine

(a)   the maximum height

(b)   the maximum range of the trajectory

12   A projectile is fired with velocity U upwards along a plane inclined at an angle of $45°$. Determine its maximum range. (The answer gives a hint.)

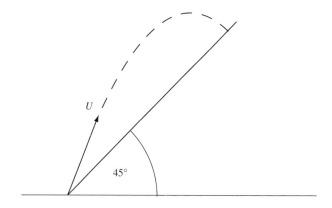

13   If the Cartesian coordinates (x, y) are transformed into polar coordinates (r, θ), establish the following results:

(a)   The components of velocity along the outward radius and tangent are respectively $\dot{r}$ and $r\dot{\theta}$

(b) $|\dot{r}|^2 = \dot{x}^2 + \dot{y}^2 = \dot{r}^2 + r^2\dot{\theta}^2$

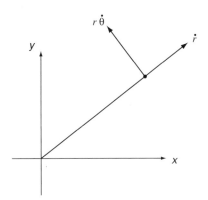

(c)   The components of the acceleration along the outward radius and tangent are respectively

$$\ddot{r} - r\dot{\theta}^2 \quad \text{and} \quad r\ddot{\theta} + 2\dot{r}\dot{\theta}$$

7.4 INTEGRATION AND APPLICATIONS

Arc Length

We know that an infinitesimal length of arc δs of a curve satisfies the relationship $(\delta s)^2 \approx (\delta x)^2 + (\delta y)^2$ (Figure 7.36). Then

$$\left(\frac{\delta s}{\delta x}\right)^2 \approx 1 + \left(\frac{\delta y}{\delta x}\right)^2$$

Let $\delta x \to 0$ then

$$\left(\frac{ds}{dx}\right)^2 = 1 + \left(\frac{dy}{dx}\right)^2$$

Hence $\dfrac{ds}{dx} = \sqrt{1 + \left(\dfrac{dy}{dx}\right)^2}$

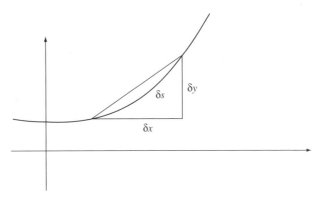

Figure 7.36 Arc length

taking the positive square root. This means the arc length of the curve $y = f(x)$ from $x = a$ to $x = b$ is

$$s = \int_a^b \left(1 + \left(\frac{dy}{dx}\right)^2\right)^{1/2} dx \tag{7.9a}$$

The sense of integration is from left to right, and only one branch of the curve can be taken at a time. Arc lengths measured in this way must be non-negative.

Example

Determine the length of arc of the curve $y = \ln x$ from $x = 1$ to $x = \sqrt{3}$
First,

$$\frac{dy}{dx} = \frac{1}{x}, \quad \text{so} \quad \frac{ds}{dx} = \left(1 + \frac{1}{x^2}\right)^{1/2}$$

We require $\displaystyle\int_1^{\sqrt{3}} \left(1 + \frac{1}{x^2}\right)^{1/2} dx$

Using the substitution $u^2 = x^2 + 1$, it can be shown that the indefinite integral is

$$(x^2 + 1)^{1/2} + \ln\left(\frac{(x^2 + 1)^{1/2} - 1}{x}\right) + C$$

so

$$\int_{1}^{\sqrt{3}} \left(1 + \frac{1}{x^2} \right)^{1/2} dx = \ln(\sqrt{2} + 1) - \frac{1}{2} \ln 3 - \sqrt{2} + 2 = 0.9179$$

∎

If the curve is described in terms of a parameter t then

$$\left(\frac{ds}{dt} \right)^2 = \left(\frac{dx}{dt} \right)^2 + \left(\frac{dy}{dt} \right)^2$$

and we have

$$s = \int_{t_1}^{t_2} \left(\left(\frac{dx}{dt} \right)^2 + \left(\frac{dy}{dt} \right)^2 \right)^{1/2} dt = \int_{t_1}^{t_2} (\dot{x}^2 + \dot{y}^2)^{1/2} \, dt \tag{7.9b}$$

The non-negative convention is again being followed.

For problems involving time-dependent movement, s represents the distance travelled and the speed $(\dot{x}^2 + \dot{y}^2)^{1/2}$ is integrated over time.

Example   Determine the length of an arch of the cycloid $x = \theta - \sin\theta, y = 1 - \cos\theta$
First

$$\frac{dx}{d\theta} = 1 - \cos\theta, \qquad \frac{dy}{d\theta} = \sin\theta$$

Then $\left(\dfrac{dx}{d\theta} \right)^2 + \left(\dfrac{dy}{d\theta} \right)^2 = 2(1 - \cos\theta) = 4\sin^2\left(\dfrac{\theta}{2} \right),$ so $\dfrac{ds}{d\theta} = 2\sin\left(\dfrac{\theta}{2} \right)$

An arch is described as θ varies from 0 to 2π, i.e. $s = 2\displaystyle\int_{0}^{2\pi} \sin\left(\frac{\theta}{2} \right) d\theta = 8.$ ∎

Polar Coordinates

Arc length can be found straightforwardly in polar coordinates.
If $x = r\cos\theta$ and $y = r\sin\theta$ then

$$\frac{dx}{d\theta} = \frac{dr}{d\theta}\cos\theta - r\sin\theta, \qquad \frac{dy}{d\theta} = \frac{dr}{d\theta}\sin\theta + r\cos\theta$$

so that $\left(\dfrac{dx}{d\theta}\right)^2 + \left(\dfrac{dy}{d\theta}\right)^2 = r^2 + \left(\dfrac{dr}{d\theta}\right)^2$

Then $\quad s = \displaystyle\int_a^b \left(1 + \left(\frac{dy}{dx}\right)^2\right)^{1/2} dx = \int_{\theta_a}^{\theta_b} \left(1 + \left(\frac{dy}{dx}\right)^2\right)^{1/2} \frac{dx}{d\theta} d\theta$

$$= \int_{\theta_a}^{\theta_b} \left(r^2 + \left(\frac{dr}{d\theta}\right)^2\right)^{1/2} d\theta \tag{7.9c}$$

Example

Find the length of the spiral $r = e^{2\theta}$ from $\theta = 0$ to $\theta = 2\pi$. The curve is shown in Figure 7.37.

$$\frac{dr}{d\theta} = 2e^{2\theta} \qquad r^2 + \left(\frac{dr}{d\theta}\right)^2 = e^{4\theta} + 4e^{4\theta} = 5e^{4\theta}$$

Then $\quad s = \sqrt{5} \displaystyle\int_0^{2\pi} e^{2\theta}\, d\theta = \frac{1}{2}\sqrt{5}(e^{4\pi} - 1)$ ∎

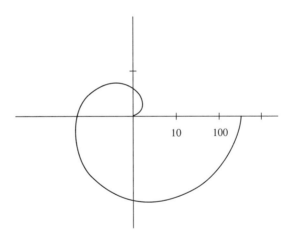

Figure 7.37 Spiral: note the logarithmic scale

The area under a polar curve is the sum of the areas of infinitesimal isosceles triangles of equal side r and included angle $\delta\theta$. Each area $\approx \dfrac{1}{2} r^2 \delta\theta$, since $\sin(\delta\theta) \cong \delta\theta$, for small angles. Then from Figure 7.38 we have that

$$\text{Shaded area} = \int_\alpha^\beta r^2\, d\theta \tag{7.10}$$

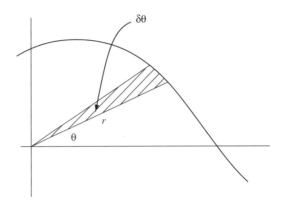

Figure 7.38   Area in polar coordinates

Example

The curve $r = a \cos 3\theta$ is sketched in Figure 7.39. Determine the area of one of its loops. The curve is contained between

$$-\frac{\pi}{6} < \theta < \frac{\pi}{6} \quad \text{and} \quad \frac{5\pi}{6} < \theta < \frac{7\pi}{6}$$

The area enclosed is

$$a^2 \int_{-\pi/6}^{\pi/6} \cos^2 3\theta \, d\theta = \frac{a^2}{2} \left[\theta + \frac{\sin 6\theta}{6} \right]_{-\pi/6}^{\pi/6} = \frac{\pi a^2}{6} \qquad \blacksquare$$

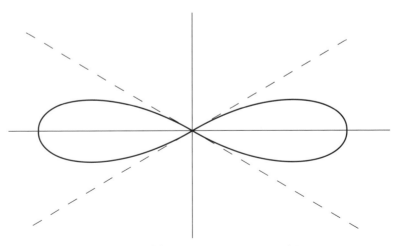

Figure 7.39   The curve $r = a \cos 3\theta$

Exercise 7.4

1   Find the length of arc of the following curves:

(a) $y = \ln(\cos x)$ from $x = 0$ to $x = \dfrac{\pi}{3}$

(b) $24xy = x^4 + 48$ from $x = 2$ to $x = 4$

2   Find the length of arc of the parabola $y^2 = 12x$ cut off between the vertex and the line $x = 3$ and lying above the x axis.

3   Find the length of arc of the following parametric curves:

(a) $x = 4t, y = t^2 - 2 \ln t$ from $t = 1$ to $t = 3$

(b) $x = t^2, y = t^3$ from $t = 0$ to $t = 4$

4   It can be proved that the curve of a heavy uniform cable takes the form of a **catenary**. If the ends of the cable are held at equal height, the catenary equation is

$$y = c \cosh \left(\frac{x}{c} \right) \qquad -a \leq x \leq a$$

Prove that the length of the cable is $2c \sinh \left(\dfrac{a}{c} \right)$

5   A point $P(x, y)$ moves across the x–y plane according to the rule

$$x = \frac{1}{2} t^2 \qquad y = \frac{1}{\sqrt{3}} (2t + 3)^{3/2}$$

where t is measured in seconds. How far does it travel in the first four seconds? (x, y are in metres)

6   For the curve $3ay^2 = x(x - a)^2$ determine

(a)   the length of arc from $x = 0$ to $x = a$ $(y \geq 0)$

(b)   the volume of revolution obtained when this curve is rotated through 2π radians about the x-axis between $x = 0$ and $x = a$

(c)   the centroid of that volume

7   Find the polar equation corresponding to

$$(x^2 + y^2)^2 = x^2 - y^2$$

Sketch the curve and determine the enclosed area for $x \geq 0$.

8   Determine the equation of the curve $x^2 + y^2 = 2y$ in polar coordinates and hence find the area enclosed by it.

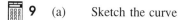 **9**   (a)   Sketch the curve

$$y^2 = \frac{1 - x^2}{1 + x^2}$$

and prove that it has two tangents of gradient ± 1 at the origin.

(b)   Determine the area enclosed by one of its loops.

(c)   If the loop $(x > 0)$ is rotated through $360°$ about the x-axis, determine the volume enclosed.

(d)   Determine the coordinates of the centroid of the volume.

10   A curve has the equation

$$(x^2 + y^2)^2 = 4xy^2$$

Determine the area enclosed by one of its loops.

11* (a)   Use polar coordinates to determine the area of the ellipse $x^2 + xy + y^2 = 1$.

(b)   Write down expressions for the integrals which give the coordinates of the centroid of the portion lying in the first quadrant. Do not attempt to evaluate the integrals.

(c)   Draw an illustrative sketch.

12   Determine the mean value of the function

$$\frac{5}{2 - x - 3x^2}$$

lying between

$$x = -\frac{1}{3} \quad \text{and} \quad x = \frac{1}{3}$$

 **13** Determine

(a)   the mean   (b)   the RMS

of the electric current function $i(t) = 20 + 100 \sin 100\pi t$ between $t = 0$ and $t = 0.02$.

14 An electric current is given by

$$i(t) = \frac{E}{R} + I \sin \omega t$$

where E, R, I, ω are constants. E represents applied d.c. voltage, R resistance, I the mean a.c. current and ω a frequency parameter. Determine the RMS current over a cycle, i.e.

$$0 \le t \le \frac{2\pi}{\omega}$$

15 The centre of pressure on an area of vertical cross-section within a fluid, is that point $(\bar{x}, \bar{y})$ where a concentrated force of magnitude F would yield the same moment with respect to any horizontal (vertical) line as the distributed forces. It can be proved that

$$\bar{x} = \frac{w}{2F} \int_c^d yx^2 \, dy \qquad \bar{y} = \frac{w}{F} \int_c^d y^2 x \, dy$$

If the surface concerned is bounded by the parabola $y = x^2$, determine $(\bar{x}, \bar{y})$.

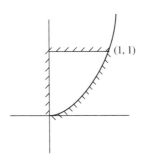

16 Write down suitable parametric coordinates to represent the ellipse $3x^2 + 4y^2 = 4$. Show that the area enclosed is $\pi \dfrac{\sqrt{3}}{2}$, and the perimeter has length

$$4 \int_0^{\pi/2} \left(1 - \frac{1}{4} \sin^2 t \right)^{1/2} dt$$

Determine its value to 5 d.p. using Simpson's rule.

17* The surface area of a solid of revolution, generated by rotating the curve $y = f(x)$ through 2π radians between $x = a$ and $x = b$ about the x-axis, is given by

$$2\pi \int_a^b y \, ds = 2\pi \int_a^b y \frac{ds}{dx} dx$$

Determine the surface areas of the following rotated volumes:

(a) $y^2 = 12x$ from $x = 0$ to $x = 3$ about the x-axis

(b) $y^2 + 4x = 2 \ln y$ from $y = 1$ to $y = 3$ about the y-axis

(c)   one arch of the cycloid $x = \dfrac{1}{4}(\theta - \sin\theta),\ y = \dfrac{1}{4}(1 - \cos\theta)$ about the x-axis.

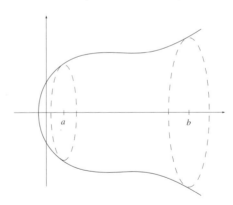

18* The work done by a constant force F over a directed distance s in a straight line is Fs units. If the force is a variable function of the distance, then $F(x)\,\delta x$ represents the work done in moving from x to $x + \delta x$. Over a distance $[a, b]$ this is $\int_b^a F(x)dx$

Determine the work done in the following cases.

(a)   Winding up a weight W (i.e. downward force W) by 10 m over a smooth pulley.

(b)   Winding in a heavy uniform chain of length L and weight W per unit length. Note that the tension (i.e. force) is equal to the weight of the unwound cable.

(c)   A combination of (a) and (b) in which the heavy chain runs over a pulley to carry the weight.

19   A particle of unit mass moves through the force field defined by

(i) $12(1 - x)^2$ between $x = 0$ and $x = 1$

(ii) $2\left(x - \dfrac{5}{2}\right)^2$ between $x = 1$ and $x = 4$

(iii) $x - 4$ between $x = 4$ and $x = 7$

(iv) $3(10 - x)$ between $x = 7$ and $x = 10$

Determine the total work done.

20*   A **curve of pursuit** is modelled as follows. A fox at $(1, 0)$ chases a rabbit at $(0, 0)$. The rabbit runs along the y-axis at half the speed of the fox. The fox follows a curved path always aiming at the rabbit, as illustrated.

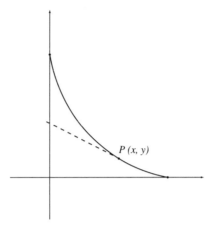

Verify that the slope of the pursuit curve satisfies

$$y' = \frac{1}{2}\left(\sqrt{x} - \frac{1}{\sqrt{x}}\right)$$

Show that the fox catches the rabbit at the point $\left(0, \frac{2}{3}\right)$

7.5 DEVELOPMENT IN THREE DIMENSIONS

We are now able to put together the ideas and concepts developed in this chapter with the three-dimensional concepts of vectors and planes.

Surfaces

The **plane** is just a flat surface, i.e. one whose normal is in a fixed direction. It has the general equation

$$ax + by + cz = \mathrm{d} \qquad \text{(Cartesians)}$$
$$(r - r_0).n = 0 \qquad \text{(vectors)}$$

(7.11)

where $n = ai + bj + ck$ is a normal and r_0 is the position vector of a fixed point P_0 in the plane (Figure 7.40).

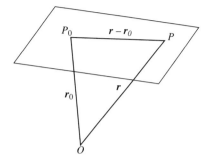

Figure 7.40   A plane

A more general surface takes the form $f(x, y, z) = 0$, where f is a functional relationship connecting the three coordinate variables. In particular, a **sphere** of radius a centred at the origin has equation

$$x^2 + y^2 + z^2 = a^2 \qquad \text{(Cartesians)}$$
$$\boldsymbol{r}.\boldsymbol{r} = a^2 \qquad\qquad \text{(vectors)} \tag{7.12}$$

the result is evident from the three-dimensional version of Pythagoras's theorem.

A **cone** is generated by rotating a vector about a fixed axis, e.g. z-axis (Figure 7.41). For a given value of z, $(x^2 + y^2)^{1/2}$ is proportional to z, so $z^2 = k(x^2 + y^2)$, $k > 0$ represents a cone.

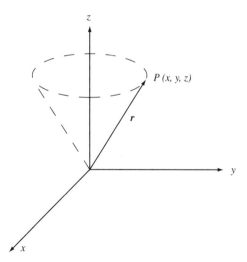

Figure 7.41   A cone

This is an example of a surface of revolution; another example is the paraboloid of revolution, formed by rotating a parabola about a coordinate axis, e.g. z (Figure 7.42):

$$z = k(x^2 + y^2) \qquad (k > 0 \text{ as drawn})$$

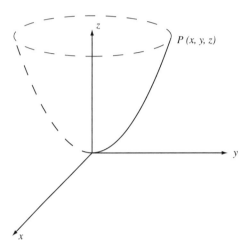

Figure 7.42 Paraboloid of revolution

Example

Describe the following surfaces and determine the equations of the curves of intersection with the given planes.

(a) $(x-5)^2 + (y-6)^2 + (z-7)^2 = 110;$ $z = 0$

(b) $x^2 + y^2 + 9(z-5) = 0;$ $z = 1$

Solution

(a) A sphere is centred at $(5, 6, 7)$ with radius $\sqrt{110}$ (Figure 7.43). When $z = 0$, the equation becomes

$$(x-5)^2 + (y-6)^2 = 61$$

i.e. a circle of radius $\sqrt{61}$. This equation reduces to

$$x^2 + y^2 = 10x + 12y$$

so the circle passes through the origin, as does the sphere.

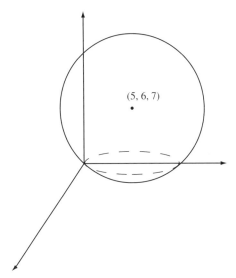

Figure 7.43 A sphere

(b) A paraboloid of revolution whose vertex is at $z = 5$. The outward normal $\boldsymbol{n}$ to the surface is upwards (Figure 7.44). When $z = 1$, $x^2 + y^2 = 36$, so the intersection is a circle of radius 6 centred on the z-axis.

We can calculate the volume and surface area of simple surfaces and we can determine the position of their centroids.

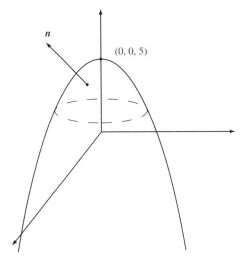

Figure 7.44 Paraboloid of revolution with vertex at $(0,0,5)$ ■

Example

(a) Determine the area of a spherical cap of a sphere of radius a which subtends a semi-angle α at the surface of the sphere.

(b) Determine the volume and centroid of the paraboloid of revolution $x^2 + y^2 = 9(z - 5)$ lying above the plane $z = 0$.

Solution

(a) Consider the circle $x^2 + y^2 = a^2$. We need to rotate the curve about the axis, i.e. to determine the area of the spherical cap (Figure 7.45).

$$\text{Area} = 2\pi \int y \, ds = 2\pi \int_{a \cos \alpha}^{a} (1 + y'^2)^{1/2} \, dx$$

$$2\pi \int_{a \cos \alpha}^{a} y \left(1 + \left(\frac{x}{y}\right)^2\right)^{1/2} dx = 2\pi a^2 (1 - \cos \alpha)$$

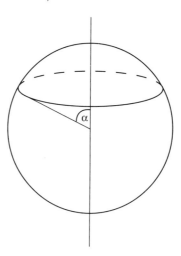

Figure 7.45 Spherical cap

From this you can easily see that the surface area of a sphere must be $4\pi a^2$ (put $\alpha = \pi$).

(b) We can represent the paraboloid as the parabola $y^2 = 9(x - 5)$ rotated about the x-axis (Figure 7.46).

$$\text{Volume of revolution} = \pi \int_{0}^{5} y^2 \, dx$$

$$= 9\pi \int_{0}^{5} (5 - x) dx = \frac{225\pi}{2}$$

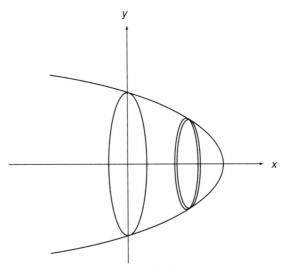

Figure 7.46

The x-coordinate of the centroid lies at

$$9\pi \int_0^5 x(5-x)dx \div \frac{225\pi}{2} = \frac{135\pi}{2} \div \frac{225\pi}{2} = \frac{5}{3}$$

The centroid of the paraboloid is thus at $\left(0, 0, \frac{5}{3}\right)$ ∎

Curves

We can best represent a three-dimensional curve (Figure 7.47) in terms of a parameter, e.g.

$$\boldsymbol{r} = x(t)\boldsymbol{i} + y(t)\boldsymbol{j} + z(t)\boldsymbol{k}$$

The straight line passing through $\boldsymbol{r}_0 = (x_0, y_0, z_0)$ in the direction (a, b, c) has the equation

$$\boldsymbol{r} = \boldsymbol{r}_0 + t\boldsymbol{u} \quad \text{where} \quad \boldsymbol{u} = a\boldsymbol{i} + b\boldsymbol{j} + c\boldsymbol{k}$$

In Cartesians the equation becomes

$$\frac{x - x_0}{a} = \frac{y - y_0}{b} = \frac{z - z_0}{c} \tag{7.13}$$

A curve of special interest is the circular helix (Figure 7.48). This conforms to the shape of the classical light spring and we can depict it as regularly enveloping the **circular cylinder** $x^2 + y^2 = a^2$ (z is unspecified since the cross-section is the same, whatever the value of z).

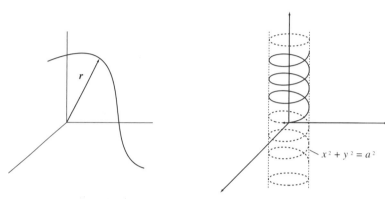

Figure 7.47 A three-dimensional curve

Figure 7.48 A circular helix may be considered as coiling around a circular cylinder

The parametric form of the helix is

$$x = a \cos t \qquad y = a \sin t \qquad z = bt$$

Example

Determine the following points of intersection:

(a) the cone $z = 2(x^2 + y^2)$ with the straight line

$$\frac{x}{1} = \frac{y}{-1} = \frac{z+1}{3} = t$$

(b) the circular helix $x = 2 \cos t$, $y = 2 \sin t$, $z = 3t$ and the plane $z = 2\pi$

Solution

(a) In parametric form $x = t, y = -t, z = -1 + 3t$, so by substituting into the equation of the cone we get

$$-1 + 3t = 2(t^2 + t^2) = 2t^2$$

then $2t^2 - 3t + 1 = 0$

so $t = 1$ or $\dfrac{1}{2}$

The points of intersection are $(1, -1, 2)$ and $\left(\dfrac{1}{2}, \dfrac{-1}{2}, \dfrac{1}{2}\right)$

(b) If $3t = 2\pi$ then $t = \dfrac{2\pi}{3}$ and $a = 2, b = 3$. The point of intersection is $(-1, \sqrt{3}, 2\pi)$

We can find arc length using parameters. In three dimensions, $(\delta s)^2 = (\delta x)^2 + (\delta y)^2 + (\delta z)^2$. This leads to

$$\left(\frac{ds}{dt}\right)^2 = \left(\frac{dx}{dt}\right)^2 + \left(\frac{dy}{dt}\right)^2 + \left(\frac{dz}{dt}\right)^2 \tag{7.14}$$

We can determine quantities such as distance and speed, e.g.

$$(\text{speed})^2 = |\dot{r}|^2 = \dot{x}^2 + \dot{y}^2 + \dot{z}^2 \tag{7.15}$$

Example

Determine the length of a complete cycle of the circular helix

$$x = a \cos t \qquad y = a \sin t \qquad z = bt$$

A complete cycle takes t from 0 to 2π, so

$$\dot{x} = -a \sin t \qquad \dot{y} = a \cos t \qquad \dot{z} = b$$

Therefore

$$s = \int_0^{2\pi} \{a^2 \sin^2 t + a^2 \cos^2 t + b^2\}^{1/2} \, dt = 2\pi(a^2 + b^2)^{1/2} \qquad \blacksquare$$

Exercise 7.5

1   The cone $z^2 = x^2 + y^2$ intersects the sphere $x^2 + y^2 + z^2 = 8$ in the plane $z = b$. Determine b and the equation for the curve of intersection.

2   A straight line passing through the origin in the direction $i + 2j + k$ intersects the cylinder $x^2 + z^2 = 4$ in two points. Determine their coordinates.

3   A straight line passes throught the point $(0, 0, 3)$ in the direction $(1, 1, 1)$. Find where it meets

(a)   the paraboloid of revolution $z = x^2 + y^2$

(b)   the **quadric surface** $z^2 - 2y^2 + xy + 2yz + 1 = 0$

4   The frustrum of a cone has the dimensions illustrated.

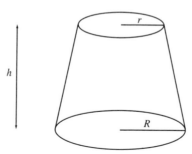

(a)   Show that the volume is $\frac{1}{3}\pi h(r^2 + rR + R^2)$

(b)   How far above the base is the centroid?

5   (a)   Determine the volume of the spherical cap in the example on page 422.

(b)   A core of radius 1 cm is drilled through the middle of a sphere of radius 3 cm. What volume remains? What is its total surface area?

6   Given that $r = xi + yj + zk$ represents the position vector $r(t)$, r is its modulus and $\dot{r} = \dot{x}i + \dot{y}j + \dot{z}k$ is its derivative.

(a)   Establish the following results:

(i) $\dfrac{d}{dt}(\mathbf{a}.\mathbf{r}) = \mathbf{a}.\dot{\mathbf{r}}$ a constant

(ii) $\dfrac{d}{dt}(r^2) = 2\mathbf{r}.\dot{\mathbf{r}}$

(b)   Determine derivatives of

(i) $r^2\mathbf{r} + (\mathbf{a}.\mathbf{r})\mathbf{b}$ b constant

(ii) $\dfrac{\mathbf{r}}{r^2} + \dfrac{r\mathbf{b}}{\mathbf{a}.\mathbf{r}}$

(iii) $\dfrac{1}{2}m\dot{r}^2$ $(\dot{r} = |\dot{\mathbf{r}}|)$

7   Determine $|\dot{\mathbf{r}} \times \ddot{\mathbf{r}}|$ where $\ddot{\mathbf{r}} = \dfrac{d\dot{\mathbf{r}}}{dt}$ for the curves

(a) $r = e^t i + e^{-t}j + \sqrt{2}tk$

(b) $r = 3ti + 3t^2j + 2t^3k$

8 (a) Determine the lengths of the curve

$$r = (3t - t^3)i + 3t^2j + (3t + t^3)k$$

lying between $(0, 0, 0)$ and $(2, 3, 4)$.

(b) At what points does the curve pass through the plane $x = 0$?

9 Calculate the work done on a particle of unit mass in the form $\int F.dr$ in the following cases:

(a) $r = e^t i + e^{-t} j + e^{2t} k,$ $F = zi + x^2 j + yk$
from $t = 0$ to $t = \ln 2$

(b) $r = \left(2 \cos t, 2 \sin t, \dfrac{t}{2\pi}\right),$ $F = g(0, 0, -1)$, g constant; from $(2, 0, 1)$ to $(2, 0, 0)$.

This represents a particle sliding down a smooth helical wire under the influence of gravity.

10 The force exerted by a large mass M, upon a small mass m in space is given by

$$F = -\frac{GMmr}{r^3} \qquad G, M \text{ constant}$$

Calculate the work done by M in moving m from $r = b$ to $r = a$ inwards along a straight path towards M. Assume that r is parallel to dr, i.e. $r.dr = r \, dr$, where $r = |r|$

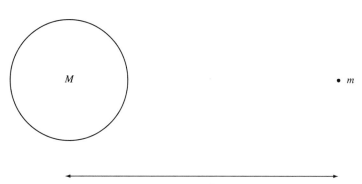

SUMMARY

- **Parabola**: $y^2 = 4ax$; focus $(a, 0)$; directrix $x = -a$; tangent at (x_1, y_1) is $yy_1 = 2a(x + x_1)$; parametric form $x = at^2, y = 2at$

- **Ellipse**: $\dfrac{x^2}{a^2} + \dfrac{y^2}{b^2} = 1, b^2 = a^2(1 - e^2)$, foci $(\pm ae, 0)$; directrices $x = \pm\dfrac{a}{e}$;

  tangent at (x_1, y_1) is $\dfrac{xx_1}{a^2} + \dfrac{yy_1}{b^2} = 1$;

  parametric form $x = a\cos\theta, y = b\sin\theta$

- **Hyperbola**: $\dfrac{x^2}{a^2} - \dfrac{y^2}{b^2} = 1, b^2 = a^2(e^2 - 1)$, foci $(\pm ae, 0)$; directrices

  $x = \pm\dfrac{a}{e}$; asymptotes $y = \pm\dfrac{b}{a}x$; tangent at (x_1, y_1) is $\dfrac{xx_1}{a^2} - \dfrac{yy_1}{b^2} = 1$;

  parametric form $x = a\cosh\theta, y = b\sinh\theta$; the rectangular hyperbola is $xy = c^2$

- **Translation of axes**: to translate the origin to (h, k) use the transformation $x' = x - h, y' = y - k$

- **Rotation of axes**: to rotate the axes anticlockwise by θ use the transformation $x = x'\cos\theta - y'\sin\theta, y = x'\sin\theta + y'\cos\theta$

- **Radius of curvature** at a point on the curve $y = f(x)$ is given by

$$\rho = \frac{\left(1 + \left(\dfrac{dy}{dx}\right)^2\right)^{3/2}}{\dfrac{d^2y}{dx^2}}$$

- **Cartesian components**: displacement $r = x\mathbf{i} + y\mathbf{j}$; velocity $\dot{r} = \dot{x}\mathbf{i} + \dot{y}\mathbf{j}$; acceleration $\ddot{r} = \ddot{x}\mathbf{i} + \ddot{y}\mathbf{j}$

- **Radial components**: velocity $\dot{r}$; acceleration $\ddot{r} - r\dot{\theta}^2$

- **Transverse components**: velocity $r\dot{\theta}$; acceleration $r\ddot{\theta} + 2\dot{r}\dot{\theta}$

- **Length of arc** along $y = f(x)$ from $x = a$ to $x = b$ is

$$\int_a^b \left(1 + \left(\frac{dy}{dx}\right)^2\right)^{1/2} dx$$

- **Area**: the area under a polar curve is $\int_\alpha^\beta \frac{1}{2} r^2\theta \, d\theta,$

  between $\theta = \alpha$ and $\theta = \beta$

- **Curved surface area** of a solid formed by rotating the curve $y = f(x)$ from

  $x = a$ to $x = b$ about the x-axis is $2\pi \int_a^b y \frac{ds}{dx} dx$, where

$$\frac{ds}{dx} = \left(1 + \left(\frac{dy}{dx}\right)^2\right)^{1/2}$$

- **Sphere**: the equation of a sphere radius a and centre the origin is

  $x^2 + y^2 + z^2 = a^2$.

Answers

Exercise 7.1

1 (a) Circle, centre $(-g, -f)$, radius $(f^2 + g^2 - c)^{1/2}$

(b) Two coincident straight lines, the y-axis

(c) Parabola

(d) Hyperbola

2 (a) $(0, 0)$, $y = 0$, $(2, 0)$, $x = -2$

(b) $(0, 0)$, $x = 0$, $\left(0, \dfrac{1}{4}\right)$, $y = -\dfrac{1}{4}$

(c) $(1, -5)$, $x = 1$, $\left(1, -\dfrac{19}{4}\right)$, $y = -\dfrac{21}{4}$

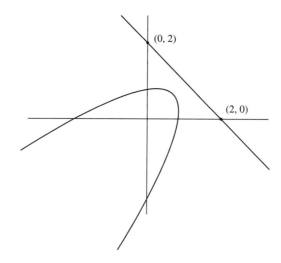

4 Vertex at $\left(\dfrac{1}{2}, \dfrac{1}{2}\right)$; axes crossed at $(-1 \pm \sqrt{2}, 0)$ and $(0, -1 \pm \sqrt{2})$

5 $\triangle QPS$ is isosceles, so $Q\hat{P}S = P\hat{Q}S = T\hat{P}M$

6 G: $(a(t^2 + 2), 0)$, Q: $(a(t^2 - 2), 4at)$ where $a = \dfrac{1}{4}$

7 (b) $y = x + a$

(c)   Tangents are

$$y = mx + \frac{a}{m}$$

$$y = -\frac{x}{m} - ma$$

Equate and cancel the factor $m + \dfrac{1}{m}$

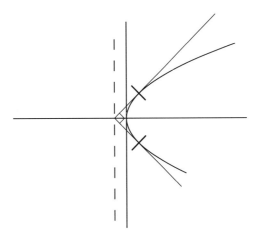

8

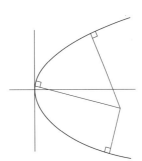

9 (a) $6, 4$; $\dfrac{\sqrt{5}}{3}, (\pm\sqrt{5}, 0)$; $x + \pm\dfrac{9}{\sqrt{5}}$

(b) $4, 6$; $\dfrac{\sqrt{5}}{3}, (0, \pm\sqrt{5})$; $y = \pm\dfrac{9}{\sqrt{5}}$

(c) $\dfrac{(x-1)^2}{25} + \dfrac{(y-2)^2}{4} = 1$ $10, 4$; $\dfrac{\sqrt{21}}{5}$; $(1 \pm \sqrt{21}, 2)$; $x = 1 \pm \dfrac{25}{\sqrt{21}}$

10

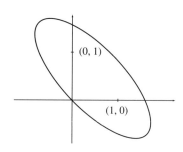

11 Tangent $25x + 6y = 137$
Normal $6x - 25y + 20 = 0$

12 (c)

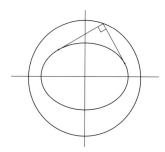

14 (a) $x = \pm 2y$; $\dfrac{\sqrt{5}}{2}$; $(\pm\sqrt{5}, 0)$; $x = \pm\dfrac{4}{\sqrt{5}}$

(b) $2x = \pm 3y$; $\dfrac{\sqrt{13}}{3}$; $(\pm\sqrt{13}, 0)$; $x = \pm\dfrac{9}{\sqrt{13}}$

16 (c) The director circle reduces to the origin and the perpendicular tangents become the asymptotes.

17 Tangent $x - 2y + 1 = 0$
Normal $2x + y + 7 = 0$

18 (a) $x + t_1 t_2 y = c(t_1 + t_2)$

(b) $x + t^2 y = 2ct$

(c) $t^2 x - y = c\left(t^3 - \dfrac{1}{t}\right)$

Exercise 7.2

1 $\dfrac{X^2}{4} + Y^2 = 1$, ellipse

2 (c)

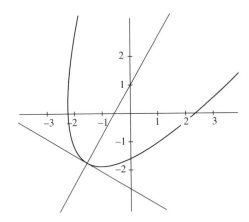

3 (a) $(x-1)^2 - (y+1)^2 = 2$

$x = 1 + \sqrt{2}\cosh t, \qquad y = -1 + \sqrt{2}\sinh t$

(b) $x = a(1 - t^2), \qquad y = 2at$

(c) $x = 4(t-2), \qquad y = \dfrac{2}{t}(2 - t)$

$(x + 8)(y + 2) = 16$

4 (a) $x = \dfrac{t}{1+t^3}, \qquad y = \dfrac{t^2}{1+t^3}$

(b) $x = \dfrac{1-t^2}{t^2}, \qquad y = \dfrac{1-t^2}{t}$

(c) $x = \dfrac{t}{1+t^3}, \qquad y = \dfrac{t^3}{1+t^3}$

(d) $x = \dfrac{1}{1+\sin t}, \qquad y = \dfrac{\cos t}{1+\sin t}$

5 (a) $x^2 + y^2 = 25$, circle centre origin, radius 5

(b) $(x-2)^2 + (y+2)^2 = 9$, circle centre $(2, -2)$, radius 3

(c) $y^2 = 2x$, parabola, $a = \dfrac{1}{2}$

(d) $y^2 = -x$, parabola on negative x-axis, $a = -\dfrac{1}{4}$

(e) $x^2 = 4y$, parabola on positive y-axis, $a = 1$

(f) $\dfrac{(x-1)^2}{4} = 3 - y$, parabola vertex $(1, 3)$, $a = -1$

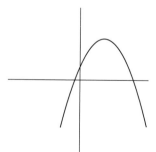

(g)   Ellipse $\dfrac{x^2}{4} + \dfrac{y^2}{9} = 1$, foci on y-axis

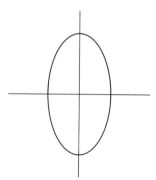

(h)   Ellipse $\dfrac{(x-1)^2}{4} + y^2 = 1$

Centre $(1, 0)$ $a = 2, b = 1$

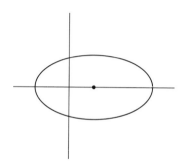

(i) $xy = 4$

(j) $xy = -1$

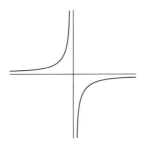

6 (a) $x = 2\left(\dfrac{y-3}{4}\right)^3$

(b) $3xy - x + y - 3 = 0$

(c) $\left(\dfrac{x+y}{2}\right)^3 = \left(\dfrac{y-x}{2}\right)$

(d) $y = \dfrac{e^x}{2e^{2x} - 1}$

7   $y - 3 - 4t = \dfrac{2}{3t^2}(x - 2t^3)$, tangent

8   $m = -\dfrac{1}{t_1 t_2}$ $c = \dfrac{a(t_1 + t_2)}{t_1 t_2}$

9   $\left(x - \dfrac{at_1}{2}\right)\left(y - \dfrac{a}{2t_1}\right) = \dfrac{a^2}{4}$

10

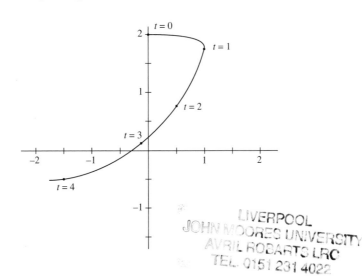

11

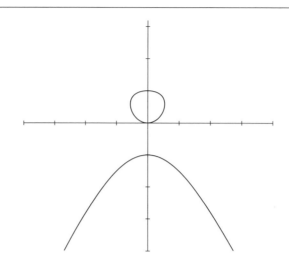

12 $x + y + e^x + e^y = 0$

Quadrant I must be excluded as $e^x, e^y > 0$ for all x, y. Also $y' = \dfrac{-e^x}{1 + e^y} < 0$

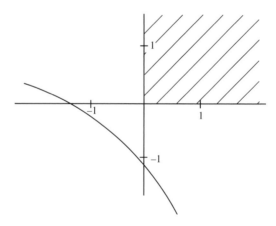

13

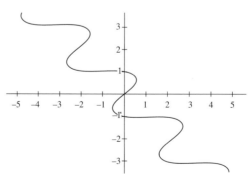

Exercise 7.3

1 $-\tan\theta$

2 $1, -\dfrac{2}{a}$

3 $-\dfrac{3}{2}$

4 $x^2 - 6ax + y^2 = 8a^2$

5 (a) $\sqrt{3}$

  (b) (i) $t = \pi$
    (ii) $t = 0, 2\pi$ $2(1 - \cos t) = 4\sin^2\left(\dfrac{t}{2}\right)$

6 -2

7 (a) $-4\sqrt{2}$

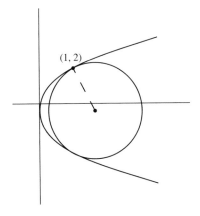

(1, 2)

Concave downwards, hence
the minus sign

(b) $-\dfrac{5}{2}\sqrt{5}$

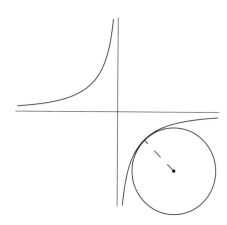

9   (b) $\boldsymbol{r.\dot{r}} = -\dfrac{1}{2}(a^2 - b^2)\sin 2t, \quad = 0,$ at the end of the axes only

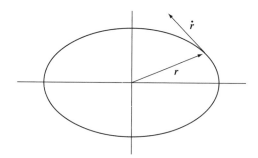

(c) $(\text{speed})^2 = a^2 \sin^2 t + b^2 \cos^2 t$

(i)   maximum at ends of minor axes a

(ii) minimum at ends of major axes b

10   (a) $r = 2ti + \dfrac{2}{t}j \qquad t < 0$

(b) $r = (1 + \cos t)\boldsymbol{i} + \sin t\,\boldsymbol{j} \qquad -\dfrac{\pi}{2} < t \le \dfrac{\pi}{2}$

11   (a) $U^2/2g$ \qquad\qquad\qquad (b) U^2/g

12   Use parabola of safety (Note that $y = x$)

$$\text{Range} = \dfrac{U^2}{g}\dfrac{\sqrt{2}}{(1 + \sqrt{2})}$$

Exercise 7.4

1 (a) $\ln(2 + \sqrt{3})$ (b) $\dfrac{17}{6}$

2 The arc is from $(0, 0)$ to $(3, 6)$ and its length is $6(\sqrt{2} + \ln(1 + \sqrt{2}))$

3 (a) $2(4 + \ln 3)$

(b) $\dfrac{8}{27}(37^{3/2} - 1)$

5 20 m

6 (a) $\dfrac{2a}{\sqrt{3}}$ (b) $\dfrac{\pi a^4}{36}$ (c) $\left(\dfrac{2a}{5}, 0\right)$

7

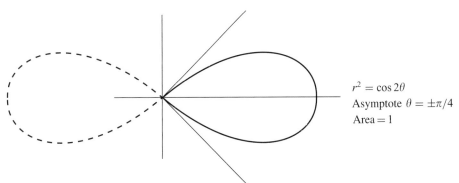

$r^2 = \cos 2\theta$
Asymptote $\theta = \pm\pi/4$
Area $= 1$

8 $r = 2\sin\theta$; circle centre $(0, 1)$, radius 1 and area π

9 (a)

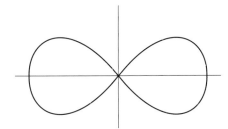

(b) Area $= \dfrac{\pi}{2} - 1$

(c) Volume $= \dfrac{5\pi}{3} - \dfrac{\pi^2}{2}$

(d) $\bar{x} = \dfrac{\ln 2 - \dfrac{1}{2}}{\text{volume}} \approx 0.641.$ $\bar{y} = 0$

10 $\dfrac{\pi}{4}$

11 (a) $\dfrac{4\pi}{\sqrt{3}}$

(b) $\displaystyle\int_0^{\pi/2} \dfrac{\cos\theta \, d\theta}{(1+\cos\theta\sin\theta)^{3/2}}, \qquad \int_0^{\pi/2} \dfrac{\sin\theta \, d\theta}{(1+\cos\theta\sin\theta)^{3/2}}$

(c)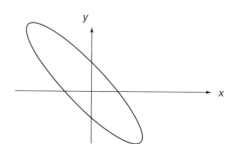

12 $\dfrac{3}{2}\ln 6$

13 73.485

14 $\left(\dfrac{E^2}{R^2}+\dfrac{1}{2}I^2\right)^{1/2}$

15 $\bar{x}=\dfrac{w}{6F} \qquad \bar{y}=\dfrac{2w}{7F}$

16 1.467 46

17 (a) $24(2\sqrt{2}-1)\pi$ (b) $\dfrac{32\pi}{3}$ (c) $\dfrac{4\pi}{3}$

18 (a) $10W$ (b) $\dfrac{WL^2}{2}$ (c) $\dfrac{L}{W}+\dfrac{WL}{2}$

19 Total work $=4+\dfrac{9}{2}+\dfrac{9}{2}+\dfrac{9}{2}=14.5$

Exercise 7.5

1 $b=4$; circle $x^2+y^2=4$; $z=4$

2 $\pm\sqrt{2}(1,2,1)$

3 (a) $(-1,-1,2)$ and $\left(\dfrac{3}{2},\dfrac{3}{2},\dfrac{9}{2}\right)$

(b) $(-1,-1,2)$ and $(-5,-5,-2)$

4   (b) $\dfrac{1}{4}\dfrac{(r^3 + r^2 R + r R^2 + R^3)}{r^2 + rR + R^2}$

5   (a) $\pi a^3 \left\{ \dfrac{2}{3} - \cos \alpha \left(1 - \dfrac{1}{3} \cos^2 \alpha \right) \right\}$

    (b) $\dfrac{64\sqrt{2}\pi}{3}, 32\sqrt{2}\pi$

6   (b)   (i) $2 r \dot{r} \mathbf{r} + r^2 \dot{\mathbf{r}} + (\mathbf{a}.\dot{\mathbf{r}})\mathbf{b}$

        (ii) $\dfrac{\dot{\mathbf{r}}}{r^2} - \dfrac{2\dot{r}\mathbf{r}}{r^3} + \dfrac{\dot{r}\mathbf{b}}{(\mathbf{a}.\mathbf{r})} - \dfrac{r(\mathbf{a}.\dot{\mathbf{r}})\mathbf{b}}{(\mathbf{a}.\mathbf{r})^2}$

        (iii) $m\dot{\mathbf{r}}.\ddot{\mathbf{r}}$

7   (a) $2\sqrt{2}\cosh t$ (b) $18(4t^4 + 4t^2 + 1)^{1/2}$

8   (a) $4\sqrt{2}$

    (b) $(0, 0, 0), (0, 9, 6\sqrt{3}), (0, 9, -6\sqrt{3})$

9   (a) $\dfrac{10}{3}$ (b) $-g$

10 $GMm \left(\dfrac{1}{a} - \dfrac{1}{b} \right)$

8 MATRICES

Introduction

Large systems of simultaneous linear equations are a feature of many problems in engineering and science. Matrices make their solution more efficient. Many other features of linear systems can be more readily understood by expressing them in terms of matrices.

Objectives

After working through this chapter you should be able to

- Identify different types of matrix
- Find the sum and difference of two matrices
- Multiply a matrix by a scalar
- Multiply two matrices together and note that the order of multiplication usually matters
- Relate matrices with two-dimensional transformations of the plane
- Interpret the eigenvalues and eigenvectors of a transformation matrix
- Calculate the eigenvalues and eigenvectors of a 2×2 matrix
- Find the determinant of a 2×2 matrix
- Use Cramer's rule to solve a system of two linear simultaneous equations
- Find the inverse of a 2×2 matrix, when it exists
- Find the determinant of a 3×3 matrix
- Use Cramer's rule to solve a system of three linear simultaneous equations.

8.1 DEFINITION AND BASIC PROPERTIES

A **matrix** is a rectangular or square **array of numbers**:

$$M = \begin{bmatrix} 6 & 5 & 7 & -1 & 3 \\ -2 & 12 & -1 & -5 & -4 \\ -1 & -3 & 0 & 2 & 0 \\ 8 & 2 & -7 & -1 & 3 \end{bmatrix}$$

M is a matrix with 4 rows and 5 columns. This array of numbers may be the basis of a table of data which can be organised into a spreadsheet.

In general, a matrix has m rows and n columns and is normally written as

$$A = \begin{bmatrix} a_{11} & a_{12} & \cdots & a_{1n} \\ \vdots & & & \vdots \\ a_{m1} & a_{m2} & \cdots & a_{mn} \end{bmatrix}$$

with the number (or **element**) in the ith row and jth column written as a_{ij} where $1 \leq k \leq m$, $1 \leq j \leq n$. The matrix A is said to have **order** $m \times n$. We sometimes write $A = [a_{ij}]$.

Types of matrix

(a) $\begin{bmatrix} 1 \\ 2 \\ 3 \end{bmatrix}$ is a **column matrix** or **column vector** or order 3×1

(b) $[-2 \quad 3 \quad 0 \quad 4]$ is a **row matrix** or **row vector** of order 1×4

(c) $\begin{bmatrix} 1 & 2 & 3 & 4 \\ -1 & -2 & -4 & 6 \\ 7 & 0 & 14 & 9 \\ -3 & 8 & -9 & 10 \end{bmatrix}$ is a **square** matrix of order 4

Note that for the square matrix above $a_{11} = 1$, $a_{13} = 3$, $a_{32} = 0$, $a_{43} = -9$, etc.

(d) If the rows and columns of a matrix are interchanged, the matrix is **transposed**; the **transpose of a matrix A** is written $\mathbf{A}^{\mathrm{T}}$. For example

$$A = \begin{bmatrix} 3 & 1 \\ 2 & -1 \\ 4 & -7 \end{bmatrix} \qquad A^{\mathrm{T}} = \begin{bmatrix} 3 & 2 & 4 \\ 1 & -1 & -7 \end{bmatrix}$$

Square matrices are particularly important; they are divided into several types. We consider matrices of order 3 and 4 but the properties apply to all orders.

If A is a square matrix then

(a) A is **symmetric** if $a_{ij} = a_{ji}$, i.e. $A = A^T$:

$$\begin{bmatrix} a & h & g \\ h & b & f \\ g & f & c \end{bmatrix} \text{ is symmetric}$$

(b) A is **skew-symmetric** if $a_{ij} = -a_{ji}$, i.e. $A = -A^T$:

$$\begin{bmatrix} 0 & -a & -b \\ a & 0 & c \\ b & -c & 0 \end{bmatrix} \text{ is skew-symmetric}$$

The elements on the leading diagonal must be zero.

(c) A is a **diagonal matrix** if the only non-zero elements are found on the leading diagonal:

$$\begin{bmatrix} a & 0 & 0 & 0 \\ 0 & b & 0 & 0 \\ 0 & 0 & c & 0 \\ 0 & 0 & 0 & d \end{bmatrix} \text{ is diagonal}$$

(d) A is a **unit matrix** or **identity matrix** if A is diagonal and the diagonal elements are all unity, i.e. they are equal to 1:

$$\begin{bmatrix} 1 & 0 \\ 0 & 1 \end{bmatrix} = I_2 \quad \text{is the unit matrix of order 2}$$

$$\begin{bmatrix} 1 & 0 & 0 \\ 0 & 1 & 0 \\ 0 & 0 & 1 \end{bmatrix} = I_3 \quad \text{is the unit matrix of order 3}$$

Matrix operations

Multiplication by a scalar

In the context of matrices, a scalar is just a number, as in the context of vectors. For example

$$3 \times \begin{bmatrix} 2 & 1 & 3 \\ 5 & 6 & -1 \end{bmatrix} = \begin{bmatrix} 6 & 3 & 9 \\ 15 & 18 & -3 \end{bmatrix}$$

i.e. each element in turn is multiplied by the scalar. In general, we write

$$kA = k[a_{ij}] = [ka_{ij}]$$

Addition and subtraction of matrices

Two matrices can be added or subtracted if and only if they have the **same order**. The corresponding numbers or elements in each matrix are added or subtracted, as required:

$$\begin{bmatrix} 1 & 2 \\ 3 & 4 \end{bmatrix} + \begin{bmatrix} -1 & -3 \\ 0 & 3 \end{bmatrix} = \begin{bmatrix} 0 & -1 \\ 3 & 7 \end{bmatrix} \qquad \begin{bmatrix} 1 & 2 \\ 3 & 4 \end{bmatrix} - \begin{bmatrix} -1 & -3 \\ 0 & 3 \end{bmatrix} = \begin{bmatrix} 2 & 5 \\ 3 & 1 \end{bmatrix}$$

In general

$$A + B = [a_{ij}] + [b_{ij}] = [a_{ij} + b_{ij}]$$
$$A - B = [a_{ij}] - [b_{ij}] = [a_{ij} - b_{ij}]$$

Multiplication of matrices

Now we break from the parallels with arithmetic. Matrices multiply together in a special predetermined manner and must be of compatible orders. Consider the product

$$\begin{bmatrix} 1 & 2 & 3 \\ -1 & 0 & 4 \end{bmatrix} \begin{bmatrix} -3 & 2 \\ 1 & 5 \\ -1 & 2 \end{bmatrix} = \begin{bmatrix} -4 & 17 \\ -1 & 6 \end{bmatrix}$$

For example, the element in row 1, column 1 of the product is obtained as follows:

$$1 \times (-3) + 2 \times 1 + 3 \times (-1) = -4$$

The elements of the first row of the matrix on the left multiply the corresponding elements of the first column on the right to form the element in the first row and first column in the product.

We have effectively taken the scalar product of the first row (vector) of the left-hand matrix with the first column (vector) of the right-hand matrix.

In the same way, the element in the first row and second column of the product is found by taking the scalar product of the first row of the left-hand matrix with the second column of the right-hand matrix.

i.e. $1 \times 2 + 2 \times 5 + 3 \times 5 = 17$

and so on.

The product can be symbolised as $A \times B = C$ or, more simply, as $AB = C$. If the product exists then the number of columns of A is equal to the number of rows of B. We saw that if A is of order (2×3) and B is of order (3×2) then $AB = C$ is of order

$$(2 \times 3) \times (3 \times 2) = 2 \times 2$$

The middle two numbers on the left-hand side must be the same and the outer two numbers indicate the order of the product. The formation of the elements in row 2 and column 2 of the product may be indicated diagrammatically as

$$\begin{bmatrix} a_{11} & a_{12} & a_{13} \\ a_{21} & a_{22} & a_{23} \end{bmatrix} \begin{bmatrix} b_{11} & b_{12} \\ b_{21} & b_{22} \\ b_{31} & b_{32} \end{bmatrix}$$

and symbolically as $c_{22} = \sum_{k=1}^{3} a_{2k} b_{k2}$

The general result is therefore

$$\begin{array}{ccc} A & \times & B & = & C \\ [m \times p] & & [p \times n] & & [m \times n] \end{array}$$

where $c_{ij} = \sum_{k=1}^{p} a_{ik} b_{kj}$. Schematically

Example

Using the matrices A and B previously described find the product BA

$$BA = \begin{bmatrix} -3 & 2 \\ 1 & 5 \\ -1 & 2 \end{bmatrix} \begin{bmatrix} 1 & 2 & 3 \\ -1 & 0 & 4 \end{bmatrix}$$

$$= \begin{bmatrix} -3 \times 1 + 2 \times (-1) & -3 \times 2 + 2 \times 0 & -3 \times 3 + 2 \times 4 \\ 1 \times 1 + 5 \times (-1) & 1 \times 2 + 5 \times 0 & 1 \times 3 + 5 \times 4 \\ -1 \times 1 + 2 \times (-1) & -1 \times 2 + 2 \times 0 & -1 \times 3 + 2 \times 4 \end{bmatrix}$$

$$= \begin{bmatrix} -5 & -6 & -1 \\ -4 & 2 & 23 \\ -3 & -2 & 5 \end{bmatrix} \qquad \blacksquare$$

You see here that $BA \neq AB$, a result which is usually true. This is completely different to the behaviour of numbers: ab and ba are *always* equal.

In the case of matrix multiplication, one or both of the products AB and BA may not be possible. If both products are possible, they may be of different order. Even if they are of the same order, they may be different.

We say that **matrices**, unlike numbers, do not generally commute; they are **non-commutative**. In other words, they may not multiply left to right and right to left interchangeably, and the order of the product left to right *does* matter.

Equality and nullity

Two matrices of the same order are **equal** if and only if corresponding elements are equal. Equality is impossible if the orders are different.

$$A = B \Leftrightarrow a_{ij} = b_{ij} \qquad \text{all } i, j$$

If $A = B$ then $A - B = 0$, a matrix of the same order as A and B with all of its elements equal to zero. Whatever the order 0 is called the **null matrix** or **zero matrix**.

Matrix products

Systematic proofs are not given for the results quoted below, though you should verify the properties for the matrices

$$A = \begin{bmatrix} 1 & 2 & 3 \\ -1 & 0 & 4 \end{bmatrix} \qquad B = \begin{bmatrix} -3 & 2 \\ 1 & 5 \\ -1 & 2 \end{bmatrix} \qquad C = \begin{bmatrix} -1 & 2 \\ 1 & 1 \end{bmatrix} \qquad D = \begin{bmatrix} 6 & 1 \\ -1 & 1 \\ 2 & 4 \end{bmatrix}$$

(i) $(AB)^{\mathrm{T}} = B^{\mathrm{T}}A^{\mathrm{T}}$: the transpose of a product is the product of the transposes in the reverse order.

(ii) If C is a square matrix then $C^{\mathrm{T}}C$ and CC^{T} are both symmetric. Does $CC^{\mathrm{T}} = C^{\mathrm{T}}C$?

(iii) $A(BC) = (AB)C$, the **associative property**, i.e. the order of multiplying out the matrix product component, AB first or BC first, is unimportant provided the left-to-right procedure is maintained.

(iv) $A(B + D) = AB + AD$, the **distributive property**.

Example

To estimate the total volume of water contained in a rectangular-shaped reservoir of uneven depth, soundings are taken at grid points 50 m apart horizontally and vertically. The following table is produced.

$$A = \begin{bmatrix} 0 & 0.2 & 0.4 & 0.8 & 0 & 0 \\ 0.4 & 1.0 & 2.3 & 3.1 & 0.2 & 0 \\ 0.6 & 3.1 & 5.9 & 6.9 & 3.5 & 0.6 \\ 0.6 & 3.4 & 4.7 & 11.1 & 7.9 & 0.8 \\ 0.4 & 0.5 & 1.7 & 3.5 & 4.7 & 1.0 \end{bmatrix}$$

The best way of estimating the volume is to estimate the amount contained in each grid square $\begin{matrix} a & b \\ c & d \end{matrix}$, i.e. $50^2 \times \frac{1}{4} \times (a + b + c + d)$, which is area $\times$ average depth.

Each corner element is counted once, each side element twice and each internal element four times. In matrix terminology we are computing

$$[1 \quad 2 \quad 2 \quad 2 \quad 1] A \begin{bmatrix} 1 \\ 2 \\ 2 \\ 2 \\ 2 \\ 1 \end{bmatrix} \quad \text{i.e.} \quad x^{\mathrm{T}}Ay \quad \text{where} \quad x = \begin{bmatrix} 1 \\ 2 \\ 2 \\ 2 \\ 1 \end{bmatrix} \quad \text{and} \quad y = \begin{bmatrix} 1 \\ 2 \\ 2 \\ 2 \\ 2 \\ 1 \end{bmatrix}$$

and A is the grid matrix.

Compute Ay first of all, then $x^{\mathrm{T}}Ay$. The dimensions of $x^{\mathrm{T}}Ay$ are $(1 \times 5) \times (5 \times 6) \times (6 \times 1) = (1 \times 1)$, i.e. a single element, the volume. The total volume is $152\,125\,\mathrm{m}^3$. ■

Exercise 8.1

1 Evaluate the matrices $A + B, AB$ and BA, if they are defined, where

(a) $A = \begin{bmatrix} 2 & 1 \\ 3 & -4 \end{bmatrix}$ $\quad B = \begin{bmatrix} 3 & 0 \\ 1 & 2 \end{bmatrix}$

(b) $A = \begin{bmatrix} 1 & 0 \\ 0 & m \end{bmatrix}$ $B = \begin{bmatrix} 0 & 1 \\ m & 0 \end{bmatrix}$

(c) $A = \begin{bmatrix} 0 & 4 & 2 \\ -1 & 1 & 3 \\ 2 & 0 & -2 \end{bmatrix}$ $B = \begin{bmatrix} 1 & -3 & 5 \\ 2 & 0 & -4 \\ 3 & 2 & 0 \end{bmatrix}$

(d) $A = \begin{bmatrix} 1 & 2 & 3 \\ -7 & 0.5 & 0 \end{bmatrix}$ $B = \begin{bmatrix} 3 & 2 \\ 5 & 6 \\ 1 & -7 \end{bmatrix}$

(e) $A = \begin{bmatrix} 1 & 2 & 3 & 4 \end{bmatrix}$ $B = \begin{bmatrix} 1 \\ 2 \\ 3 \\ 4 \end{bmatrix}$

2   Evaluate the matrices given by the products

(a) $\begin{bmatrix} 2 & 1 \\ -1 & 0 \\ 2 & 3 \end{bmatrix} \begin{bmatrix} 3 & 0 \\ 0 & 1 \end{bmatrix} \begin{bmatrix} -1 & 2 & 3 \\ 4 & 0 & 1 \end{bmatrix}$

(b) $\begin{bmatrix} 1 & 2 & 3 & 4 \\ -1 & -3 & -5 & -6 \\ 6 & 1 & 7 & 5 \end{bmatrix} \begin{bmatrix} 5 & 9 \\ 6 & 10 \\ 7 & 12 \\ 8 & 11 \end{bmatrix} \begin{bmatrix} 2 & 3 \\ 1 & 2 \end{bmatrix}$

3   Verify the associative and distributive properties of matrix products, where

$$A = \begin{bmatrix} 0 & 4 & 2 \\ -1 & 1 & 3 \\ 2 & 0 & -2 \end{bmatrix} \quad B = \begin{bmatrix} 1 & -3 & 5 \\ 2 & 0 & -4 \\ 3 & 2 & 0 \end{bmatrix} \quad C = \begin{bmatrix} 2 & 1 \\ 4 & 0 \\ 0 & 1 \end{bmatrix}$$

4   If $A = \begin{bmatrix} 1 & 1 \\ 0 & 1 \end{bmatrix}$ we define $A^2 = AA = \begin{bmatrix} 1 & 1 \\ 0 & 1 \end{bmatrix}\begin{bmatrix} 1 & 1 \\ 0 & 1 \end{bmatrix}$

(a)   Determine A^2
(b)   Verify the associative property $A^2 A = A A^2 = A^3$
(c)   What is $A^4 = AAAA = AA^3 = AA^2A = A^2A^2 = A^3A$?
(d)   Write down a form for

$$A^n = A \times \cdots \times A \qquad (n \text{ times})$$

$$A^{n+1} = A \times \cdots \times A \qquad (n+1 \text{ times})$$

Now show that $A^{n+1} = AA^n = A^nA$. By this stage you will have proved by mathematical induction the form for A^n.

(e) Prove that $A^m A^n = A^n A^m = A^{m+n}$, i.e. that positive integer powers of A commute. Note that such powers of a square matrix always commute.

5 If A is the matrix in the previous exercise and $X = \begin{bmatrix} a & b \\ 0 & c \end{bmatrix}$

(a) Determine X if $XA = I$, the identity matrix.

(b) Verify that X and A commute, i.e.

$$XA = AX = I$$

(c) Find X^2 and X^3 and write down a form for X^n

(d) What is the relationship between X^n and A^n?

 6 A quarry with a rectangular base is to be drawn out of the side of a hill. It is known that the start line is 200 m long, straight and level. The depth of rock above this line is estimated from an Ordnance Survey map. Depths of rock are given over a square grid mesh of side 40 m.

Quarrying 160 m into the side of the hill gives the following estimates of depth

0	0	0	0	0	0
8	9	11	17	6	5
15	20	23	30	11	9
24	31	39	50	30	17
35	44	52	61	40	29

Use the formulation of the reservoir example (page 450) to calculate the total amount of rock excavated.

7 Write each of the following as a pair of simultaneous equations and solve them:

(a) $\begin{bmatrix} 3 & 4 \\ -2 & 3 \end{bmatrix} \begin{bmatrix} x \\ y \end{bmatrix} = \begin{bmatrix} 11 \\ 4 \end{bmatrix}$

(b) $\begin{bmatrix} a & b \\ -2b & 3a \end{bmatrix} \begin{bmatrix} 3 \\ 2 \end{bmatrix} = \begin{bmatrix} 5 \\ 0 \end{bmatrix}$

8 $A = \begin{bmatrix} a & 0 \\ 0 & b \end{bmatrix}$ and $B = \begin{bmatrix} c & 0 \\ 0 & d \end{bmatrix}$ are diagonal matrices of order 2.

(a) Prove that A and B commute, i.e. $AB = BA$.

(b) Write down symbolically two diagonal matrices of order 3. Do they commute?

(c) Do diagonal matrices always commute?

9 Given the matrices

$$A = \begin{bmatrix} -3 & 1 & 0 \\ 2 & -3 & 2 \\ 0 & 1 & -3 \end{bmatrix} \quad B = \begin{bmatrix} 1 & 1 & -1 \\ 2 & 0 & 2 \\ 1 & -1 & -1 \end{bmatrix} \quad D = \begin{bmatrix} a & 0 & 0 \\ 0 & b & 0 \\ 0 & 0 & c \end{bmatrix}$$

Determine a, b, c if

$$AB + BD = 0$$

where 0 is the zero matrix.

10 The matrix $R(\theta)$ is defined as a function of θ by

$$R(\theta) = \begin{bmatrix} \cos\theta & -\sin\theta \\ \sin\theta & \cos\theta \end{bmatrix}$$

(a) Prove that $R(\theta)R(\phi) = R(\theta + \phi) = R(\phi)R(\theta)$, i.e.

$$\begin{bmatrix} \cos\theta & -\sin\theta \\ \sin\theta & \cos\theta \end{bmatrix} \begin{bmatrix} \cos\phi & -\sin\phi \\ \sin\phi & \cos\phi \end{bmatrix} = \begin{bmatrix} \cos(\theta + \phi) & -\sin(\theta + \phi) \\ \sin(\theta + \phi) & \cos(\theta + \phi) \end{bmatrix}$$

(b) Write down $R(\theta)$ and deduce that $R(\theta)R(-\theta) = I$

(c) Deduce that $\{R(\theta)\}^2 = R(\theta) \times R(\theta) = R(2\theta)$. What is $\{R(\theta)\}^n$?

(d) Write down

(i) $R(0)$ (ii) $R\left(\dfrac{\pi}{6}\right)$ (iii) $R\left(\dfrac{\pi}{4}\right)$ (iv) $R\left(\dfrac{\pi}{3}\right)$ (v) $R\left(\dfrac{\pi}{2}\right)$

(e) What is the value of $R\left(\dfrac{5\pi}{12}\right)$?

8.2 TRANSFORMATIONS IN TWO DIMENSIONS

Transformations

Consider Figure 8.1. We will refer to the vector $r = OP = xi + yj$ as $\begin{bmatrix} x \\ y \end{bmatrix}$ in matrix notation and the 2×2 matrices which convert one vector into another as **transformation**

matrices. As an example, the matrix statement

$$\begin{bmatrix} 2 & 1 \\ -1 & 3 \end{bmatrix}\begin{bmatrix} 1 \\ 2 \end{bmatrix} = \begin{bmatrix} 4 \\ 5 \end{bmatrix}$$

means that the matrix $\begin{bmatrix} 2 & 1 \\ -1 & 3 \end{bmatrix}$ converts the vector $i + 2j$ to the vector $4i + 5j$

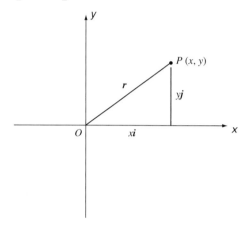

Figure 8.1   Two-dimensional vectors

Reflection in the x-axis

The matrix $\begin{bmatrix} 1 & 0 \\ 0 & -1 \end{bmatrix}$ transforms the vector $\begin{bmatrix} 3 \\ 1 \end{bmatrix}$ to $\begin{bmatrix} 3 \\ -1 \end{bmatrix}$ and the vector $\begin{bmatrix} 7 \\ 2 \end{bmatrix}$ to $\begin{bmatrix} 7 \\ -2 \end{bmatrix}$, i.e. it maps the x-coordinate onto itself while changing the sign of the y-coordinate (Figure 8.2).

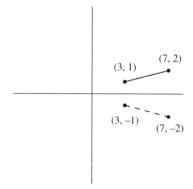

Figure 8.2   Reflection in the x-axis

In general, the vector $\begin{bmatrix} x \\ y \end{bmatrix}$ is mapped to

$$\begin{bmatrix} 1 & 0 \\ 0 & -1 \end{bmatrix}\begin{bmatrix} x \\ y \end{bmatrix} = \begin{bmatrix} x \\ -y \end{bmatrix}$$

One-way stretch

The matrix $\begin{bmatrix} 1 & 0 \\ 0 & 2 \end{bmatrix}$ transforms $\begin{bmatrix} 1 \\ 2 \end{bmatrix}$ into $\begin{bmatrix} 1 \\ 4 \end{bmatrix}$ and $\begin{bmatrix} 3 \\ 4 \end{bmatrix}$ into $\begin{bmatrix} 3 \\ 8 \end{bmatrix}$, so it doubles the y-coordinates (Figure 8.3). In general, the vector $\begin{bmatrix} x \\ y \end{bmatrix}$ is mapped to

$$\begin{bmatrix} 1 & 0 \\ 0 & 2 \end{bmatrix}\begin{bmatrix} x \\ y \end{bmatrix} = \begin{bmatrix} x \\ 2y \end{bmatrix}$$

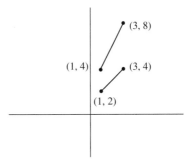

Figure 8.3 One-way stretch

Rotation anticlockwise by 90°

The matrix $\begin{bmatrix} 0 & -1 \\ 1 & 0 \end{bmatrix}$ moves $\begin{bmatrix} 2 \\ 1 \end{bmatrix}$ to $\begin{bmatrix} -1 \\ 2 \end{bmatrix}$ and $\begin{bmatrix} 4 \\ 4 \end{bmatrix}$ to $\begin{bmatrix} -4 \\ 4 \end{bmatrix}$ (Figure 8.4). In general, the vector $\begin{bmatrix} x \\ y \end{bmatrix}$ is mapped to

$$\begin{bmatrix} 0 & -1 \\ 1 & 0 \end{bmatrix}\begin{bmatrix} x \\ y \end{bmatrix} = \begin{bmatrix} -y \\ x \end{bmatrix}$$

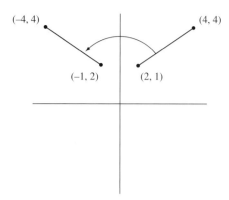

Figure 8.4   Rotation

Translation

The vector $\begin{bmatrix} 1 \\ 2 \end{bmatrix}$ is translated to the vector $\begin{bmatrix} 5 \\ 3 \end{bmatrix}$ and the vector $\begin{bmatrix} 2 \\ 5 \end{bmatrix}$ to the vector $\begin{bmatrix} 6 \\ 6 \end{bmatrix}$. Adding $\begin{bmatrix} 4 \\ 1 \end{bmatrix}$ to the vector performs the task (Figure 8.5).

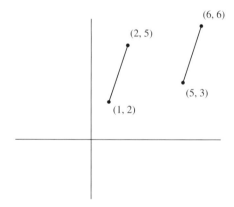

Figure 8.5   Translation

General transformations

Reflection, one-way stretch and rotation involved matrix multiplication; translation involved addition. This is because the first three are changes in relative position with reference to the origin whereas translation entails movement to a new origin. Generally speaking, a **transformation** is a combination of **translation** and **relative movement**.

Examples

1. Find the transformations which map the shaded triangle (Figure 8.6) into each of the other triangles. $\begin{bmatrix} x \\ y \end{bmatrix}$ represents the position vector of a vertex of the shaded triangle.

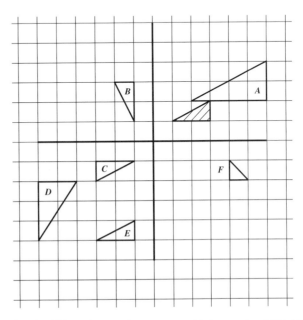

Figure 8.6 The shaded triangle can be mapped into each of the lettered triangles by combining simple transformations

(a) *A*: direct enlargement by a factor 2. The matrix of transformation is $2I = \begin{bmatrix} 2 & 0 \\ 0 & 2 \end{bmatrix}$
   Then

$$\begin{bmatrix} 2 & 0 \\ 0 & 2 \end{bmatrix} \begin{bmatrix} x \\ y \end{bmatrix} = \begin{bmatrix} 2x \\ 2y \end{bmatrix}$$

(b) *B*: 90° rotation followed by a move by one unit to the left. The transformation is defined as follows:

$$\begin{bmatrix} 0 & -1 \\ 1 & 0 \end{bmatrix} \begin{bmatrix} x \\ y \end{bmatrix} + \begin{bmatrix} -1 \\ 0 \end{bmatrix} = \begin{bmatrix} -(y+1) \\ x \end{bmatrix}$$

(c) C: rotation by 180°, coordinates change sign. The transformation is

$$\begin{bmatrix} -1 & 0 \\ 0 & -1 \end{bmatrix}\begin{bmatrix} x \\ y \end{bmatrix} = \begin{bmatrix} -x \\ -y \end{bmatrix}$$

(d) D: use C as the starting-point, enlarge in the y-direction, then translate as indicated in Figure 8.7.

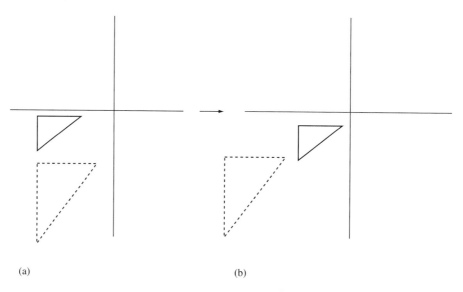

(a)   (b)

Figure 8.7 Transformation D

The enlargement matrix is $\begin{bmatrix} 1 & 0 \\ 0 & 3 \end{bmatrix}$ and the translation vector is $\begin{bmatrix} -3 \\ 1 \end{bmatrix}$. The full operation is thus

$$\begin{bmatrix} 1 & 0 \\ 0 & 3 \end{bmatrix}\begin{bmatrix} -1 & 0 \\ 0 & -1 \end{bmatrix}\begin{bmatrix} x \\ y \end{bmatrix} + \begin{bmatrix} -3 \\ 1 \end{bmatrix}$$

$$= \begin{bmatrix} -1 & 0 \\ 0 & -3 \end{bmatrix}\begin{bmatrix} x \\ y \end{bmatrix} + \begin{bmatrix} -3 \\ 1 \end{bmatrix} = \begin{bmatrix} -(x+3) \\ -3y+1 \end{bmatrix}$$

The matrix for the first transformation is placed immediately to the left of $\begin{bmatrix} x \\ y \end{bmatrix}$. The matrix for the second transformation is placed immediately to the left of $\begin{bmatrix} -1 & 0 \\ 0 & -1 \end{bmatrix}\begin{bmatrix} x \\ y \end{bmatrix}$. Successive transformations are applied by matrix multiplication on the left.

(e) E is just a translation of the original: the transformation is

$$\begin{bmatrix} x \\ y \end{bmatrix} + \begin{bmatrix} -4 \\ -6 \end{bmatrix} \quad \text{or} \quad \begin{bmatrix} x \\ y \end{bmatrix} - \begin{bmatrix} 4 \\ 6 \end{bmatrix} = \begin{bmatrix} x - 4 \\ y - 6 \end{bmatrix}$$

(f) F can be obtained by first reflecting E in the y-axis, then contracting the x-coordinate by $\dfrac{1}{2}$, followed by a translation (Figure 8.8).

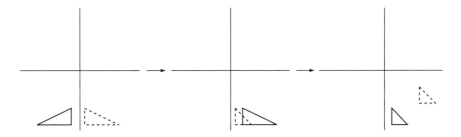

Figure 8.8 Transformation to F: starting from triangle E

The full operation is therefore

$$\begin{bmatrix} \dfrac{1}{2} & 0 \\ 0 & 1 \end{bmatrix} \begin{bmatrix} -1 & 0 \\ 0 & 1 \end{bmatrix} \begin{bmatrix} x - 4 \\ y - 6 \end{bmatrix} + \begin{bmatrix} 3\dfrac{1}{2} \\ 3 \end{bmatrix}$$

$$\begin{bmatrix} -\dfrac{1}{2} & 0 \\ 0 & 1 \end{bmatrix} \begin{bmatrix} x - 4 \\ y - 6 \end{bmatrix} + \begin{bmatrix} 3\dfrac{1}{2} \\ 3 \end{bmatrix}$$

i.e.

$$\begin{bmatrix} -\dfrac{1}{2} & 0 \\ 0 & 1 \end{bmatrix} \begin{bmatrix} x \\ y \end{bmatrix} + \begin{bmatrix} 5\dfrac{1}{2} \\ -3 \end{bmatrix} = \begin{bmatrix} -\dfrac{1}{2}x + \dfrac{11}{2} \\ y - 3 \end{bmatrix}$$

At this point you can see that we could have started by reflecting the original triangle in the y-axis, contracting it in the x-direction then translating it (Figure 8.9).

2. Find a matrix which reflects points in the line $y = x$. This reflection interchanges the coordinates, no matter what they are (Figure 8.10). Let the matrix be $\begin{bmatrix} a & b \\ c & d \end{bmatrix}$, then

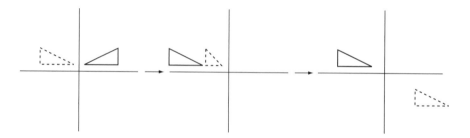

Figure 8.9 Transformation to F: starting from the shaded triangle

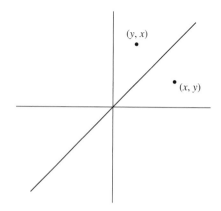

Figure 8.10 Reflection in the line $y = x$

we require that $\begin{bmatrix} a & b \\ c & d \end{bmatrix} \begin{bmatrix} x \\ y \end{bmatrix} = \begin{bmatrix} y \\ x \end{bmatrix}$, whatever x, y. Choose x and y to be whatever you wish:

e.g. $x = 1, y = 0$ means $a \times 1 + b \times 0 = 0 \Rightarrow a = 0, c \times 1 + d \times 0 = 1 \Rightarrow c = 1$
and $x = 0, y = 0$ means $b = 1, d = 0$

so the matrix is $\begin{bmatrix} 0 & 1 \\ 1 & 0 \end{bmatrix}$. Try it and see!

3. The rotation matrix

$$R(\theta) = \begin{bmatrix} \cos\theta & -\sin\theta \\ \sin\theta & \cos\theta \end{bmatrix}$$

rotates the vector $\begin{bmatrix} x \\ y \end{bmatrix}$ through an angle θ anticlockwise, e.g.

$$R\left(\frac{\pi}{2}\right) = \begin{bmatrix} 0 & -1 \\ 1 & 0 \end{bmatrix} \quad \text{takes} \quad \begin{bmatrix} 1 \\ 0 \end{bmatrix} \quad \text{to} \quad \begin{bmatrix} 0 \\ 1 \end{bmatrix}$$

$$\begin{bmatrix} 0 \\ 1 \end{bmatrix} \quad \text{to} \quad \begin{bmatrix} -1 \\ 0 \end{bmatrix}$$

is rotation through $\frac{\pi}{2}$ anticlockwise. ■

It was proved in the previous exercises that matrices of the type $R(\theta)$ are commutative and that

$$R(\theta)R(\phi) = R(\phi)R(\theta) = R(\theta + \phi)$$

and $R(\theta)R(-\theta) = I$

Rotation matrices maintain the magnitude of a vector. Referring to Figure 8.11, the point (x, y) is moved to the point

$$(x', y') \quad \text{where} \quad \begin{bmatrix} x' \\ y' \end{bmatrix} = \begin{bmatrix} \cos\theta & -\sin\theta \\ \sin\theta & \cos\theta \end{bmatrix}\begin{bmatrix} x \\ y \end{bmatrix}$$

i.e.

$$x' = x\cos\theta - y\sin\theta \qquad y' = x\sin\theta + y\cos\theta$$

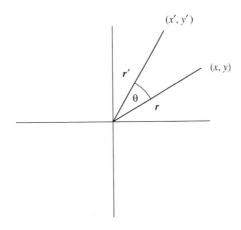

Figure 8.11 Rotation

Note that

$$x'^2 + y'^2 = (x\cos\theta - y\sin\theta)^2 + (x\sin\theta + y\cos\theta)^2$$
$$= x^2\cos^2\theta - 2xy\cos\theta\sin\theta + y^2\sin^2\theta + x^2\sin^2\theta$$
$$+ 2xy\sin\theta\cos\theta$$
$$+ y^2\cos^2\theta$$
$$= x^2 + y^2$$

General matrices tend to rotate vectors and alter their magnitudes as well.

Examples   1. Consider the following examples, where the matrix multiplies each of the given vectors in turn.

(a) $\begin{bmatrix} 2 & 1 \\ 1 & 2 \end{bmatrix}$ $\begin{bmatrix} 1 \\ 1 \end{bmatrix}, \begin{bmatrix} 1 \\ -1 \end{bmatrix}, \begin{bmatrix} -1 \\ 6 \end{bmatrix}$

(b) $\begin{bmatrix} 1 & 2 \\ -2 & 4 \end{bmatrix}$ $\begin{bmatrix} 1 \\ 1 \end{bmatrix}, \begin{bmatrix} 1 \\ -2 \end{bmatrix}, \begin{bmatrix} 2 \\ -1 \end{bmatrix}$

(c) $\begin{bmatrix} 2 & -1 \\ 1 & 2 \end{bmatrix}$ $\begin{bmatrix} 1 \\ 0 \end{bmatrix}, \begin{bmatrix} 1 \\ 1 \end{bmatrix}, \begin{bmatrix} 3 \\ 2 \end{bmatrix}$

Solution

(a) (i) $\begin{bmatrix} 2 & 1 \\ 1 & 2 \end{bmatrix} \begin{bmatrix} 1 \\ 1 \end{bmatrix} = \begin{bmatrix} 3 \\ 3 \end{bmatrix} = 3\begin{bmatrix} 1 \\ 1 \end{bmatrix}$

(ii) $\begin{bmatrix} 2 & 1 \\ 1 & 2 \end{bmatrix} \begin{bmatrix} 1 \\ -1 \end{bmatrix} = \begin{bmatrix} 1 \\ -1 \end{bmatrix}$

(iii) $\begin{bmatrix} 2 & 1 \\ 1 & 2 \end{bmatrix} \begin{bmatrix} -1 \\ 6 \end{bmatrix} = \begin{bmatrix} 4 \\ 11 \end{bmatrix}$

In cases (i) and (ii) the direction is unchanged and the magnification factors are 3 and 1 respectively. In case (iii) the direction is changed and the magnification factor is

$$\sqrt{\frac{(4^2 + 11^2)}{((-1)^2 + 6^2)}} = \sqrt{\frac{137}{37}} = 1.924$$

This situation arises because $\begin{bmatrix} 1 \\ 1 \end{bmatrix}$ and $\begin{bmatrix} 1 \\ -1 \end{bmatrix}$ have a special status for the matrix $\begin{bmatrix} 2 & 1 \\ 1 & 2 \end{bmatrix}$ in that their directions are unchanged. They are called **eigenvectors** which is German for characteristic vectors. The magnification factors 3 and 1 are called **eigenvalues** or characteristic values. You will notice too that $\begin{bmatrix} -1 \\ 6 \end{bmatrix}$ did change direction and was magnified, i.e. stretched by a factor lying between 1 and 3.

A 2×2 matrix can have at most **two** eigenvalues and if the matrix is symmetric, as $\begin{bmatrix} 2 & 1 \\ 1 & 2 \end{bmatrix}$ happens to be, the magnitude of its 'stretching power' always lies between the magnitude of the eigenvalues.

(b) (i) $\begin{bmatrix} 1 & 2 \\ -2 & -4 \end{bmatrix} \begin{bmatrix} 1 \\ 1 \end{bmatrix} = \begin{bmatrix} 3 \\ -6 \end{bmatrix}$

(ii) $\begin{bmatrix} 1 & 2 \\ -2 & -4 \end{bmatrix} \begin{bmatrix} 2 \\ -1 \end{bmatrix} = \begin{bmatrix} 0 \\ 0 \end{bmatrix}$

(iii) $\begin{bmatrix} 1 & 2 \\ -2 & -4 \end{bmatrix} \begin{bmatrix} 1 \\ -2 \end{bmatrix} = \begin{bmatrix} -3 \\ 6 \end{bmatrix} = -3 \begin{bmatrix} 1 \\ -2 \end{bmatrix}$

In case (i) the vector is not an eigenvector and the stretching factor is $3\sqrt{\dfrac{5}{2}} = 4.743$. For case (ii) the vector $\begin{bmatrix} 2 \\ -1 \end{bmatrix}$ is compressed into the zero vector **0**.

This is because $\begin{bmatrix} 2 \\ -1 \end{bmatrix}$ is an eigenvector with eigenvalue 0. Any matrix with 0 as an eigenvalue is said to be **singular**, a topic to which we will return in the next section. Note, however, that for matrices the equation

$$AX = 0 \text{ does not imply that either } A = 0 \text{ or } X = 0$$

but either $A = 0$ or $X = 0$ does imply that $AX = 0$.

In case (iii), $\begin{bmatrix} 1 \\ -2 \end{bmatrix}$ is an eigenvector and -3 is the eigenvalue. But note that $\begin{bmatrix} 1 & 2 \\ -2 & -4 \end{bmatrix}$ is not symmetric, so the stretching power of the matrix is not between 0 and 3.

(c) (i) $\begin{bmatrix} 2 & -1 \\ 1 & 2 \end{bmatrix} \begin{bmatrix} 1 \\ 0 \end{bmatrix} = \begin{bmatrix} 2 \\ 1 \end{bmatrix}$

(ii) $\begin{bmatrix} 2 & -1 \\ 1 & 2 \end{bmatrix} \begin{bmatrix} 1 \\ 1 \end{bmatrix} = \begin{bmatrix} 1 \\ 3 \end{bmatrix}$

(iii) $\begin{bmatrix} 2 & -1 \\ 1 & 2 \end{bmatrix} \begin{bmatrix} 3 \\ 2 \end{bmatrix} = \begin{bmatrix} 4 \\ 7 \end{bmatrix}$

None of the vectors is an eigenvector, and for this matrix no real eigenvalues exist. The stretching of the vector is constant $= \sqrt{5}$ in each case. This is because the matrix can be written as

$$\sqrt{5} \begin{bmatrix} \dfrac{2}{\sqrt{5}} & \dfrac{-1}{\sqrt{5}} \\ \dfrac{1}{\sqrt{5}} & \dfrac{2}{\sqrt{5}} \end{bmatrix}$$

which you will recognise as a multiple ($\sqrt{5}$), of the rotation matrix with $\cos\theta = \dfrac{2}{\sqrt{5}}$, $\sin\theta = \dfrac{1}{\sqrt{5}}$ or $\theta = \tan^{-1}\left(\dfrac{1}{2}\right)$. Not surprisingly, every vector is rotated anticlockwise by this angle and increased in length by a factor $\sqrt{5}$.

2. Consider the matrix $\begin{bmatrix} 1 & 2 \\ -2 & -4 \end{bmatrix}$, part (b) in the previous example. We can determine its maximum stretching power by taking a vector of unit length and determining the maximum magnification. Let the unit vector be $\boldsymbol{r} = \cos\theta\boldsymbol{i} + \sin\theta\boldsymbol{j}$ or, in matrix terminology, $\begin{bmatrix} \cos\theta \\ \sin\theta \end{bmatrix}$ (Figure 8.12). First

$$\begin{bmatrix} 1 & 2 \\ -2 & -4 \end{bmatrix}\begin{bmatrix} \cos\theta \\ \sin\theta \end{bmatrix} = \begin{bmatrix} \cos\theta + 2\sin\theta \\ -2\cos\theta - 4\sin\theta \end{bmatrix}$$

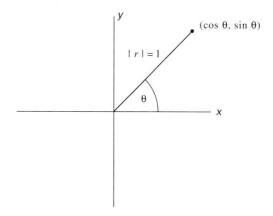

Figure 8.12   Finding the maximum stretching power of a matrix

The stretching factor is

$$\frac{\{(\cos\theta + 2\sin\theta)^2 + (-2\cos\theta - 4\sin\theta)^2\}^{1/2}}{\{\cos^2\theta + \sin^2\theta\}^{1/2}} = \sqrt{5}|\cos\theta + 2\sin\theta|$$

since $\cos^2\theta + \sin^2\theta \equiv 1$

The maximum value of this occurs when

$$\frac{d}{d\theta}(\cos\theta + 2\sin\theta) = 0$$

i.e. $-\sin\theta + 2\cos\theta = 0$ or $\tan\theta = 2$, i.e. $\theta = \tan^{-1}2$

so either $\sin\theta = \dfrac{2}{\sqrt5}$ $\cos\theta = \dfrac{1}{\sqrt5}$

or $\sin\theta = \dfrac{-2}{\sqrt5}$ $\cos\theta = \dfrac{-1}{\sqrt5}$

The maximum value is thus

$$\sqrt5 \left| \frac{1}{\sqrt5} + 2 \times \frac{2}{\sqrt5} \right| = \sqrt5 \times \frac{5}{\sqrt5} = 5 \qquad\blacksquare$$

Exercise 8.2

1   Write down the transformation matrices which perform the following:

(a)   a reflection in the y-axis

(b)   a stretching by a factor of 4 in the x-direction coupled with a contraction by a factor of 3 in the y-direction

(c)   a clockwise rotation of $90°$

(d)   a translation by 6 units to the right in the x-direction and 2 units down in the y-direction

2   Draw a triangle ABC in the first quadrant and sketch its image under the transformations

(a) $\begin{bmatrix} 1 & 0 \\ 0 & -1 \end{bmatrix}$ (b) $\begin{bmatrix} 0 & 1 \\ -1 & 0 \end{bmatrix}$

(c) $\begin{bmatrix} 2 & 0 \\ 0 & 1 \end{bmatrix}$ (d) $\begin{bmatrix} 2 & 0 \\ 0 & 1 \end{bmatrix}\begin{bmatrix} x \\ y \end{bmatrix} + \begin{bmatrix} 2 \\ 1 \end{bmatrix}$

where $\begin{bmatrix} x \\ y \end{bmatrix}$ or $r = xi + yj$ is the position vector of any of the vertices of the triangle

(e) $\dfrac{1}{\sqrt2}\begin{bmatrix} 1 & -1 \\ 1 & 1 \end{bmatrix}$

3 Determine transformations which map the shaded triangle in the diagram into triangles *A* to *F*.

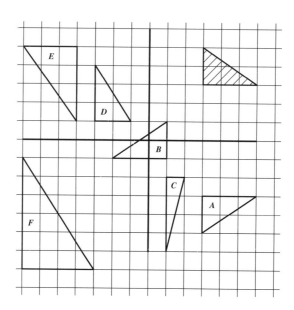

4 Determine transformation matrices which carry out the following:

(a) a reflection in the line $y = -x$

(b) a reflection in the line $y = 2$

5 Determine the matrix $\begin{bmatrix} a & b \\ c & d \end{bmatrix}$ which transforms the line segment *AB* into the line segment *CD*, as illustrated.

Start by transforming *A* into *C*, then transform *B* into *D*.

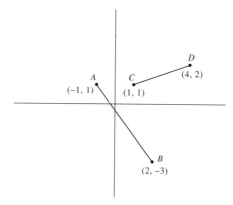

6 Prove that the transformation comprising of a reflection in the x-axis followed by a reflection in the y-axis is the same as that in which the reflections are carried out in the reverse order and is equivalent to a rotation of $180°$.

7 Determine the resultant transformation matrix for the following:

   (a) a reflection in the line $y = x$ followed by a rotation of $45°$ anticlockwise

   (b) a reflection in the y-axis followed by an enlargement by a factor of 2 in the y-direction only, followed by a reflection in the x-axis

   (c) a rotation of $90°$ anticlockwise followed by a reflection in the x-axis and a further $90°$ rotation anticlockwise

   Do any of the matrices in the above transformation commute, i.e. can any of the operations be taken in a different order?

8 Determine the vectors which arise as a result of the matrix products given below. Identify any eigenvectors, eigenvalues and singular matrices.

   (a) $\begin{bmatrix} 2 & -1 \\ 0 & 1 \end{bmatrix}$ with $\begin{bmatrix} 1 \\ 0 \end{bmatrix}, \begin{bmatrix} 1 \\ -1 \end{bmatrix}, \begin{bmatrix} 1 \\ 1 \end{bmatrix}$

   (b) $\begin{bmatrix} -6 & 2 \\ 3 & -1 \end{bmatrix}$ with $\begin{bmatrix} 1 \\ 3 \end{bmatrix}, \begin{bmatrix} 3 \\ 1 \end{bmatrix}, \begin{bmatrix} -2 \\ 1 \end{bmatrix}$

   (c) $\begin{bmatrix} 1 & 0 \\ 1 & a \end{bmatrix}$ with $\begin{bmatrix} 1 \\ 0 \end{bmatrix}, \begin{bmatrix} 0 \\ 1 \end{bmatrix}, \begin{bmatrix} -1 \\ 1 \end{bmatrix}$ $(a \neq 1)$

   (d) $\begin{bmatrix} 4 & 1 \\ 2 & 3 \end{bmatrix}$ with $\begin{bmatrix} 1 \\ 2 \end{bmatrix}, \begin{bmatrix} 1 \\ 1 \end{bmatrix}, \begin{bmatrix} 1 \\ -2 \end{bmatrix}$

9 A general result of matrix theory is that the matrices A and A^2 have the same eigenvectors. Verify this result for the matrix $A = \begin{bmatrix} 1 & 2 \\ -2 & -4 \end{bmatrix}$, with eigenvectors $\begin{bmatrix} 2 \\ -1 \end{bmatrix}, \begin{bmatrix} 1 \\ -2 \end{bmatrix}$. Verify in this case the general result that the eigenvalues of A^2 are the squares of the eigenvalues of A.

10 Determine the maximum stretching power of the matrix $\begin{bmatrix} -6 & 2 \\ 3 & -1 \end{bmatrix}$

8.3 PROPERTIES OF SQUARE MATRICES

In this section we look at some of the properties of square matrices and how they relate to the solution of simultaneous linear equations. We confine our attention to 2×2 matrices, but in the next section we extend some of the concepts to 3×3 matrices. Many of these concepts extend to square matrices of any order.

Determinants

For every square matrix A there exists a number called the **determinant** of A written as $\det A$ or $|A|$. If

$$A = \begin{bmatrix} a & b \\ c & d \end{bmatrix} \quad \text{then} \quad \det A = |A| = \begin{vmatrix} a & b \\ c & d \end{vmatrix} = ad - bc$$

The determinant is therefore a combination of the elements of the matrix. The significance of $|A|$ will emerge when we apply matrices to the solution of equations.

Examples

1. Consider the matrix $A = \begin{bmatrix} 1 & 3 \\ -1 & 2 \end{bmatrix}$ and the vectors $x = \begin{bmatrix} x_1 \\ x_2 \end{bmatrix}$ and $b = \begin{bmatrix} 7 \\ 3 \end{bmatrix}$. The matrix equation $Ax = b$, i.e.

$$\begin{bmatrix} 1 & 3 \\ -1 & 2 \end{bmatrix} \begin{bmatrix} x_1 \\ x_2 \end{bmatrix} = \begin{bmatrix} 7 \\ 3 \end{bmatrix}$$

can be expanded by multiplying out Ax and comparing each element to its equivalent in b. This gives the pair of simultaneous equations

$$x_1 + 3x_2 = 7$$
$$-x_1 + 2x_2 = 3$$

Solving the equations gives $x_1 = 1, x_2 = 2$. Note that $|A| = 1 \times 2 - (-1) \times 3 = 5$.

2. Consider the similar problem with $A = \begin{bmatrix} 1 & 2 \\ -2 & -4 \end{bmatrix}, b = \begin{bmatrix} 3 \\ 5 \end{bmatrix}$. The equivalent equations for x_1 and x_2 are

$$x_1 + 2x_2 = 3$$
$$-2x_1 - 4x_2 = 5$$

Adding twice the first equation to the second, in order to eliminate x_1, has the effect of also eliminating x_2 and leads to the equation $6 + 5 = 0$, an absurd result. What has happened? ∎

The equations can be represented by two parallel lines (Figure 8.13). There is no point of intersection, so the equations have **no solution**.

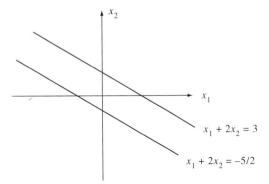

Figure 8.13 Parallel lines

In the first example the lines met at the solution point $(1, 2)$. However, in the second case $|A| = 1 \times (-4) - (-2) \times 2 = -4 = 0$. And recalling the previous section, notice that A has an eigenvalue equal to 0. The two statements are equivalent but the real significance arises in the following theorem.

The determinant of a square matrix is non-zero, i.e. $|A| \neq 0$ if, and only if, the matrix equation $Ax = b$ has a unique solution.

The determinant of a square matrix is zero, i.e. $|A| = 0$ if, and only if, the matrix equation $Ax = b$ does not have a unique solution.

Consider the equations

$$\begin{bmatrix} a & b \\ c & d \end{bmatrix}\begin{bmatrix} x_1 \\ x_2 \end{bmatrix} = \begin{bmatrix} p \\ q \end{bmatrix}$$

i.e. $\quad ax_1 + bx_2 = p \quad$ (i)
$\quad\quad cx_1 + dx_2 = q \quad$ (ii)

where a, b, c, d and p, q are all known. We solve them by multiplying equation (i) by d and equation (ii) by b, and eliminating x_2 by subtraction. We obtain first the equations

$$adx_1 + bdx_2 = pd$$
$$bcx_1 + bdx_2 = bq$$

subtracting gives

$$(ad - bc)x_1 = pd - bq$$

If ad and bc were equal, x_1 would also have been eliminated along with x_2. The same would have happened had we tried to eliminate the variables in the other order.

On the other hand, if $ad \neq bc$, dividing by $ad - bc$ gives

$$x_1 = \frac{pd - bq}{ad - bc} \quad \text{and} \quad x_2 = \frac{aq - pc}{ad - bc} \qquad \text{(a unique solution)}$$

It is clear therefore that matrices A with $|A| = 0$ have a very special significance. On the other hand, note that when $|A| \neq 0$ we can write

$$x_1 = \frac{pd - ba}{ad - bc} = \frac{\begin{vmatrix} p & b \\ q & d \end{vmatrix}}{\begin{vmatrix} a & b \\ c & d \end{vmatrix}} = \frac{|A_1|}{|A|} \qquad x_2 = \frac{aq -\!- pc}{ad - bc} = \frac{\begin{vmatrix} q & p \\ c & q \end{vmatrix}}{\begin{vmatrix} a & b \\ c & d \end{vmatrix}} = \frac{|A_2|}{|A|}$$

Therefore, x_1 and x_2 are expressible as a ratio of determinants. The matrices A_1 and A_2 are formed from A by replacing the first and second columns by the vector elements of the right-hand side. This result extends to matrices of higher order than 2×2 and is known as **Cramer's rule**.

Example

(a) If $\begin{vmatrix} x & 2x \\ x-1 & x+1 \end{vmatrix} = -10$, determine the values of x.

(b) If $\begin{vmatrix} x+a & b \\ b & x+a \end{vmatrix} = 0$, determine the values of x.

(c) Show that the determinant of the matrix

$$\begin{bmatrix} x & 2x-1 \\ 2 & 4 \end{bmatrix}$$

is independent of x and is non-zero.

(d) Find the determinant of the matrix

$$\begin{bmatrix} 4x & 3(x-1) \\ 8(y+1) & 6y \end{bmatrix}$$

What is the relationship between x and y if the determinant is zero?

Solution

(a) The determinant has the value $x(x+1) - 2x(x-1) = -x^2 + 3x$, so

$$-x^2 + 3x = -10 \quad \text{or} \quad x^2 - 3x - 10 = 0$$

This equation has solutions $x = -2$ or 5

(b) The equation is $(x + a)^2 - b^2 = 0$, so $x = b - a$ or $x = -(a + b)$

(c) $|A| = 4x - 2(2x - 1) = 2$, which is independent of x and non-zero.

(d) The value of the determinant is $24xy - 24(x - 1)(y + 1) \equiv 24(y - x - 1)$. Hence $y - x - 1 = 0$ or $y = 1 + x$, if the determinant is zero.

2. The value of the determinant of a 2×2 matrix is unaltered if a multiple of one row is added to another row, or if a multiple of one column is added to another column. For example, adding a multiple k of row 1 to row 2 in the determinant $\begin{vmatrix} a & b \\ c & d \end{vmatrix}$ gives the determinant

$$\begin{vmatrix} a & b \\ c + ka & d + kb \end{vmatrix}$$

The value of the second determinant is

$$a(d + kb) - b(c + ka) = ad - bc$$

which is the value of the original determinant. Using specific values for a, b, c, d and k,

(i) Subtracting twice row 1 from row 2,

$$\begin{vmatrix} 1 & 3 \\ 2 & 4 \end{vmatrix} = \begin{vmatrix} 1 & 3 \\ 2 - 2 \times 1 & 4 - 2 \times 3 \end{vmatrix} = \begin{vmatrix} 1 & 3 \\ 0 & -2 \end{vmatrix} = -2$$

(ii) Subtracting three times column 1 from column 2

$$\begin{vmatrix} 11 & 33 \\ 3 & 10 \end{vmatrix} = \begin{vmatrix} 11 & 33 - 3 \times 11 \\ 3 & 10 - 3 \times 3 \end{vmatrix} = \begin{vmatrix} 11 & 0 \\ 3 & 1 \end{vmatrix} = 11$$

This property of the determinant being **invariant**, i.e. unchanged, by the addition or subtraction of multiples of rows or columns holds for higher orders of determinant and is also of enormous use in simplifying the arithmetic involved in evaluating them.

3. Use Cramer's rule to write down the solution of the equations

$$3x + 4y = 13$$
$$-2x + 5y = -1$$

Cramer's rule gives

$$x = \frac{\begin{vmatrix} 13 & 4 \\ -1 & 5 \end{vmatrix}}{\begin{vmatrix} 3 & 4 \\ -2 & 5 \end{vmatrix}} = \frac{\begin{vmatrix} 0 & 69 \\ -1 & 5 \end{vmatrix}}{23} = \frac{69}{23} = 3$$

$$y = \frac{\begin{vmatrix} 0 & 13 \\ -2 & -1 \end{vmatrix}}{23} = \frac{23}{23} = 1$$

We can check by direct substitution in the original equations that these values are correct. ∎

The inverse matrix

Consider the system of equations

$$Ax = b$$

For a general system of two equations in two unknowns this becomes

$$\begin{bmatrix} a & b \\ c & d \end{bmatrix} \begin{bmatrix} x_1 \\ x_2 \end{bmatrix} = \begin{bmatrix} p \\ q \end{bmatrix}$$

Solving these equations (e.g. by Cramers' rule) we obtained

$$x_1 = \frac{pd - bq}{ad - bc}$$

$$x_2 = \frac{aq - pc}{ad - bc}$$

In matrix form this is

$$x = \begin{bmatrix} x_1 \\ x_2 \end{bmatrix} = \frac{1}{ad - bc} \begin{bmatrix} d & -b \\ -c & a \end{bmatrix} \begin{bmatrix} p \\ q \end{bmatrix} = \frac{1}{|A|} \begin{bmatrix} d & -b \\ -c & a \end{bmatrix} b$$

The process of solving the equations

$$Ax = b$$

to obtain the solution vector x is equivalent to multiplying the entire matrix system by a matrix B on the left, i.e.

$$BAx = Bb$$

to give the equation $Ix = x = Bb$. Therefore, $BA = I$.

We already know the matrix B; it is

$$B = \frac{1}{|A|} \begin{bmatrix} d & -b \\ -c & a \end{bmatrix}$$

We confirm this by finding the product BA.

$$BA = \frac{1}{|A|} \begin{bmatrix} d & -b \\ -c & a \end{bmatrix} \begin{bmatrix} a & b \\ c & d \end{bmatrix} = \frac{1}{|A|} \begin{bmatrix} |A| & 0 \\ 0 & |A| \end{bmatrix} = I$$

It is also easily proved that B and A commute, so $AB = BA = I$. B is unique and is called the **inverse** of A, written A^{-1}. If a matrix does not possess an inverse, its determinant is zero and it is called a **singular** matrix. If a matrix does possess a (unique) inverse, its determinant is non-zero and it is called **non-singular**.

If the inverse is known, it can be used to solve the simultaneous equations $Ax = b$:

$$A^{-1}(Ax) = Ix = x = A^{-1}b$$

Example   Solve the system of equations $\begin{bmatrix} 3 & -1 \\ 1 & 2 \end{bmatrix} \begin{bmatrix} x \\ y \end{bmatrix} = \begin{bmatrix} 8 \\ 5 \end{bmatrix}$

Matrix $A = \begin{bmatrix} 3 & -1 \\ 1 & 2 \end{bmatrix}$ has inverse $A^{-1} = \frac{1}{7} \begin{bmatrix} 2 & 1 \\ -1 & 3 \end{bmatrix}$, so that

$$\begin{bmatrix} x \\ y \end{bmatrix} = \frac{1}{7} \begin{bmatrix} 2 & 1 \\ -1 & 3 \end{bmatrix} \begin{bmatrix} 8 \\ 5 \end{bmatrix} = \frac{1}{7} \begin{bmatrix} 21 \\ 7 \end{bmatrix} = \begin{bmatrix} 3 \\ 1 \end{bmatrix}$$

i.e. $x = 3, y = 1$.   ∎

The most important applications of the inverse matrix A^{-1} rely on the fact that it totally inverts the operation performed by matrix A.

Example

Refer back to Figure 8.6 and the accompanying example (page 457), Figure 8.14 shows a shaded triangle and three other triangles which are the results of mappings A, C and F.

(a) A is an enlargement by a factor 2, i.e. the matrix of transformation is $A = 2I$. Hence $A^{-1} = \frac{1}{2}I$, which represents a contraction by a factor 2. The vertices of the triangle A, $\begin{bmatrix} x \\ y \end{bmatrix}$, map to $\begin{bmatrix} \frac{x}{2} \\ \frac{y}{2} \end{bmatrix}$, i.e. the vertices of the original triangle under A^{-1}.

(b) C is a rotation of $180°$ (or is alternatively an enlargement/contraction by a factor -1, i.e. $C = \begin{bmatrix} -1 & 0 \\ 0 & -1 \end{bmatrix} = -I$. Then $C^{-1} = -I$ also, and $C^2 = I$. C is therefore its own inverse.

(c) In this transformation, $\begin{bmatrix} x \\ y \end{bmatrix}$ is mapped to

$$\begin{bmatrix} -\frac{1}{2} & 0 \\ 0 & 1 \end{bmatrix}\begin{bmatrix} x \\ y \end{bmatrix} + \begin{bmatrix} 5\frac{1}{2} \\ -3 \end{bmatrix} = \begin{bmatrix} \frac{1}{2}(11 - x) \\ y - 3 \end{bmatrix}$$

The operation is composite in that translations are involved. It could be carried out by applying in turn the following transformations:

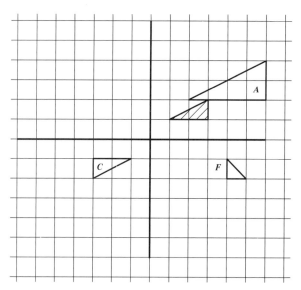

Figure 8.14 Finding inverses for three of the transformations in Figure 8.6

F_1: subtract the vector $\begin{bmatrix} 4 \\ 6 \end{bmatrix}$; this gives the vector $\begin{bmatrix} x-4 \\ y-6 \end{bmatrix}$

F_2: multiply by the matrix $\begin{bmatrix} -1 & 0 \\ 0 & 1 \end{bmatrix}$; this gives the vector $\begin{bmatrix} 4-x \\ y-6 \end{bmatrix}$

F_3: multiply by the matrix $\begin{bmatrix} \frac{1}{2} & 0 \\ 0 & 1 \end{bmatrix}$; this gives the vector $\begin{bmatrix} 2-\frac{1}{2}x \\ y-6 \end{bmatrix}$

F_4: add the vector $\begin{bmatrix} 3\frac{1}{2} \\ 3 \end{bmatrix}$; this gives the vector $\begin{bmatrix} 5\frac{1}{2}-x \\ y-3 \end{bmatrix}$

The inverse process is performed with the operations of F reversed and taken in reverse order.

First we subtract the vector $\begin{bmatrix} 3\frac{1}{2} \\ 3 \end{bmatrix}$. Next we multiply by the matrix

$$\begin{bmatrix} \frac{1}{2} & 0 \\ 0 & 1 \end{bmatrix}^{-1} = \begin{bmatrix} 2 & 0 \\ 0 & 1 \end{bmatrix}$$

Then we multiply by the matrix

$$\begin{bmatrix} -1 & 0 \\ 0 & 1 \end{bmatrix}^{-1} = \begin{bmatrix} -1 & 0 \\ 0 & 1 \end{bmatrix}$$

Finally we add the vector $\begin{bmatrix} 4 \\ 6 \end{bmatrix}$

The process may be summarised as

$$\begin{bmatrix} x \\ y \end{bmatrix} \rightarrow \begin{bmatrix} -1 & 0 \\ 0 & 1 \end{bmatrix}\begin{bmatrix} 2 & 0 \\ 0 & 1 \end{bmatrix}\begin{bmatrix} x-3\frac{1}{2} \\ y-3 \end{bmatrix} + \begin{bmatrix} 4 \\ 6 \end{bmatrix} = \begin{bmatrix} 11-2x \\ y+3 \end{bmatrix}$$

$\blacksquare$

The eigenvalue problem

We have seen already that certain vectors do not change their direction, merely their magnitude when multiplied by a 2×2 matrix. If this is so, when the vector x is multiplied by the matrix A, we may write

$$Ax = \lambda x \quad \text{(where λ is a scalar)}$$

i.e. $(A - \lambda I)x = 0$.

The vector x is called an **eigenvector** of the matrix and the scalar λ is its associated **eigenvalue**. If $B = A - \lambda I$ so that $Bx = 0$, then two possibilities arise:

(i) $|B| \neq 0$; then $x = B^{-1}0 = 0$, in other words the zero vector is the only solution.
(ii) $x \neq 0$, solutions can only exist if $|B| = 0$, i.e. $|A - \lambda I| = 0$.

If $A = \begin{bmatrix} a & b \\ c & d \end{bmatrix}$,

then $|A - \lambda I| = \begin{vmatrix} a - \lambda & b \\ c & d - \lambda \end{vmatrix} = \lambda^2 - (a+d) + (ad - bc) = \lambda^2 - (a+d) + |A| = 0.$

Real eigenvalues exist if this quadratic equation has real solutions, i.e.

$$(a + d)^2 \geq 4|A|$$

Therefore $a^2 + 2ad + d^2 \geq 4ad - 4bc$ or $(a - d)^2 + 4bc \geq 0$.

Since $(a - d)^2 \geq 0$, it is sufficient to require that $bc \geq 0$, i.e. the off-diagonal elements are of the same sign. An alternative criterion is that $|A| < 0$. These conditions are *sufficient* but not *necessary*.

Example

Determine the eigenvalues and eigenvectors of the matrices

(a) $\begin{bmatrix} 3 & -2 \\ -1 & 4 \end{bmatrix}$

(b) $\begin{bmatrix} 1 & 6 \\ 5 & 2 \end{bmatrix}$

Solution

(a) The equation $|A - \lambda I| = 0$ becomes $\begin{vmatrix} 3 - \lambda & -2 \\ -1 & 4 - \lambda \end{vmatrix} = 0$

$$(3 - \lambda)(4 - \lambda) - 2 = 0$$
$$\lambda^2 - 7\lambda + 10 = 0$$

The solutions, i.e. the eigenvalues, are $\lambda = 2$ and $\lambda = 5$.
For $\lambda = 2$ the equation $(A - \lambda I)x = 0$ becomes

$$\begin{bmatrix} 1 & -2 \\ -1 & 2 \end{bmatrix} \begin{bmatrix} x_1 \\ x_2 \end{bmatrix} = 0$$

i.e. $x_1 + 2x_2 = 0$ and $-x_1 - 2x_2 = 0$

The two equations are effectively the same, as we should have expected.

The best we can achieve is $x_1 = 2x_2$. A suitable solution vector is $\begin{bmatrix} x_1 \\ x_2 \end{bmatrix} = \begin{bmatrix} 2 \\ 1 \end{bmatrix}$, which is an eigenvector of the matrix. Note that only the direction of the eigenvectors can be found, therefore $\begin{bmatrix} 2 \\ 1 \end{bmatrix}$ and $\begin{bmatrix} 1 \\ \frac{1}{2} \end{bmatrix}$ are also possible eigenvectors. In effect we have a family of eigenvectors and we choose a suitable representative; any other member of the family can be obtained from the representative by multiplying both its components by the same number. Multiplying by the matrix scales up the chosen eigenvector by the factor 2, i.e. the eigenvalue.

For $\lambda = 5$ the equation $(A - \lambda I)x = 0$ becomes

$$\begin{bmatrix} -2 & -2 \\ -1 & -1 \end{bmatrix}\begin{bmatrix} x_1 \\ x_2 \end{bmatrix} = 0$$

Both equations are equivalent to $x_1 = -x_2$, so $\begin{bmatrix} 1 \\ -1 \end{bmatrix}$ is the other eigenvector of the matrix. Incidentally, in the matrix A, b and c are not of the same sign and $|A| = 10$, which is not negative. Even so, the eigenvalues are real.

(b) This time b and c are of the same sign and $|A| = -28$, so the eigenvalues must be real.

The equation we solve is $\begin{vmatrix} 1 - \lambda & 6 \\ 5 & 2 - \lambda \end{vmatrix} = 0$

i.e. $(1 - \lambda)(2 - \lambda) - 30 = 0$

hence $\lambda^2 - 3\lambda - 28 = 0$

Therefore the eigenvalues are $\lambda = 7$ and $\lambda = -4$.

For $\lambda = 7$ we solve $\begin{bmatrix} -6 & 6 \\ 5 & -5 \end{bmatrix}\begin{bmatrix} x_1 \\ x_2 \end{bmatrix} = 0$.

The solution is $x_1 = x_2$, so $\begin{bmatrix} 1 \\ -1 \end{bmatrix}$ is an eigenvector.

For $\lambda = -4$ we solve $\begin{bmatrix} 5 & 6 \\ 5 & 6 \end{bmatrix}\begin{bmatrix} x_1 \\ x_2 \end{bmatrix} = 0$.

The solution is $5x_1 + 6x_2 = 0$, so $\begin{bmatrix} 6 \\ -5 \end{bmatrix}$ is an eigenvector.

Eigenvalues and eigenvectors have enormous importance in science and engineering. We look at another widely-used result. We start from the equation

$$Ax = \lambda x$$

Multiplying both sides by A gives

$$A^2 x = A(Ax) = A(\lambda x) = \lambda(Ax) = \lambda^2 x$$

i.e. if λ is an eigenvalue of A with eigenvector x, then λ^2 is an eigenvalue of A^2 with the same eigenvector x. You can easily prove this for any power of A.

If we multiply by A^{-1}, if it exists, we obtain

$$A^{-1}(Ax) = Ix = \lambda A^{-1}x$$

or $\quad A^{-1}x = \lambda^{-1}x$

so λ^{-1} is an eigenvalue of A^{-1}, with the same eigenvector.

The (quadratic) equation in λ which emerges from the determinant equation $|A - \lambda I| = 0$ is called the **characteristic equation** of the matrix A.

For matrix $A = \begin{bmatrix} 3 & 2 \\ -1 & 4 \end{bmatrix}$ the characteristic equation is $\lambda^2 - 7\lambda + 10 = 0$. Now

$$(A^2 - 7A + 10I)x = A^2x - 7Ax + 10x$$
$$= \lambda^2 x - 7\lambda x + 10x$$
$$= (\lambda^2 - 7\lambda + 10)x = 0$$

It can be proved that $A^2 - 7A + 10I = 0$ by the **Cayley–Hamilton theorem**, which states that a square matrix satisfies its own characteristic equation.

Note that, in the characteristic equation, λ^2 has been replaced by A^2, λ by $A, 10 = 10 \times 1$ by $10I$ and 0 by $\mathbf{0}$. These last two replacements are necessary in order to produce an equation in which all terms are matrices; in order that they are all 2×2 matrices, the unit and zero matrices must also be 2×2 matrices.

If the inverse exists, then premultiplying the matrix equation by A^{-1} gives

$$A - 7I + 10A^{-1} = 0$$

$$A^{-1} = \frac{1}{10}(7I - A)$$

Exercise 8.3

1 Prove that

$$\begin{vmatrix} a + kc & b + kd \\ b & d \end{vmatrix} = \begin{vmatrix} a + kc & c \\ b + kd & d \end{vmatrix} = \begin{vmatrix} a & b \\ c & d \end{vmatrix}$$

2 Evaluate the following using Question 1 as appropriate.

(a) $\begin{vmatrix} 2 & 4 \\ 6 & 3 \end{vmatrix}$

(b) $\begin{vmatrix} 9 & 10 \\ 11 & 12 \end{vmatrix}$

(c) $\begin{vmatrix} 100 & 101 \\ 99 & 102 \end{vmatrix}$

(d) $\begin{vmatrix} 1001 & 999 \\ 1000 & 1004 \end{vmatrix}$

3   Determine x in the following equations:

(a) $\begin{vmatrix} x & x+1 \\ x+3 & x+2 \end{vmatrix} = 0$

(b) $\begin{vmatrix} x(x-2) & x+4 \\ 2 & 1 \end{vmatrix} = -4$

(c) $\begin{vmatrix} x+1 & x-1 \\ x+1 & x-5 \end{vmatrix} = \begin{vmatrix} x & 6x-1 \\ 1 & 3 \end{vmatrix}$

(d) $\begin{vmatrix} x+a & b \\ c & x+d \end{vmatrix} = 0$ given $\begin{vmatrix} a & b \\ c & d \end{vmatrix} = 0$

4   It can be shown that if two square matrices A and B of the same order are multiplied together then $|AB| = |A\|B| = |BA|$, even if A and B do not commute.

(a)   Verify this property for $A = \begin{bmatrix} 2 & 4 \\ 6 & 3 \end{bmatrix}$, $B = \begin{bmatrix} 9 & 10 \\ 11 & 12 \end{bmatrix}$

(b)   If $A_1, A_2, \ldots, A_n$ are matrices of the same order, one of which is singular, what is the determinant of the product of the n matrices taken in any order? What happens if more than one matrix is singular?

(c)   If C is a 2×2 matrix, with determinant $|C|$, what is

(i) $|kC|$, where k is a scalar   (ii) $|C^{-1}|$

5   Use Cramer's rule to solve the following sets of equations:

(a) $6x - y = 8$ (b) $19x - 17y = -15$ (c) $5x - 3y = 7$
       $2x + 3y = 6$ $11x + 5y = 21$ $4x - y = 5$

6   Solve the equations in Question 5, this time using the inverse matrix.

7   Determine inverse transformations to the following:

(a) $B: \begin{bmatrix} x \\ y \end{bmatrix} \rightarrow \begin{bmatrix} 0 & -1 \\ 1 & 0 \end{bmatrix} \begin{bmatrix} x \\ y \end{bmatrix} + \begin{bmatrix} -1 \\ 0 \end{bmatrix} = \begin{bmatrix} -y-1 \\ x \end{bmatrix}$

(b) $D: \begin{bmatrix} x \\ y \end{bmatrix} \rightarrow \begin{bmatrix} 1 & 0 \\ 0 & 3 \end{bmatrix} \begin{bmatrix} -1 & 0 \\ 0 & -1 \end{bmatrix} \begin{bmatrix} x \\ y \end{bmatrix} + \begin{bmatrix} -3 \\ 1 \end{bmatrix} = \begin{bmatrix} -x-3 \\ -3y+1 \end{bmatrix}$

8   (a)   If A and B are square matrices of the same order prove that

$$(AB)^{-1} = B^{-1}A^{-1}$$

(b)   If C is a third compatible square matrix, what is $(ABC)^{-1}$?

9   (a)   If A and B are as in Question 8 and $A^2 = (AB)^2 = B^2 = I$, prove that A and B commute.

(b)   If C is a third matrix such that $C^2 = I$, and C commutes with A and B, simplify

(i) $A(BC)^{-2}(AC^{-1}B)^{-1}$

(ii) $(ABC)^2$

(iii) $(AC^{-1}B)C(AC^{-1}B)$

10 Prove that the product of two symmetric 2×2 matrices is symmetric.

11 Determine the eigenvalues and eigenvectors of the matrices

(a) $\begin{bmatrix} 1 & 1 \\ -2 & 4 \end{bmatrix}$ (b) $\begin{bmatrix} 1 & 3 \\ 6 & 4 \end{bmatrix}$ (c) $\begin{bmatrix} 6 & 1 \\ -2 & 3 \end{bmatrix}$

12 Using the characteristic equation of each matrix in Question 11 along with the Cayley–Hamilton theorem, deduce each inverse matrix in the form

$$A^{-1} = \alpha A + \beta I$$

8.4 3×3 MATRICES

Determinants

The determinant of a 3×3 matrix A, written $|A|$, has exactly the same significance as the determinant of a 2×2 matrix but is a little more awkward to define. To help explain the defintion, we use the idea of a **permutation**.

The digits 1, 2 and 3 can be re-ordered in six different ways:

(1 2 3)   (3 1 2)   (2 3 1)   (1 3 2)   (2 1 3)   (3 2 1).

You will see that the orderings of the first three are **cyclic**, i.e. $1 \to 2 \to 3 \to 1 \to 2$ as one moves left to right and back to left again. The digits in the last three orderings are not cyclic. Regardless of this, every permutation has a sign, positive or negative. We can find this easily by the 'linking-of-arms' check, with reference to the natural ordering 1 2 3, i.e.

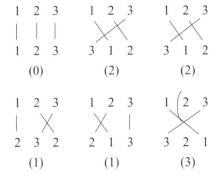

Every crossing is kept separate and the number is counted (shown in parentheses). If the number is even (and that includes zero) the permutation is **even**, i.e. positive; if the number is odd, the permutation is **odd**, i.e. negative. You can see straight away that there are equal numbers of each sign in the above example.

We now consider a **cross-column product** for a 3×3 matrix. A, where

$$A = \begin{bmatrix} a & b & c \\ d & f & g \\ h & k & l \end{bmatrix}$$

An example of a cross-column product is bgh. We have started with column 2, then moved to column 3 and finally moved to column 1, going down each row in turn. Notice there are six possible choices of cross-column product, i.e. 3 choices for the first column × 2 choices for the second × 1 choice for the third; remember $6 = 3!$

$$(+) \qquad\qquad (+) \qquad\qquad (+)$$

$$(-) \qquad\qquad (-) \qquad\qquad (-)$$

The $(+)$ and $(-)$ represent the sign of the permutation of column numbers taken in order. We can now make the following definition:

$$|A| = \sum (\text{cross column product}) \times (\text{sign of permutation})$$

This result is true for matrices of any order n.

Example

Find the determinant of the following matrices:

(a) $\begin{bmatrix} 2 & 3 \\ 1 & 5 \end{bmatrix}$ (using permutation and cross column products)

(b) $\begin{bmatrix} 1 & 0 & 1 \\ 0 & 2 & 1 \\ 1 & 2 & 3 \end{bmatrix}$ (c) $\begin{bmatrix} 0 & 1 & 0 & 0 \\ 0 & 0 & 0 & 1 \\ 0 & 0 & 1 & 0 \\ 1 & 2 & 3 & 4 \end{bmatrix}$

Solution

(a) For a 2×2 matrix, using cross-column products, we have one of each sign.

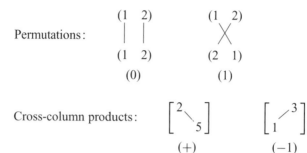

Permutations:

Cross-column products:

therefore $|A| = 2 \times 5 - 3 \times 1 = 7$.

(b) Only three out of the six possible products are non-zero, i.e.

so that $|A| = 1 \times 2 \times 3 - 1 \times 1 \times 2 - 1 \times 2 \times 1 = 2$.

(c) Carrying forward the idea to a 4×4 matrix and its determinant, notice that

is the only non-zero cross-column product. The permutation is $(2\,4\,3\,1)$ with a positive sign

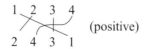

because there are four crossings. Then $|A| = 1 \times 1 \times 1 \times 1 = 1$. ∎

A quick way to find $|A|$ for a 3×3 matrix is as follows. Write down the elements then repeat the first two columns, i.e.

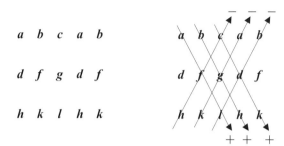

Draw in the arrows and append the signs as shown. Remember this method works for 3×3 determinants only. Each product of elements along one of the six lines is multiplied by the sign attached to the line and the results added. This gives

$$\{afl + bgh + cdk\} - \{cfh + agk + bdl\}$$

We will often be able to simplify the determinant, i.e. reduce the number of non-zero products, by adding or subtracting a multiple of any row or column to any row or column, e.g.

$$\begin{vmatrix} a & b+ma & c \\ d & f+mb & g \\ h & k+mh & l \end{vmatrix} = \begin{vmatrix} a & b & c \\ d & f & g \\ h & k & l \end{vmatrix}$$

The proof is left as an exercise. Two important corollaries emerge.

1. Interchanging any two rows or columns of the matrix changes the sign of its determinant.
2. If any two rows or columns of a matrix are equal, or one is a constant multiple of the other, then $|A| = 0$.

Example

Find the following determinants:

(a) $\begin{vmatrix} 101 & 203 & 303 \\ 115 & 262 & 345 \\ 179 & 285 & 537 \end{vmatrix}$ (b) $\begin{vmatrix} 10 & 11 & 12 \\ 20 & 14 & 15 \\ 30 & 25 & 36 \end{vmatrix}$ (c) $\begin{vmatrix} 5 & 6 & 7 \\ 3 & 5 & 8 \\ 7 & 11 & 16 \end{vmatrix}$

Solution

(a) Note that column $3 = 3 \times$ column 1, so $|A| = 0$.

(b) $\begin{vmatrix} 10 & 11 & 12 \\ 20 & 14 & 15 \\ 30 & 25 & 36 \end{vmatrix} = \begin{vmatrix} 10 & 11 & 12 \\ 0 & -8 & -9 \\ 0 & -8 & 0 \end{vmatrix}$ (row 2 $-$ 2 $\times$ row 1)
(row 3 $-$ 3 $\times$ row 1)

$$= -[10 \times (-9) \times (-8)] = -720$$

(c) $\begin{vmatrix} 5 & 6 & 7 \\ 3 & 5 & 8 \\ 8 & 11 & 16 \end{vmatrix} = \begin{vmatrix} 5 & 6 & 7 \\ 3 & 5 & 8 \\ 0 & 0 & 1 \end{vmatrix}$ (row 3 $-$ row 1 $-$ row 2)

so that $|A| = 5 \times 5 \times 1 - 6 \times 3 \times 1 = 7$ ∎

Solving equations

Equations of the form $Ax = b$ can be solved for a 3×3 matrix A provided that $|A| \neq 0$. We can also find A^{-1} so that $x = A^{-1}b$. However, the determination of A^{-1} is a much more complicated for 3×3 matrices than for 2×2 matrices, and the solution of $Ax = b$ can be much more easily found without knowing A^{-1}. Cramer's rule can provide solutions but it requires much more effort than in the 2×2 case. For the system

$$\begin{bmatrix} a & b & c \\ d & f & g \\ h & k & l \end{bmatrix} \begin{bmatrix} x_1 \\ x_2 \\ x_3 \end{bmatrix} = \begin{bmatrix} p \\ q \\ r \end{bmatrix}$$

we have

$$x_1 = \frac{|A_1|}{|A|} \qquad x_2 = \frac{|A_2|}{|A|} \qquad x_3 = \frac{|A_3|}{|A|}$$

where

$$|A_1| = \begin{vmatrix} p & b & c \\ q & f & g \\ r & k & l \end{vmatrix}$$

i.e. replacing the first column by the right-hand vector. $|A_2|$ and $|A_3|$ are similarly defined; the right-hand side replaces the second and third columns respectively. The evaluation of four 3×3 determinants is markedly simpler than finding A^{-1}.

Examples

1. Solve the equations

$$\begin{array}{ll} x_1 + 2x_2 + 3x_3 = 14 & \text{(i)} \\ -2x_1 + x_2 + x_3 = 3 & \text{(ii)} \\ x_1 - x_2 + 2x_3 = 5 & \text{(iii)} \end{array}$$

using (a) Cramer's rule and (b) variable elimination.

Solution

(a) Note that

$$|A| = 16 \qquad |A_1| = 16 \qquad |A_2| = 32 \qquad |A_3| = 48$$

Then

$$x_1 = 1 \qquad x_2 = 2 \qquad x_3 = 3$$

(b) We can eliminate x_1 from the second equation by adding twice the first equation to the second. We can eliminate x_1 from the third equation by subtracting the first equation from the third. We then have

$$
\begin{aligned}
x_1 + 2x_2 + 3x_3 &= 14 \quad &\text{(i)} \\
5x_2 + 7x_3 &= 31 \quad &\text{(iv)} \\
-3x_2 - x_3 &= -9 \quad &\text{(v)}
\end{aligned}
$$

Now we estimate x_2 from equation (v) by adding $\dfrac{3}{5}$ of equation (iv) to equation (v). This gives the reduced set of equations

$$
\begin{aligned}
x_1 + 2x_2 + 3x_3 &= 14 \quad &\text{(i)} \\
5x_2 + 7x_3 &= 31 \quad &\text{(iv)} \\
\frac{16}{5}x_3 &= \frac{48}{5} \quad &\text{(vi)}
\end{aligned}
$$

Then, **back-substituting** for x_3, x_2, x_1 in turn gives

$$
\begin{aligned}
x_3 &= 3 \quad &\text{from (vi)} \\
x_2 &= \frac{31 - 7 \times 3}{5} = 2 \quad &\text{from (iv)} \\
x_1 &= 14 - 3 \times 3 - 2 \times 2 = 1 \quad &\text{from (i)}
\end{aligned}
$$

This method of elimination, in which we systematically eliminate variables then back-substitute, is called **Gaussian elimination**. It works when $|A| \neq 0$, i.e. when each of the three equations comprises a totally separate and non-conflicting piece of information.

2. Equations which do not have a unique solution

(a)
$$
\begin{aligned}
x + y + z &= 3 \\
2x + 3y + 4z &= 9 \\
3x + 4y + 5z &= 12
\end{aligned}
$$

(b)
$$
\begin{aligned}
x + y + z &= 1 \\
x + y + z &= 2 \\
2x - 3y + 5z &= 12
\end{aligned}
$$

In system (a) you see that $x = y = z = 1$ is one solution, but the third equation is just the sum of the first two and does not give an independent restriction on the variables. There are **infinitely many solutions**. Gaussian elimination would start by subtracting twice the first equation from the second, and three times the first equation from the third. In either case we would obtain the equation $y + 2z = 3$, showing clearly the duplication. Verify this for yourself.

In system (b) the first two equations are incompatible, so **no solutions** are possible; the system is **inconsistent**. In both (a) and (b) the determinant of the matrix of coefficients on the left-hand side is zero.

Rank

When $|A| = 0$ we rely on the concept of the **rank** of a matrix to determine whether no solution to a set of simultaneous equations exists, or if a solution does exist, to provide the best information possible about it. We illustrate the concept of rank by an example.

Example

Solve, as far as possible, the sets of equations

(a) $\quad x + y = 2$
$\quad\quad 2x + 3y = 5$

(b) $\quad x + y = 2$
$\quad\quad 2x + 2y = 4$

(c) $\quad x + y = 2$
$\quad\quad 2x + 2y = 5$

Solution

(a) Putting the equations in the form $Ax = b$, we obtain

$$\begin{bmatrix} 1 & 1 \\ 2 & 3 \end{bmatrix} \begin{bmatrix} x \\ y \end{bmatrix} = \begin{bmatrix} 2 \\ 5 \end{bmatrix}$$

Here $\quad |A| = 1$ and $A^{-1} = \begin{bmatrix} 3 & -1 \\ -2 & 1 \end{bmatrix}$

Therefore $x = A^{-1}b = \begin{bmatrix} 3 & -1 \\ -2 & 1 \end{bmatrix} \begin{bmatrix} 2 \\ 5 \end{bmatrix} = \begin{bmatrix} 1 \\ 1 \end{bmatrix}$

The *unique solution* is $x = y = 1$.

The matrix A is non-singular, and since $|A|$ is a non-zero 2×2 determinant, we say that the rank of the matrix A is 2, written rank$(A) = 2$.

(b) It is easy to check that $x = y = 1$ is one solution of these equations. But the second equation is simply the first equation multiplied by 2, so any pair of values x, y which satisfy the equation $x + y = 2$ is a solution of the second equation also. The equations are consistent.

For the matrix $A = \begin{bmatrix} 1 & 1 \\ 2 & 2 \end{bmatrix}$, $|A| = 0$. We form the **minors** of A by deleting the row and column containing each element in turn and taking the determinant of the remaining elements. Such a determinant is called a **subdeterminant** of A. In this case each deletion produces 1×1 subdeterminant, i.e. a single number. The process can be illustrated as follows:

$$\begin{bmatrix} 1 & 1 \\ 2 & 2 \end{bmatrix}, \begin{bmatrix} 1 & 1 \\ 2 & 2 \end{bmatrix}, \begin{bmatrix} 1 & 1 \\ 2 & 2 \end{bmatrix}, \begin{bmatrix} 1 & 1 \\ 2 & 2 \end{bmatrix}$$

The subdeterminants, i.e. the numbers 2, 2, 1, 1, are all non-zero, so we say that $\text{rank}(A) = 1$.

The existence of a solution depends on the right-hand side of the set of equations, represented by the vector b, as we shall see in case (c). We really need to look at the augmented matrix $A:b$, in this case the 2×3 matrix $\begin{bmatrix} 1 & 1 & : & 2 \\ 2 & 2 & : & 4 \end{bmatrix}$

We can obtain 2×2 minors by deleting each of its columns in turn:

i.e. $\qquad \begin{vmatrix} 1 & 2 \\ 2 & 4 \end{vmatrix}, \begin{vmatrix} 1 & 2 \\ 2 & 4 \end{vmatrix}, \begin{vmatrix} 1 & 1 \\ 2 & 2 \end{vmatrix}$

Each of these subdeterminants is zero and all the matrix elements are non-zero; we write $\text{rank}(A:b) = 1$, since the largest subdeterminants which are non-zero are 1×1. We summarise by stating that, because $\text{rank}(A) = \text{rank}(A:b)$ $(=1)$, the set of equations is consistent, i.e. solutions exist.

(c) Once again $A = \begin{bmatrix} 1 & 1 \\ 2 & 2 \end{bmatrix}$ but now $b = \begin{bmatrix} 2 \\ 5 \end{bmatrix}$. We already know that $\text{rank}(A) = 1$; now we look at the augmented matrix $A:b$. Its minors are $\begin{vmatrix} 1 & 2 \\ 2 & 5 \end{vmatrix}, \begin{vmatrix} 1 & 2 \\ 2 & 5 \end{vmatrix}, \begin{vmatrix} 1 & 1 \\ 2 & 2 \end{vmatrix}$

The first two minors, both of which contain the elements of b, are non-zero, so $\text{rank}(A:b) = 2$. In this case, $\text{rank}(A:b) > \text{rank}(A)$ and the equations are inconsistent, yielding no solution. ∎

The general results are now stated for a system of n simultaneous linear equations in n unknowns, where we are solving the equations $Ax = b$. The following possibilities exist:

(i) $\text{rank}(A) = n$, i.e. $|A| \neq 0$, and there is a **unique solution** $x = A^{-1}b$
(ii) $\text{rank}(A) = m < n$, i.e. $|A| = 0$, so there is no unique solution

- (a) $\text{rank}(A:b) = \text{rank}(A)$ and there is an **infinite family** of solutions involving $n - m$ independent parameters
- (b) $\text{rank}(A:b) > \text{rank}(A)$ and there are **no solutions**; the system of equations is **inconsistent**.

Note that only the zero matrix has a rank of zero.

Example

Consider again the systems of equations

$$x + y + z = 3$$
(a) $$2x + 3y + 4z = 9$$
$$3x + 4y + 5z = 12$$

(b)
$$x + y + z = 1$$
$$x + y + z = 2$$
$$2x - 3y + 5z = 12$$

In case (a) $A:b = \begin{bmatrix} 1 & 1 & 1 & 3 \\ 2 & 3 & 4 & 9 \\ 3 & 4 & 5 & 12 \end{bmatrix}$ which is a 3×4 matrix whose 3×3 minors are

$$\begin{vmatrix} 1 & 1 & 3 \\ 3 & 4 & 9 \\ 4 & 5 & 12 \end{vmatrix}, \begin{vmatrix} 1 & 1 & 3 \\ 2 & 4 & 9 \\ 3 & 5 & 12 \end{vmatrix}, \begin{vmatrix} 1 & 1 & 3 \\ 2 & 3 & 9 \\ 3 & 4 & 12 \end{vmatrix} \quad \text{and} \quad |A| = \begin{vmatrix} 1 & 1 & 1 \\ 2 & 3 & 4 \\ 3 & 4 & 5 \end{vmatrix}$$

All these determinants are zero, as can be verified. The nine 2×2 minors, formed by deleting a row and two columns from $A:b$, are non-zero. For example

$$\begin{vmatrix} 1 & 1 & 1 & 3 \\ 2 & 3 & 4 & 9 \\ 3 & 4 & 5 & 12 \end{vmatrix} \rightarrow \begin{vmatrix} 1 & 1 \\ 3 & 5 \end{vmatrix} \rightarrow 2$$

Hence rank$(A:b) = $ rank$(A) = 2$.
In case (b) rank$(A:b) = $ rank$(A) = 3$; this is left as an exercise. ∎

For matrices larger than 3×3 a large number of minors need to be examined. Fortunately, there is another method of finding rank, but that lies beyond our scope.

Eigenvalues revisited

We return to the eigenvalue problem $Ax = \lambda x$ from which we produced the equivalent form $(A - \lambda I)x = 0$, and the determining equation $|A - \lambda I| = 0$.

Examples

1. Find the eigenvalues and eigenvectors of the matrix

$$A = \begin{bmatrix} 2 & 1 & 3 \\ 1 & 2 & 3 \\ 3 & 3 & 20 \end{bmatrix}$$

The determining equation is

$$|A - \lambda I| = \begin{vmatrix} 2 - \lambda & 1 & 3 \\ 1 & (2 - \lambda) & 3 \\ 3 & 3 & 20 - \lambda \end{vmatrix} = 0$$

Multiplying out the six products gives

$$(2 - \lambda)(2 - \lambda)(20 - \lambda) + 1 \times 3 \times 3 + 3 \times 1 \times 3$$
$$- \{(2 - \lambda) \times 3 \times 3 + 1 \times 1 \times (20 - \lambda) + 3 \times (2 - \lambda) \times 3\} = 0$$

i.e. $\lambda^3 - 24\lambda^2 + 65\lambda - 42 = 0$

This is a cubic equation in λ. The remainder theorem shows readily that $\lambda = 1$ and $\lambda = 2$ are roots, and the cubic equation can be written

$$(\lambda - 1)(\lambda - 2)(\lambda - 21) = 0$$

The solutions, i.e. the eigenvectors are of the form $\begin{bmatrix} x_1 \\ x_2 \\ x_3 \end{bmatrix}$; we can find the ratios $x_1 : x_2 : x_3$. For $\lambda = 1$, $(A - I)x = 0$ becomes

$$x_1 + x_2 + 3x_3 = 0$$
$$x_1 + x_2 + 3x_3 = 0$$
$$3x_1 + 3x_2 + 19x_3 = 0$$

Multiplying either of the first two equations by 3 gives the equation

$$3x_1 + 3x_2 + 9x_3 = 0$$

To be consistent with the third, we require that

$$x_3 = 0$$

Any of the equations now yield $x_2 = -x_1$, so $\begin{bmatrix} 1 \\ -1 \\ 0 \end{bmatrix}$ is a suitable eigenvector.

For $\lambda = 2$, $(A - 2I)x = 0$ becomes

$$x_2 + 3x_3 = 0$$
$$x_1 + 3x_3 = 0$$
$$3x_1 + 3x_2 + 18x_3 = 0$$

therefore $x_1 = x_2$, by comparing the first two equations. Then $x_1 = x_2 = -3x_3$ so that $\begin{bmatrix} 3 \\ 3 \\ -1 \end{bmatrix}$ is an eigenvector.

For $\lambda = 21$, $(A - 21I)x = 0$ becomes

$$-19x_1 + x_2 + 3x_3 = 0$$
$$x_1 - 19x_2 + 3x_3 = 0$$
$$3x_1 + 3x_2 - x_3 = 0$$

From the first two equations $x_1 = x_2$, and from the third equation $x_3 = 6x_1 = 6x_2$, so

$\begin{bmatrix} 1 \\ 1 \\ 6 \end{bmatrix}$ is an eigenvector.

2. Use the Cayley–Hamilton theorem to find the inverse of the matrix in the previous example.

We know that the characteristic equation of the matrix is

$$\lambda^3 - 24\lambda^2 + 65\lambda - 42 = 0$$

Applying the Cayley–Hamilton theorem, we obtain the matrix equation

$$A^3 - 24A^2 + 65A - 42I = 0$$

We multiply throughout by A^{-1} to obtain

$$A^2 - 24A + 65I - 42A^{-1} = 0$$

Then

$$A^{-1} = \frac{1}{42}(A^2 - 24A + 65I)$$

$$= \frac{1}{42}\begin{bmatrix} 31 & -11 & -3 \\ -11 & 31 & -3 \\ -3 & -3 & 3 \end{bmatrix}$$

after much manipulation. Note that $|A| = 42$.

Exercise 8.4

1 Write down the 24 permutations of the digits 1 2 3 4 and use the linking-of-arms method to identify the 12 positive and 12 negative permutations.

2 Determine the signs of the following permutations of $1\,2\,3\,4\,5$ and $1\,2\,3\,4\,5\,6$:

(a) $(2\,3\,1\,5\,4)$ (b) $(1\,3\,5\,2\,4)$ (c) $(6\,4\,2\,3\,5\,1)$

3 These matrices are called **permutation matrices**

(a) $\begin{bmatrix} 0 & 1 & 0 \\ 0 & 0 & 1 \\ 1 & 0 & 0 \end{bmatrix}$ (b) $\begin{bmatrix} 0 & 0 & 1 & 0 \\ 0 & 0 & 0 & 1 \\ 0 & 1 & 0 & 0 \\ 1 & 0 & 0 & 0 \end{bmatrix}$

They have a single 1 in every row and every column, all other elements are zero. The determinant of permutation matrix is either 1 or -1. Find the determinants of (a) and (b).

4 Evaluate the following determinants:

(a) $\begin{vmatrix} 1 & 1 & 1 \\ 35 & 37 & 34 \\ 23 & 26 & 25 \end{vmatrix}$ (b) $\begin{vmatrix} 13 & 16 & 19 \\ 14 & 17 & 20 \\ 15 & 18 & 21 \end{vmatrix}$

(c) $\begin{vmatrix} 13 & 3 & 23 \\ 30 & 7 & 53 \\ 39 & 9 & 70 \end{vmatrix}$ (d) $\begin{vmatrix} 1 & 0 & 0 \\ 0 & \cos\theta & \sin\theta \\ 0 & -\sin\theta & \cos\theta \end{vmatrix}$

5 The following determinants are all zero. Find x in each case.

(a) $\begin{vmatrix} x+a & b & c \\ c & x+b & a \\ a & b & x+c \end{vmatrix}$ (b) $\begin{vmatrix} 1 & -4 & 1 \\ 8 & 3 & 3 \\ 3 & 23 & x \end{vmatrix}$

(c) $\begin{vmatrix} 3+x & 2(1+x) & (x-2) \\ 2x-3 & 2-x & 3 \\ 3 & 7 & -1 \end{vmatrix}$

6 Given the matrices

$$A = \begin{bmatrix} a & b \\ c & d \end{bmatrix} \qquad B = \begin{bmatrix} p & q \\ r & s \end{bmatrix} \qquad C = \begin{bmatrix} 1 & 0 & 0 \\ 0 & a & b \\ 0 & c & d \end{bmatrix} \qquad D = \begin{bmatrix} a & b & 0 \\ c & d & 0 \\ 0 & 0 & 1 \end{bmatrix}$$

$$E = \begin{bmatrix} a & b & 0 & 0 \\ c & d & 0 & 0 \\ 0 & 0 & p & q \\ 0 & 0 & r & s \end{bmatrix}$$

prove that

(a) $|C| = |A|$ (b) $|D| = |A|$ (c) $|E| = |A||B|$

Hence evaluate the determinant of the matrix

$$\begin{bmatrix} 1 & 2 & 0 & 0 \\ 4 & 3 & 0 & 0 \\ 0 & 0 & 5 & 6 \\ 0 & 0 & 8 & 7 \end{bmatrix}$$

7 Use Cramer's rule to solve the following sets of equations:

(a)
$$\begin{aligned} x_1 - x_2 + x_3 &= -6 \\ -x_1 + 2x_2 - x_3 &= 8 \\ 2x_1 - x_2 + x_3 &= -7 \end{aligned}$$

(b)
$$\begin{aligned} x_1 + 3x_2 + 4x_3 &= 14 \\ x_1 + 2x_2 + x_3 &= 7 \\ 2x_1 + x_2 + 2x_3 &= 2 \end{aligned}$$

8 If $A = \begin{bmatrix} 1 & 3 & 3 \\ 1 & 4 & 3 \\ 1 & 3 & 4 \end{bmatrix}$ prove that $A^{-1} = \begin{bmatrix} 7 & -3 & -3 \\ -1 & 1 & 0 \\ -1 & 0 & 1 \end{bmatrix}$ by showing that $AA^{-1} = A^{-1}A = I$.

Hence solve the equations

$$\begin{aligned} x + 3y + 3z &= 8 \\ x + 4y + 3z &= 9 \\ x + 3y + 4z &= 10 \end{aligned}$$

9 Solve the equations in Questions 7 and 8 using Gaussian elimination.

10 Verify that $x = 1, y = -1, z = 2$ is a solution of the equations

$$\begin{aligned} x + 2y + 3z &= 5 \\ 3x + y - z &= 0 \\ x + y + z &= 2 \end{aligned}$$

Why is it not the only solution?

11   Prove that $A = LU$ where

$$A = \begin{bmatrix} 2 & 12 & 6 \\ -1 & -3 & 3 \\ -2 & -11 & -3 \end{bmatrix} \qquad L = \begin{bmatrix} 2 & 0 & 0 \\ -1 & 3 & 0 \\ -2 & 1 & 1 \end{bmatrix} \qquad U = \begin{bmatrix} 1 & 6 & 3 \\ 0 & 1 & 2 \\ 0 & 0 & 1 \end{bmatrix}$$

Solve the system of equations

$$\begin{aligned} 2x_1 + 12x_2 + 6x_3 &= 12 \\ -x_1 - 3x_2 + 3x_3 &= 9 \\ -2x_1 - 11x_2 - 3x_3 &= -4 \end{aligned}$$

i.e. $Ax = b$ by first of all solving the subsidiary system of equations $Ly = b$, where $y = \begin{bmatrix} y_1 \\ y_2 \\ y_3 \end{bmatrix}$, using forward substitution.

Then solve for x the system of equations $Ux = y$ using back-substitution.

12   Determine the rank of the following matrices.

(a) $\begin{bmatrix} 2 & 1 \\ 4 & 2 \end{bmatrix}$ (b) $\begin{bmatrix} 0 \\ 1 \end{bmatrix}$ (c) $\begin{bmatrix} 1 & 2 & 3 \\ 4 & 5 & 6 \end{bmatrix}$

(d) $\begin{bmatrix} 1 & 2 \\ 2 & 1 \\ 6 & 5 \end{bmatrix}$ (e) $\begin{bmatrix} 0 \\ 0 \end{bmatrix}$

In the case of non-square matrices, examine the minors formed by deleting rows (or columns) in turn.

13   By determining rank(A) and rank($A:b$) determine whether the following sets of equations have a unique solution, infinitely many solutions or no solutions at all.

(a) $\begin{aligned} 2x + 3y &= 10 \\ 3x - y &= 1 \end{aligned}$ (b) $\begin{aligned} 2x + 3y &= 10 \\ 4x + 6y &= 17 \end{aligned}$ (c) $\begin{aligned} x + y + z &= 1 \\ 2x + 2y + 2z &= 2 \\ 3x + 3y + 3z &= 3 \end{aligned}$

14   For the matrices

$$A = \begin{bmatrix} 1 & 1 & 1 \\ 1 & 1 & 1 \\ 2 & -3 & 5 \end{bmatrix} \quad \text{and} \quad b = \begin{bmatrix} 3 \\ 9 \\ 12 \end{bmatrix}$$

show that rank(A) = 2, but that rank ($A:b$) = 3.

15   If $A = \begin{bmatrix} a & 0 & 0 \\ 0 & b & 0 \\ 0 & 0 & c \end{bmatrix}$, find the matrices A^2 and A^3. What is A^n? If a, b, c are all non-zero, write down A^{-1}.

16   The sum of the diagonal elements of a matrix A is called the **trace** of A and is written $\operatorname{tr} A$. A well-known result in matrix theory is that

$$\operatorname{tr} A = \text{sum of the eigenvalues of } A$$

It is known that $\lambda = 1$ is an eigenvalue of the singular matrix

$$A = \begin{bmatrix} -2 & -8 & -12 \\ 1 & 4 & 4 \\ 0 & 0 & 1 \end{bmatrix}$$

A singular matrix has a zero determinant. Show how this implies that $\lambda = 0$ is one of its eigenvalues. Determine the remaining eigenvalue of the given matrix and find its associated eigenvector.

17   Determine the eigenvalues and eigenvectors of the matrix

$$\begin{bmatrix} -3 & 1 & 0 \\ 2 & -3 & 2 \\ 0 & 1 & -3 \end{bmatrix}$$

18   Starting from the characteristic equation for the matrix in Question 17, use the Cayley–Hamilton theorem to find A^{-1} in the form

$$A^{-1} = \alpha A^2 + \beta A + \gamma I$$

SUMMARY

- **Order and transpose**: a rectangular array A of numbers in m rows and n columns has order $m \times n$; its elements are a_{ij}; if the rows and columns of A are interchanged we obtain the transpose A^{T}

- **Diagonal matrices**: a diagonal matrix has $a_{ij} = 0$ if $i \neq j$, the unit or identity matrix is diagonal and has $a_{ij} = 1$; the zero matrix has $a_{ij} = 0$ and is denoted by $\mathbf{0}$

- **Arithmetic operations**

$$kA = k[a_{ij}] = [ka_{ij}] \qquad A + B = [a_{ij}] + [b_{ij}] = [a_{ij} + b_{ij}]$$

$$C = AB \quad \text{where} \quad [c_{ij}] = \sum a_{ik}b_{kj} \qquad AB \neq BA \text{ in general}$$

- **Equality**: $A = B$ if $[a_{ij}] = [b_{ij}]$ for all i and j

- **Transformations in the plane**

$$\begin{bmatrix} x' \\ y' \end{bmatrix} = \begin{bmatrix} a & b \\ c & d \end{bmatrix} \begin{bmatrix} x \\ y \end{bmatrix}$$

- **Determinant of a square matrix**

$$\begin{vmatrix} a & b \\ c & d \end{vmatrix} = ad - bc$$

- **Cramer's rule**: used when solving a system of equations

$$ax_1 + bx_2 = p$$
$$cx_1 + dx_2 = q$$

if a solution exists, it takes the form

$$x_1 = \frac{\begin{vmatrix} p & b \\ q & d \end{vmatrix}}{\begin{vmatrix} a & b \\ c & d \end{vmatrix}} \qquad x_2 = \frac{\begin{vmatrix} a & p \\ c & q \end{vmatrix}}{\begin{vmatrix} a & b \\ c & d \end{vmatrix}}$$

- **Inverse matrix**: if $A = \begin{bmatrix} a & b \\ c & d \end{bmatrix}$ and $|A| \neq 0$ then $A^{-1} = \dfrac{1}{|A|} \begin{bmatrix} d & -b \\ -c & a \end{bmatrix}$

- **Eigenvalues and eigenvectors**: if $Ax = \lambda x$ then x is an eigenvector of A and λ is its associated eigenvalue; the characteristic equation is $|A - \lambda I| = 0$

- **3×3 matrices**

$$\begin{vmatrix} a & b & c \\ d & f & g \\ h & k & l \end{vmatrix} = (afl + bgh + cdk) - (cfh + agk + bdl)$$

- **Rank**: A is an $m \times n$ matrix; if $\text{rank}(A) = n$ then the equations represented by $Ax = b$ have a unique solution; if not, then either $\text{rank}(A:b) = \text{rank}(A) < n$ and there is an infinite number of solutions or $\text{rank}(A:b) > \text{rank}(A)$ and there are no solutions.

Answers

Exercise 8.1

1		$A + B$	AB	BA
	(a)	$\begin{bmatrix} 5 & 1 \\ 4 & -2 \end{bmatrix}$	$\begin{bmatrix} 7 & 2 \\ 5 & -8 \end{bmatrix}$	$\begin{bmatrix} 6 & 3 \\ 8 & -7 \end{bmatrix}$
	(b)	$\begin{bmatrix} 1 & 1 \\ m & m \end{bmatrix}$	$\begin{bmatrix} 0 & 1 \\ m^2 & 0 \end{bmatrix}$	$\begin{bmatrix} 0 & m \\ m & 0 \end{bmatrix}$
	(c)	$\begin{bmatrix} 1 & 1 & 7 \\ 1 & 1 & -1 \\ 5 & 2 & -2 \end{bmatrix}$	$\begin{bmatrix} 14 & 4 & -16 \\ 10 & 9 & -9 \\ -4 & -10 & 10 \end{bmatrix}$	$\begin{bmatrix} 13 & 1 & -17 \\ -8 & 8 & 12 \\ -2 & 14 & 12 \end{bmatrix}$
	(d)	undefined	$\begin{bmatrix} 16 & -7 \\ -18.5 & -11 \end{bmatrix}$	$\begin{bmatrix} -11 & 7 & 9 \\ -37 & 13 & 15 \\ 50 & -1.5 & 3 \end{bmatrix}$
	(e)	undefined	$\begin{bmatrix} 1 & 2 & 3 & 4 \\ 2 & 4 & 6 & 8 \\ 3 & 6 & 9 & 12 \\ 4 & 8 & 12 & 16 \end{bmatrix}$	$[30]$

2 (a) $\begin{bmatrix} -2 & 12 & 19 \\ 3 & -6 & -9 \\ 6 & 12 & 21 \end{bmatrix}$ (b) $\begin{bmatrix} 249 & 428 \\ -377 & -648 \\ 453 & 781 \end{bmatrix}$

4 (a) $\begin{bmatrix} 1 & 2 \\ 0 & 1 \end{bmatrix}$ (c) $A^4 = \begin{bmatrix} 1 & 4 \\ 0 & 1 \end{bmatrix}$

(d) $A^n = \begin{bmatrix} 1 & n \\ 0 & 1 \end{bmatrix}$ $A^{n+1} = \begin{bmatrix} 1 & n+1 \\ 0 & 1 \end{bmatrix}$

5 (a) $\begin{bmatrix} 1 & -1 \\ 0 & 1 \end{bmatrix}$ (c) $X^2 = \begin{bmatrix} 1 & -2 \\ 0 & 1 \end{bmatrix}$ $X^n = \begin{bmatrix} 1 & -n \\ 0 & 1 \end{bmatrix}$

(d) $X^n A^n = A^n X^n = I$

6 $688\,800\,\text{m}^3$

7 (a) $x = 1, y = 2$ (b) $a = b = 1$

8 Diagonal matrices of all orders commute.

9 $a = 1, b = 3, c = 5$

10 (a) $\begin{bmatrix} \cos\theta & \sin\theta \\ -\sin\theta & \cos\theta \end{bmatrix}$

(c) $\begin{bmatrix} \cos n\theta & -\sin n\theta \\ \sin n\theta & \cos n\theta \end{bmatrix} = R(n\theta)$

(d) (i) $\begin{bmatrix} 1 & 0 \\ 0 & 1 \end{bmatrix}$

(ii) $\dfrac{1}{2}\begin{bmatrix} \sqrt{3} & -1 \\ 1 & \sqrt{3} \end{bmatrix}$

(iii) $\dfrac{1}{\sqrt{2}}\begin{bmatrix} 1 & -1 \\ 1 & 1 \end{bmatrix}$

(iv) $\dfrac{1}{2}\begin{bmatrix} 1 & -\sqrt{3} \\ \sqrt{3} & 1 \end{bmatrix}$

(v) $\begin{bmatrix} 0 & -1 \\ 1 & 0 \end{bmatrix}$

(e) $\dfrac{1}{2\sqrt{2}}\begin{bmatrix} \sqrt{3}-1 & -\sqrt{3}-1 \\ \sqrt{3}+1 & \sqrt{3}-1 \end{bmatrix}$

Exercise 8.2

1 (a) $\begin{bmatrix} -1 & 0 \\ 0 & 1 \end{bmatrix}$ $(x, y) \rightarrow (-x, y)$

(b) $\begin{bmatrix} 4 & 0 \\ 0 & \dfrac{1}{3} \end{bmatrix}$ $(x, y) \rightarrow \left(4x, \dfrac{y}{3}\right)$

(c) $\begin{bmatrix} 0 & 1 \\ -1 & 0 \end{bmatrix}$ $(x, y) \rightarrow (y, -x)$

(d) $\begin{bmatrix} x-6 \\ y+2 \end{bmatrix}$ $(x, y) \rightarrow (x-6, y+2)$

2 The diagram shows an example.

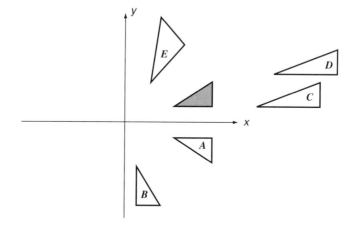

3 (a) $\begin{bmatrix} 1 & 0 \\ 0 & -1 \end{bmatrix}$ $(x, y) \rightarrow (x, -y)$

(b) $\begin{bmatrix} -1 & 0 \\ 0 & 1 \end{bmatrix}\begin{bmatrix} x \\ y \end{bmatrix} + \begin{bmatrix} 4 \\ -4 \end{bmatrix}$ $(x, y) \rightarrow (-x+4, y-4)$

(c) $\begin{bmatrix} \frac{1}{3} & 0 \\ 0 & -2 \end{bmatrix} \begin{bmatrix} x \\ y \end{bmatrix} + \begin{bmatrix} 0 \\ 4 \end{bmatrix}$ $(x, y) \rightarrow \left(\frac{x}{3}, -2y + 4\right)$

(d) $\begin{bmatrix} 0 & 1 \\ 1 & 0 \end{bmatrix} \begin{bmatrix} x \\ y \end{bmatrix} - \begin{bmatrix} 6 \\ 2 \end{bmatrix}$ $(x, y) \rightarrow (y - 6, x - 2)$

(e) $\begin{bmatrix} -1 & 0 \\ 0 & -2 \end{bmatrix} \begin{bmatrix} x \\ y \end{bmatrix} + \begin{bmatrix} -1 \\ 11 \end{bmatrix}$ $(x, y) \rightarrow (-x - 1, -2y + 11)$

(f) $\begin{bmatrix} 0 & 2 \\ 2 & 0 \end{bmatrix} \begin{bmatrix} x \\ y \end{bmatrix} - 13 \begin{bmatrix} 1 \\ 1 \end{bmatrix}$ $(x, y) \rightarrow (2y - 13, 2x - 13)$

4 (a) $\begin{bmatrix} 0 & -1 \\ -1 & 0 \end{bmatrix}$ $(x, y) \rightarrow (-y, -x)$

(b) $\begin{bmatrix} 1 & 0 \\ 0 & -1 \end{bmatrix} \begin{bmatrix} x \\ y \end{bmatrix} + \begin{bmatrix} 0 \\ 4 \end{bmatrix}$ $(x, y) \rightarrow (x, 4 - y)$

5 $\begin{bmatrix} -7 & -6 \\ -5 & -4 \end{bmatrix}$

6 $\begin{bmatrix} 1 & 0 \\ 0 & -1 \end{bmatrix} \begin{bmatrix} -1 & 0 \\ 0 & 1 \end{bmatrix} = \begin{bmatrix} -1 & 0 \\ 0 & 1 \end{bmatrix} \begin{bmatrix} 1 & 0 \\ 0 & -1 \end{bmatrix} = \begin{bmatrix} -1 & 0 \\ 0 & -1 \end{bmatrix} = \begin{bmatrix} \cos \pi & -\sin \pi \\ \sin \pi & \cos \pi \end{bmatrix}$

7 (a) $\frac{1}{\sqrt{2}} \begin{bmatrix} 1 & -1 \\ 1 & 1 \end{bmatrix} \begin{bmatrix} 0 & 1 \\ 1 & 0 \end{bmatrix} = \frac{1}{\sqrt{2}} \begin{bmatrix} -1 & 1 \\ 1 & 1 \end{bmatrix}$

(b) $\begin{bmatrix} 1 & 0 \\ 0 & -1 \end{bmatrix} \begin{bmatrix} 1 & 0 \\ 0 & 2 \end{bmatrix} \begin{bmatrix} -1 & 0 \\ 0 & 1 \end{bmatrix} = \begin{bmatrix} -1 & 0 \\ 0 & -2 \end{bmatrix}$

(c) $\begin{bmatrix} 0 & -1 \\ 1 & 0 \end{bmatrix} \begin{bmatrix} 1 & 0 \\ 0 & -1 \end{bmatrix} \begin{bmatrix} 0 & -1 \\ 1 & 0 \end{bmatrix} = \begin{bmatrix} 1 & 0 \\ 0 & -1 \end{bmatrix}$

The matrices in (b) are diagonal and all commute. Transformation is $180°$ rotation plus an enlargement in the y-direction, in either order. The matrices in (a) and (c) do not commute.

8 (a) (i) $\begin{bmatrix} 2 \\ 0 \end{bmatrix}$ $\begin{bmatrix} 1 \\ 0 \end{bmatrix}$ is an eigenvector, 2 is the eigenvalue

(ii) $\begin{bmatrix} 3 \\ -1 \end{bmatrix}$ (iii) $\begin{bmatrix} 1 \\ 1 \end{bmatrix}$ $\begin{bmatrix} 1 \\ 1 \end{bmatrix}$ is an eigenvector, 1 is the eigenvalue

(b) (i) $\begin{bmatrix} 0 \\ 0 \end{bmatrix}$ $\begin{bmatrix} 1 \\ 3 \end{bmatrix}$ is an eigenvector, 0 is the eigenvalue

(ii) $\begin{bmatrix} -16 \\ 8 \end{bmatrix} = 8 \begin{bmatrix} -2 \\ 1 \end{bmatrix}$

(iii) $\begin{bmatrix} 14 \\ -7 \end{bmatrix}$ $\begin{bmatrix} -2 \\ 1 \end{bmatrix}$ is an eigenvector, -7 is the eigenvalue

The matrix is singular.

(c) (i) $\begin{bmatrix} 1 \\ 1 \end{bmatrix}$

(ii) $\begin{bmatrix} 0 \\ a \end{bmatrix}$ $\begin{bmatrix} 0 \\ 1 \end{bmatrix}$ is an eigenvector, a is the eigenvalue

(iii) $\begin{bmatrix} -1 \\ -1+a \end{bmatrix}$

(d) (i) $\begin{bmatrix} 6 \\ 8 \end{bmatrix}$

(ii) $\begin{bmatrix} 5 \\ 5 \end{bmatrix}$ $\begin{bmatrix} 1 \\ 1 \end{bmatrix}$ is an eigenvector, 5 is the eigenvalue

(iii) $\begin{bmatrix} 2 \\ -4 \end{bmatrix}$ $\begin{bmatrix} 1 \\ -2 \end{bmatrix}$ is an eigenvector, 2 is the eigenvalue

9 $A^2 = \begin{bmatrix} -3 & -6 \\ 6 & 12 \end{bmatrix}$ has eigenvalues 0 and 9; A has eigenvalues 0 and 3.

10 $\sqrt{50}$

Exercise 8.3

2 (a) -18 (b) -2 (c) 201 (d) 6004

3 (a) $-\dfrac{3}{2}$ (b) $2(1 + \sqrt{2}), 2(1 - \sqrt{2})$

(c) $-\dfrac{5}{4}$ (d) $0, -(a+d)$

4 (a) $|A| = -18, |B| = -2$

$AB = \begin{bmatrix} 62 & 68 \\ 87 & 96 \end{bmatrix}$ $\quad |AB| = 36$

(b) $|A_1 \dots A_n| = |A_1||A_2| \dots |A_n|$
If for some j, $|A_j| = 0$, the determinant of any product involving A_j is zero. If any other matrix in the product has zero determinant, the position is unchanged.

(c) (i) $k^2|C|$ (ii) $|C^{-1}| = \dfrac{1}{|C|}$

5 (a) $x = \dfrac{3}{2}, y = 1$ (b) $x = 1, y = 2$ (c) $x = \dfrac{8}{7}, y = -\dfrac{3}{7}$

7 B^{-1}: $\begin{bmatrix} x \\ y \end{bmatrix} \rightarrow \begin{bmatrix} x \\ y \end{bmatrix} + \begin{bmatrix} 1 \\ 0 \end{bmatrix} = \begin{bmatrix} x+1 \\ y \end{bmatrix}$

$$\begin{bmatrix} 0 & 1 \\ -1 & 0 \end{bmatrix} \begin{bmatrix} x+1 \\ y \end{bmatrix} = \begin{bmatrix} y \\ -(x+1) \end{bmatrix} = \begin{bmatrix} 0 & 1 \\ -1 & 0 \end{bmatrix} \begin{bmatrix} x \\ y \end{bmatrix} + \begin{bmatrix} 0 \\ -1 \end{bmatrix}$$

D^{-1}: $\begin{bmatrix} x \\ y \end{bmatrix} \rightarrow \begin{bmatrix} x \\ y \end{bmatrix} + \begin{bmatrix} 3 \\ -1 \end{bmatrix} = \begin{bmatrix} x+3 \\ y-1 \end{bmatrix}$

$$\begin{bmatrix} -1 & 0 \\ 0 & -\frac{1}{3} \end{bmatrix} \begin{bmatrix} x+3 \\ y-1 \end{bmatrix} = \begin{bmatrix} -(x+3) \\ -\frac{1}{3}(y-1) \end{bmatrix} = \begin{bmatrix} -1 & 0 \\ 0 & -\frac{1}{3} \end{bmatrix} \begin{bmatrix} x \\ y \end{bmatrix} + \begin{bmatrix} -3 \\ \frac{1}{3} \end{bmatrix}$$

8 (b) $C^{-1}B^{-1}A^{-1}$

9 (b) (i) BC (ii) I (iii) C

11 (a) $\lambda_1 = 2, x_1 = \begin{bmatrix} 1 \\ 1 \end{bmatrix}$ $\lambda_2 = 3, x_2 = \begin{bmatrix} 1 \\ 2 \end{bmatrix}$

(b) $\lambda_1 = -2, x_1 = \begin{bmatrix} 1 \\ -1 \end{bmatrix}$ $\lambda_2 = 7, x_2 = \begin{bmatrix} 1 \\ 2 \end{bmatrix}$

(c) $\lambda_1 = 4, x_1 = \begin{bmatrix} 1 \\ -2 \end{bmatrix}$ $\lambda_2 = 5, x_2 = \begin{bmatrix} 1 \\ -1 \end{bmatrix}$

12 (a) $\frac{1}{6}(5I - A) = \frac{1}{6}\begin{bmatrix} 4 & -1 \\ 2 & 1 \end{bmatrix}$

(b) $-\frac{1}{14}(5I - A) = -\frac{1}{14}\begin{bmatrix} 4 & -3 \\ -6 & 2 \end{bmatrix}$

(c) $-\frac{1}{20}(9I - A) = \frac{1}{20}\begin{bmatrix} 3 & -1 \\ 2 & 6 \end{bmatrix}$

Exercise 8.4

1 The positive permutations are

(1 2 3 4) (1 3 4 2) (1 4 2 3)
(2 1 4 3) (2 3 1 4) (2 4 3 1)
(3 1 2 4) (3 2 4 1) (3 4 1 2)
(4 1 3 2) (4 2 1 3) (4 3 2 1)

The rest are negative.

2 (a) negative   (b) negative

(c) negative (11 crossings)

3 (a) 1 (b) -1

4 (a) 7 (b) 0
   (c) 1 (d) 1

5 (a) $x = 0$ or $-(a + b + c)$ (b) $x = -2$
   (c) $x = 3$ or $-\dfrac{1}{22}$

6 65

7 (a) $x_1 = -1, x_2 = 2, x_3 = -3$ (b) $x_1 = -2, x_2 = 4, x_3 = 1$

8 $x = -1, y = 1, z = 2$

10 $|A| = 0$

11 $x_1 = 3, x_2 = 1, x_3 = 3$

12 (a) 1 (b) 1 (c) 2 (d) 2 (e) 0

13 (a) $x = 1, y = 2$; unique solution, rank$(A) = 2$
   (b) no solution, rank$(A) = 1$, rank$(A : b) = 2$
   (c) infinitely many solutions, $x = y = z = 1$; rank$(A) = $ rank$(A : b) = 1$

15 $A^n = \begin{bmatrix} a^n & 0 & 0 \\ 0 & b^n & 0 \\ 0 & 0 & c^n \end{bmatrix}$ $A^{-1} = \begin{bmatrix} a^{-1} & 0 & 0 \\ 0 & b^{-1} & 0 \\ 0 & 0 & c^{-1} \end{bmatrix}$

16 $\lambda = 2, x = \begin{bmatrix} -2 \\ 1 \\ 0 \end{bmatrix}$

17 $\lambda_1 = -1, x_1 = \begin{bmatrix} 1 \\ 2 \\ 1 \end{bmatrix}$ $\lambda_2 = -3, x_2 = \begin{bmatrix} 1 \\ 0 \\ -1 \end{bmatrix}$ $\lambda_3 = -5, x_3 = \begin{bmatrix} -1 \\ 2 \\ -1 \end{bmatrix}$

18 $A^{-1} = -\dfrac{1}{15}(A^2 + 9A + 23I) = -\dfrac{1}{15} \begin{bmatrix} 7 & 3 & 2 \\ 6 & 9 & 6 \\ 2 & 3 & 7 \end{bmatrix}$

9 UNCERTAINTY

Introduction

Uncertainty in data is a fact of life for engineers and scientists, so it is vital to take this into account when presenting results. Besides a descriptive summary, we need to make inferences from sampled data. Probability theory provides the link between observed data and theoretical models. We make our inferences by using this link.

Objectives

After working through this chapter you should be able to:

- Represent data using appropriate pictograms
- Represent a frequency distribution by a histogram
- Calculate the mean, median and mode of a set of data in grouped or ungrouped form
- Calculate the range, interquartile range, variance and standard deviation for a set of data in grouped or ungrouped form
- Define and calculate the probability of an event
- Calculate the probability for the complement of an event and for the union and intersection of two events
- Understand conditional probability and statistical independence
- Understand the meaning of a discrete random variable and a continuous random variable
- Use the binomial, Poisson and geometric distributions
- Apply the probability density function of a continuous random variable
- Calculate probabilities from the normal distribution
- Use the normal approximation to the binomial distribution.

9.1 ANALYSIS OF DATA

Uncertainty exists at every level in the real world. In this chapter we examine the effect of uncertainty in data. When a quantity is measured it provides an item of data. Any numerate quantity determined from the data is called a **statistic**. Before looking at a specific example, we must distinguish carefully between certainty and uncertainty. **Certainty** relates to matters, results or values that are fixed; for example, the natural constants e and π are fixed for all time and space. In mathematics we speak of **deterministic problems**, i.e. problems whose results can be systematically and uniquely found to whatever accuracy is available, e.g. finding the roots of a quadratic equation. On the other hand, **statistical problems** are those whose solution can only be found within a margin of error or uncertainty, e.g. estimating the mean weight of a batch of metal bars by taking measurements on a sample. In this context we call the weight of a bar a **random variable**.

A **discrete random variable** is one which can take individual values only, e.g. 0, 1, 2, 3. As an example, consider the number of defective items in a box of 100 mass-produced components; this can take one of a finite number of values—the integers from 0 to 100 inclusive. A second example is the number of mass-produced items tested until a defective item is discovered; this variable can take any value from 0 upwards—the possible values it can take are literally infinite.

A **continuous random variable** can take a range of real values on a continuous scale. As an example, consider the time to failure for an electronic component.

Pictorial representation of data

There are several methods available for the pictorial representation of data. Pie charts, bar charts and line diagrams are normally used for discrete data whereas the histogram and frequency polygon are used for continuous data.

Pie chart

The pie chart is a circular area divided into sectors. Each sector represents a particular subset of the data and the area of the sector is proportional to the number of items of data in that subset.

Example

The population of each of the six continental areas of the world can be calculated as a fraction of the total world population. The data, based on 1994 values, are displayed in Figure 9.1.

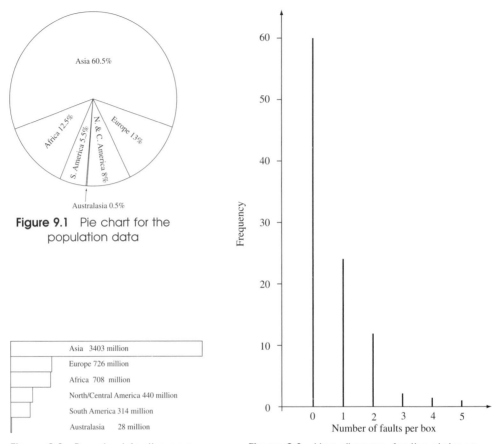

Figure 9.1 Pie chart for the population data

Figure 9.2 Bar chart for the population data

Figure 9.3 Line diagram for the data on defectives

Bar charts

The data of the previous example can alternatively be displayed in the form of a bar chart, shown in Figure 9.2, where the length of each bar is proportional to the population which it represents. The order of the bars could be in increasing order of population, in decreasing order of an alphabetical order, depending on the effect required.

Line diagrams

Vertical lines are drawn to represent subsets of data; the length of each line is proportional to the number of items in the subset.

Example

The number of defective items in boxes of 100 mass-produced components was recorded for 100 boxes. The results are displayed in Figure 9.3.

Frequency distributions

When measurements are made of a continuous random variable it is often useful to divide the data into subsets called **classes** and to work with the classes. How many classes we use is at our discretion, but a rough guide is to select a number between 6 and 12.

Example

The results displayed in Table 9.1 were obtained for the resistance of 60 electrical resistors; the resistance was measured to the nearest ohm.

Table 9.1

999	1 003	1 002	999	990	1 007
1 012	996	1 000	995	1 009	993
998	998	1 003	1 003	1 002	997
993	1 006	1 004	1 000	1 014	999
990	1 003	999	995	1 008	1 001
1 000	1 004	1 003	1 001	991	1 005
1 011	997	1 001	994	1 010	995
996	998	1 004	1 002	1 004	996
994	1 007	1 002	1 001	1 016	1 001
991	1 004	999	994	1 009	1 000

We may group the data as in Table 9.2. Such a table is known as a **frequency distribution** because the total frequency is distributed or shared out among the classes.

Table 9.2

Resistance	Class mark	Frequency	Cumulative frequency
990–992	991	4	4
993–995	994	8	12
996–998	997	8	20
999–1 001	1 000	14	34
1 002–1 004	1 003	14	48
1 005–1 007	1 006	4	52
1 008–1 010	1 009	4	56
1 011–1 013	1 012	2	58
1 014–1 016	1 015	2	60

The **class mark** is the midpoint of each class interval; each observation in that interval is effectively assumed to be taken at that point. **Frequency** is the number of times that an item has been recorded in that interval. ∎

Histograms

To construct a **histogram** from a frequency distribution we draw a set of rectangles where the midpoints of the base are the class marks and where the area of each rectangle is proportional to the class frequency.

In the example above the intervals are widened so that they adjoin. For example, the interval 990–992 is widened to 989.5–992.5, the interval 993–995 is widened to 992.5–995.5, and so on. The intervals then meet and the width of each interval is 3 ohms. The height of each rectangle is therefore proportional to the frequency of items in that interval. Note that, although the resistances have been measured to the nearest ohm, resistance is a continuous variable. The histogram for the data is shown in Figure 9.4.

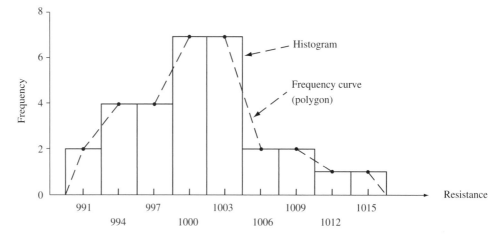

Figure 9.4 Histogram and frequency polygon for the resistance data

The histogram may be replaced by a curve consisting of a set of straight line segments joining the midpoints of the top of neighbouring rectangles. This curve is called a **frequency polygon**. When there are a large number of observations and the class widths are relatively small, the polygon can be replaced by a smooth curve passing through the midpoints of the rectangle tops; the curve is then called a **frequency curve**.

Another useful way of plotting data is to construct a **cumulative frequency diagram**. If the observations are arranged in ascending order of magnitude, it is possible to find the cumulative frequency of observations which are less than or equal to a particular value. The cumulative frequencies are easier to interpret if they are expressed as relative cumulative frequencies, by dividing each frequency by the total number of observations.

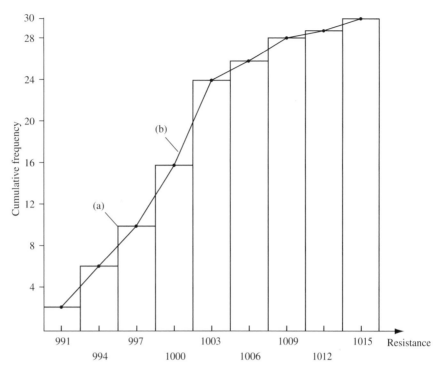

Figure 9.5 Histogram (a) and cumulative frequency polygon (b).

The relative cumulative frequency diagram is useful for identifying or estimating a mathematical function, known as a theoretical **probability distribution**, to model the behaviour of the variable. The cumulative frequency diagram for the resistance data is shown in Figure 9.5(a). Also shown is the **cumulative frequency polygon**, Figure 9.5(b), formed in the same way as the frequency polygon.

Mean and dispersion

In addition to graphical techniques it is often useful to calculate numerical values which summarise the data. We have already mentioned that such a quantity is called a statistic and this is a function of the measurements or observations. Most simple statistics may be categorised as a measure of location, e.g. *mean*, *median*, and *mode*, or a measure of dispersion, e.g. *variance*, *standard deviation* and *fractile/percentile*. We will concentrate mainly upon the mean and standard deviation.

Mean is the term used for arithmetic mean, commonly known as the average. It is sometimes called the **mathematical expectation**, the context of which will be explained during this chapter. For example, the mean of the 10 temperatures (measured in °C)

$$7, 9, 11, 4, 7, 13, 9, 6, 11, 13$$

is
$$\frac{7 + 9 + 11 + 4 + 7 + 13 + 9 + 6 + 11 + 13}{10} = 9 \, °C$$

i.e. the sum divided by the total number of items.

Note carefully that a **population**, possibly infinite, is a complete set of items or objects whereas a **sample** is a subset of items selected from a population. The mean of a population and of a sample are referred to by the symbols μ and $\bar{x}$ respectively. Whatever the context we define

$$\text{Mean} = \frac{x_1 + \cdots + x_n}{n} = \frac{\sum_{i=1}^{n} x_1}{n} = \frac{1}{n} \sum_{i=1}^{n} x_i \tag{9.1}$$

If the values $x_1, x_2, \ldots, x_n$ occur with frequencies, $f_1, f_2, \ldots, f_n$ then the mean is given by

$$\mu = \frac{f_1 x_1 + f_2 x_2 + \cdots + f_n x_n}{f_1 + f_2 + \cdots + f_n} = \frac{\sum_{i=1}^{n} f_i x_i}{\sum_{i=1}^{n} f_i} \tag{9.2}$$

Example

Calculate the mean for the set of examination results shown in Table 9.3. The marks are expressed as percentages in the column headed x_i, and the frequency of each mark is shown in the column headed f_i. The third column is the product of entries in the first two columns; the fourth column is the cumulative frequency.

Then $\mu = \dfrac{\sum f_i x_i}{\sum f_i} = \dfrac{5173}{100} = 51.73$

For individual or repeated items of data, $x_1, \ldots, x_n$, where n is large, it is often easier to estimate the mean by arranging the data in m **classes** using the formula

$$\mu = \frac{\sum_{i=1}^{m} f_i y_i}{\sum_{i=1}^{m} f_i} \tag{9.3}$$

where y_i are the class marks and f_i are the class frequencies. The calculated value of the mean will not be exact, but for large quantities of data with small class sizes the error will be small.

	Table 9.3				Table 9.4		
x_i	f_i	$f_i x_i$	F_i	Class	y_i	f_i
18	1	18	1	10–19	14.5	1
22	2	44	3	20–29	24.5	4
27	2	54	5	30–39	34.5	8
35	3	105	8	40–49	44.5	33
39	5	195	13	50–59	54.5	28
40	7	280	20	60–69	64.5	15
42	6	252	26	79–79	74.5	8
45	8	360	34	80–89	84.5	2
49	12	588	46	90–99	94.5	1
51	14	714	50			
55	14	770	74			
60	9	540	83			
64	3	192	86			
69	3	207	89			
71	2	142	91			
75	4	300	95			
79	2	158	97			
82	2	164	99			
90	1	90	100			
	$\sum f_i = 100$	$\sum f_i x_i = 5173$				

Example

Regrouping the data in Table 9.3 into 10 classes we obtain Table 9.4.

Then
$$\mu = \frac{\displaystyle\sum_{i=1}^{10} f_i y_i}{\displaystyle\sum_{i=1}^{10} f_i} = \frac{5210}{100} = 52.10$$

a close approximation obtained after a shorter calculation.

Median

This essentially means the **middle value** of an ordered distribution, i.e. the value above which and below which 50% of the data lie. If the middle two values differ then we take their arithmetic mean as the median of the distribution.

Consider the following sets of data ranked in ascending order:

(a) 3, 4, 4, 5, 6, 8, 9

(b) 2, 3, 3, 4, 5, 6, 8, 9

In case (a) there are seven items of data; the 'middle value' is the fourth, which is 5, hence the median is 5. In case (b) there are eight items; the 'middle' two values are 4 and 5; their mean is 4.5, hence the median is 4.5.

In the Example of Table 9.3 this is the value midway between the 50th and 51st examination mark; in this case, since both values are the same, the median is 51%.

If the mean lies to the left of the median, the distribution is **negatively skewed**; if it lies to the right, the distribution is **positively skewed**. If the mean and the median coincide then the distribution is **symmetric**.

Mode

The mode is the **most frequently occurring value**, or highest point on the frequency curve. In the data set above, (a) has mode 4 and (b) has mode 3. In the examination marks of Table 9.3, 14 candidates obtained each of 51% and 55% so the mode is the average of the two, i.e. 53%. If a frequency curve has two peaks the distribution is called **bimodal**.

Measures of dispersion

The dispersion of data, i.e. the spread or variation about the mean is also of statistical interest. The simplest measure is the **range**, which is the difference between the highest and the lowest values in the set of data. The most important measure of spread is the **variance**. The **standard deviation** is the square root of the variance.

Population variance is represented by σ^2 (**standard deviation** σ) whereas **sample variance** is represented by s^2 (**standard deviation** s). There is a subtle distinction between the two, as we shall soon see.

To compute the standard deviation σ for readings $x_1, x_2, \ldots, x_n$:

(a) Find the mean μ

(b) Find the deviation of each reading from the mean $d_i = x_i - \mu$.

(c) Square each deviation $d_i^2 = (x_i - \mu)^2$.

(d) Find the mean of these squared deviations; this is the variance:

$$\sigma^2 = \frac{d_1^2 + d_2^2 + \cdots + d_n^2}{n} \quad \text{i.e.} \quad \sigma^2 = \frac{\sum\limits_{i=1}^{n} (x_i - \mu)^2}{n} \tag{9.4}$$

(e) Take the square root to give the standard deviation σ.

Expanding (9.4) gives another formula for σ^2:

$$\sigma^2 = \frac{\sum_{i=1}^{n}(x_i^2 - 2x_i\mu + \mu^2)}{n}$$

$$= \frac{\left(\sum_{i=1}^{n}x_i^2\right) - 2\mu(x_1 + x_2 + \cdots + x_n) + n\mu^2}{n}$$

$$= \frac{\left(\sum_{i=1}^{n}x_i^2\right) - 2n\mu^2 + n\mu^2}{n} \qquad \left(\text{since } n\mu = \sum_{i=1}^{n}x_i\right)$$

i.e. $\qquad \sigma^2 = \dfrac{\sum_{i=1}^{n}x_i^2 - n\mu^2}{n} = \dfrac{1}{n}\sum_{i=1}^{n}x_i^2 - \mu^2 \qquad\qquad$ (9.5)

We remember this as *the mean of the squares minus the square of the mean.*

Equation (9.5) is more convenient to use for computational purposes when calculating the variance, but the two are totally equivalent.

If the data is grouped into m classes then

$$\sigma^2 = \frac{\sum_{i=1}^{n}f_i y_i^2 - n\mu^2}{\sum_{i=1}^{m}f_i} \qquad\qquad (9.6)$$

Example

The data in Table 9.5 represent the weight of a number of crates. Calculate the standard deviation of the weights given that $\mu = 67.45$ kg.

Table 9.5

Mass (kg)	x_i	$x_i - \mu$	$(x_i - \mu)^2$	f_i	$f_i(x_i - \mu)^2$
60–62	61	−6.45	41.6025	5	208.0125
63–65	64	−3.45	11.9025	18	214.2450
66–68	67	−0.45	0.2025	42	8.5050
69–71	70	2.55	6.5025	27	175.5675
72–74	73	5.55	30.8025	8	246.4200
				100	852.7500

Method 1

First $\quad \sigma^2 = \dfrac{\sum f_i(x_i - \mu)^2}{\sum f_i} = \dfrac{852.75}{100} = 8.5275 \text{ kg}^2$

then $\quad \sigma = 2.92 \text{ kg}$

Method 2

Using equation (9.5)

$\sum f_i x_i^2 = 455\,803, \qquad \mu = 67.45, \qquad \mu^2 = 4549.5025, \qquad n\mu^2 = 454\,950.25$

Then $\quad \sigma^2 \dfrac{455\,803 - 454\,950.25}{100} = 8.5275 \text{ kg}^2$

therefore $\sigma = 2.92 \text{ kg}$ ∎

The standard deviation is a common measure of variation. To develop an understanding of its meaning, you should compute the standard deviation for various sets of data. Of course, the standard deviation is always positive or zero. If the readings are close together the standard deviation will be small, but if the readings are spread out the standard deviation will be large.

The difference between σ^2 and s^2 (or σ and s) for population and sample variance is as follows:

$$\sigma^2 = \frac{\sum\limits_{i=1}^{n} (x_i - \mu)^2}{n} \qquad s^2 = \frac{\sum\limits_{i=1}^{n} (x_i - \bar{x})^2}{n-1}$$

where μ and $\bar{x}$ are respectively the population and sample means. There are equivalent grouped frequency forms as well. The $n-1$ divisor removes bias in a sample and is informally, though not rigorously, justified by arguing that estimating the sample mean, using $\bar{x}$, takes up one degree of freedom. Most calculators refer to σ and s as σ_n and σ_{n-1}, depending upon the divisor used. For large values of n the difference is small and for the current example

$$\sigma_n = 2.920, \qquad \sigma_{n-1} = 2.935$$

Fractiles are extensions of the median and divide the data into subsets of equal size. They are extensively used in economic and social statistics. The distribution of data according to range of value is split into fractions of the distribution lying under the frequency curve. The most commonly used forms are as follows:

Quartile: in Figure 9.6 the data is grouped into four subsets of equal size. Note that the first quartile is also called the **lower quartile**, Q1, and the third quartile is called the

upper quartile, Q3. Quartiles split into the area under the curve into quarters. The midway mark, or ordinate, i.e. median separates the second and third quarters; the convention is to take first to second to third to fourth, from left to right.

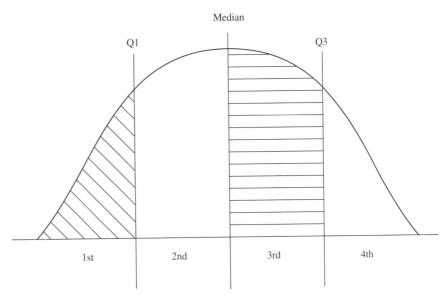

Figure 9.6 Quartiles of a distribution

Percentiles and deciles: deciles split the area under the curve into 10 equal parts; the first decile, second decile, etc., are taken from left to right. The range marks separating them are often called the 10 percentile, 20 percentile, etc. In like fashion the ranges separating the first and second and the third and fourth quarters are called the 25th percentile (written 25% ile) and the 75th percentile. The range of values between them is called the **interquartile range**. As you have doubtless guessed, **percentiles** divide the area under the curve into one hundred parts.

Example

Find the quartile and decile range marks for the examination results in Table 9.3. What is the nature of the skewness?

(a) As $\sum f_i = 100$, there must be 25 candidates in each quartile. Twenty-five candidates obtained 42 or less with six obtaining 42. So 42 is the lower quartile. The median, or 50th percentile, is 51, known already.

   Twenty-six candidates obtained 60 or more with nine obtaining 60 itself. So 60 is the upper quartile.

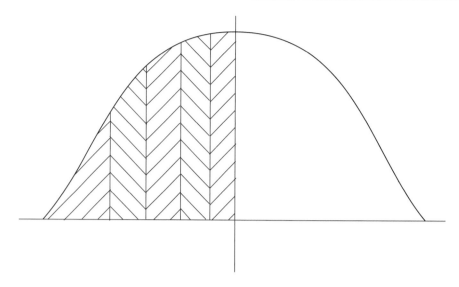

Figure 9.7 Deciles of a distribution

(b) The 10th percentile, 20th percentile, etc., obtained from the data are

$$39, 40, 45, 49, 51, 51, 55, 60, 71$$

For example, in the sixth decile (10 students) all obtained 51 and the top 10% obtained 71 or more.

(c) As the median is less than the mean, i.e. $51 < 51.73$, the distribution is positively skewed, though only slightly. ■

Note that in the example the calculation of the percentiles was straightforward. Should percentiles lie within class intervals their precise location should be found by apportioning the appropriate frequency proportions on either side.

Exercise 9.1

 1 The number N of cars parked in a private inner-city car park, maximum capacity 100, was observed at four-hour intervals during the day for a seven-day period. The observations are given below.

	Mon	Tue	Wed	Thu	Fri	Sat	Sun
0800	19	34	32	37	46	5	0
1200	86	91	93	94	81	13	5
1600	71	59	85	90	61	6	2
2000	17	11	9	14	10	0	0

Assume that each observation represents a class mark value of the average number of cars parked in four-hour intervals over the period 0600–2200 for seven days and draw a pie chart to represent the proportions of time in which the car park is 0–20%, 20–40%, 40–60% and 80–100% full.

2 The table below represents the distribution of weights to the nearest 1 kg of a sample of 1000 boxes handled by a delivery firm.

Weight	Class mark	Frequency
28–31	29.5	1
32–35	33.5	14
36–39	37.5	56
40–43	41.5	172
44–47	45.5	245
48–51	49.5	263
52–55	53.5	156
56–59	57.5	67
60–63	61.5	23
64–67	65.5	3

(a) Draw a histogram of frequency against weight and mark in the frequency curve.

(b) Determine the cumulative frequencies and plot the cumulative frequency polygon.

3 These data items were obtained by measuring the frequencies (in kilohertz) of 60 tuned circuits. Construct a frequency distribution for the classes 12.24–12.26, 12.27–12.29 kHz, etc., hence draw a histogram to represent this distribution.

12.37	12.29	12.40	12.41	12.31	12.35	12.37	12.35	12.33
12.36	12.32	12.36	12.40	12.38	12.33	12.35	12.30	12.30
12.34	12.39	12.43	12.32	12.27	12.32	12.41	12.40	12.37
12.40	12.35	12.34	12.38	12.43	12.36	12.35	12.26	12.28
12.36	12.24	12.42	12.39	12.44	12.42	12.28	12.25	12.34
12.33	12.32	12.39	12.38	12.27	12.35	12.35	12.34	12.36
12.36	12.32	12.31	12.35	12.29	12.30			

4 For the weights of the 1000 boxes in Question 2, determine the sample mean and standard deviation ($\bar{x}$ and s). Determine also the quartile range marks, interpolating as necessary.

5 Calculate the mean and standard deviation for the following frequency distribution:

Diameter (mm)	Frequency
11.46	1
11.47	4
11.48	12
11.49	15
11.50	11
11.51	6
11.52	3
11.53	1

6 The table shows a frequency distribution for the lifetime of cathode-ray tubes. Calculate the mean and standard deviation.

Lifetime (hours)	400–499	500–599	600–699	700–799	800–899
Frequency	14	50	82	46	8

7 The table below shows the distribution of maximum loads supported by certain cables manufactured by a steel wire company.

Maximum load (kN)	Number of cables
84–88	4
89–93	10
94–98	24
99–103	34
104–108	28
109–113	12
114–118	6
119–123	2

Calculate the mean and standard deviation for this distribution, and the quartile range marks.

8   The following table shows the age distribution of heads of families in the United States during the year 1957.

(a)   Find the median age.

(b)   Why is the median a more suitable measure of central tendency than the mean in this case?

Age of head of family (years)	Number (in millions)
Under 25	2.22
25–29	4.05
30–34	5.08
35–44	10.45
45–54	9.47
55–64	6.63
65–74	4.16
75 and over	1.66
Total	43.72

9   Find the mean, median and mode for these sets of numbers:

(a)   7, 4, 10, 9, 15, 12, 7, 9, 7

(b)   8, 11, 4, 3, 2, 5, 10, 6, 4, 1, 10, 8, 12, 6, 5, 7

10   The table below shows a frequency distribution of examination marks.

Mark	Number of students
90–100	9
80–89	32
70–79	43
60–69	21
50–59	11
40–49	3
30–39	1
Total	120

(a)   Find the quartiles of the distribution.

(b)   Interpret clearly the significance of each.

11 Find

(a) the second decile (b) the fourth decile

(c) the 90th percentile (d) the 68th percentile

for the data of Question 7, intepreting clearly the significance of each.

9.2 PROBABILITY THEORY AND THE MODELLING OF UNCERTAINTY

Probability theory is concerned with the modelling of real-life situations in which an action or experiment takes place with an outcome that is uncertain. Such an experiment is called a **random experiment**. Games of chance are classic examples of random experiments. The outcome of throwing a die is not known with certainty before the throw. Firing a rocket is an example of performing a random experiment; it could result in success or failure.

Associated with any random experiment is a set S of all its possible outcomes; this set S is called the **sample space** of the random experiment. Each outcome in a sample space S is called a **sample point**. An **event** is a subset of a sample space and contains those sample points for which the event occurs.

Examples

1. Three items are selected at random from the output of a manufacturing process. Each item is inspected and classified defective (D) or non-defective (N). The sample space is

$$S = \{\text{NNN, NDN, DNN, NND, DDN, DND, NDD, DDD}\}$$

where, for example, DNN means that the first item is defective, the second and third items are non-defective.

Let A be the event 'the number of defectives is greater than one', then

$$A = \{\text{DDN, DND, NDD, DDD}\}$$

2. An electronic component is placed on test and we are interested in the time to failure of the component in hours. The sample space is $S = \{t: 0 \leq t < \infty\}$.

Let A be the event 'the component fails before 5 hours have elapsed', then

$$A = \{t: 0 \leq t < 5\}$$

3. An experiment consists of tossing a coin then tossing it a second time if a head (H) occurs on the first toss. If a tail (T) occurs on the first toss, a die is thrown. Let (T,2) represent a tail and a 2, etc. The sample space is

$$S = \{(H, H), (H, T), (T, 1), (T, 2), (T, 3), (T, 4), (T, 5), (T, 6)\}$$

Let A be the event 'a number less than 4 occurred on the die', then

$$A = \{(T, 1), (T, 2), (T, 3)\}$$

the die is not thrown if a head appears on the first throw.   ■

Operations on events

The **intersection** of two events A and B, denoted by the symbol $A \cap B$, is the event containing all the sample points that are common to both A and B.

The **union** of two events A and B, denoted by the symbol $A \cup B$, is the event containing all the sample points that belong to A or to B or to both.

The **complement** of an event A with respect to a sample space S is the set of all sample points of S that are not in A. Then complement of A is denoted by A'.

These new events correspond to the ideas of 'and', 'or' and 'not' respectively; see Figure 9.8. The pictorial representations are known as **Venn diagrams**.

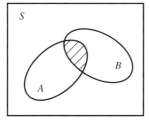
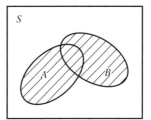
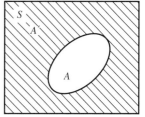

Figure 9.8 Venn diagram: the shaded areas indicate (a) $A \cap B$, intersection, (b) $A \cup B$, union and (c) A', complement

Example   An electric circuit with two components R_1 and R_2 in series can function only when both R_1 and R_2 function (Figure 9.9).

Figure 9.9 Series electric circuit

We assume that R_1, R_2 can be in two possible states, 1 or 0, denoting operative or defective respectively. The sample space S for the circuit is

$$S = \{(1, 1), (0, 1), (1, 0), (0, 0)\}.$$

If E_1 is the event 'the entire circuit is operative' then $E_1 = \{(1, 1)\}$ and if E_2 is the event 'at least one of R_1 and R_2 is operative' then $E_2 = \{(1, 1), (0, 1), (1, 0)\}$.

We notice that E_1 and E_2 are *not* mutually exclusive since the sample point $(1, 1)$ belongs to both events. ∎

Example   An electronic component is placed on test. Let A be the event 'the component fails before 5 hours' and B the event 'the component fails before 10 hours'. Now the sample space is $S = \{t: 0 \leq t < \infty\}$. Also

$$A \cap B = \{t: 0 \leq t < 5\}$$
$$A \cup B = \{t: 0 \leq t < 10\}$$
$$A' = \{t: 5 \leq t < \infty\} \qquad B' = \{t: 10 \leq t < \infty\}$$

Note that $A \subset B$, i.e. A is a subset of B and $B' \subset A'$. ∎

Mutually exclusive events

In certain statistical experiments we may define two events A and B that cannot occur simultaneously. The events A and B are then said to be **mutually exclusive** and have no sample points in common, i.e. $A \cap B = \varnothing$, the **null set**.

Example   Suppose a die is thrown, let A be the event 'an even number is face up' and B is the event 'an odd number is face up'. Then $A = \{2, 3, 6\}$, $B = \{1, 3, 5\}$ and $A \cap B = \varnothing$, the null set, hence A and B are mutually exclusive events. ∎

Probability of an event

Historically, the probability of an event has been explained by one of three methods: the relative frequency method, the equally likely outcome method and the subjective degree-of-belief method. The results obtained by these methods have led to the formulation of an axiomatic definition of probability which avoids many of the pitfalls inherent in these historical methods.

The mathematical theory of probability for finite sample spaces provides a set of numbers called **weights**, ranging from 0 to 1, which are a means of evaluating the likelihood of occurrence for events resulting from a statistical experiment. To every point in the sample space we assign a weight such that the sum of all the weights is 1.

If we have a reason to believe that a certain sample point is quite likely to occur when the experiment is conducted, the weight assigned should be close to 1. On the other hand, a weight closer to zero is assigned to a sample point that is unlikely to occur. In many experiments all the sample points have the same chance of occurring and are assigned equal weights.

The **probability of an event** A is the **sum of the weights of all sample points in** A. If we denote the probability of event A by $P(A)$ then

$$0 \le P(A) \le 1 \qquad P(\varnothing) = 0 \qquad P(S) = 1$$

Uncertainty is therefore quantified on the range $[0, 1]$ between impossibility, 0, and certainty, 1.

Example

A die is biased in such a way that an even number is twice as likely to occur as an odd number. If E is the event that a number less than 4 occurs on a single throw of the die, find $P(E)$.

The sample space $S = \{1, 2, 3, 4, 5, 6\}$ and we assign a weight w to the odd numbers and $2w$ to the even numbers. Since the sum of the weights, $9w$, must total 1, then $w = \dfrac{1}{9}$.

Now $E = \{1, 2, 3\}$ so

$$P(E) = \frac{1}{9} + \frac{2}{9} + \frac{1}{9} = \frac{4}{9}$$

∎

Equally likely outcomes

If an experiment is of such a nature that we can assume equal weights for the sample points of S, then the probability of any event A is the ratio of the number of elements in A to the number of elements in S.

Example

A bag contains the five tags marked with $1, 2, 3, 4$ and 5. Two tags are drawn at random, the first tag being replaced before the second is drawn. Find the probabilities of the events $A =$ 'the same tag is drawn twice' and $B =$ 'the second number drawn is greater than the first number drawn'.

The sample space contains the 25 ordered pairs (i, j), $1 \le i \le 5$, $1 \le j \le 5$, where i indicates the first number drawn and j indicates the second number drawn. So each ordered pair is equally likely to occur with probability $\dfrac{1}{25}$.

Now $A = \{(1.1), (2, 2), (3, 3), (4, 4), (5, 5)\}$ so that $P(A) = \dfrac{5}{25}$

and $B = \{(1, 2), (1, 3), (1, 4), (1, 5), (2, 3), (2, 4), (2, 5), (3, 4), (3, 5), (4, 5)\}$

so that $P(B) = \dfrac{10}{25}$ ■

Figure 9.10 illustrates the events.

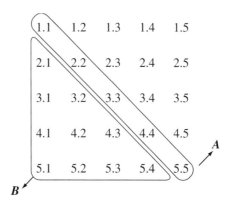

Figure 9.10 Drawing two tags from a bag of five tags, with replacement

Probability laws

We can calculate the probability of the union of two events A and B knowing the individual probabilities of A and B.

If A and B are any two events in a sample space S then

$$P(A \cup B) = P(A) + P(B) - P(A \cap B) \tag{9.7}$$

To indicate the proof of this result we use a Venn diagram (Figure 9.11). Remember that $P(A \cup B)$ is the sum of the weights of the sample points in $A \cup B$. Now $P(A) + P(B)$ is the sum of all the weights in A plus the sum of all the weights in B. We have added the weights in $A \cap B$ twice. Hence $P(A \cup B) = P(A) + P(B) - P(A \cap B)$.

Corollary 1   If A and B are mutually exclusive, i.e. $A \cap B = \varnothing$ so that $P(A \cap B) = 0$ then

$$P(A \cup B) = P(A) + P(B) \tag{9.8}$$

Hence for the union of exclusive events we add probabilities.

Example   Calculate the probability of getting a total of 7 or 11 when a pair of dice are thrown.

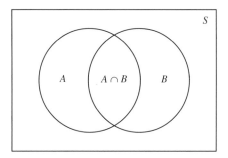

Figure 9.11   Venn diagram for the union of two events

Let A be the event that 7 occurs, then $A = \{(1, 6), (2, 5), (3, 4), (4, 3), (5, 2), (6, 1)\}$. Let B be the event that 11 occurs, then $B = \{(5, 6), (6, 5)\}$. The sample space S has 36 equally likely sample points, so

$$P(A) = \frac{6}{36} \qquad P(B) = \frac{2}{36} \qquad P(A \cap B) = 0$$

Then $P(A \cup B) = P(A) + P(B)$ since A and B are mutually exclusive events. Therefore

$$P(A \cup B) = \frac{8}{36} = \frac{2}{9} \qquad\qquad\qquad\blacksquare$$

Corollary 2   If A and A' are complementary events then

$$P(A') = 1 - P(A) \tag{9.9}$$

Since $S = A \cup A'$ and $A \cap A' = \varnothing$

then $P(S) = P(A) + P(A')$

but $P(S) = 1$ so $P(A') = 1 - P(A)$

In the last example the probability of getting a total other than 7 or 11 is $1 - \dfrac{2}{9} = \dfrac{7}{9}$

Conditional probability and independence

The probability of an event B occurring when it is known that some other event A has already occurred is called a **conditional probability** and is denoted by $P(B/A)$, read as 'the probability of B given A'.

Consider the event B of getting a score which is perfect square when a biased die is tossed – the even numbers on the die being twice as likely to occur as the odd numbers.
Now

$$B = \{1, 4\} \quad \text{and} \quad P(B) = \frac{1}{9} + \frac{2}{9} = \frac{1}{3}$$

The calculation may be understood using a **probability tree**, showing the probability of all individual events (Figure 9.12).

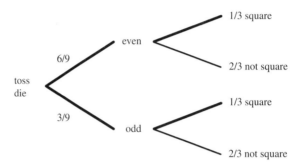

Figure 9.12   A probability tree

Summing the probabilities along the branches in bold, we obtain the probability of a square

$$P(B) = \left(\frac{6}{9} \times \frac{1}{3} \right) + \left(\frac{3}{9} \times \frac{1}{3} \right) = \frac{2}{9} + \frac{1}{9} = \frac{1}{3}$$

Suppose it is known that the toss of the die resulted in a number greater than 3. We are now dealing with a reduced sample space $A = \{4, 5, 6\}$, which is a subset of the sample space $S = \{1, 2, 3, 4, 5, 6\}$.

To find the probability that B occurs relative to the space A, we need to assign new weights to the elements of A in proportion to their original weights such that their sum is 1.

Assigning a weight w to the odd number in A and a weight $2w$ to each of the two even numbers then $w = \frac{1}{5}$ (Figure 9.13). The event $B/A = \{4\}$, $P(B/A) = \frac{2}{5}$ and $P(A \cap B) = \frac{2}{9}$

This example illustrates that events may have different probabilities when considered relative to different sample spaces. We note that

$$P(B/A) = \frac{2}{9} \div \frac{8}{9}$$

where $P(A)$ and $P(A \cap B)$ are found from the original sample space S.

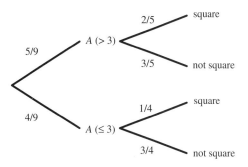

Figure 9.13 Conditional probability

The **conditional probability** of B, given A, denoted $P(B/A)$ is defined by

$$P(B/A) = \frac{P(A \cap B)}{P(A)}$$

(9.10)

given that $P(A) > 0$.
  We may also write

$$P(A \cap B) = P(A) \times P(B/A) = P(B) \times P(A/B) = P(B \cap A)$$

In other words, the probability of the joint event A and B, or B and A, is the product of the probability of one of them occurring and the probability of the other occurring given that the first has already occurred. The result may be summarised using a probability tree (Figure 9.14).

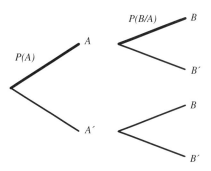

Figure 9.14 Probability tree for conditional probability

Multiplying along the branches in bold gives

$$P(A \cap B) = P(A) \times P(B/A)$$

Examples

1. A company owns two factories that produce similar items. Factory 1 produces 1000 items, 100 of which are defective, and Factory 2 produces 4000 items, 200 of which are defective. An item is chosen at random from the production of the company and found to be defective. What is the probability that it came from Factory 1?

Let A be the event 'the item chosen is defective' and B the event 'the item chosen came from Factory 1'. We require $P(B/A)$. Now

$$P(A \cap B) = P \text{ (item chosen is defective and from Factory 1)} = \frac{100}{5000}$$

$$\text{and } P(A) = P \text{ (item chosen is defective)} = \frac{300}{5000}$$

Then $\quad P(B/A) = \dfrac{P(A \cap B)}{P(A)} = \dfrac{1}{3}$

Alternatively, the calculation may be performed using a probability tree (Figure 9.15)

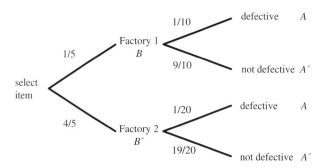

Figure 9.15 Probability tree for the two factories

Again, $\quad P(B/A) = \dfrac{P(A \cap B)}{P(A)}$

Multiplying probabilities along the relevant branches gives

$$P(B/A) = \frac{\dfrac{1}{10} \times \dfrac{1}{5}}{\left(\dfrac{1}{10} \times \dfrac{1}{5}\right) + \left(\dfrac{1}{20} \times \dfrac{4}{5}\right)} = \frac{1}{50} \times \frac{50}{3} = \frac{1}{3}$$

2. A fuse box contains 20 fuses of which 5 are defective. If 2 fuses are selected at random, without replacement, what is the probability that both fuses are defective?

   Let A be the event 'the first fuse is defective'; and B the event 'the second fuse is defective'. $A \cap B$ represents the joint event that both fuses are defective.

   Since the selection of a defective fuse leaves a box of 19 fuses of which 4 are defective, then

$$P(A) = \frac{5}{20} = \frac{1}{4} \quad \text{and} \quad P(B/A) = \frac{4}{19}$$

Hence

$$P(A \cap B) = \left(\frac{1}{4}\right) \times \left(\frac{4}{19}\right) = \frac{1}{19}$$

If, in the second example, the first fuse is replaced and the fuses thoroughly rearranged before the second fuse is removed, the probability of drawing a defective fuse on the second selection is exactly as it was on the first selection, i.e. $\frac{1}{4}$

In this case $P(B/A) = P(B)$, so the events are **independent**. In general, the events A and B are independent if and only if

$$P(A \cap B) = P(A) \times P(B) \tag{9.11}$$

Hence for independent events we multiply probabilities.

In the second example with replacement

$$P(A \cap B) = P(A) \times P(B) = \frac{1}{4} \times \frac{1}{4} = \frac{1}{16}$$

This is slightly greater than without replacement, as you might expect.   ∎

Exercise 9.2

1 An experiment involves throwing a pair of dice and recording the numbers that are face up.

   (a)   List the elements of the sample space S.

   (b)   List the elements of event A 'the sum is less than 5'.

   (c)   List the elements of event B 'a 6 occurs on at least one die'.

   (d)   List the elements of event C 'a 2 is face up on the first die'.

2 Let the sample space S consist of all real numbers and define the events

$$A = \{x: x > 0\} \qquad B = (x: -1 < x < 2\} \qquad C = \{x: x < 0\}$$

(a) Describe and sketch on the real line the following events:

$$A \cup C, \qquad A \cap C, \qquad A', \qquad B \cap C, \qquad A \cap B', \qquad A \cup B$$

(b) Express the following events in terms of A, B and C

$$D = \{x: 0 < x < 2\}, \qquad E = \{x: x = 0\}, \qquad F = \{x: x \leq -1\}$$

3 An assembly of electronic equipment consists of three components arranged in the illustrated series–parallel circuit. Each component is either operative or fails under load. The entire assembly fails only if the path from A to B is broken. Let the sample space S consists of the eight possible combinations of operative or inoperative components. Let E_1 be the event 'the assembly is operative'; let E_2 be the event 'R_2 has failed but the assembly is operative'; let E_3 be the event 'R_3 has failed but the assembly is operative'.

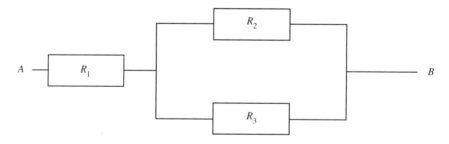

(a) List the sample points of S, E_1, E_2 and E_3.

(b) Investigate whether E_1, E_2 and E_3 are mutually exclusive.

4* A coin is tossed until a head is obtained. Describe the appropriate sample space. If each point in the sample space requiring n tosses has probability $\left(\dfrac{1}{2}\right)^n$ what is the probability that the first head is obtained on an even-numbered toss?

5 A fair coin is tossed five times and each outcome is equally likely. What is the probability of obtaining a sequence of at least three consecutive heads?

6   A coin is tossed and a die rolled. Calculate the probabilities of

   (a)   a tail and a 5   (b)   a head and an even number.

7   What is the probability of rolling two dice to obtain the sum 7 and/or the number 3 on at least one die?

8   A box contains five white, three red and four blue balls. Three are drawn at random without replacement. Find the probability that:

   (a)   no ball is red   (b)   exactly one is red

   (c)   at least one is blue   (d)   no two are the same colour.

9   Three female and two male candidates for a public appointment are short-listed for interview. The interviewers know the candidates by their surname only and interview them individually in random order. Calculate the probabilities of the following events:

   (a)   the first interviewee is female

   (b)   the second interviewee is female given that the first is male

   (c)   the first two interviewees are of different gender.

10   In a noisy communications system transmitting binary digits (0's and 1's) the sender may send a digit (e.g. 1) and the receiver may receive the other digit (0). Let A be the event 'a 1 is sent' and let B be the event 'a 1 is received'. Hence A' is the event 'a 0 is sent' and B' is the event 'a 0 is received'.

   Assume that $P(B'/A) = 0.01, P(B/A') = 0.01$ and $P(A) = 0.5$, then determine the following probabilities:

   (a) $P(A/B)$ (b) $P(A'/B)$

   (c) $P(A/B')$ (d) $P(A'/B')$

These conditional probabilities are useful to the receiver in interpreting the incoming messages.

11   Modern electronic systems often comprise durable components connected in series or parallel. The **reliability** (i.e. survival probability) of two completely independent components connected in this way obeys one of the laws illustrated in the figure.

Two systems (a) and (b) are each made up from four components whose reliabilities over 3000 hours running are shown. Calculate the reliability of each system as a whole over 3000 hours running.

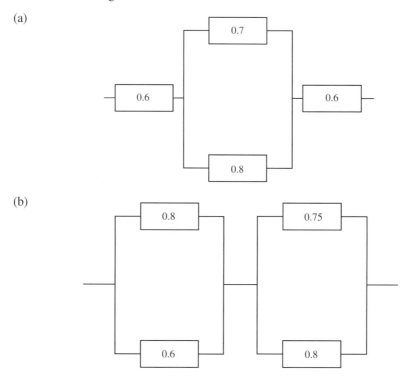

12 A factory worker travels to work either by bus or by bicycle depending upon weather and season. He uses the bus one-third of the time and the bicycle two-thirds of the time. When going by bus he arrives home by 6 pm 75% of the time and when cycling he is home by 6 pm 70% of the time. On a particular evening he is home after 6 pm; what is the probability that he cycled?

13 One per cent of a population suffer from a certain disease. A new diagnostic test gives a positive indication 97% of the time when an individual has the disease, and a negative response 95% of the time if an individual does not have the disease. A person selected at random reacts positively to the test. What is the probability that this person actually has the disease? Calculate the probability that the test yields the correct diagnosis for an individual chosen at random.

14 A lie detector given to a suspect is known to be 90% reliable when the person is guilty and 99% reliable when the person is innocent. If a suspect is selected from a group of which only 5% have ever committed a crime and the detector indicates guilt, what is the probability that the suspect is innocent?

9.3 DISCRETE RANDOM VARIABLES

In the context of data and statistics a **random variable** is a measurable attribute or number X whose value cannot be predicted before an experiment to measure it. When applied to random experiments or theoretical models, X is a real number assigned to each point in the sample space S.

Consider the random experiment of tossing a coin twice. The sample space is

$$S = \{HH, HT, TH, TT\}$$

A random variable X can be defined for each element of S as the number of heads achieved. Hence

$$X\,(\mathrm{HH}) = 2, \qquad X\,(\mathrm{HT}) = 1, \qquad X\,(\mathrm{TH}) = 1, \qquad X\,(\mathrm{TT}) = 0$$

The values which X can take are $\{0, 1, 2\}$, a discrete or countable set, so that X is a **discrete random variable**. We use the word 'variable' because X can take more than one value, the word 'random' because we cannot be certain which value X will take before the coin is tossed and the word 'discrete' because the values it takes form a discrete, or countable, set.

Now consider the time to failure, X, of piece of equipment; this is a **continuous random variable**, since the set of values it can take, i.e. any value greater than or equal to zero, is infinite and uncountable.

In both cases we will use X for the name of the random variable and x for the values it takes.

Discrete probability distributions

Let X take the values $x_1, \ldots, x_n$, with probabilities

$$P(X = x_1) = p_1 \qquad P(X = x_2) = p_2 \quad \ldots \quad P(X = x_n) = p_n$$

where $p_i \geq 0$, for each i. Then $\sum_{i=1}^{n} p_i = 1$, and n is either finite or infinite. The set of values of X and the probabilities associated with them define a **discrete probability distribution**. We now consider some of the most important examples.

The binomial distribution

The binomial distribution arises when a random experiment, or trial, in which the outcomes can be classified under two headings, e.g. success and failure, is repeated a fixed number of times. The probability of success on a particular trial is assumed to be constant. Each of the trials is assumed to be independent of the others.

The sample spaces of outcomes for the number of trials $n = 1$ to 3 are shown in Table 9.6 with X being the number of successes in each case.

Table 9.6

n	Sample space	X
1	S, F	$1, 0$
2	SS, SF, FS, FF	$2, 1, 1, 0$
3	$SSS, SSF, SFS, FSS,$	$3, 2, 2, 2,$
	FFS, FSF, SFF, FFF	$1, 1, 1, 0$

If we let $P(S) = p$ and $P(F) = q = 1 - p$, then $P(SF) = pq$, $P(FSF) = qpq = pq^2$, etc. Furthermore $P(X = x)$ and $X = x$ corrrespond uniquely, as shown in Figure 9.16.

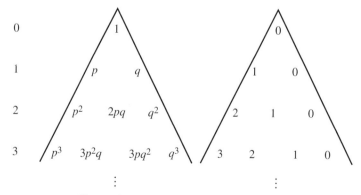

Figure 9.16 Binomial probabilities

You will notice that coefficients in the left-hand triangle are those of Pascal's triangle and that the sum of the probabilities in any row is 1, e.g. for $n = 3$ we have

$$
\begin{array}{ccccccc}
p^3 & + & 3p^2q & + & 3pq^2 & + & q^3 = (p+q)^3 = 1^3 = 1 \\
\| & & \| & & \| & & \| \\
P(X = 3) & + & P(X = 2) & + & P(X = 1) & + & P(X = 0) = 1
\end{array}
$$

It is because of this property, which operates irrespective of n, that the binomial distribution gets its name. The applications for it are considerable.

Examples

1. A production line supervisor tests video cassette recorders in batches of 10 each day. Production records show that 5% of all the VCRs produced have at least some small fault.

   (a) What is the probability that on a given day an entire batch passes the tests?
   (b) If more than two defectives are found in a day's batch, the whole batch is rejected. What is the probability of this happening?

   Let p denote the probability of an item being defective and let n denote the number of trials. Then $p = 0.05$ and $n = 10$. Let X denote the number of defective items in a batch.

   (a) $P(X = 0) = (1 - p)^{10} = (0.95)^{10} = 0.5987$
   The entire batch passes the testing procedure on approximately 3 days per working week of 5 days.

   (b) P (rejection of the batch) $= P(X > 2)$

   $$
   = 1 - \{P(X = 0) + P(X = 1) + P(X = 2)\}
   $$

   $$
   P(X = 1) = 10p(1 - p)^9 = 10 \times 0.05 \times (0.95)^9 = 0.3151
   $$

   $$
   P(X = 2) = \frac{10 \times 9}{2} \times p^2 \times (1 - p)^8
   $$

   $$
   = \frac{10 \times 9}{2} \times (0.05)^2 \times (0.95)^8 = 0.0746
   $$

   so that $P(X > 2) = 0.0116$, or just over 1% of batches are rejected outright.

2. The Stalin Organ used by the Red Army in World War II consisted of parallel pipes, like organ pipes through each of which an identical rocket was fired at a target. If each rocket had a probability of 60% of achieving a hit, how many pipes would the organ need to be 95% certain of hitting its target?

Each rocket firing is a trial and the number of trials is equal to the number of pipes. Call this n. Let p be the probability that a rocket hits, i.e. $p = 0.6$ and let X be the number of hits. The organ system fails to hit the target only when all rockets miss. Therefore

$$P \text{ (at least one hit)} = 1 - P \text{ (all rockets miss)}$$
$$= 1 - P(X = 0)$$
$$= 1 - (1 - p)^n \geq 0.95$$

hence $(1 - p)^n \leq 0.05$ where n is the smallest integer for this to be so.

Taking logarithms (noting that $(1 - p)^n$ decreases as n increases) gives

$$n \geq \frac{\ln 0.05}{\ln 0.4} = 3.27$$

so that four pipes per organ are needed.

The probability distribution is

$$P \text{ (0 hits)} = (0.4)^4 = 0.026$$
$$P \text{ (1 hit)} = 4 \times 0.6 \times (0.4)^3 = 0.154$$
$$P \text{ (2 hits)} = 6 \times 0.6^2 \times 0.4 = 0.346$$
$$P \text{ (3 hits)} = 4 \times 0.6^3 \times 0.4 = 0.346$$
$$P \text{ (4 hits)} = (0.6)^4 = 0.130$$

The line diagram for the distribution is shown in Figure 9.17. ■

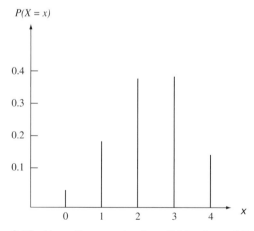

Figure 9.17 Line diagram for the distribution of Example 2

For the general binomial distribution X takes the values $0, 1, 2, \ldots, n$ and the sample point corresponding to r successes in n trials is of the form

$$\{SFSSFFFSFSFF\ldots SFS\}$$

where there are r successes occurring with probability p^r and $(n - r)$ failure occuring with probability q^{n-r}. There are also $\binom{n}{r}$ orderings of the S's and F's, so that

$$P(X = r) = \binom{n}{r} p^r q^{n-r} \qquad (9.12)$$

Furthermore, $\sum_{r=0}^{n} P(X = r) = (p + q)^n = 1$, where $P(X = r)$ is a typical term in the binomial expansion of $(p + q)^n$.

The geometric distribution

Another discrete distribution of interest is where a sequence of independent success and failure trials (i.e. binomial) are conducted but the trials stop as soon as a success occurs. The sample space is easy to write down

$$\{S\} \qquad \text{1 trial} \qquad \text{success first time}$$
$$\{FS\} \qquad \text{2 trials} \qquad \text{success second time}$$
$$\{FFS\} \qquad \text{3 trials} \qquad \text{success third time}$$
$$\vdots$$
$$\{FFF\ldots FS\} \quad n \text{ trials} \quad \text{success } n\text{th time}$$

In theory the trials could go on forever.

Let $P(S) = p, P(F) = 1 - p = q$, as before. Then

$$P(X = 1) = p$$
$$P(X = 2) = pq$$
$$P(X = 3) = pq^2$$
$$\vdots$$
$$P(X = n) = pq^{n-1}$$

The total probability is given by

$$\sum_{r=1}^{\infty} P(X = r) = \sum_{r=1}^{\infty} pq^{r-1} = \frac{p}{1 - q} = \frac{p}{p} = 1$$

using for formula for the sum of a geometric series, hence the name of the distribution.

Example

Three players A, B, C throw a die before starting a board game. A throws first, followed by B then C, then A again, etc, and the first to throw a six starts the game. What are the respective probabilities of A, B and C starting the game?

The throwing of the die follows a geometric distribution with

S: a six is thrown F: a six is not thrown

So $P(S) = \dfrac{1}{6}$ $P(F) = \dfrac{5}{6}$

The throwing ceases as soon as a six appears.

If X is the random variable and $X = n$ denotes a six being thrown on the nth occasion, then A starts the game if the six occurs on the first throw, or the fourth throw, or the seventh throw, etc.

B starts the game if the six occurs on the second throw, or the fifth throw, or the eighth throw, etc.

C starts the game if the six occurs on the third throw, or the sixth throw, or the ninth throw, etc.

In summary, if $P(A)$ is the probability that A starts the game, etc., then

$$P(A): P(X = 1) + P(X = 4) + P(X = 7) + \ldots$$
$$P(B): P(X = 2) + P(X = 5) + P(X = 8) + \ldots$$
$$P(C): P(X = 3) + P(X = 6) + P(X = 9) + \ldots$$

Now the probability that A throws a six on the first throw is $\dfrac{1}{6}$. For A to throw the six on the fourth throw, A, B and C must fail to throw a six at their first attempts then A must succeed at the second attempt. The probability of this occuring is

$$\frac{5}{6} \times \frac{5}{6} \times \frac{5}{6} \times \frac{1}{6} = \left(\frac{5}{6}\right)^3 \times \frac{1}{6} = \frac{1}{6} \times \left(\frac{5}{6}\right)^3$$

Similarly, for A to throw the six on the seventh throw, A, B and C must fail twice and A must succeed at the third attempt; the probability of this is

$$\frac{1}{6} \times \left(\frac{5}{6}\right)^6$$

Therefore

$$P(A) = \frac{1}{6}\left(1 + \left(\frac{5}{6}\right)^3 + \left(\frac{5}{6}\right)^6 + \ldots\right)$$

Similar arguments show that

$$P(B) = \frac{1}{6} \times \frac{5}{6} \times \left(1 + \left(\frac{5}{6}\right)^3 + \left(\frac{5}{6}\right)^6 + \ldots\right)$$

and that

$$P(C) = \frac{1}{6} \times \left(\frac{5}{6}\right)^2 \times \left(1 + \left(\frac{5}{6}\right)^3 + \left(\frac{5}{6}\right)^6 + \ldots\right)$$

You could sum the geometric series in each case, but note that the ratios $P(A) : P(B) : P(C)$ are

$$1 : \frac{5}{6} : \frac{25}{36} \quad \text{i.e.} \quad \frac{36}{36} : \frac{30}{36} : \frac{25}{36} \quad \text{or} \quad 36 : 30 : 25$$

and since one of A, B and C must win eventually then $P(A) + P(B) + P(C) = 1$, so

$$P(A) = \frac{36}{91} \qquad P(B) = \frac{30}{91} \qquad P(C) = \frac{25}{91} \qquad \blacksquare$$

The Poisson distribution

A very important discrete distribution is the Poisson distribution. It models rare events where the average rate at which the events occur can be assumed constant.

If a Poisson random variable X represents the number of successes occurring in a given time interval, then the probability that it takes the value x is given by

$$P(X = x) = \frac{e^{-\lambda}\lambda^x}{x!} \tag{9.13}$$

where λ is the average number of events occurring in the given interval. This is assumed to be constant.

Example

In a laboratory experiment the average number of radioactive particles passing through a counter in one millisecond is 4. The particles arrive at random. What is the probability that 6 particles enter the counter in a particular millisecond?

Using the Poisson distribution with $x = 6, \lambda = 4$ we find that

$$P(X = 6) = \frac{e^{-4} \times 4^6}{6!} = 0.1042 \qquad \blacksquare$$

The binomial and Poisson distribution are closely related in that if n is large and p is very small then the binomial probabilities can be approximated well by Poisson probabilities where $\lambda = np \sim 0(1)$.

Example

Rework the example on page 535 replacing the binomial probabilities by Poisson approximations.

The probability of a defective item is $p = 0.05$. The batch size is $n = 10$. Then $\lambda = np = 0.5$.

(a) $P(X = 0) = e^{-\lambda} = 0.6065$

(b) $P(X > 2) = 1 - \{P(X = 0) + P(X = 1) + P(X = 2)\}$

$$= 1 - \left(e^{-\lambda} + \frac{\lambda}{1!} e^{-\lambda} + \frac{\lambda^2}{2!} e^{-\lambda} \right)$$

$$= 1 - e^{-\lambda} \left(1 + \frac{\lambda}{1!} + \frac{\lambda^2}{2!} \right)$$

$$= 1 - 0.6065(1 + 0.5 + 0.125)$$

$$= 0.0144$$

The value in (a) is close to the binomial value and that in (b) is of the right order of magnitude. Table 9.7 compares the binomial and Poisson probabilities for $n = 10, p = 0.05$ and $\lambda = np = 0.5$.

Table 9.7

X	0	1	2	3
Binomial	0.5987	0.3151	0.0746	0.0100
Poisson	0.6065	0.3032	0.0758	0.0126

If we have to calculate a succession of Poisson probabilities we can use the **recurrence relation**

$$P(X = x) = \frac{e^{-\lambda} \times \lambda^x}{x!} = \frac{e^{-\lambda} \times \lambda \times \lambda^{x-1}}{x \times (x-1)!} = \frac{\lambda}{x} \times P(X = x - 1)$$

If we let p_r be an alternative notation for $P(X = r)$ then the recurrence relation may be written

$$p_r = \frac{\lambda}{r} p_{r-1} \tag{9.14}$$

Expected value and variance of discrete random variables

The **mean** or **expected value** of a discrete random variable X which takes the values $x_1, \ldots, x_n$, where n may be finite or infinite, is defined as

$$\mu = E[X] = \sum_{i=1}^{n} \{x_i \times P(X = x_i)\}$$

As with the descriptive analysis of data, the mean is a measure of central tendency.

The **variance** of X is defined to be

$$\sigma^2 = \text{Var}[X] = E[(X - \mu)^2]$$
$$= \sum_{i=1}^{n} \{(x_i - \mu)^2 \times P(X = x_i)\}$$

This measures the dispersion about the mean and, as with the analysis of data, is a measure of spread. The standard deviation σ is the square root of the variance.

Note that the variance can be calculated more simply, and often more accurately, by the formula

$$\sigma^2 = E[X^2] - \mu^2 \tag{9.15}$$

For the distributions we have looked at so far, the mean and variance are as follows:

$$\begin{aligned}
&\text{Binomial} &&\mu = np, &&\sigma^2 = npq \\
&\text{Geometric} &&\mu = \frac{1}{p}, &&\sigma^2 = \frac{q}{p^2} \\
&\text{Poisson} &&\mu = \lambda, &&\sigma^2 = \lambda
\end{aligned}$$

You can see that the mean and variance of the Poisson distribution are numerically equal. If a binomial distribution is such that n is large and p is small then np is approximately equal to npq and the Poisson distribution with $\lambda = np$ is a reasonable approximation to the binomial distribution.

Example

(a) A binomial distribution with $n = 10, p = 0.3$ has mean $\mu = 3$, i.e. with a 30% chance of success in a trial then 3 successes are expected in 10 trials.

(b) A geometric distribution. In throwing a die until a six appears, $p = \frac{1}{6}$, then we expect that over a large number of occasions six throws are required on average. ■

The proofs behind the determination of mean and variance involves series expansions. One of the more straightforward demonstrations is to find the mean of the geometric distribution.

Note first the binomial expansion

$$\frac{1}{(1-x)^2} = 1 + 2x + 3x^2 + 4x^3 + \dots \qquad |x| < 1$$

Then for the geometric distribution

$$\begin{aligned}
\mu &= \sum_{r=1}^{\infty} \left(r \times P(X = r) \right) \\
&= 1 \times p + 2 \times pq + 3 \times pq^2 + 4 \times pq^3 + \dots \\
&= p(1 + 2q + 3q^2 + 4q^3 + \dots) \\
&= \frac{p}{(1-p)^2} = \frac{p}{p^2} = \frac{1}{p}
\end{aligned}$$

The variance is harder to find, as are many of the others. More important at this stage is that you know of the results and their significance for the models which they represent.

Exercise 9.3

1 A box contains four black balls and two green balls. Three balls are drawn in succession, each ball being replaced in the box before the next draw is made. Find the probability distribution for the random variable X representing the number of green balls drawn.

2 A coin is biased so that a head is twice as likely to occur as a tail. If the coin is tossed four times, find the probability distribution for the random variable X representing the number of heads.

3 A certain class of 60 W light bulbs has a probability of 0.98 of being satisfactory. What is the probability of getting two or more defectives in a sample of size 5?

4 Packets of food are filled automatically and the proportion of packets in a large batch which are underweight is p. A sample size n is selected randomly from the batch and the probability that the sample contains exactly r defective packets ($r = 0, 1, \dots, n$) is required. Write down the probability that the sample contains exactly r defective packets. For one particular process, it has been found in the past that 2% of the packets are underweight. An inspector takes a random sample of 10 packets.

(a)   Calculate the expected number of packets in the sample which are underweight.

(b)   Calculate the probability that no packets in the sample are underweight.

(c)   Calculate the probability that two or more of the packets in the sample are underweight.

5   Suppose that the number of dust particles per unit volume in a coal mine is randomly distributed with a Poisson distribution and that the average density is μ particles per litre. A sampling apparatus collects a one-litre sample and counts the number of particles in it. If the true value of μ is 6, what is the probability of getting a reading less than 2?

6   It has been found in the past that 1% of electronic components produced in a certain factory are defective. A sample of size 100 is drawn from each day's production. Calculate to 6 decimal places, the probability that the sample contains fewer than two defectives by using

(a)   the binomial distribution

(b)   the Poisson approximation to the binomial distribution.

7   A and B alternately cut a pack of 52 cards and the pack is shuffled after each cut. If A starts the game and the game is continued until one cuts a diamond, find the probability of A winning. If B starts the game, what is the probability of A winning?

8   A game ends with three equal winners A, B, C. To decide on the overall victor a fair coin is tossed. A calls first then B and C follow in turn, repeatedly. What are the probabilities of each winning the play-off?

9.4 CONTINUOUS RANDOM VARIABLES

Suppose cars arrive at a supermarket with an average overall rate of μ cars per minute, although each arrival is random. The arrival pattern might look like Figure 9.18, where each blob denotes an arrival at a given time. We already know that X, the number of cars arriving in a given minute, is a discrete random variable following a Poisson distribution with mean μ. But what of T, the time between arrivals or inter-arrival time, which takes real values in the infinite interval $[0, \infty)$? This is an example of a **continuous random variable** whose probability distribution might be estimated by observing arrivals over a long period of time and plotting a histogram of relative frequency against time between arrivals, so that the total area under the histogram is equal to 1. As more and more observations are taken, the class interval is made smaller and the histogram of relative

Figure 9.18 Arrival of cars at a supermarket

frequencies tends to a smooth curve called the **probability density curve**. The height of the curve at some point t is usually denoted by $f_T(t)$ and is called the **probability density function** or **p.d.f.** for the distribution (Figure 9.19). You can see from the diagrams that the probability density curve closely parallels the frequency curve of data analysis. The matching of the two is the principal aim of fitting a probability model to real-life data.

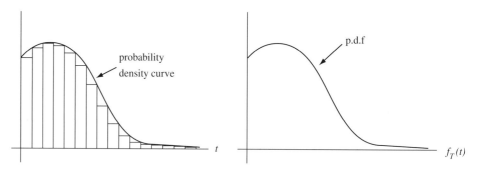

Figure 9.19 Probability density function

For the probability density function $f_T(t)$

(a) $f_T(t) \geq 0$ all t

(b) $\displaystyle\int_{\text{Range of } T} f_T(t)dt = 1$

Probabilities can be represented as areas under the curve of the p.d.f. (Figure 9.20). Hence the two results

(c) $P(t < T < t + \delta t) \approx f_T(t)\delta t$ for small δt

(d) $P(t_1 < T < t_2) = \displaystyle\int_{t_1}^{t_2} f_T(t)dt$

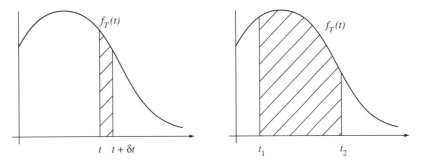

Figure 9.20 Probabilities as areas

We cannot measure the probability that T will take an exact value, only the probability that it takes a value in a given range.

(e) $P(T = t)$ is infinitesimally small and can be regarded as zero

Note that f is used for a continuous probability density function. To distinguish between the f's we write f_T and f_X for the p.d.f.'s corresponding to the random variables T and X, respectively.

Example

It can be proved that the p.d.f. for the random variable T representing the inter-arrival time for a Poisson arrival process is

$$f_T(t) = \begin{cases} \mu e^{-\mu t} & t \geq 0 \\ 0 & t < 0 \end{cases}$$

(a) Verify properties (a) and (b) above.

(b) Determine $P\left(\dfrac{1}{\mu} < T < \dfrac{2}{\mu} \text{ given } T > \dfrac{1}{\mu}\right)$

Refer to Figure 9.21.

(a) $e^{-\mu t} > 0$ for all t, so that $f_T(t) \geq 0$ since $\mu > 0$, and

$$\int_0^\infty \mu e^{-\mu t}\, dt = \left[\mu \times \left(-\frac{1}{\mu}\right) e^{-\mu t}\right]_0^\infty = 1$$

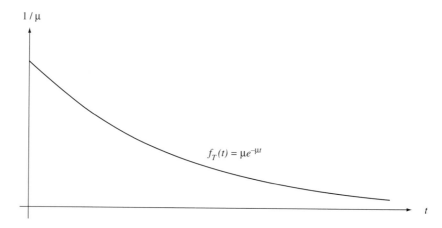

Figure 9.21   Graph of $f_T(t) = \mu e^{-\mu t}, t \geq 0$

(b) Denote the events E and F as

$$E: \frac{1}{\mu} < T < \frac{2}{\mu}$$

$$F: T > \frac{1}{\mu}$$

We require

$$P(E/F) = \frac{P(E \cap F)}{P(F)}.$$

Now

$$E \cap F = \frac{1}{\mu} < T < \frac{2}{\mu} \cap T > \frac{1}{\mu}$$

$$= \frac{1}{\mu} < T < \frac{2}{\mu} = E$$

$$P(E) = \int_{1/\mu}^{2/\mu} \mu e^{-\mu t} dt = \left[-e^{-\mu t}\right]_{1/\mu}^{2/\mu} = \frac{1}{e} - \frac{1}{e^2} = \frac{1}{e}\left(1 - \frac{1}{e}\right)$$

$$P(F) = \int_{1/\mu}^{\infty} \mu e^{-\mu t} dt = \left[-e^{-\mu t}\right]_{1/\mu}^{\infty} = \frac{1}{e}$$

$$P(E/F) = \frac{1/e(1 - 1/e)}{1/e} = 1 - \frac{1}{e}$$

∎

If you wish, think of this result as the area of curve between $\dfrac{1}{\mu}$ and $\dfrac{2}{\mu}$ as a proportion of that which is to the right of $\dfrac{1}{\mu}$. ∎

Expected value and variance of continuous distributions

Let X be a continuous random variable with p.d.f. $f_X(x)$ defined on $[a, b]$. Then the expected value is

$$E[X] = \int_a^b xf_X(x)dx \qquad\qquad (9.16)$$

$E[\ ]$ indicates the multiplier in the integrand. The variance is

$$\mathrm{Var}[X] = E[(x - \mu)^2] = \int_a^b (x - \mu)^2 f_X(x)dx, \quad \text{where} \quad \mu = E[X] \qquad (9.17)$$

Note that

$$\int_a^b (x - \mu)^2 f_X(x)dx = \int_a^b (x^2 - 2\mu x + \mu^2)f_X(x)dx$$

$$= \int_a^b x^2 f_X(x)dx - 2\mu \underbrace{\int_a^b xf_X(x)dx}_{\mu} + \mu^2 \underbrace{\int_a^b f_X(x)dx}_{1}$$

$$= \int_a^b x^2 f_X(x)dx - 2\mu^2 + \mu^2 = E[X^2] - \mu^2$$

Hence $\mathrm{Var}[X] = E[X^2] - \mu^2$

Examples

1. The mean of the exponential distribution is

$$E[T] = \int_0^\infty t\mu e^{-\mu t}\, dt$$

$$= [-te^{-\mu t}]_0^\infty + \int_0^\infty e^{-\mu t}\, dt$$

$$= \left[\frac{-e^{-\mu t}}{\mu}\right]_0^\infty$$

$$= \frac{1}{\mu}$$

The variance of the exponential distribution is

$$\text{Var}[T] = \int_0^\infty \left(t - \frac{1}{\mu}\right)^2 \mu e^{-\mu t}\, dt$$

$$= \int_0^\infty t^2 \mu e^{-\mu t}\, dt - 2\int_0^\infty te^{-\mu t}\, dt + \frac{1}{\mu}\int_0^\infty e^{-\mu t}\, dt$$

$$= \left[-t^2 e^{-\mu t}\right]_0^\infty + 2\int_0^\infty te^{-\mu t}\, dt - 2\int_0^\infty te^{-\mu t}\, dt + \frac{1}{\mu^2}$$

$$= \frac{1}{\mu^2}$$

2. A continuous random variable has p.d.f. $f_X(x)$ defined by

$$f_X(x) = \begin{cases} cx^2(3-x) & 0 \le X \le 3 \\ = 0 & \text{elsewhere} \end{cases}$$

(a) Determine cM

(b) Find $E[X]$

(c) Find $\text{Var}[X]$

(d) Find $P(X > 2/X > 1)$

Solution

(a) $\displaystyle\int_0^3 f_X(x)dx = 1$,   i.e. $\displaystyle c\int_0^3 x^2(3-x)dx = 1$

Now $\displaystyle\int_0^3 x^2(3-x)dx = \left[x^3 - \frac{x^4}{4}\right]_0^3$

$$= 27\left(1 - \frac{3}{4}\right) = \frac{27}{4},   \text{so}   c = \frac{4}{27}$$

(b) $\displaystyle E[X] = \frac{4}{27}\int_0^3 x \times x^2(3-x)dx$

$$= \frac{4}{27}\left[\frac{3x^4}{4} - \frac{x^5}{5}\right]_0^3 = \frac{4}{27} \times \frac{3^5}{20} = \frac{9}{5} = 1.8$$

(c) $\text{Var}[X] = \dfrac{4}{27} \displaystyle\int_0^3 x^2 \times x^2(3-x)dx - (1.8)^2$

$= \dfrac{4}{27}\left[\dfrac{3x^5}{5} - \dfrac{x^6}{6}\right]_0^3 - (1.8)^2$

$= \dfrac{4}{27} \times \dfrac{3^6}{30} - (1.8)^2$

$= 3.6 - (1.8)^2 = 0.36$

(d) $P(X > 1) = \dfrac{4}{27}\displaystyle\int_1^3 x^2(3-x)dx = \dfrac{4}{27}\left[x^3 - \dfrac{x^4}{4}\right]_1^3 = \dfrac{24}{27}$

$P(X > 2) = \dfrac{4}{27}\left[x^3 - \dfrac{x^4}{4}\right]_2^3 = \dfrac{11}{27}$

$P(X > 2/X > 1) = \dfrac{P(X > 2 \cap X > 1)}{P(X > 1)} = \dfrac{P(X > 2)}{P(X > 1)} = \dfrac{11}{24}$ ∎

Mode, median and percentiles

A **mode** occurs at a maximum on the density curve, i.e. $f_X'(x) = 0, f_X''(x) < 0$, i.e. at the point x_m shown in Figure 9.22; $f_X(x)$ may have more than one mode.

The **median** is the point x_M on either side of which the area under the curve is divided into two equal parts, i.e.

$$\int_a^{x_M} f_X(x)dx = \int_{x_M}^b f_X(x)dx = \dfrac{1}{2}$$

The **mean** $\mu = E[X]$ is usually different as was shown above.

The **pth percentile** x_p is the point to the left side of which there lies $p\%$ of the area under the curve:

$$\int_a^{x_p} f_X(x)dx = \dfrac{p}{100}$$

In practical terms the mode (or modes) are generally straightforward to find for many of the classical distributions. The percentiles on the other hand are often not so amenable to explicit determination and numerical equation-solving techniques, e.g. iteration, are often needed.

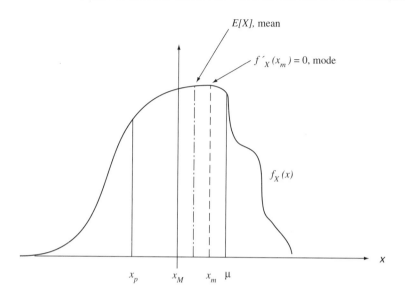

Figure 9.22 Mode and median of a continuous random variable

The normal distribution

The most important continuous distribution is the **normal distribution** (Figure 9.23). Its symmetric bell-shaped curve models many data distributions such as examination marks, heights and weights of a human population.

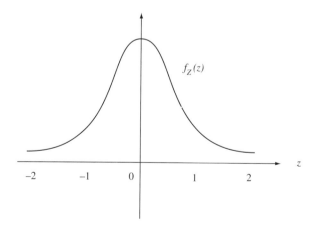

Figure 9.23 Standard normal distribution

The random variable Z is said to be distributed as a **standard normal variable** or $N(0, 1)$ when its p.d.f. is

$$f_Z(z) = \frac{1}{\sqrt{2\pi}} e^{-z^2/2} \qquad (-\infty < z < \infty)$$

$f_Z(z)$ cannot be integrated by analytical means, so the area under the curve is tabulated in statistical tables. However, mathematical analysis can prove that the factor of $\dfrac{1}{\sqrt{2\pi}}$ ensures that $\displaystyle\int_{-\infty}^{\infty} f_Z(z)dz = 1$

It follows that the mean, mode and median are coincident at $Z = 0$, and the variance is 1, i.e.

$$\mu = E[Z] = 0 = z_m = z_M$$
$$\sigma^2 = \text{Var}[Z] = 1 \qquad (\text{standard deviation } \sigma = 1)$$

Tables of $N(0, 1)$ exist in a variety of forms, the most common of which tabulates

$$\int_0^x f_Z(z)dz$$

Refer to Table 9.8 and Figure 9.24.

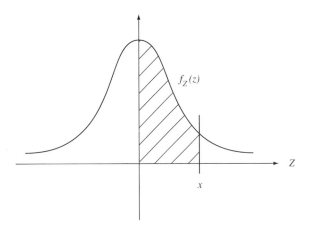

Figure 9.24 Area under the standard normal distribution curve

Table 9.8 Standard normal distribution[a]

$\dfrac{x-\mu}{\sigma}$	0	1	2	3	4	5	6	7	8	9
0	0000	0040	0080	0120	0160	0199	0239	0279	0319	0359
0.1	0398	0438	0478	0517	0557	0596	0636	0675	0714	0753
0.2	0793	0832	0871	0909	0948	0987	1026	1064	1103	1141
0.3	1179	1217	1255	1293	1331	1368	1406	1443	1480	1517
0.4	1555	1591	1628	1664	1700	1736	1772	1808	1844	1879
0.5	1915	1950	1985	2019	2054	2088	2123	2157	2190	2224
0.6	2257	2291	2324	2357	2389	2422	2454	2486	2517	2549
0.7	2580	2611	2642	2673	2703	2734	2764	2794	2822	2852
0.8	2881	2910	2939	2967	2995	3023	3051	3078	3106	3133
0.9	3159	3186	3212	3238	3264	3289	3315	3340	3365	3389
1.0	3413	3438	3461	3485	3508	3531	3554	3577	3599	3621
1.1	3643	3665	3686	3708	3729	3749	3770	3790	3810	3830
1.2	3849	3869	3888	3907	3925	3944	3962	3980	3997	4015
1.3	4032	4049	4066	4082	4099	4115	4131	4147	4162	4177
1.4	4192	4207	4222	4236	4251	4265	4279	4292	4306	4319
1.5	4332	4345	4357	4370	4382	4394	4406	4418	4429	4441
1.6	4452	4463	4474	4484	4495	4505	4515	4525	4535	4545
1.7	4554	4564	4573	4582	4591	4599	4608	4616	4625	4633
1.8	4641	4649	4656	4664	4671	4678	4686	4693	4699	4706
1.9	4713	4719	4726	4732	4738	4744	4750	4756	4761	4767
2.0	4772	4778	4783	4788	4793	4798	4803	4808	4812	4817
2.1	4821	4826	4830	4834	4838	4842	4846	4850	4854	4857
2.2	4861	4865	4868	4871	4875	4878	4881	4884	4887	4890
2.3	4893	4896	4898	4901	4904	4906	4909	4911	4913	4916
2.4	4918	4920	4922	4925	4927	4929	4931	4932	4934	4936
2.5	4938	4940	4941	4943	4946	4947	4948	4949	4951	4952
2.6	4953	4955	4956	4957	4959	4960	4961	4962	4963	4964
2.7	4965	5966	4967	4968	4969	4970	4971	4972	4973	4974
2.8	4974	4975	4976	4977	4977	4978	4979	4979	4980	4981
2.9	4981	4982	4982	4983	4984	4984	4985	4985	4986	4986
3.0	4987	4990	4993	4995	4997	4998	4998	4999	4999	4999

[a] The normal probability integral is

$$\frac{1}{\sigma\sqrt{2\pi}}\int_{\mu}^{x}\exp\left\{-\frac{(x-\mu)^2}{2\sigma^2}\right\}dx = \frac{1}{\sqrt{2\pi}}\int_{0}^{(x-\mu)/\sigma} e^{-z^2/2}\,dz$$

The area is easily read off. In particular

$$P(0 < Z < 1) = 0.3413$$
$$P(-1 < Z < 1) = 0.6826 \qquad \text{(by symmetry)}$$
$$P(-1.96 < Z < 1.96) \approx 0.95$$

In other words just over 68% of the area under the curve is within one standard deviation of the mean.

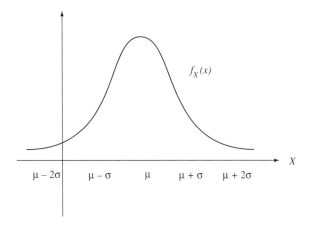

Figure 9.25 General normal distribution

A **general normal variable** (Figure 9.25) is distributed $N(\mu, \sigma^2)$, i.e. with mean μ and variance σ^2 and has p.d.f.

$$f_X(x) = \frac{1}{\sqrt{2\pi}\sigma} \exp\left\{\frac{(x - \mu)^2}{2\sigma^2}\right\}$$

It can be proved that

$$E[X] = \mu = x_m = x_M$$
$$\text{Var}[X] = \sigma^2$$

If data are being fitted to a normal distribution of the form X, then the transformation

$$Z = \frac{X - \mu}{\sigma}$$

converts them to $N(0, 1)$. Consider the following example.

Example   A large number of students sat a public examination. Analysis of the results indicated a normal distribution of marks with a mean of 56% and standard deviation 12%. With the pass mark set at 40%, a credit level at 60% and distinction at 70% find out what proportion fell into each category. Note first that

$$z_i = \frac{x_i - \mu}{\sigma}$$

$$z_1 = \frac{40 - 56}{12} = -1.333, \qquad P(Z < z_1) = 0.5 - 0.4087 = 0.0913$$
$$(40\%)$$

$$z_2 = 0.333, \qquad P(Z < z_2) = 0.5 + 0.1306 = 0.6306$$
$$(60\%)$$

$$z_3 = 1.167, \qquad P(Z < z_3) = 0.5 + 0.3784 = 0.8784$$
$$(70\%)$$

$$P(z_1 < Z < z_2) = 0.6306 - 0.0913 = 0.5393$$
$$P(z_2 < Z < z_3) = 0.8784 - 0.6306 = 0.2478$$
$$P(Z > z_3) = 1 - 0.8784 = 0.1216$$

i.e. approximately 12% obtain a distinction, 25% obtain a credit, 54% pass and 9% fail. ∎

Normal approximation to the binomial distribution

When n is large and neither p nor q is small, the behaviour of a binomial random variable x approaches the behaviour of a normal random variable.

It can be proved that the discrete binomial random variable X distributed with mean np and variance npq approaches a normal random variable with the same mean and variance. We write $X \sim N(np, npq)$ as $n \to \infty$. In other words

$$Z = \frac{X - np}{\sqrt{npq}} \sim N(0, 1)$$

We can use this property to estimate binomial probabilities when n is large.

Example Estimate from the normal distribution the following probabilities from 100 tosses of a fair coin:

(a) exactly 50 heads (b) more than 60 heads.

Solution

(a) Since we are trying to model a precise discrete value, i.e. exactly 50 heads, with a value taken from a continuous distribution, we need to allow for this by estimating the area under the Normal curve between $x_1 = 49.5$ and $x_2 = 50.5$. The equivalent values of the standard Normal variable Z, noting $\mu = np = 5$, $\sigma = \sqrt{npq} = 5$, are

$$z_1 = \frac{49.5 - 50}{5} = -0.1, \quad \text{and} \quad z_2 = \frac{50.5 - 50}{5} = 0.1$$

so that $P(50 \text{ heads}) = P(0.1 \leq Z \leq 0.1) = 0.0796$.

In other words there is barely an 8% chance of exactly 50 heads in 100 tosses. This is partly because 48, 49, 51, 52 heads, etc., are almost equally likely.

(b) To find the probability of more than 60 heads, we need to exclude exactly 60 and determine $P(X \geq 60.5)$.

i.e. $\quad P\left(Z \geq \frac{60.5 - 50}{5}\right) = P(Z \geq 2.1) \simeq 0.017\,86$ ∎

Exercise 9.4

 1 The random variable X has p.d.f.

$$f_X(x) = \begin{cases} \dfrac{3x^2}{2} & -1 < x < 1 \\ 0 & \text{elsewhere} \end{cases}$$

Calculate

(a) $P\left(|X| < \dfrac{1}{2}\right)$

(b) $P\left(-\dfrac{1}{4} < X < \dfrac{2}{3}\right)$

(c) $E[X]$

(d) $\text{Var}[X]$

(e) x_m, the mode

(f) x_M, the median

(g) x_{90}, the location of the 90th percentile

Sketch the graph of $f_X(x)$.

2 The lifetime in hours of a certain piece of equipment is a continuous random variable X with p.d.f.

$$f_X(x) = \begin{cases} \dfrac{xe^{-x/10}}{100} & x > 0 \\ 0 & \text{elsewhere} \end{cases}$$

(a) Show that $\int_0^\infty f_X(x)(x)dx = 1$

(b) Compute the probability that the lifetime exceeds 20 hours.

3 A plane carrying bombs flies directly above a straight road. When a bomb is dropped, its distance from the road (measured perpendicular to the road) is distributed with p.d.f.

$$f_X(x) = \begin{cases} \dfrac{50 + x}{2500}, & -50 \le x \le 0 \\ \dfrac{50 - x}{2500}, & 0 < x \le 50 \end{cases}$$

where x is measured in metres and is taken positive on one side of the road and negative on the other. If a bomb falls within 20 metres of the road, the road will be damaged. What is the probability that the road will be damaged by a bomb?

4 If two bombs are dropped in Question 3 what is the probability that the road is damaged?

5 The random variable X has p.d.f.

$$f_X(x) = \frac{6x}{(1 + 3x^2)^2}$$

Determine

(a) the mode x_m

(b) the location of the median x_M

(c) the set of values of X which lie within the interquartile range.

 6 A model of a small hatchback car is designed to travel a mean distance of 200 miles on a tank of petrol. The population of such cars is observed to have their travel distance normally distributed with standard deviation 9 miles.

(a) Calculate the proportion which achieve 210 miles

(b) Of those that achieve 210 miles, what proportion achieve 220 miles?

7 A random variable has p.d.f.

$$f_X(x) = ke^{-x/\theta} \qquad 0 \le x \le \theta$$

Find k and the mean value of X.

8 For a standard normal variable Z determine

(a) $P(0 \le Z \le 1)$ (b) $P(-2 < Z < 2)$

(c) $P(0.3 \le Z < 3.2)$ (d) $P(|Z| > 0.3)$

(note that $P(Z = z) = 0$), where z is an exact value, so $P(a < Z < 6) = P(a \le Z \le b)$

9 The time taken to change a wheel on a certain kind of racing car is precisely regulated and known to be normally distributed with a mean time of 29 seconds and a standard deviation of 2 seconds. Determine the probability that a particular wheel change is more than 4 seconds outside the mean.

10 A fair coin is tossed 200 times. Determine the probability that the tosses produce

(a)   between 80 and 120 tails

(b)   less than 90 heads

(c)   less than 85 and more than 115 heads

(d)   exactly 100 heads

11 For the following numbers of tosses, n, of a fair coin, estimate the probabilities as indicated.

(a) $n = 10$ (b) $n = 100$

         (i) $P(x = 5)$ (ii) $P(x > 6)$ (i) $P(X = 50)$ (ii) $P(X > 60)$

         Use the binomial distribution.

(c) $n = 1000$

         (i) $P(X = 500)$ (ii) $P(X > 600)$

What do you observe?

12 Following a general election it is known that 37% of the electorate voted for party A. Before the result was known a sample of 1000 people, assumed to be random and fully representative, were interviewed for a TV exit poll. Determine the probability that more than 400 claimed to have voted for party A.

SUMMARY

- **Pictorial representation of data**: methods include pie charts, bar charts, line diagrams, histograms
- **Frequency distribution**: a set of observations $x_1, x_2, \ldots, x_n$ and the frequencies $f_1, f_2, \ldots, f_n$ with which they occur
- **Mode and median**: the mode is the most commonly occurring value; the median splits the data in two equally sized sets, one above and one below.
- **Mean**: the mean of n items of data, $x_1, x_2, \ldots, x_n$ is commonly known as the average and is given by

$$\frac{x_1 + x_2 + \cdots + x_n}{n}$$

- **Range and interquartile range**: measures of dispersion given by

$$\text{Range} = \text{highest value} - \text{lowest value}$$

$$\text{Interquartile range} = \text{third quartile} - \text{first quartile}$$

where a quartile is one of four sets into which the data is separated; each quartile contains the same number of data items

- **Variance and standard deviation**: the standard deviation (σ) is the square root of the variance; the variance is given by

$$\sigma^2 = \frac{x_1^2 + x_2^2 + \cdots + x_n^2}{n} - \mu^2$$

where μ is the mean of the data

- **Grouped data**: $\text{mean} = \dfrac{\sum_{i=1}^{s} f_i x_1}{\sum_{i=1}^{s} f_i}$ $\qquad$ $\text{variance} = \dfrac{\sum_{i=1}^{s} f_i x_i^2}{\sum_{i=1}^{s} f_i} - (\text{mean})^2$

- **Probability of event A**

$$\frac{\text{number of outcomes in } A}{\text{total number of outcomes}}$$

if outcomes are equally likely; $0 \leq P(A) \leq 1$

- **Empty set and universal set**: if $\emptyset$ is the empty set and S represents the set of all possible outcomes, then $P(\emptyset) = 0$ and $P(S) = 1$
- **Combined events**:

$$P(A') = 1 - P(A), \qquad P(A \cup B) = P(A) + P(B) - P(A \cap B).$$

- **Conditional probability**:

$$P(B/A) = \frac{P(A \cap B)}{P(A)}$$

but if events A and B are independent then $P(A \cap B) = P(A) \times P(B)$
- **Discrete probability distributions**

binomial $\qquad P(X = r) = \binom{n}{r} p^r q^{n-r} \qquad \mu = np \qquad \sigma^2 = npq$

Poisson $\qquad P(X = r) = \dfrac{e^{-\lambda} \lambda^r}{r!} \qquad \mu = \lambda \qquad \sigma^2 = \lambda$

geometric $\qquad P(X = r) = pq^{r-1} \qquad \mu = \dfrac{1}{p} \qquad \sigma^2 = \dfrac{q}{p^2}$

- **Continuous random variable**: for a continuous random variable with probability density function $f_T(t)$ we have

probability $\qquad P(t_1 < T < t_2) = \int_{t_1}^{t_2} f_T(t) dt$

expected value $\qquad E(T) = \int t f_T(t) dt$

variance $= E(T^2) - (E(T))^2$

- **Normal distribution**: if x is distributed $N(\mu, \sigma^2)$ then $z = \dfrac{x - \mu}{\sigma}$ is the standard normal variable, distributed $N(0, 1)$, for which numerical values are tabulated. The normal approximation to a binomial distribution puts $\mu = np$, $\sigma^2 = npq$.

Answers

Exercise 9.1

1

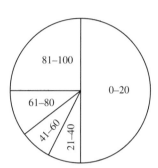

2 (a)

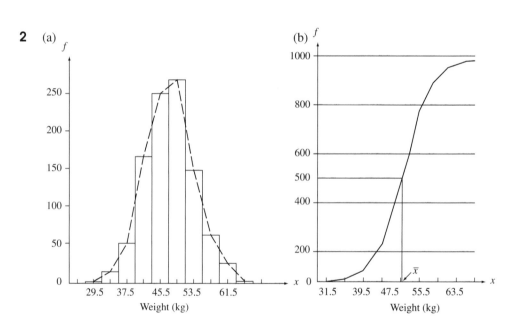

(b)

3

kHz	Frequency
12.24–12.26	3
12.27–12.29	6
12.30–12.32	10
12.33–12.35	15
12.36–12.38	12
12.39–12.41	9
12.42–12.44	5

4 $\bar{x} = 47.71, s = 6.291$

Quartiles 43.61, 47.68, 51.48

5 $\mu = 11.49\,\text{mm}, \sigma = 0.015\,\text{mm}$

6 $\mu = 642\,\text{h}, \sigma = 95.58\,\text{h}$

7 $\mu = 101.92\,\text{kN}, \sigma = 7.33\,\text{kN}$; interquartile range (96, 106)

8 (a)   Median $= 45.1$

(b) $\mu = 46.6$

The lowest and highest class marks were taken to be 23 and 80. These are intrinsically inaccurate so the median is a much better estimate.

9 (a)   Mean $= 8.9$, median $= 7$

(b)   Mean $= 6.4$, median $= 6$

Since each of the numbers 4, 5, 6, 8 and 10 occurs twice, we can consider that they are the five modes. However, it is more reasonable to conclude in this case that no mode exists.

10 (a)   Lower quartile $= 67$, middle quartile $=$ median $= 75$, upper quartile $= 83$

(b)   25% scored 67 or lower (or 75% scored 67 or higher), 50% scored 75 or lower (or 50% scored 75 or higher), 75% scored 83 or lower (or 25% scored 83 or higher).

11 (a)   32.4   (b)   40.9   (c)   68.5   (d)   53.4

Exercise 9.2

1 (a) $S = \{(x, y): 1 \leq x \leq 6, 1 \leq y \leq 6\}$ i.e. 36 ordered pairs

(b) $A = \{(1, 1), (1, 2), (1, 3), (2, 1), (2, 2), (3, 1)\}$

(c) $B = \{(1, 6), (2, 6), (3, 6), (4, 6), (5, 6), (6, 6), (6, 1), (6, 2), (6, 3), (6, 4), (6.5)\}$

(d) $C = \{(2, 1), (2, 2), (2, 3), (2, 4), (2, 5), (2, 6)\}$

2 (a) $A \cup C = \mathbb{R} \backslash \{0\}$, i.e., the set of real numbers excluding zero.
$A \cap C = \varnothing$
$A' = \{x: -\infty < x \leq 0\}, B \cap C = \{x: -1 < x < 0\}$
$A \cap B' = \{x: 2 \leq x < \infty\}, A \cup B = \{x: -1 < x < \infty\}$

(b) $D = A \cap B, E = A' \cap C', F = C \cap B'$.

3 (a) Let 0 denote failure, 1 denote operative
$S = \{(0, 0, 0), (0, 0, 1), (0, 1, 0), (1, 0, 0), (0, 1, 1), (1, 0, 1), (1, 1, 0), (1, 1, 1)\}$
$E_1 = \{(1, 1, 1), (1, 1, 0), (1, 0, 1)\}; E_2 = \{(1, 0, 1)\}; E_3 = \{(1, 1, 0)\}$

(b) E_2 and E_3 are mutually exclusive.

4 P (first head on an even numbered toss) $= \dfrac{1}{3}$

5 $\dfrac{1}{4}$

6 (a) $\dfrac{1}{12}$ (b) $\dfrac{1}{4}$

7 $\dfrac{5}{12}$

8 (a) 21/55 (b) 27/55 (c) 41/55 (d) 3/11

9 (a) 3/5 (b) 3/4 (c) 3/5

10 (a) $P(A/B) = 0.99$ (b) $P(A'/B) = 0.01$
(c) $P(A/B') = 0.01$ (d) $P(A'/B') = 0.99$

11 (a) Reliability $= 0.3384$ (b) Reliability $= 0.874$

12 12/17

13 0.164; 0.9502

14 0.1743

Exercise 9.3

1

X	0	1	2	3
Probability	$\dfrac{8}{27}$	$\dfrac{12}{27}$	$\dfrac{6}{27}$	$\dfrac{1}{27}$

2

X	0	1	2	3	4
Probability	$\dfrac{1}{81}$	$\dfrac{8}{81}$	$\dfrac{24}{81}$	$\dfrac{32}{81}$	$\dfrac{16}{81}$

3   0.0038

4   (a)   0.2   (b)   0.817   (c)   0.0162

5   0.0174

6   (a)   0.735 762   (b)   0.735 759

7   If A starts, $P(A \text{ wins}) = \dfrac{4}{7}$

If B starts, $P(A \text{ wins}) = \dfrac{3}{7}$

8 $P(A) = \dfrac{4}{7}$, $P(B) = \dfrac{2}{7}$, $P(C) = \dfrac{1}{7}$

Exercise 9.4

1   (a) $\dfrac{1}{8}$ (b)   0.156   (c)   0

(d)   0.6   (e)   no mode x_m (f)   0

(g)   0.928

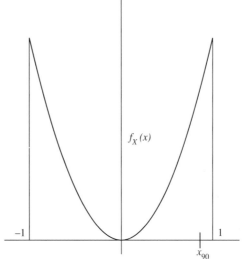

2   (b)   0.406

3 $\dfrac{16}{25}$

4 $\dfrac{554}{625}$; the road is undamaged only when both bombs miss.

5 (a) $\dfrac{1}{3}$ (b) $\dfrac{1}{\sqrt{3}}$

(c) interquartile range $\left(\dfrac{1}{3}, 1\right)$

6 (a) 13.3% (b) 10%

7 $k = \dfrac{1}{\theta(e-1)}$; mean is $\dfrac{\theta(e-2)}{(e-1)}$

8 (a) 0.3413 (b) 0.9545
(c) 0.3814 (d) 0.7642

9 0.9545

10 (a) 0.9962 (b) 0.0687
(c) 0.0284 (d) 0.0564

11 (a) (i) 0.2461 (ii) 0.3680
(b) (i) 0.0796 (ii) 0.0179
(c) (i) 0.0258 (ii) 0.0008

The probability of an exact number decreases with n as does the relative variation away from the mean.

12 2.48%

10 INTEGER VARIABLE

Introduction

Data is normally transmitted as a sequence of digits. Many signals are discrete samples of a continuous process. In order to analyse this information it is necessary to understand the behavior of sequences and related concepts.

Objectives

After working through this chapter you should be able to

- Understand that sequences may be convergent, divergent, bounded or oscillatory
- Classify a sequence and identify its main features, including its limit, if it exists
- Obtain the partial sums of a series and discuss the behaviour of the series
- Use the comparison and ratio tests to determine whether the series converges
- Understand how to use recursion
- Understand how to use mathematical induction
- Understand that convergence of a power series in x depends upon the magnitude of x
- Obtain simple Maclaurin series for functions from first principles or by combining standard series.

10.1 SEQUENCES AND SERIES

Sequences

A sequence of numbers is a set of numbers that can be counted. Written as $\{a_n\} = \{a_1, a_2, a_3, \ldots\}$, the sequence may be finite or infinite.

Examples   (a) $\left\{ \dfrac{1}{2}, \dfrac{2}{3}, \dfrac{5}{7}, \dfrac{3}{7}, \dfrac{2}{11}, \dfrac{1}{15}, -\dfrac{2}{21}, \ldots \right\}$

(b) $\{0.23, 0.023, 0.0023, 0.00023, \ldots\}$

Both of these sets of numbers are sequences. In (a) there is no obvious pattern but in (b) there is a pattern in that $a_n = f(n)$ can be defined as a function of n, i.e. $f(n) = 0.23 \times (0.1)^{n-1}, n \geq 1$.   ∎

When $a_n = f(n)$ can be defined and when $f(n)$ tends to a finite limit L as $n \to \infty$, we say that the sequence $\{a_n\} \to L$, **converges** to L or is **convergent** to L. On the other hand, if $f(n) \to \pm \infty$ as $n \to \infty$ we say that $\{a_n\}$ **diverges** or is **divergent**.

Examples   Which of the following sequences are convergent, divergent or neither?

(a) $\left\{ \dfrac{1}{n} \right\}$ (b) $\left\{ 1 - \dfrac{1}{n} \right\}$

(c) $\{n^2\}$ (d) $\{1 + (-1)^n\}$ $(n \geq 1)$

Solution

(a) $\left\{ \dfrac{1}{n} \right\} = \left\{ 1, \dfrac{1}{2}, \dfrac{1}{3}, \dfrac{1}{4}, \dfrac{1}{5}, \ldots \right\} \to 0$ as $n \to \infty$

When $a_{n+1} < a_n$, $\{a_n\}$ is **monotonic decreasing**. Furthermore, $a_n > 0$ for all n. This sequence converges to the limit 0.

(b) $\left\{ 1 - \dfrac{1}{n} \right\} = \left\{ 0, \dfrac{1}{2}, \dfrac{2}{3}, \dfrac{3}{4}, \dfrac{4}{5}, \ldots \right\}$

$\{a_n\}$ is **monotonic increasing** when $a_{n+1} > a_n$ for all n, but $a_n < 1$. This sequence converges to the limit 1.

(c) $\{n^2\} = \{1, 4, 9, 16, 25, \ldots\}$. Note that $n^2 > n$ for all $n > 1$, so $\{a_n\} \to \infty$; the sequence diverges.

(d) $\{1 + (-1)^n\} = \{0, 2, 0, 2, 0, 2, \ldots\}$

Because every odd term is 0 and every even term is 2, convergence cannot take place to a single limit. We say that the sequence $\{a_n\}$ is **bounded**, i.e. the terms are constrained within fixed finite values and the sequence does not diverge. This particular sequence is also **periodic**, as it repeats itself (in this case every second term). ∎

In examining convergence or divergence, we reasonably assume that as $n \to \infty$

(a) $n^k \to \infty$ if $k > 0$ $n^k \to 0$ if $k < 0$

(b) $f(n)g(n) \to 0$ if $f(n) \to 0$ and $g(n)$ is bounded, e.g. $f(n) = \dfrac{1}{n}$, $g(n) = (-1)^n$

Example

Examine the following sequences for convergence. If a limit exists, determine its value. The nth terms of the sequence are

(a) $\dfrac{3n - 5}{(n + 2)(n + 1)}$ (b) $(-1)^n \dfrac{n}{n^2 + 1}$ (c) $\cos \dfrac{n\pi}{2}$

Write out the first five terms in (b) and (c).

(a) For any rational function of n, divide by whatever power of n is required to make n appear everywhere as a reciprocal. Dividing top and bottom by n^2 in this example gives

$$\frac{3n - 5}{(n + 2)(n + 1)} = \frac{\left(\dfrac{3}{n} - \dfrac{5}{n^2}\right)}{\left(1 + \dfrac{2}{n}\right)\left(1 + \dfrac{1}{n}\right)}$$

Removing a factor of $\dfrac{1}{n}$ from the numerator

$$\frac{\left(3 - \dfrac{5}{n}\right)}{n\left(1 + \dfrac{2}{n}\right)\left(1 + \dfrac{1}{n}\right)}$$

You can see that $3 - \dfrac{5}{n} \to 3$, $1 + \dfrac{2}{n} \to 1$, $1 + \dfrac{1}{n} \to 1$, so $a_n \approx \dfrac{3}{n}$ when n is large, hence $a_n \to 0$. You could also use the binomial theorem, ignoring terms of $O\left(\dfrac{1}{n^2}\right)$.

Since $\left(1 + \dfrac{1}{n}\right)^{-1} \approx 1 - \dfrac{1}{n}$

then $a_n \approx \dfrac{3}{n}\left(1 - \dfrac{5}{3n}\right)\left(1 - \dfrac{2}{n}\right)\left(1 - \dfrac{1}{n}\right)$

$$= \frac{3}{n} + O\left(\frac{1}{n^2}\right)$$

(b) $\dfrac{n}{n^2 + 1} = \dfrac{1}{n} + O\left(\dfrac{1}{n^2}\right)$; hence $\dfrac{n}{n^2 + 1} \to 0$ as $n \to \infty$.

Also, $(-1)^n \dfrac{n}{n^2 + 1} \to 0$, using the assumptions stated earlier.

$$\{a_n\} = \left\{-\frac{1}{2}, \frac{2}{5}, -\frac{3}{10}, \frac{4}{17}, -\frac{5}{26}, \ldots\right\}$$

(c) $\left\{\cos\dfrac{n\pi}{2}\right\} = \{0, 1, 0, -1, 0, \ldots\}$

This sequence is both periodic and **oscillating** (in this case between 1 and -1). ■

A sequence $\{a_n\}$ can be generated forwards if each successive term is a function of some or all of its predecessors, i.e.

$$a_{n+1} = f(a_n, a_{n-1}, a_{n-2}, \ldots)$$

The following are two examples:

(a) $a_{n+1} = \dfrac{1}{2}(1 + a_n) \qquad n \geq 1$

(b) $a_{n+1} = a_n^2 - 2a_{n-1} + a_{n-2}^3 \qquad n \geq 3$

Example

Verify that the sequence which satisfies $a_{n+1} = \dfrac{1}{2}(1 + a_n)$, and for which $a_1 = 0$, converges. State its limit.
  The terms of the sequence are

$$\left\{0, \frac{1}{2}, \frac{3}{4}, \frac{7}{8}, \frac{15}{16}, \ldots\right\}$$

If the limit is L, then L must satisfy the functional relation identically, i.e.

$$L = \frac{1}{2}(1 + L) \quad \text{so that} \quad L = 1$$

The sequence clearly converges to this limit. ■

You can repeat this example with any other starting value a_1. The limit is 1, although the process of convergence may take longer, especially if $|a_1|$ is large.
  One might ask whether the choice of a_1 is important at all, or whether a sequence actually converges to a limit it appears to possess. The answers are

Choice of a_1: usually this must be in a finite interval $[a, b]$, but sometimes it can be any real number.
Limit L: Must satisfy the functional relationship.

A general analysis of convergence is beyond our scope but an inkling comes from the process of fixed-point interation, namely

$$x_{n+1} = g(x_n)$$

in which convergence to the root α of $f(x) = x - g(x) = 0$ requires x_1 to be close to α and $|g'(\alpha)| < 1$.

Example

Identify potential limits to the sequence $\{a_n\}$ which satisfies the relationship

$$a_{n+1} = a_n^2 - 2a_{n-1} + a_{n-2}^3$$

and examines its convergence.

Solution

L must satisfy the relationship

$$L = L^2 - 2L + L^3 \qquad \text{i.e. } L(L^2 + L - 3) = 0$$

Hence $L = 0, \dfrac{\pm\sqrt{13} - 1}{2}$, i.e. 0, 1.303, -2.303

Choosing $a_1 = a_2 = a_3 = 0$ gives an identically zero sequence. Any other choice, e.g. $a_1 = 0, a_2 = 1, a_3 = 2$, produces a divergent sequence. You can verify that each of the limit points is a **focus of repulsion** by taking close starting-values and observing that the ensuing sequence is repelled away. ∎

Series

When the terms of a sequence are summed we refer to the sum as a **series**, i.e.

$$S_n = a_1 + a_2 + \cdots + a_n = \sum_{r=1}^{n} a_r$$

The sums taken after one term, two terms, three terms, etc. form a sequence $\{S_n\}$, called the **sequence of partial sums**. The convergence of the series $\sum a_r$ as $n \to \infty$ is therefore equivalent to the convergence of the sequence $\{S_n\}$ to a finite limit L.

Before proceeding, we quote without proof the result that a bounded monotonic sequence is convergent. Consider the example in Figure 10.1:

if $S_n < K$ (all n) and $S_{n+1} \geq S_n$ (monotonic increasing)
then $\{S_n\} \to L \ (\leq K)$

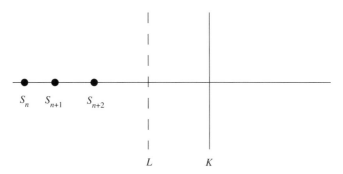

Figure 10.1 A bounded monotonic sequence converges

K acts as an upper bound to the forward progress of $\{S_n\}$, so the results seems reasonable. An equivalent argument follows for a monotonic decreasing sequence. The sequences $\left\{1 - \dfrac{1}{n}\right\}$ and $\left\{\dfrac{1}{n}\right\}$ are good examples, converging to 1 and 0 respectively.

For the moment, we restrict ourselves to series of non-negative terms. Consider the partial sum

$$S_n = \sum_{r=1}^{n} a_r \qquad a_r \geq 0$$

It follows that $S_{n+1} \geq S_n$. If $S_{n+1} \leq K$ for all n, where K is a fixed constant, then the series $\sum_{r=1}^{\infty} a_r$ converges. Otherwise it diverges.

Example

The geometric series satisfies

$$a_r = x^{r-1} \qquad 0 \leq x < 1$$

Then $\quad S_n = \sum_{r=1}^{n} a_r = 1 + x + x^2 + \cdots + x^{n-1}$

$$= \frac{1 - x^n}{1 - x} \qquad \text{(proved in \textit{Foundation Mathematics}, Chapter 6)}$$

$$< \frac{1}{1 - x}$$

So $\quad \{S_n\} \to L \quad \left(= \dfrac{1}{1 - x}, \text{ because } x^n \to 0\right)$ ∎

The comparison test

If we know that a particular series of positive terms converges, then so does any other series of positive terms when the corresponding terms are smaller in each case. Therefore suppose that $a_r \geq 0$, and

$$S_n = \sum_{r=1}^{n} a_r \to L \quad \text{as} \quad n \to \infty$$

If $0 \leq b_r \leq a$, for all r, then $T_n = \sum_{r=1}^{n} b_r \leq \sum_{r=1}^{n} a_r = S_n$.

$\{T_n\}$ is monotonic increasing and $T_n \leq S_n \leq L$, so that $\{T_n\} \to M$, a finite limit, with $M \leq L$.

Conversely, if $T_n = \sum_{r=1}^{n} b_r$, $b_r \geq 0$, represents a known divergent series, i.e. $\{T_n\} \to \infty$, then any series whose corresponding terms are larger will also diverge, i.e. if $0 \leq b_r \leq a_r$ all r then $T_n = \sum_{r=1}^{n} b_r \leq \sum_{r=1}^{n} a_r = S_n$

Because $\{T_n\} \to \infty$ then $S_n \to \infty$. We can summarise this as follows: If

$$0 \leq b_r \leq a_r \quad \text{(all } r\text{)} \qquad \begin{cases} \sum a_r \text{ converges} \Rightarrow \sum b_r \text{ converges} \\ \sum a_r \text{ diverges} \Leftarrow \sum b_r \text{ diverges} \end{cases}$$

Examples

1. Consider $S_n = \sum_{r=1}^{n} \dfrac{1}{r^2} = 1 + \dfrac{1}{2^2} + \dfrac{1}{3^2} + \cdots + \dfrac{1}{n^2}$

So $\quad S_n \leq 1 + \dfrac{1}{1 \times 2} + \dfrac{1}{2 \times 3} + \cdots + \dfrac{1}{n(n-1)}$

$\qquad = 1 + \sum_{r=1}^{n-1} \dfrac{1}{r(r+1)} = 2 - \dfrac{1}{n} \qquad$ (see Section 10.3)

Then $S_n \leq 2$, for all n, a constant upper bound. Therefore $S_n \to L \leq 2$. Actually the limit L is $\dfrac{\pi^2}{6}$ (≈ 1.645)

2. Consider the harmonic series

$$S_n = \sum_{r=1}^{n} \frac{1}{r} = 1 + \frac{1}{2} + \frac{1}{3} + \frac{1}{4} + \cdots + \frac{1}{n}$$

At first sight it seems reasonable that $S_n \to L$, a finite limit, as $n \to \infty$, but suppose we look at the first 2^n terms rather than the first n terms

$$S_{2^n} = 1 + \left(\frac{1}{2}\right) + \left(\frac{1}{3} + \frac{1}{4}\right) + \left(\frac{1}{5} + \frac{1}{6} + \frac{1}{7} + \frac{1}{8}\right) + \cdots$$

$$+ \left(\frac{1}{2^{n-1}+1} + \cdots + \frac{1}{2^n}\right)$$

$$\geq 1 + \left(\frac{1}{2}\right) + \left(\frac{1}{4} + \frac{1}{4}\right) + \left(\frac{1}{8} + \frac{1}{8} + \frac{1}{8} + \frac{1}{8}\right) + \cdots + \left(\frac{1}{2^n} + \cdots + \frac{1}{2^n}\right)$$

$$= 1 + \left(\frac{1}{2}\right) + \left(\frac{1}{2}\right) + \cdots + \left(\frac{1}{2}\right) \qquad (n \text{ times})$$

$$= 1 + \frac{n}{2}$$

Given n, 2^n is a very much larger number, though still finite, but even so, $S_{2^n} \geq 1 + \dfrac{n}{2}$ whatever the value of n. Thus S_{2^n} is unbounded as $n \to \infty$, and the harmonic series diverges. The rate of divergence is very slow and logarithmic in nature. For example, the sum of the first 1000 terms is close to 7.5 only.

(It can be proved that $S_n - \ln n \to 0.5772$, **Euler's constant** as $n \to \infty$). ∎

We can use $\sum x^n$, $\sum \dfrac{1}{n}$ and $\sum \dfrac{1}{n^2}$ as a basis of comparison for other series, noting that the ultimate behaviour of $\{a_n\}$ as $n \to \infty$ is the determining factor.

Example   Examine the convergence of the series whose nth terms are

(a) $\dfrac{3n + 1}{(n + 2)(5n + 3)}$

(b) $\dfrac{n^2 + 1}{(n + 2)(3n^3 - 4)}$

(c) $\dfrac{1}{2^n(3 + (-1)^n)}$

Solution

(a) Divide by n^2 top and bottom to obtain

$$a_n = \frac{3\left(1 + \dfrac{1}{n}\right)}{5n\left(1 + \dfrac{2}{n}\right)\left(1 + \dfrac{3}{5n}\right)}$$

Therefore $a_n \sim \dfrac{3}{5n}$ if n is very large. The $\sim$ sign, known as tilde, sometimes as twiddle, means 'behaves like'. Since $\sum \dfrac{1}{n}$ diverges so does $k \sum \dfrac{1}{n}$ for any constant k.

The series therefore diverges by comparison with the harmonic series.

(b) $a_n = \dfrac{1 + \dfrac{1}{n^2}}{n^2\left(1 + \dfrac{2}{n}\right)\left(3 - \dfrac{4}{n^3}\right)}$

$$a_n \sim \frac{1}{3n^2} \qquad \text{(large } n\text{)}$$

The series converges by comparison with $\sum \dfrac{1}{n^2}$. As in (a) the constant factor, $\dfrac{1}{3}$, does not affect the result.

(c) The nth terms of the series are

$$\frac{1}{2^{n+2}} \quad \text{when } n \text{ is even}$$

or $\quad \dfrac{1}{2^{n+1}} \quad$ when n is odd

In either case, $a_n < \dfrac{1}{2^n}$. By comparison with the geometric series, convergence is assured. ∎

Ratio test

We have so far restricted ourselves to examining the convergence of non-negative series. In the exercises we look at further applications, e.g. the Cauchy condensation test, and at oscillating series, but the convergence of general series is beyond our scope. There is an exception and that rests on our knowledge of the geometric series below.

$$\sum ax^r = a + ax + ax^2 + \cdots + ax^{n-1} = a\frac{(1 - x^n)}{1 - x}$$

provided that $x \neq 1$, where a is a given real number.

If $|x| < 1$ then $a\sum x^r$ converges to $\dfrac{a}{1 - x}$, and we note that the result is also true even if $x < 0$ and the terms oscillate in sign.

If we consider the general power series $\sum a_r x^r$, where the ratio of successive coefficients $\dfrac{a_{r+1}}{a_r} \to L$ as $r \to \infty$, then if r is large

$$a_{r+1} \simeq La_r$$
$$a_{r+2} \simeq L^2 a_r$$
$$\vdots$$
$$a_{r+q} \simeq L^q a_r$$

In other words, each successive term in the series is approximately Lx times its predecessor. The power series thus converges if $|Lx| = |L||x| < 1$, i.e. if

$$|x| < \frac{1}{|L|} = R, \text{ say}$$

We call this quantity R the **radius of convergence**.

For a non-power series, i.e. a power series with $x = 1$, we merely consider the magnitude of the ratio of successive coefficients: $\left|\dfrac{a_{r+1}}{a_r}\right|$. If the limit is L then convergence or divergence follows by comparison with the geometric series. This result is known as the **D'Alembert ratio test** and is defined as follows.

The series $\sum a_r$

(a) converges if $\left|\dfrac{a_{r+1}}{a_r}\right| \to L < 1$

(b) diverges if $\left|\dfrac{a_{r+1}}{a_r}\right| \to L > 1$

The test gives no conclusion when $\left|\dfrac{a_{r+1}}{a_r}\right| \to 1$

You will note that both $\sum\dfrac{1}{n}$ and $\sum\dfrac{1}{n^2}$ come into category (c). However, some useful results can be found by applying the test.

Example

Examine the convergence of the series whose nth terms are

(a) $\dfrac{3^n}{n^3}$

(b) $\dfrac{(-1)^n n^2}{2^n}$

(c) $\dfrac{(-1)^{n-1} x^{2(n-1)}}{(2n)!}$ $(a_1 = 1)$

What is the radius of convergence in case (c)?

(a) Ratio $= \dfrac{3^{n+1}(n+1)^3}{3^n n^3} = 3\left(1 + \dfrac{1}{n}\right)^3 \rightarrow 3$, the series is divergent.

(b) Ratio $= -\dfrac{1}{2}\left(\dfrac{n}{n+1}\right)^2 \rightarrow -\dfrac{1}{2}$, the series is convergent.

(c) Ratio $= \dfrac{(-1)x^2}{(2n+2)(2n+1)} \rightarrow 0$ as $n \rightarrow \infty$, for a fixed x.

The series (c) converges for all values of x, i.e. R is infinitely large. In fact it is the Maclaurin series for $\cos x$, i.e.

$$1 - \frac{x^2}{2!} + \frac{x^4}{4!} - \frac{x^6}{6!} + O(x^8)$$

■

Exercise 10.1

1 Write down the first six terms of the following sequences $\{a_n\}$

(a) $\dfrac{1}{n^2}$

(b) $\dfrac{3n - 2}{n + 1}$

(c) $\dfrac{(2n + 1)(n - 1)}{(n + 1)(n + 2)}$

(d) $\dfrac{n^3 + 2}{3n + 1}$

(e) $\dfrac{n + 1}{(n + 2)(n - 5)}$

(f) $\dfrac{(n + 5)(n - 3)}{(n + 1)(n + 2)(n + 4)}$

Leave the answers for (b) to (f) in ratio and/or product form.

2 Write down the limit to which each sequence in Question 1 converges, as appropriate. You can formalise proof of convergence by dividing numerator/denominator by n, n^2, etc., and noting the behaviour of a_n for large n. If the sequence does not converge, say so.

3 Establish that

$$\frac{2n - 1}{n + 1} = \left(2 - \frac{1}{n}\right)\left(1 - \frac{1}{n} + O\left(\frac{1}{n^2}\right)\right)$$

using the binomial theorem.

Given $a_n = \dfrac{2n-1}{n+1}$ observe that $a_n = 2 + O\left(\dfrac{1}{n}\right)$ for large n and write down the limit of the sequence.

4 Examine the sequences below and state whether they are convergent or divergent. State also whether they are periodic, oscillating or monotonic.

(a) $\dfrac{(-1)^{n+1}}{n+1}$

(b) $(-3)^{n-1}$

(c) $\dfrac{(-1)^{n-1}}{2^n + (-1)^{n-1}}$

(d) $\sin\dfrac{n\pi}{4}$

(e) $\dfrac{1}{n!}$

(f) $\dfrac{1}{n}\cos\dfrac{n\pi}{2}$

In each case write down the first six terms.

5 The $(n+1)$th term of each following sequence is defined in terms of its predecessors. Write down the first six terms of each of the sequences given below and examine the progressive behaviour. (The answers give the limit if it exists.)

(a) $a_{n+1} = \dfrac{1}{2}(1 + a_n)$, $a_1 = 3$

(b) $a_{n+1} = \sqrt{1 + a_n}$, $a_1 = 8$

(c) $a_{n+1} = \dfrac{2 + a_n}{2 + 3a_n}$, $a_1 = 2$

(d) $a_{n+1} = \dfrac{1}{4}(4 - a_n^2)$, $a_1 = 1$

(e) $a_{n+1} = 5 + 6a_n - a_n^2$, $a_1 = 1$

(f) $F_{n+1} = F_n + F_{n-1}$, $F_0 = 0, F_1 = 1$ (the Fibonacci sequence)

(g) By taking a value of a_1 of your own choice, positive or negative, establish that the sequence in (a) converges to the same limiting value.

6 The classical method of determining the square root of a number N is based upon the Newton–Raphson method and sets up a sequence

$$x_{n+1} = \dfrac{1}{2}\left(x_n + \dfrac{N}{x_n}\right)$$

It is known to converge rapidly if the initial approximation x_0 is close.

(a) For $N = 2$ verify that $x_0 = 1.4$ establishes a sequence which rapidly converges to $\sqrt{2}$

(b) Take another $x_0 > 0$ and verify that $\sqrt{2}$ be the limit.

(c) What happens if $x_0 < 0$?

7   Determine the limits of the following sequence by considering each to be a fixed-point iteration. The term a_{n+1} is equal to

(a) $\dfrac{3 + 4a_n}{3 + 2a_n}$ (b)* $\exp(-a_n)$ (c) $\dfrac{1}{4}(2 - 3a_n + a_n^2)$

Note that a good initial approximation is usually needed to ensure convergence. Examine the flexibility in the choice of a_1 that can be permitted in each of the above cases.

8   Prove that

$$a_{n+1} = \frac{1}{6}\left(2a_n + a_{n-1} + \frac{3}{a_{n-1}}\right)$$

has two possible limits. By taking both of a_1, a_2 close to each limit in turn, verify that the suggested limits are both valid.

9   By comparison with the series $\sum \dfrac{1}{n}$ and $\sum \dfrac{1}{n^2}$ test for convergence the series whose nth terms are as follows:

(a) $\dfrac{1}{\sqrt{n}}$ (b) $\dfrac{n + 1}{n(n + 3)}$ (c) $\dfrac{n^2 + 5}{(n^4 + 4)^{2/3}}$

(d) e^{-n} (e) $\dfrac{\ln n}{n}$ (f) $\dfrac{1}{n!}$

In this question and in the ones that follow, the symbol $\sum$ represents the infinite series $\displaystyle\sum_{n=1}^{\infty}$ unless otherwise stated.

10   (a)   Regroup the series $\sum \dfrac{1}{n^2}$ as

$$S = 1 + \left(\frac{1}{2^2} + \frac{1}{3^2}\right) + \left(\frac{1}{4^2} + \frac{1}{5^2} + \frac{1}{6^2} + \frac{1}{7^2}\right) + \left(\frac{1}{8^2} + \cdots + \frac{1}{(15)^2}\right) + \cdots$$

By considering the content of each bracket, deduce that

$$S \leq 1 + \left(\frac{1}{2}\right) + \left(\frac{1}{4}\right) + \left(\frac{1}{8}\right) + \cdots$$

Hence prove that $S \leq 2$

(b)   Now repeat the argument for

$$S_N = \sum_{n=1}^{N} \frac{1}{n^k} \qquad k > 1, N \text{ large}$$

Prove that

$$S_N \leq 1 + 2^{1-k} + 4^{1-k} + \cdots$$

Deduce that $\sum \dfrac{1}{n^k}$ converges for $k > 1$. Determine a bound for its sum.

(c) What happens to $\sum \dfrac{1}{n^k}$ if $k \leq 1$?

11 (a) Given that $a_n > 0$ and that $\sum a_n$ is convergent, say what conclusions can be deduced concerning

(i) $\sum a_n^2$ (ii) $\sum \sqrt{a_n}$

(b) What happens if $\sum a_n$ is divergent?

12 Use the D'Alembert ratio test to establish the convergence or divergence of the series whose nth terms are

(a) $\dfrac{2^n}{n^2}$ (b) $\dfrac{1}{n3^n}$ (c) $\dfrac{n^5}{n!}$

(d) $\dfrac{n!}{(2n)!}$ (e) $nx^{n+1}, \quad 0 < x < 1$ (f) $\dfrac{100^n}{n!}$

13 The Maclaurin series for e^x is $\sum \dfrac{x^r}{r!}$. Prove that it must converge if $x > 0$.

14 A series of the form $\sum (-1)^{n-1} a_n$ is called an **oscillating series**, i.e. $a_1 - a_2 + a_3 - a_4 + \text{etc.}, a_n \geq 0$.

If the sequence $\{a_n\}$ also decreases monotonically to the limit zero, i.e. $a_n \geq a_{n+1} > 0$ and $a_n \to 0$, then the oscillating series converges.

(a) Prove that $1 - \dfrac{1}{2} + \dfrac{1}{3} - \dfrac{1}{4} + \dfrac{1}{5} \cdots = \sum \dfrac{(-1)^{n-1}}{n}$ converges.

(Its sum is $\ln 2$.)

(b) For which values of k does $\sum \dfrac{(-1)^{n-1}}{n^k}$ converge?

(c) Establish that $\sqrt{n+1} - \sqrt{n} = \dfrac{1}{\sqrt{n+1} + \sqrt{n}}$ and prove that the series

$$\sum (-1)^n (\sqrt{n+1} - \sqrt{n})$$

converges.

15   The series $\dfrac{1}{2} - \dfrac{2}{5} + \dfrac{3}{10} - \dfrac{4}{17} + \cdots$ can be written as $\sum \dfrac{(-1)^{n-1}n}{n^2+1}$. Establish that

$$\frac{n}{n^2+1} > \frac{n+1}{(n+1)^2+1}$$

and prove that the series is convergent.

16   Observe that

$$1 - \frac{1}{2^2} + \frac{1}{3^2} - \frac{1}{4^2} + \cdots = 1 + \frac{1}{2^2} + \frac{1}{3^2} + \frac{1}{4^2} + \cdots - 2\left(\frac{1}{2^2} + \frac{1}{4^2} + \cdots\right)$$

If $\sum \dfrac{1}{n^2} = \dfrac{\pi^2}{6}$, what is the value of $\sum \dfrac{(-1)^{n-1}}{n^2}$?

17   If $\{a_n\}$ is monotonically decreasing to 0 then $\sum a_n$ is convergent or divergent whenever $\sum 2^n a_{2^n}$ is convergent or divergent, a result known as the **Cauchy condensation test**. For example, since

$$1 + \frac{1}{2} + \frac{1}{3} + \frac{1}{4} + \cdots + \frac{1}{8} \cdots$$

diverges then

$$1 + \frac{2}{2} + \frac{4}{4} + \frac{8}{8} + \cdots$$

also diverges.

(a)   Use the test to prove that

$$\sum \frac{1}{n^k} \text{ converges if } k > 1 \text{ and diverges if } k \le 1$$

(b)   For which values of k are the following convergent or divergent?

(i) $\sum \dfrac{1}{n(\ln n)^k}$ $(n \ge 2)$ (ii) $\sum \dfrac{1}{n \ln n(\ln \ln n)^k}$ $(n \ge 3)$

18   (a)   Observe that

$$1 + \frac{1}{2} + \frac{1}{3} + \frac{1}{4} + \frac{1}{5} + \cdots < 1 + \frac{1}{2} + \frac{1}{2} + \frac{1}{4} + \frac{1}{4} + \cdots$$

Hence prove for the harmonic series, $\sum \dfrac{1}{n}$, that $S_{1000} < 10$ (S_{1000} is the sum of the first 1000 terms).

(b)   Use a computer to find to 4 d.p.

(i) S_{1000} (ii) S_{10^6} (if possible)

(c)   Determine $S_n - \ln n$ for the largest feasible value of n. What do you observe?

19 By taking S_{10}, S_{100}, etc., observe the slow convergence of the following series; the sum is shown in brackets.

(a) $\sum \dfrac{(-1)^{n-1}}{n}$ $(\ln 2)$

(b) $\sum \dfrac{(-1)^{n-1}}{(2n-1)}$ $\left(\dfrac{\pi}{4}\right)$

20 Use a computer to find to 3 d.p. the sum to infinity of the series

(a) $\dfrac{1}{1.3} - \dfrac{3}{5.7} + \dfrac{5}{9.11} - \dfrac{7}{11.13} + \cdots$

(b) $1 - \dfrac{1}{\sqrt{3}} + \dfrac{1}{\sqrt{5}} - \dfrac{1}{\sqrt{7}} + \cdots$

(c) $\dfrac{1}{\ln 2} - \dfrac{1}{\ln 3} + \dfrac{1}{\ln 4} - \dfrac{1}{\ln 5} + \cdots$

(d) $(2-1) - (\sqrt{3}-1) + (4^{1/3}-1) - (5^{1/4}-1) + \cdots$

10.2 RECURSION

We have already met the Fibonacci sequence in which successive terms are generated as the sum of their two predecessors by the formula

$$F_{n+1} = F_n + F_{n-1}$$

This is a **recursive** property and its use to calculate successive terms of the sequence is called **recursion**.

In this sequence we start with F_0. With $F_0 = 0$ and $F_1 = 1$, it follows that

$$F_2 = F_1 + F_0 = 1$$
$$F_3 = F_2 + F_1 = 2$$

and the terms are $\{0, 1, 1, 2, 3, 5, 8, 13 \dots\}$. We do not yet possess a formula for F_n which represents it as a function of n alone.

A more practical illustration of a recursive relationship arises in financial investment. Suppose £Q_0 is invested at a fixed interest rate $p\%$ for n years and by that time it has increased to £Q_n. We can see how the capital increases.

$$\text{After one year} \quad Q_1 = \left(1 + \frac{p}{100}\right)Q_0$$

$$\text{After two years} \quad Q_2 = \left(1 + \frac{p}{100}\right)Q_1 = \left(1 + \frac{p}{100}\right)^2 Q_0$$

Each Q_k, the capital value at k years, where $k = 1, 2, \dots, n$, is being recursively generated by its predecessor Q_{k-1}, the capital value at $(k-1)$ years, by the relation

$$Q_k = \left(1 + \frac{p}{100}\right)Q_{k-1}$$

This is true for all values of k because the interest rate is constant at $p\%$. It is clear that

$$Q_n = \left(1 + \frac{p}{100}\right)^n Q_0$$

and in the next section we prove such results formally.

Example

Now suppose that £100 is added annually to the total amount invested and that $p = 5$. Find a recurrence relation for Q_k, the amount in pounds sterling after k years, and use a geometric series to find the value after 10 years.

First

$$Q_k = (1.05)Q_{k-1} + 100 \qquad (k = 1, 2, \dots, 10)$$

But

$$\begin{aligned}
Q_n &= (1.05)Q_{n-1} + 100 \\
&= (1.05)\{(1.05)Q_{n-2} + 100\} + 100 \\
&= (1.05)\{(1.05)((1.05)Q_{n-3} + 100) + 100\} + 100, \text{ etc.}
\end{aligned}$$

We can see that continuing forward we obtain

$$Q_n = (1.05)^n Q_0 + 100\{(1.05)^{n-1} + (1.05)^{n-2} + \cdots\}$$

with every additional £100 invested for $1, 2, \dots, (n-1)$ years respectively.

Therefore

$$Q_{10} = (1.05)^{10}Q_0 + 100\{(1.05)^9 + (1.05)^8 + \cdots + 1\}$$

$$= (1.05)^{10}Q_0 + 100\left\{\frac{(1.05)^{10} - 1}{0.05}\right\}$$

$$= 1628.89 + 1257.79 = 2526.68$$

Hence the value after 10 years is £2526.68 ◼

We know that a sequence $\{a_n\}$ is recursive if every a_n can be determined by a straightforward function of a fixed number of its predecessors. For example

$$a_{n+1} = 2a_n + 1 \qquad \text{(i.e. } a_n = 2a_{n-1} + 1)$$
$$a_{n+2} + 3a_{n+1} + 2a_n = 0 \quad \text{(i.e. } a_n = -2a_{n-1} - a_{n-2})$$
$$a_{n+2} + 2na_{n+1} - a_n^2 = 0 \quad \text{(i.e. } a_n = -2(n-2)a_{n-1} - a_{n-2}^2)$$

The recursive equations so formed are called **difference equations**. They have many applications in economics and engineering. The investment examples above involved first-order linear difference equations of the type

$$a_{n+1} = ka_n \quad \text{or} \quad a_{n+1} = ka_n + r \qquad (k, r \text{ constants})$$

for which the prime solution is of the type $a_n = Ab^n$, where A and b are constants. In fact, we can try to fit such a solution to a three-term linear difference equation.

Example   Obtain the general solution to the difference equation

$$a_{n+2} + 3a_{n+1} + 2a_n = 0$$

Let $a_n = Ab^n$, so that

$$a_{n+2} = Ab^{n+2}, \qquad a_{n+1} = Ab^{n+1} \quad \text{and} \quad Ab^n(b^2 + 3b + 2) = 0$$

Now $A = 0$ or $b = 0$ would give perfectly valid solutions, but they would be of little interest since then $a_n = 0$ for all n. However, if $b^2 + 3b + 2 = 0$, then $b = -1$ or -2 and $A(-1)^n$ or $B(-2)^n$ are valid solutions, where B is another constant. In general

$$a_n = A(-1)^n + B(-2)^n = (-1)^n(A + B \times 2^n)$$

If, say, $a_0 = 0$ and $a_1 = 1$, then

$$a_0 = (-1)^0(A + B) = A + B = 0$$
$$a_1 = (-1)^1(A + 2B) = -A - 2B = 1$$

so $A = 1, B = -1$

Hence

$$a_n = (-1)^n(1 - 2^n)$$

so $a_0 = 0, a_1 = 1, a_2 = -3, a_3 = 7$, etc.

Note that $a_3 + 3a_2 + 2a_1 = 0$ since $7 - 3 \times 3 + 2 = 0$ ∎

You can try to find a_n, $n \geq 4$, by both difference equation and formula and check they are the same.

Note also that $b^2 + 3b + 2 = 0$, called the **auxiliary quadratic**, possesses real roots; this means that solutions of the type Ab^n are evident. Where the roots are complex, there is a way forward, but we will leave this to Chapter 11.

Recursion is strongly-related to iteration. The Newton–Raphson method determines the square root of a number $a > 0$ by using the sequence of iterates $\{x_n\}$ where

$$x_{n+1} = \frac{1}{2}\left(x_n + \frac{a}{x_n}\right) \qquad (n \geq 0)$$

so x_{n+1} is recursively defined. The initial term x_0 is the first approximation to $\sqrt{a}$. Note that $\{x_n\}$ converges to $\sqrt{a}$ whereas $\{a_n\}$ diverges in the example above.

Example   The recursive iteration $x_{n+1} = x_n(2 - ax_n)$ converges to $\dfrac{1}{a}$, $a > 0$. For $a = 2$ the formula becomes $x_{n+1} = 2x_n(1 - x_n)$

Taking $x_0 = 0.8$, we can compute recursively

$$x_1 = 2(0.8)(1 - 0.8) = 0.32$$
$$x_2 = 0.4352$$
$$x_3 = 0.4916$$
$$x_4 = 0.4999, \text{ etc.}$$

∎

With suitable starting conditions, the terms of a difference equation are easily generated recursively. A first-order equation needs one condition, a second-order equation two, etc. Recursion can also be applied to functions.

Example

The Chebychev polynomials used in approximation theory are defined to be

$$T_n(x) = \cos(n \cos^{-1} x)$$

It is not obvious that $T_n(x)$ is a polynomial, but observe that

$$T_0(x) = \cos(0) = 1$$
$$T_1(x) = \cos(\cos^{-1} x) = x$$

Also if $\theta = \cos^{-1} x$, then using the identity

$$\cos(n+1)\theta + \cos(n-1)\theta \equiv 2 \cos \theta \cos n\theta$$

and writing $T_n(x) = \cos n\theta$, it follows that

$$T_{n+1}(x) + T_{n-1}(x) = 2x T_n(x)$$

or rearranging

$$T_{n+1}(x) = 2x T_n(x) - T_{n-1}(x)$$

which is a recurrence relation for $\{T_n(x)\}$.

But why is $T_n(x)$ a polynomial? Now, $T_0(x) = 1$ and $T_1(x) = x$ are polynomials of degree 0 and 1 respectively. Assume that $T_n(x)$ is a polynomial of degree n, for some $n > 1$, and that $T_{n-1}(x)$ is of degree $n - 1$. Then using the recursive relation

$$T_{n+1}(x) = 2x T_n(x) - T_{n-1}(x)$$

we see that

$$T_{n+1}(x) = 2x \times (\text{polynomial of degree } n) - (\text{polynomial of degree } (n-1))$$
$$= (\text{polynomial of degree } (n+1)) - (\text{polynomial of degree } (n-1))$$
$$= (\text{polynomial of degree } (n+1))$$

In other words, if $T_n(x)$ is a polynomial of degree n, then $T_{n+1}(x)$ is a polynomial of degree $n + 1$. This argument is a special case of mathematical induction, which we look at in the next section. We know here that $T_0(x)$ and $T_1(x)$ are true, so the result is true for all n, stepping forward. For example

$$T_2(x) = 2x^2 - 1$$
$$T_3(x) = 2x(2x^2 - 1) - x = 4x^3 - 3x$$

∎

Exercise 10.2

 1 (a) A sum of money £S_0 is invested at $p\%$ per annum. Write down the difference equation for S_n and S_{n-1} where the terms represent the value of the investment after n and $(n - 1)$ years respectively.

(b) What is the value of the investment after exactly three years?

(c) A sum of £100 is invested at a fixed annual interest rate of 5%.

(i) What is the value of the investment after 5 years?
(ii) What is the value after 7 years and 7 months, assuming that interest is paid *pro rata* for each completed month in the eighth year?
(iii) How long will it be before the original investment is doubled?

 2 A building society pays interest on an investment of £I_0 at an annual rate which can change at any time. Assuming that the interest is added on for every day of the investment and that the annual interest rate is $p_n\%$ on the nth day, write down a recurrence relation linking I_{n+1} and I_n, the values of the investment on the nth and $(n + 1)$th days respectively. How does I_n relate to I_0? (Remember that n may reach several thousand.) You may assume that the investment increase daily by a proportion equal to the annual interest rate divided by 365.

 3 (a) A hire purchase debt of £D is being repaid by fixed monthly instalments of £R starting at the end of the first month. If the debt attracts a fixed annual rate of interest of $p\%$, taken as $(p/12)\%$ per month, write down a recurrence relation linking the debt at the nth and $(n - 1)$th months.

(b) If £D_n denotes the outstanding debt after n months, prove that

$$D_n = \left(1 + \frac{p}{1200}\right)^n D - \frac{R\left\{\left(1 + \dfrac{p}{1200}\right)^n - 1\right\}}{\dfrac{p}{1200}}$$

(c)   If the debt is totally discharged after N months, i.e. $D_N = 0$, show that

(i) $R = \dfrac{\dfrac{p}{1200}\left(1 + \dfrac{p}{1200}\right)^N D}{\left(1 + \dfrac{p}{1200}\right)^N - 1}$

(ii) $N = \dfrac{\ln\left(\dfrac{1}{1 - \dfrac{pD}{1200\,R}}\right)}{\ln\left(1 + \dfrac{p}{1200}\right)}$

(iii) $R > \dfrac{pD}{1200}$, for the repayments to reduce the outstanding debt.

(d)   An outstanding hire purchase debt of £400 on a video cassette recorder is to be repaid by monthly instalments over a two-year period at 11% per year and set at 11/12% per month. At what rate must the monthly repayment be fixed?

(e)   What does 11/12% per month interest actually equate to per year?

4   A mortgage of £M is repaid by fixed monthly instalments of £S over N years at an agreed fixed monthly rate of $i\%$. Prove that

$$S = M\left[\frac{i}{1 - (1 + i)^{-12N}}\right]$$

If $M = 30\,000$, $N = 25$, $i = 7\%$ per year, determine S.

5   The sequence $\{a_n\}$ satisfies the recurrence relation

$$a_{n+2} = 5a_{n+1} - 6a_n$$

(a)   Given $a_0 = 0$ and $a_1 = 1$, compute a_2, a_3, a_4 and a_5.

(b)   Write the recurrence relation as a difference equation

$$a_{n+2} - 5a_{n+1} + 6a_n = 0$$

and show that $a_n = 2^n$ satisfies the equation.

Assuming a general solution of the form $a_n = Ab^n$, obtain the auxiliary quadratic equation and show that the general solution is of the form

$$a_n = A \times 2^n + B \times 3^n$$

Hence use the property $a_0 = 0$ and $a_1 = 1$ to evaluate A and B.

6   The Fibonacci sequence satisfies the recurrence relation

$$F_{n+1} = F_n + F_{n-1}$$

Assuming a solution of the form $F_n = Ar^n$, show that $F_n = Ar_1^n + Br_2^n$ where

$$r_1, r_2 = \frac{1}{2}(1 \pm \sqrt{5})$$

If $F_0 = 0$ and $F_1 = 1$, prove that

$$F_n = \frac{1}{\sqrt{5}} \left\{ \left(\frac{1 + \sqrt{5}}{2} \right)^n - \left(\frac{1 - \sqrt{5}}{2} \right)^n \right\}$$

and verify that it is correct for $n = 2$ and $n = 3$.

7   The following linear difference equations are given together with their solutions:

(a) $y_{n+2} - 2y_{n+1} + 2y_n = 0$, $y_1 = 1$, $y_2 = 2$; $y_n = 2^{n/2} \sin \frac{n\pi}{4}$

(b) $y_{n+2} - 6y_{n+1} + 9y_n = 0$, $y_0 = 1$, $y_1 = 6$; $y_n = (n + 1)3^n$

Verify that the equations are satisfied for $n \le 4$

(c) Show that the general solution to the equation

$$y_{n+2} - 3y_{n+1} + 2y_n = 4 \times 3^n \quad \text{is} \quad y_n = c_1 + c_2 2^n + 2 \times 3^n$$

8   For the non-linear difference equation given in the text, namely

$$a_{n+2} + 2na_{n+1} - a_n^2 = 0$$

obtain a_n, $n \le 6$, given $a_0 = 0$, $a_1 = 1$

9 The existence of a limit l of a sequence of iterates $\{x_n\}$ can always be validated by putting x_n, x_{n+1}, etc., in the iterative formula equal to l; hence $l = \sqrt{a}$ gives

$$x_{n+1} = \frac{1}{2}\left(x_n + \frac{a}{x_n}\right) \Rightarrow \sqrt{a} = \frac{1}{2}\left(\sqrt{a} + \frac{a}{\sqrt{a}}\right)$$

Verify that $\dfrac{1}{\sqrt{a}}$ is a limit of the squence

$$x_{n+1} = \frac{1}{2}x_n(3 - ax_n^2)$$

Starting with $x_0 = 0.3$, obtain the first five iterates and the limit of the sequence

$$x_{n+1} = \frac{1}{2}x_n(3 - 4x_n^2)$$

10 If $R_n = \dfrac{F_{n+1}}{F_n}$ where F_n and F_{n+1} are the nth and $(n+1)$th Fibonacci numbers, obtain $R_n, n \leq 10$. From Question 7, accepting that $|r_1| > 1$ and $|r_2| < 1$, write down an approximate form for F_n if n is large. What limit does R_n approach?

11* The functions $S_n(x)$ are defined as

$$S_n(x) = \sin(n \sin^{-1} x) \qquad n = 0, 1, 2, \ldots$$

Determine $S_n(x), n \leq 3$. Which of the $S_n(x)$ are polynomials of degree n?

12* The Chebychev polynomials are said to be **orthogonal** over the range $[-1, 1]$ with the weighting factor $w = \dfrac{1}{\sqrt{1 - x^2}}$. In other words

$$\int_{-1}^{1} \frac{T_m(x)T_n(x)}{\sqrt{1 - x^2}}\,dx = \begin{cases} 0 & m \neq n \\ \dfrac{\pi}{2} & m = n \neq 0 \\ \pi & m = n = 0 \end{cases}$$

where $T_m(x)$ and $T_n(x)$ are the polynomials of degree m and n respectively. By changing the integration variable to θ via $x = \cos\theta$, establish the orthogonal property for m and $n \leq 2$.

13*   Determine the following:

(a) $P_1(x) = \dfrac{1}{2}\dfrac{d}{dx}(x^2 - 1)$

(b) $P_2(x) = \dfrac{1}{2^2}\dfrac{1}{2!}\dfrac{d^2}{dx^2}((x^2 - 1)^2)$

(c) $P_3(x) = \dfrac{1}{2^3!}\dfrac{1}{3}\dfrac{d^3}{dx^3}((x^2 - 1)^3)$

If $\dfrac{d^n}{dx^n}$ denotes differentiation n times, explain why

$$P_n(x) = \dfrac{1}{2^n n!}\dfrac{d^n}{dx^n}((x^2 - 1)^n)$$

must be a polynomial of degree n (the **Legendre polynomial** of degree n).

14*   For $m, n = 0, 1, 2$ prove the following results for the Legendre polynomials of degree m and n:

(a) $\displaystyle\int_{-1}^{1} P_m(x)P_n(x)dx = \begin{cases} 0 & m \neq n \\ \dfrac{2}{2n+1} & m = n \end{cases}$

assuming that $P_0(x) = 1$.

And for $n = 0, 1, 2$ prove

(b) $(2n + 1)P_n(x) = (n + 1)P_{n+1}(x) + nP_{n-1}(x)$

(c) $x\dfrac{dP_n(x)}{dx} - \dfrac{dP_{n-1}(x)}{dx} = nP_n(x)$

15*   Given that $y_n = \displaystyle\int_0^1 \dfrac{x^n}{x^2 + x + 1}dx$ prove that

$$y_{n+2} + y_{n+1} + y_n = \dfrac{1}{n+1}$$

Show that $y_0 = \dfrac{\pi\sqrt{3}}{9}$ and $y_1 = \dfrac{1}{2}\left(\ln 3 - \dfrac{\pi\sqrt{3}}{9}\right)$ and determine $y_n, n \leq 6$.

10.3 MATHEMATICAL INDUCTION

The arithmetic series

$$S_n = a + (a + d) + (a + 2d) + \cdots + (a + (n - 1)d)$$

can be succinctly expressed as

$$s_n = \sum_{r=0}^{n-1}(a + rd) = \frac{n}{2}(2n + (n - 1)d)$$

In *Foundation Mathematics* 1, Chapter 8, we proved this by taking S_n as a sum first from left to right, then from right to left, adding the results and dividing by 2.

Now if $a = 1, d = 1$ the formula gives

$$1 + 2 + 3 + \cdots + n = \sum_{r=0}^{n-1}(r + 1) = \sum_{r=1}^{n} r = \frac{1}{2}n(n + 1)$$

This compact formula for the sum of the first n positive integers can be shown to be true using mathematical induction. Mathematical induction is a very important theorem-proving method in integer variable analysis, we will summarise it here.

The principle of mathematical induction

A sequence of propositions $\{p_n\}$ is such that each in turn implies its successor, i.e. $p_1 \Rightarrow p_2, p_2 \Rightarrow p_3, \ldots, p_n \Rightarrow p_{n+1}, \ldots$. Furthermore, if it is known that p_1 is true, then $p_1 \Rightarrow p_n$, for all n. The method works as follows:

(i) p_n is assumed to be true
(ii) p_{n+1} is proved to be true when p_n is true, i.e. $p_n \Rightarrow p_{n+1}, n \geq 1$
(iii) p_1 is proved to be true by a direct method

It follows then that $p_1 \Rightarrow p_2 \Rightarrow p_3 \Rightarrow \cdots \Rightarrow p_n$.

Example   Prove by induction that $S_n = \sum_{r=1}^{n} r^2 = \frac{1}{6}n(n + 1)(2n + 1)$; this is p_n.

Then

$$S_{n+1} = \sum_{r=1}^{n+1} r^2 = S_n + (n+1)^2$$

$$= \frac{1}{6} n(n+1)(2n+1) + (n+1)^2$$

$$= \frac{n+1}{6} \{n(2n+1) + 6(n+1)\}$$

$$= \frac{n+1}{6} (2n^2 + 7n + 6)$$

$$= \frac{1}{6}(n+1)(n+2)(2n+3)$$

$$= \frac{1}{6}(n+1)[(n+1)+1][2(n+1)+1]$$

This is the formula for S_n with n replaced by $n+1$. Hence $p_n \Rightarrow p_{n+1}$. Also, $S_1 = \frac{1}{6} \times 1 \times 2 \times 3 = 1 = 1^2$, i.e. p_1 is true; therefore p_n is true. ∎

Induction can *validate* the compact formulae for $\sum r$ and $\sum r^2$ but it *does not construct* them. Now consider the two series, each of n terms:

$$Y = 1 \times 2 \times 3 + 2 \times 3 \times 4 + \cdots + n(n+1)(n+2)$$

and $$Z = 0 \times 1 \times 2 + 1 \times 2 \times 3 + \cdots + (n-1)n(n+1)$$

i.e. $$Y = \sum_{r=1}^{n} r(r+1)(r+2) \quad \text{and} \quad Z = \sum_{r=0}^{n-1} r(r+1)(r+2)$$

By cancelling common terms, we see that $Y - Z = n(n+1)(n+2)$. But if we subtract the first term of Z from the first term of Y, then subtract the second term of Z from the second term of Y, and so on, we obtain

$$1 \times 2 \times (3-0) + 2 \times 3 \times (4-1) + \cdots = 3(1 \times 2 + 2 \times 3 + \cdots)$$

$$= 3 \sum_{r=1}^{n} r(r+1)$$

Therefore

$$\sum_{r=1}^{n} r(r+1) = \frac{1}{3}n(n+1)(n+2) = \sum_{r=1}^{n} r^2 + \sum_{r=1}^{n} r$$

If $\displaystyle\sum_{r=1}^{n} r$ is known, then $\displaystyle\sum_{r=1}^{n} r^2$ can be easily found.

An exactly parallel analysis can prove that

$$\sum_{r=1}^{n} r(r+1)(r+2) = \frac{1}{4}n(n+1)(n+2)(n+3)$$

from which we can prove that

$$\sum_{r=1}^{n} r^3 = \frac{1}{4}n^3(n+1)^3$$

Example

Find a compact form to n terms for

(a) $1 \times 4 + 2 \times 7 + 3 \times 10 + \cdots$ 　　　　(b) $9^2 + 11^2 + 13^2 + \cdots$

Solution

(a) $\displaystyle S_n = \sum_{r=1}^{n} r(3r+1) = 3\sum_{r=1}^{n} r^2 + \sum_{r=1}^{n} r$

$\displaystyle \quad = 3\frac{n(n+1)(2n+1)}{6} + \frac{n(n+1)}{2}$

$\displaystyle \quad = \frac{n(n+1)}{2}(2n+1+1)$

$\displaystyle \quad = \frac{n(n+1) \times 2(n+1)}{2}$

$\displaystyle \quad = n(n+1)^2$

(b) $S_n = \sum\limits_{r=1}^{n} (2r + 7)^2$

$= \sum\limits_{r=1}^{n} (4r^2 + 28r + 49)$

$= 4 \sum\limits_{r=1}^{n} r^2 + 28 \sum\limits_{r=1}^{n} r + 49 \sum\limits_{r=1}^{n} 1$

$= \dfrac{4}{6} n(n+1)(2n+1) + \dfrac{28}{2} n(n+1) + 49n$

$= \dfrac{n}{3} [2(n+1)(2n+1) + 42(n+1) + 147]$

$= \dfrac{n}{3} (4n^2 + 48n + 191)$ ∎

There are generally no compact formulae to describe $\sum\limits_{r=1}^{n} r^k$, where k is a negative integer, but we can find compact expressions for certain sums using partial fractions.

Example   Consider

$$\sum_{r=1}^{n} \frac{1}{r(r+1)} = \frac{1}{1 \times 2} + \frac{1}{2 \times 3} + \cdots + \frac{1}{n(n+1)}$$

Note that

$$\frac{1}{r(r+1)} = \frac{1}{r} - \frac{1}{r+1}$$

so

$$\frac{1}{1 \times 2} + \frac{1}{2 \times 3} + \cdots + \frac{1}{n(n+1)} = \left(1 - \frac{1}{2}\right) + \left(\frac{1}{2} - \frac{1}{3}\right) + \left(\frac{1}{n} - \frac{1}{n+1}\right)$$

$$= 1 - \frac{1}{2} + \frac{1}{2} - \frac{1}{3} + \frac{1}{3} - \cdots - \frac{1}{n} + \frac{1}{n} - \frac{1}{n+1}$$

$$= 1 - \frac{1}{n+1}$$

cancelling out adjacent terms. ∎

This result is of great importance in mathematical analysis because it proves that

$$S_n = \sum_{r=1}^{n} \frac{1}{r(r+1)} = 1 - \frac{1}{n+1}$$

converges to 1 as $n \to \infty$, since $\dfrac{1}{n+1} \to 0$. Hence

$$\frac{1}{2} + \frac{1}{6} + \frac{1}{12} + \frac{1}{20} + \cdots = 1$$

Proof by mathematical induction is particularly useful for validating a suggested result.

Example

$$\frac{1}{4 \times 7} + \frac{1}{7 \times 10} + \frac{1}{10 \times 13} + \cdots + \frac{1}{(3n+1)(3n+4)} = \frac{n}{4(3n+4)}$$

Let

$$S_n = \frac{1}{4 \times 7} + \frac{1}{7 \times 10} + \frac{1}{10 \times 13} + \cdots + \frac{1}{(3n+1)(3n+4)}$$

Then

$$S_{n+1} = S_n + \frac{1}{(3n+4)(3n+7)}$$

$$= \frac{n}{4(3n+4)} + \frac{1}{(3n+4)(3n+7)} = \frac{1}{(3n+4)} \left(\frac{n}{4} + \frac{1}{3n+7} \right)$$

$$= \frac{1}{(2n+4)} \left(\frac{3n^2 + 7n + 4}{4(3n+7)} \right) = \frac{1}{(3n+4)} \frac{(3n+4)(n+1)}{4(3n+7)}$$

$$= \frac{n+1}{4(3(n+1)+7)}$$

In other words, $(n+1)$ has replaced n in the formula. Furthermore,

$$S_1 = \frac{1}{4 \times 7} = \frac{1}{4 \times (3+4)}$$

and the result is therefore proved. ∎

Mathematical induction can be used to prove the binomial theorem, i.e.

$$p_n : (1+x)^n = \sum_{r=0}^{n} \binom{n}{r} x^r \qquad (n \text{ a positive integer})$$

Assume the expansion is true and consider $(1+x)^{n+1} = (1+x)^n(1+x)$. The coefficients of x^r in the expansion of $(1+x)^{n+1}$ is formed by adding from the $(1+x)^n$ expansion:

$$1 \times (\text{coeff. of } x^r) + x \times (\text{coeff. of } x^{r-1}) = \binom{n}{r} + \binom{n}{r-1} = \binom{n+1}{r}$$

which is the addition property of Pascal's triangle, already proved. As we know, this is just the coefficient of x^r with $n \rightarrow n+1$, i.e.

$$p_{n+1}: (1+x)^{n+1} = \sum_{r=0}^{n+1} \binom{n+1}{r} x^r$$

Also $p_1: (1+x) = 1+x$, so the theorem is proved. The argument can be followed in exactly the same way for the expansion of $(a+b)^n$.

We can also use mathematical induction to differentiate a function to any order n.

Example

Prove that

$$\frac{d^n}{dx^n}(\ln x) = \frac{(-1)^{n-1}(n-1)!}{x^n}$$

Assume

$$p_n: \frac{d^n}{dx^n}(\ln x) = \frac{(-1)^{n-1}(n-1)!}{x^n}$$

Now

$$\frac{d}{dx}\left\{\frac{(-1)^{n-1}(n-1)!}{x^n}\right\} = (-1)^{n-1} \times (n-1)! \frac{n}{x^{n+1}}$$

$$= \frac{(-1)^n n!}{x^{n+1}}, \quad \text{i.e. } p_{n+1}, \quad \text{since } n \rightarrow n+1, n-1 \rightarrow n$$

Also $p_1: \dfrac{d}{dx}(\ln x) = \dfrac{1}{x}$ is true, $(0! = 1)$ ∎

Combining nth order differentiation with the binomial theorem, we obtain **Leibniz' theorem** for the nth derivative of the product of two functions of x, namely u and v. The theorem is

$$\frac{d^n}{dx^n}(uv) = \sum_{r=0}^{n} \binom{n}{r} u^{(r)} v^{(n-r)}$$

where $u^{(r)} = \dfrac{d^r u}{dx^r}$, etc.; $u^{(0)} = u$.

The proof is exactly the same as for the binomial theorem with $(a+b)^n$ except that derivatives replace powers. The Pascal's triangle equivalent is

$$uv$$
$$u'v + uv'$$
$$u''v + 2u'v' + uv''$$
$$u'''v + 3u''v' + 3u'v'' + uv'''$$

In general $u^{(n)}v + \binom{n}{1}u^{(n-1)}v' + \cdots + \binom{n}{r}u^{(n-r)}v^{(r)} + \cdots + uv^{(n)}$

Example   Determine the nth derivatives of

(a) $x^2 e^x$ (b) $(1-x^2)y$

where y is a function of x.

Solution

(a) Take $u = e^x, v = x^2$. The third and higher derivatives of x^2 all vanish and all derivatives of e^x are e^x. We obtain

$$e^x\left\{x^2 + \binom{n}{1} \times 2x + \binom{n}{2} \times 2\right\} = e^x(x^2 + 2nx + n(n-1))$$

(b) Note that $\dfrac{d}{dx}(1-x^2) = -2x$, $\dfrac{d^2}{dx^2}(1-x^2) = -2$ and that higher derivatives vanish. We therefore obtain

$$(1-x^2)y^{(n)} + \binom{n}{1} \times (-2x) \times y^{(n-1)} + \binom{n}{2} \times (-2) \times y^{(n-2)}$$

$$= (1-x^2)y^{(n)} - 2nxy^{(n-1)} - n(n-1)y^{(n-2)} \qquad \blacksquare$$

Exercise 10.3

1   Prove by mathematical induction that

(a) $1 + 3 + 5 + \cdots + (2n-1) = n^2$

(b) $1 \times 2 + 2 \times 5 + 3 \times 8 + \cdots + n(3n-1) = n^3 + n^2$

(c) $\dfrac{1}{2 \times 3} + \dfrac{1}{3 \times 4} + \dfrac{1}{4 \times 5} + \cdots + \dfrac{1}{(n+1)(n+2)} = \dfrac{n}{2(n+2)}$

2   Given that $\displaystyle\sum_{r=1}^{n} r = \dfrac{1}{2}n(n+1)$, it is obvious that the right-hand side is an integer because 2 divides exactly one of n or $n+1$. Justify that the right-hand side of the following formula is an integer

$$\sum_{r=1}^{n} r^2 = \frac{1}{6}n(n+1)(2n+1)$$

3   By considering

$$X = 1 \times 2 \times 3 \times 4 + 2 \times 3 \times 4 \times 5 + \cdots + n(n+1)(n+2)(n+3)$$

prove that

$$\sum_{r=1}^{n} r(r+1)(r+2) = \frac{1}{4}n(n+1)(n+2)(n+3)$$

Hence show that $\displaystyle\sum_{r=1}^{n} r^3 = \left(\dfrac{1}{2}n(n+1)\right)^2$, i.e. the sum of the cubes of the first n positive integers, is equal to the square of their sum.

4   What is the largest integer that always divides $n(n+1)(n+2)(n+3)$?

5   Verify the results in Question 1 by using the known results of $\displaystyle\sum_{r=1}^{n} r$ and $\displaystyle\sum_{r=1}^{n} r^2$

6   Without using induction prove that the sum to n terms of the series given below are as follows:

(a) $1 \times 3 + 2 \times 5 + 3 \times 7 + \cdots = \dfrac{1}{6}n(n+1)(4n+5)$

(b) $1 \times 1^2 + 4 \times 3^2 + 7 \times 5^2 + \cdots = \dfrac{5}{6}(18n^3 - 4n^2 - 9n + 1)$

7   Prove the following:

(a) $\dfrac{1}{1 \times 3} + \dfrac{1}{3 \times 5} + \dfrac{1}{5 \times 7} + \cdots + \dfrac{1}{(2n-1)(2n+1)} = \dfrac{n}{2n+1}$

(b) $\dfrac{1}{4 \times 7} + \dfrac{1}{7 \times 10} + \cdots + \dfrac{1}{(3n+1)(3n+4)} = \dfrac{n}{4(3n+4)}$

8 Use mathematical induction to prove that $f(n) = 3^{2n+1} - 2^{2n-1}$ is divisible by 5.

9 Prove that the series in Question 7 are convergent by considering their sums if n is large. What are the sums as $n \to \infty$?

10 Write down the first three terms of $\sum_{r=2}^{n} \ln\left(1 - \frac{1}{r}\right)$ and find their sum. Hence determine the sum of the series.

11 Assuming that $\frac{d}{dx}(ax + b) = a$, use mathematical induction to prove that

$$\frac{d}{dx}(ax + b)^n = na(ax + b)^{n-1},$$

where n is a positive integer.

12 A function satisfies

$$f(x + y) = f(x)f(y) \qquad (x, y \text{ real})$$
$$f(1) = e \qquad f(0) = 1$$

Use mathematical induction to prove that $f(n) = e^n$ for n a positive integer. Deduce that $f(-n) = e^{-n}$ and $f\left(\frac{1}{n}\right) = e^{1/n}$. Further deduce that $f\left(\frac{p}{q}\right) = e^{p/a}$ where p and q are integers and, assuming $f(x)$ to be continuous, write down its apparent form.

13 Given that the indefinite integral of x is $\frac{1}{2}x^2$ plus an arbitrary constant use integration by parts, together with mathematical induction, to prove that

$$\int x^n \, dx = \frac{x^{n+1}}{n + 1} + C$$

14 Use mathematical induction to prove that $\frac{d^n}{dx^n}(\sin x) = \sin\left(x + \frac{n\pi}{2}\right)$

15 (a) If k is real use mathematical induction to prove that

$$\frac{d^n}{dx^n}(x^k) = k(k - 1)\cdots(k - n + 1)x^{k-n},$$

where n is a positive integer, assuming

that $\frac{d}{dx}(x^k) = kx^{k-1}$

(b)   What happens if (i) k is a positive integer $< n$ and (ii) $k = n$?

(c)   Write down the nth derivative of $\sqrt{x}$

16   Obtain nth derivatives of the following functions:

(a) $x^2 \sin x$ $(u = \sin x; v = x^2; v^{(r)}(x) = 0, r \geq 3)$

(b) $x^n(1 + x^2)$ (some n, integer)

(c) $x^3 e^x$

(d) $x^2 \ln x$

17 (a)   If y is a function of x, write down the nth derivatives of
          (i) y' (ii) $y^{(5)}$

(b)   Use Leibniz theorem to determine nth derivatives of
          (i) xy (ii) $(2x + 3)y'$ (iii) $x^2 y''$

(c)   A differential equation for $y(x)$ takes the form

$$(1 - x^2)y'' - xy' - y = 0$$

By differentiating each of the three terms separately n times, prove that

$$(1 - x^2)y^{(n+2)}(x) - (2n + 1)xy^{(n+1)}(x) - (n^2 + 1)y^{(n)}(x) = 0$$

18   If $y = \sin^2 x$ prove that $y'' + 4y = 2$, and that

$$y^{(n+2)}(x) + 4y^{(n)}(x) = 0, \qquad n \geq 1$$

Write down $y(0), y^{(r)}(0), 1 \leq r \leq 6$. What can you say about $y^{(r)}(0)$ in general when r is odd?

19 (a)   Given the matrix $A = \begin{bmatrix} 1 & 1 \\ 0 & 1 \end{bmatrix}$, prove by induction that $A^n = \begin{bmatrix} 1 & n \\ 0 & 1 \end{bmatrix}$. Determine A^{-1} and prove that $(A^n)^{-1} = (A^{-1})^n$

(b)   For the matrix $B = \begin{bmatrix} 1 & 0 \\ 1 & 2 \end{bmatrix}$, generate powers of B and aim to identify a form for B^n. Prove by induction that it is valid and that the form also holds for B^{-1}. Note that B^0 is the identity matrix.

20 The rotation matrix $\boldsymbol{R}(\theta)$ can be written in the form

$$\boldsymbol{R}(\theta) = \begin{bmatrix} \cos\theta & \sin\theta \\ -\sin\theta & \cos\theta \end{bmatrix}$$

(a) Prove by induction that

$$\boldsymbol{R}(n\theta) = (\boldsymbol{R}(\theta))^n$$

(b) What happens if n is a negative integer?

10.4 POWER SERIES

An infinite series of the form

$$f(x) = \sum_{r=0}^{\infty} a_r x^r \equiv a_0 + a_1 x + a_2 x^2 + \cdots$$

is called a **power series**. We have already encountered the concept in Chapter 5 with Maclaurin series but we must be aware that the Maclaurin series is only a special example. At first sight a power series looks like an infinite polynomial, but the function $f(x)$ to which it converges is not a polynomial unless the series is finite. In other words, $f(x)$ does not generally obey the fundamental theorem of algebra, e.g. by having infinitely many roots. It may not possess any roots at all.

First of all a power series needs to converge to be meaningful, and the chances of this happening diminish as x becomes larger in magnitude. A threshold may be reached beyond which the series cannot converge. We call this the **radius of convergence** and denote it by R. This property is best illustrated by an example.

Example Determine a power series valid for small $|x|$ which converges to $f(x) = \dfrac{1}{1+x}$. The binomial expansion for $f(x)$ is $f(x) = 1 - x + x^2 - x^3 + O(x^4)$, i.e. $\sum_{n=0}^{\infty} (-1)^n x^n$. The Maclaurin series for $f(x)$ is identical.

You can verify easily that $R = 1$ using the D'Alembert ratio test. Hence, if $|x| < 1$ then the series converges. ∎

This demonstrates a fundamental property of power series, quoted without proof, that a power series is **unique**. It therefore does not matter how we find it.

Furthermore, $f(x) = \dfrac{1}{1+x}$ is not a polynomial and is never zero for any $x \in \mathbb{R}$, let alone for $-1 < x \le 1$. However, if x is very small, perhaps so that (x^3) could be neglected, then $1 - x + x^2$ would approximate $f(x)$ very closely.

The radius of convergence acts as a barrier to the outward application of $f(x)$ for larger values of $|x|$. We can see that if $f(x) = \dfrac{1}{1+x}$, then $f(1) = \dfrac{1}{2}$ but the series does not converge at $x = 1$ even though it is bounded. However, $f(-1)$ is infinite and the series rather obviously diverges when $x = -1$. On the other hand $f(3) = \dfrac{1}{4}$, though quite evidently the power series diverges, but we can write instead

$$f(x) = \frac{1}{1+x} = \frac{1}{x\left(1 + \dfrac{1}{x}\right)}$$

This time we can expand $\left(1 + \dfrac{1}{x}\right)^{-1}$ by the binomial theorem, valid for $\left|\dfrac{1}{x}\right| < 1$, i.e. $|x| > 1$, so a unique power series in reciprocal powers of x results. Such a device can be used for other functions too, and parallel arguments can be applied for the determination of radii of convergence.

It may be true that a power series exists but to find it may involve much algebra.

Example

Use computer algebra to determine $e^{\cos x - 1}$ as far as $O(x^8)$.

The problem here is to find the expansion in powers of x, i.e. for the expression

$$\frac{1}{e}\left(1 + e^{\cos x} + \frac{e^{2\cos x}}{2!} + O(e^{3\cos x}) + \cdots\right)$$

Each exponent has its own expansion. It can be shown that the series is

$$1 - \frac{x^2}{2} + \frac{x^4}{6} - \frac{31x^6}{720} + \frac{379x^8}{40\,320} + O(x^{10}).$$ ∎

The series above is in powers of x^2, rather than x, but this is evident from the evenness of $\cos x$ and thus $e^{\cos x}$. A power series expansion of another function might be in different powers of x.

Example

Find the first five terms of the power series expansions for

(a) $\dfrac{\sin \sqrt{x}}{x\sqrt{x}}$ 　　　　　　 (b) $\ln\left(\dfrac{x+2}{x}\right), x \gg 0$ 　　　 (c) $e^{1/\sqrt{x}}$

For which ranges of x are they valid? ($x \gg 0$ means x is very large and positive)

(a) If $y = \sqrt{x}$, then

$$\sin y = y - \frac{y^3}{6} + \frac{y^5}{120} + O(y^7)$$

Substituting $y = \sqrt{x}$ and dividing by $x\sqrt{x}$, we obtain the series

$$\frac{1}{x} - \frac{1}{6} + \frac{x}{120} - \frac{x^3}{5040} + \frac{x^5}{362\,880}$$

(b) $\ln\left(\frac{x+2}{x}\right) = \ln\left(1 + \frac{2}{x}\right)$. Since $x \gg 0$, we can safely assume that $0 < \frac{2}{x} < 1$. We set $y = \frac{2}{x}$, then use the Maclaurin expansion for $\ln(1 + y)$, i.e.

$$\ln\left(\frac{x+2}{x}\right) = \frac{2}{x} - \frac{2}{x^2} + \frac{8}{3x^3} - \frac{4}{x^4} + \frac{32}{5x^5} + O(x^6)$$

(c) If $y = \frac{1}{\sqrt{x}}$ then, using the expansion of e^x, e^y can be expanded as

$$e^{1/\sqrt{x}} = 1 + \frac{1}{x^{1/2}} + \frac{1}{2x} + \frac{1}{6x^{3/2}} + \frac{1}{24x^2} + O(x^{-5/2})$$

In (a) the series is valid for $x > 0$, not $x = 0$, but for $x < 0$ it converges to

$$\frac{\sin\sqrt{-x}}{x\sqrt{-x}}$$

In (b) because $R = 1$ for $\ln(1 + y)$ then $\frac{2}{x} < 1$, i.e. $x > 2$. In (c), as with (a), the series is valid for $x > 0$, not valid for $x = 0$, but $e^{1/\sqrt{(-x)}}$ can be similarly expanded if $x < 0$. ∎

Power series expansions can be very helpful in approximating functions more simply if x is suitably constrained.

Example (a) Given $y = \sinh(\sin^{-1} x)$ prove that

$$(1 - x^2)y'' - xy' - y = 0$$

(b) Show that upon differentiating n times,>
$$(1 - x^2)y^{(n+2)} - (2n + 1)xy^{(n+1)} - (n^2 + 1)y^{(n)} = 0$$

(c) Show that $y^{(n+2)}(0) = (n^2 + 1)y^{(n)}(0)$ and determine the Maclaurin series for $y(x)$ to $O(x^7)$. What is its radius of convergence?

(d) Compute $\sinh(\sin^{-1} 0.3)$ from the series and compare it with the answer from your calculator.

Solution

(a)
$$y' = \frac{\cosh(\sin^{-1} x)}{\sqrt{(1 - x^2)}}$$

$$y'' = \frac{\sinh(\sin^{-1} x)}{(1 - x^2)} + \frac{x\cosh(\sin^{-1} x)}{(1 - x^2)^{3/2}}$$

$$= \frac{y}{1 - x^2} + \frac{xy'}{1 - x^2}$$

i.e. $(1 - x^2)y'' - xy' - y = 0$

(b) Differentiating n times, using Leibniz' theorem, gives

$$(1 - x^2)y^{(n+2)} - 2nxy^{(n+1)} - \frac{2n(n - 1)}{2}y^{(n)} - xy^{(n+1)} - ny^{(n)} - y^{(n)} = 0$$

$$(1 - x^2)y^{(n+2)} - x(2n + 1)y^{(n+1)} + (-n^2 + n - n + 1)y^{(n)} = 0$$

$$(1 - x^2)y^{(n+2)} - (2n + 1)xy^{(n+1)} - (n^2 + 1)y^{(n)} = 0$$

(c) When $x = 0$ we obtain

$$y^{(n+2)}(0) = (n^2 + 1)y^{(n)}(0)$$

Now $y(0) = 0$ $y'(0) = 1$ $y''(0) = 0$

Since $y''(0) = 0$ $y^{(2n)}(0) = 0$ (all n)

Also $y^{(3)}(0) = 2$ $y^{(5)}(0) = 20$ $y^{(7)}(0) = 520$

Therefore

$$\sinh(\sin^{-1} x) = x + \frac{x^3}{3} + \frac{x^5}{6} + \frac{13x^7}{126} + O(x^9)$$

The ratio of successive odd-power terms is

$$\frac{(n^2 + 1)x^2}{(n+2)^2 + 1} \quad \text{which converges to } x^2 \text{ as } n \to \infty$$

Hence convergence of the series occurs if $x^2 < 1$, i.e. $|x| < 1$. The radius of convergence is therefore $R = 1$.

(d) The series at $x = 0.3$ has the value $0.309\ 427$ and $\sinh(\sin^{-1}(0.3)) = 0.309\ 429$, so they agree to within 2×10^{-6}. ■

Table 10.1 shows some Maclaurin series and their radius of convergence. Notice that $R = \infty$ for the basic exponential (i.e. hyperbolic) and trigonometric functions and that $R = 1$ for ratio functions (tan, tanh), inverse functions and logarithmic functions.

The radius of convergence of the Maclaurin series of simple modifications of these functions can be found straightforwardly.

Example

Determine the radius of convergence of the Maclaurin series for

(a) e^{-2x} (b) $\ln(1 - ax)$ (c) $(1 + x - x^2)^{1/2}$

(d) $(a + bx)^3$ (e) $(a + bx)^{-3}$

Solution

(a) Comparison with the standard exponential series is made by replacing x by $(-2x)$. For convergence we require that $|-2x| < \infty$. Now

$$|-2x| < \infty \Rightarrow |x| < \infty \quad \text{so} \quad R = \infty.$$

(b) Here x is replaced in the logarithmic series by $(-ax)$. Then the requirement becomes $|-ax| < 1$

$$\text{i.e.} \quad |x| < \frac{1}{|a|}, \quad \text{so} \quad R = \frac{1}{|a|}$$

Table 10.1

$f(x)$	Type	Series	Validity	R
e^x	Exponential	$1 + x + \dfrac{x^2}{2!} + \dfrac{x^3}{3!} + \cdots$	$x \in \mathbb{R}$	∞
$\ln(1+x)$	Logarithmic	$x - \dfrac{x^2}{2} + \dfrac{x^3}{3} - \cdots$	$\lvert x \rvert < 1$	1
$\dfrac{1}{1+x}$	Inverse polynomial	$1 - x + x^2 - x^3 + \cdots$	$\lvert x \rvert < 1$	1
$\sin x$	Trigonometric	$x - \dfrac{x^3}{3!} + \dfrac{x^5}{5!} - \cdots$	$x \in \mathbb{R}$	∞
$\cos x$	Trigonometric	$1 - \dfrac{x^2}{2!} + \dfrac{x^4}{4!} - \cdots$	$x \in \mathbb{R}$	∞
$\tan x$	Trigonometric	$x + \dfrac{x^3}{3} + \dfrac{2x^5}{15} + \cdots$	$\lvert x \rvert < 1$	1
$\sin^{-1} x$	Inverse trigonometric	$x + \dfrac{x^3}{6} + \dfrac{3x^5}{40} + \cdots$	$\lvert x \rvert < 1$	1
$\tan^{-1} x$	Inverse trigonometric	$x - \dfrac{x^3}{3} + \dfrac{x^5}{5} - \cdots$	$\lvert x \rvert < 1$	1
$\sinh x$	Hyperbolic	$x + \dfrac{x^3}{3!} + \dfrac{x^5}{5!} + \cdots$	$x \in \mathbb{R}$	∞
$\cosh x$	Hyperbolic	$1 + \dfrac{x^2}{2!} + \dfrac{x^4}{4!} + \cdots$	$x \in \mathbb{R}$	∞
$\tanh x$	Hyperbolic	$x - \dfrac{x^3}{3} + \dfrac{2x^5}{15} - \cdots$	$\lvert x \rvert < 1$	1
$\sinh^{-1} x$	Inverse hyperbolic	$x - \dfrac{x^3}{6} + \dfrac{3x^5}{40} - \cdots$	$\lvert x \rvert < 1$	1
$\tanh^{-1} x$	Inverse hyperbolic	$x + \dfrac{x^3}{3} + \dfrac{x^5}{5} + \cdots$	$\lvert x \rvert < 1$	1

(c) In the binomial series x is replaced by $x - x^2 \equiv x(1 - x)$, so

$$\lvert x(1-x) \rvert < 1 \Rightarrow \frac{1}{2}(1 - \sqrt{5}) < x < \frac{1}{2}(1 + \sqrt{5}), \quad \text{i.e.} \quad -0.618 < x < 1.618$$

We take $R = 0.618$ so that convergence is guaranteed for all $\lvert x \rvert < R$.

(d) $(a + bx)^3 \equiv a^3 + 3a^2 bx + 3ab^2 x^2 + b^3 x^3$, a cubic polynomial which is valid for all x. Then $R = \infty$, as it must always be for any finite polynomial.

(e) $(a + bx)^{-3} = a^{-3}\left(1 + \dfrac{bx}{a}\right)^{-3}$. The Maclaurin (binomial) series converges if

$$\left|\frac{bx}{a}\right| < 1 \quad \text{i.e.} \quad |x| < \left|\frac{a}{b}\right| \quad \text{so} \quad R = \left|\frac{a}{b}\right| \qquad \blacksquare$$

When we deal with the Maclaurin series of composite functions, the overall radius of convergence must be equal to the minimum of each of the series components, so

$$R = \min(R_1, R_2) \qquad \text{for two series}$$
$$R = \min(R_1, R_2, R_3) \text{ for three series}$$

Example

Write out Maclaurin series for the following functions, where they exist, and determine the radius of convergence where possible. $O(x^p)$ operates when applicable

(a) $\dfrac{\sin(2x + 1)}{3x + 1}$, $\qquad O(x^3)$

(b) $\ln(1 - 5x)\exp(-x(1 + x))\tan^{-1}(2x)$, $\qquad O(x^4)$

(c) $(\tan^{-1} x)^{1/3}$, $\qquad O(x^4)$

(d) $\ln\left(\dfrac{\sin x}{x}\right)$, $\qquad O(x^4)$

Solution

(a) $\sin(2x + 1) = \sin 2x \cos(1) + \cos 2x \sin(1)$ so that $\cos(1)$ and $\sin(1)$ will be multipliers in the series. The expressions for $\sin 2x$ and $\cos 2x$ exist for all $x \in \mathbb{R}$.

$(1 + 3x)^{-1}$ expands binomially, provided that $|3x| < 1$ i.e. $|x| < \dfrac{1}{3}$
The expansion is

$$\cos(1)\left(2x - 6x^2 + \frac{50x^3}{3}\right) + \sin(1)(1 - 3x + 7x^2 - 21x^3) + O(x^4) \text{ and } R = \frac{1}{3}$$

(b) First, $\ln(1 - 5x)$ is valid if $|5x| < 1$, i.e. $|x| < \dfrac{1}{5}$. Then, $\exp(-x(1 + x)) = e^{-x} \times e^{-x^2}$
each valid for all $x \in \mathbb{R}$. Then, $\tan^{-1}(2x)$ is valid for $|2x| < 1$, i.e. $|x| < \dfrac{1}{2}$
The expansion is

$$-10x^2 - 15x^3 - 40x^4 + O(x^5) \qquad R = \min\left(\frac{1}{5}, \infty, \frac{1}{2}\right) = \frac{1}{5}$$

(c) $(\tan^{-1} x)^{1/3} = \left(x - \dfrac{x^3}{3} + \dfrac{x^5}{5} - \cdots \right)^{1/3}$. There is no Maclaurin expansion. However, if the factor $x^{1/3}$ is removed, we have

$$x^{1/3}\left(1 - \dfrac{x^2}{2} + \dfrac{x^4}{5} - \cdots \right)^{1/3} = x^{1/3}\left(1 - \dfrac{x^2}{9} + \dfrac{22x^4}{405} - \cdots \right)$$

It is not possible to determine R easily for the series on the right, although it gives good approximations for $|x| < 0.5$. Note that $(\tan^{-1} 0.5)^{1/3} = 0.7740$, and $(0.5)^{1/3} \times$ (Series to $O(x^4)$) $= 0.7743$.

(d) First, $\ln\left(\dfrac{\sin x}{x} \right) = \ln\left(1 - \dfrac{x^2}{3!} + \dfrac{x^4}{5!} \cdots \right)$. This does possess a Maclaurin series, i.e.

$$-\dfrac{x^2}{6} - \dfrac{x^4}{180} + O(x^6)$$

It is not possible to determine R easily but approximations appear good for $|x| < 1$. To 4 d.p. we have

$$\ln(\sin 1) = -0.1726, \qquad \text{series value} = -0.1722 \qquad \blacksquare$$

Exercise 10.4

1   Assuming the series for $\ln(1 + x)$, $\sin x$ and $\cos x$, as far as $O(x^4)$ find the series for

(a)   (i) $\ln\left(\dfrac{\sin x}{x} \right)$ (ii) $\ln(\cos x)$

(b)   Show that if x is small

$$\tan x \approx x] \exp\left(\dfrac{x^2}{3} \right)$$

2   (a)   Write down a series of negative powers of x which is an expansion of $e^{1/x}$.

(b)   No matter how small x happens to be, this series must exist. Why?

(c)   Put $x = 1$ and calculate values to 4 d.p. Compare with your calculator value for e^{-1}.

(d)   Repeat (c) for $\sqrt{e}$.

3 Expand $\cos(x + h)$ and use the Maclaurin series for $\sin h$ and $\cos h$ to calculate $\cos 61°$ and $\cos 59°$ to 4 decimal places. Assume that $\cos 60° = 0.5$, $\sin 60° = \dfrac{\sqrt{3}}{2}$. Verify the result with your calculator.

 4 Use computer algebra to find the Maclaurin series of the following functions. Unless otherwise stated, work to $O(x^5)$.

   (a) $e^{\sin x}$ (b) $\cos(\sqrt{1 + x^2})$ (c) $\dfrac{\sinh x + (1 + x)^{1/3}}{\sqrt{1 - x^2}}$

   (d) $\cos\left(3x + \dfrac{\pi}{2}\right)\tan\left(x + \dfrac{\pi}{4}\right)$, $O(x^4)$

   (e) $\sqrt{1 + x}\ln(1 + \sqrt{x})e^x$, $O(x^3)$

   Note that in (e) the series is in powers of $\sqrt{x}$.

5 If $y = \dfrac{\sin^{-1} x}{(1 - x^2)^{1/2}}$ show that

   (a) $(1 - x^2)y' - xy = 1$ (b) $y^{(n+2)}(0) = (n + 1)^2 y^{(n)}(0)$

   Hence show that

$$y = x + \frac{2x^3}{3} + \frac{8x^5}{15} + O(x^7)$$

6 Consider $y = \tanh^{-1} x$

   (a)   Show that $(1 - x^2)\dfrac{dy}{dx} = 1$

   (b)   Show that $(1 - x^2)y^{(n+1)} - 2nxy^{(n)} - n(n - 1)y^{(n-1)} = 0$

   (c)   Show that $\tanh^{-1} x = \displaystyle\sum_{n=1}^{\infty} \frac{x^{2n-1}}{2n - 1}$, $|x| < 1$

7 Consider $y = \exp(\sin^{-1} x)$

   (a)   Show that

$$(1 - x^2)y'' - xy' - y = 0$$

   (b)   By differentiating n times, show that $(1 - x^2)y^{(n+2)} - (2n + 1)xy' - (n^2 + 1)y = 0$

   (c)   Hence prove that the first five terms in the Maclaurin expansion are

$$1 + x + \frac{x^2}{2} + \frac{x^3}{3} + \frac{5x^4}{24} + O(x^5)$$

8 (a) If $f(x) = x^m$ where m is a non-negative integer, prove that the Maclaurin series of x^m is x^m itself.

(b) What is the Maclaurin series of $g(x) = P_n(x)$, where $P_n(x)$ is a polynomial of degree n in x?

9 (a) Write down the power series for

$$f(x) = e^{-x^2/2}$$

Integrate it term by term and prove that the **error function**

$$\text{erf}(x) = \frac{1}{\sqrt{2\pi}} \int_0^x e^{-t^2/2}\, dt$$

$$= \frac{1}{\sqrt{2\pi}} \left(x - \frac{x^3}{6} + \frac{x^5}{40} - \frac{x^7}{336} + O(x^9) \right)$$

Compare erf(1) calculated from the series with the true value of 0.3413

(b) By integrating the power series expansion for $f(x) = \dfrac{1}{1+x^2}$ obtain a power series for $\tan^{-1} x$. Hence write down a power series that converges to the sum 1.

10 When analysing the stability of dynamic systems there often arises a differential equation of the form

$$\frac{d^2x}{dt^2} + n^2 x = 0 \qquad (n\text{ constant})$$

This is known as the **simple harmonic motion** (SHM) equation and the frequency of oscillation is $\left(\dfrac{2\pi}{n}\right)$. Throughout, x is a small displacement.

Very often SHM equation comes in a disguised form, i.e.

$$\frac{d^2x}{dt^2} + f(x) = 0$$

where $f(0) = 0, f'(0) > 0$ and $f(x)$ has a power series expansion in x. If x is small, then the frequency of oscillation is $\dfrac{2\pi}{\sqrt{f'(0)}}$. Calculate n in the following cases:

(a) $\dfrac{d^2x}{dt^2} + \dfrac{8(1 - \cos x)}{x} = 0$

(b) $\dfrac{d^2x}{dt^2} + \sin(x + \sin x) = 0$

(c) $\quad \dfrac{d^2x}{dt^2} + \dfrac{\left(e^x - \left(1 + x + \dfrac{x^2}{2}\right)\right)}{\sin^2 x} = 0$ (d) $\quad \dfrac{7}{5}(R - r)\dfrac{d^2\theta}{dt^2} = -g\sin\theta$

Equation (d) describes the period of oscillation of a small sphere of radius r placed near the bottom of a spherical bowl of radius R; g is the acceleration due to gravity.

11 The displacement of a small particle as a function of time is given by

$$x(t) = e^{-t}(1 + 3t)$$

Determine $x(t)$ to $O(t^2)$ when t is small. What happens to $x(t)$ when t is large?

12 A hollow cylindrical metal tube with internal and external radii a and b respectively is rolled down a straight slope of length x inclined at an angle θ. If the time is t then a, b, x, θ and t are connected by the relationship

$$b = a\left(3 - \dfrac{gt^2 \sin\theta}{x}\right)^{1/2}$$

If all the parameters except θ are known determine θ in terms of the others given that it is very small.

13 Planck's formula for the emissive power E_λ of a black body at wavelength λ is

$$E_\lambda = \dfrac{8\pi hc}{\lambda^5}\dfrac{1}{e^{-hc/\lambda kT} - 1}$$

If $O((hc)^2)$ can be neglected, determine an approximate form for E_λ.

14 Determine the radius of convergence of the Maclaurin series for

(a) $\quad (1 + x + 2x^2)^{1/2}$ (b) $\quad \sin(2x + 1) \times \exp(x + x^2)$
(c) $\quad e^{-2x}(1 + 2x)^{-1/4}$ (d) $\quad (1 + 2x)^{1/2}(1 - 3x)^2$
(e) $\quad (1 + 2x)^{1/2}(1 - 3x)^{-2}$ (f) $\quad \sin(x^2 + 1) \times (1 + x + x^2)^{-1}$

15 It is known that the series for $\ln(1 + x)$ converges when $x = 1$ but not when $x = -1$. Find the sums of the series

(a) $\quad 1 - \dfrac{1}{2} + \dfrac{1}{3} - \dfrac{1}{4} + \dfrac{1}{5} - \dfrac{1}{6} + \cdots$ (b) $\quad 1 + \dfrac{1}{2} + \dfrac{1}{3} + \dfrac{1}{4} + \dfrac{1}{5} + \dfrac{1}{6} + \cdots$

SUMMARY

- **Sequences**

  convergent if $\{a_n\} \to L$ as $n \to \infty$

  divergent if $\{a_n\} \to \pm\infty$ as $n \to \infty$

  monotonic increasing if $a_{n+1} > a_n$ for all n

- **Series**: converges (to a finite sum) if the sequence of its partial sums converges

- **Comparison test**: if all terms in both series are positive then if $0 \le b_r \le a_r$

  $$\sum a_r \text{ converges } \Rightarrow \sum b_r \text{ converges}$$

  $$\sum a_r \text{ diverges } \Leftarrow \sum b_r \text{ diverges}$$

- **Ratio test**

  $$\sum a_r \text{ converges if } \left| \frac{a_{r+1}}{a_r} \right| \to L < 1$$

  $$\sum a_r \text{ diverges if } \left| \frac{a_{r+1}}{a_r} \right| \to L > 1$$

  $$\text{No conclusion when } \left| \frac{a_{r+1}}{a_r} \right|$$

- **Recursion**: the general term of a sequence is a function of one or more of its predecessors

- **Mathematical induction**: a sequence of propositions $\{p_n\}$ is such that each in turn implies its successor, i.e. $p_1 \Rightarrow p_2, p_2 \Rightarrow p_3, \ldots, p_{n-1} \Rightarrow p_n, \ldots$; if it is known that p_1 is true, then $p_1 \Rightarrow p_n$, for all n

- **Power series**: an infinite series

  $$\sum_{r=0}^{\infty} a_r x^r \equiv a_0 + a_1 x + a_2 x^2 + \cdots$$

  a power series converges if $|x| < R$, where R is the radius of convergence

- **Maclaurin series**

$$e^x = 1 + x + \frac{x^2}{2!} + \frac{x^3}{3!} + \cdots$$

$$\sin x = x - \frac{x^3}{3!} + \frac{x^5}{5!} - \cdots$$

$$\cos x = 1 - \frac{x^2}{2!} + \frac{x^4}{4!} - \cdots$$

$$\ln(1 + x) = x - \frac{x^2}{2} + \frac{x^3}{3} - \cdots$$

valid for all x except the series of $\ln(1 + x)$, which is valid for $-1 < x \leq 1$.

Answers

Exercise 10.1

1 (a) $1, \dfrac{1}{4}, \dfrac{1}{9}, \dfrac{1}{16}, \dfrac{1}{25}, \dfrac{1}{36}$

(b) $\dfrac{1}{2}, \dfrac{4}{3}, \dfrac{7}{2}, 2, \dfrac{13}{6}, \dfrac{16}{7}$

(c) $0, \dfrac{5\times1}{3\times4}, \dfrac{7\times2}{4\times5}, \dfrac{9\times3}{5\times6}, \dfrac{11\times4}{6\times7}, \dfrac{13\times5}{7\times8}$

(d) $\dfrac{3}{4}, \dfrac{10}{7}, \dfrac{29}{10}, \dfrac{66}{13}, \dfrac{127}{16}, \dfrac{218}{19}$

(e) $\dfrac{2}{3\times(-4)}, \dfrac{3}{4\times(-3)}, \dfrac{4}{5\times(-2)}, \dfrac{5}{6\times(-1)}, \infty, \dfrac{7}{8\times1}$

(f) $\dfrac{6(-2)}{2\times3\times5}, \dfrac{7(-1)}{3\times4\times6}, 0, \dfrac{9\times1}{5\times6\times9}, \dfrac{10\times2}{6\times7\times10}, \dfrac{11\times3}{7\times8\times11}$

2 (a) 0   (b) 3   (c) 2   (d) $\sim \dfrac{n^2}{3} \to \infty$

(e) 0   (f) 0

3 2

4 (a) $\dfrac{1}{2}, -\dfrac{1}{3}, \dfrac{1}{4}, -\dfrac{1}{5}, \dfrac{1}{6}, -\dfrac{1}{7}$
   Oscillating, converges to 0 from both sides

(b) $1, -3, 9, -27, 243, -729$
   Oscillating, diverges to $\pm\infty$

(c) $\dfrac{1}{3}, -\dfrac{1}{3}, \dfrac{1}{9}, -\dfrac{1}{15}, \dfrac{1}{33}, -\dfrac{1}{63}$
   Oscillating, converges to 0 from both sides

(d) $\dfrac{1}{\sqrt{2}}, 1, \dfrac{1}{\sqrt{2}}, 0, -\dfrac{1}{\sqrt{2}}, -1$
   Periodic, not convergent

(e) $1, \dfrac{1}{2}, \dfrac{1}{6}, \dfrac{1}{24}, \dfrac{1}{120}, \dfrac{1}{720}$
   Monotonic, decreasing to limit 0

(f) $\dfrac{1}{\sqrt{2}}, 0, -\dfrac{1}{3\sqrt{2}}, -\dfrac{1}{4}, -\dfrac{1}{5\sqrt{2}}, 0$
   Limit 0, even though some terms are 0

5 (a) $3, 2, \dfrac{3}{2}, \dfrac{5}{4}, \dfrac{9}{8}, \dfrac{17}{16}, \dots;$ $L = 1$

(b) $8, 3, 2, \sqrt{3} (= 1.732), 1.6529, 1.6288;$ $L = 1.6180 = \dfrac{1}{2}(1 + \sqrt{5})$
   L satisfies $L = \sqrt{1 + L}$

(c) $2, \dfrac{1}{2}, \dfrac{5}{7}, \dfrac{19}{29}, \dfrac{77}{115}, \dfrac{307}{461}$

$= 2, 0.5, 0.7143, 0.6552, 0.6696, 0.6659$

$L = \dfrac{2}{3}$, satisfying $L = \dfrac{2+L}{2+3L}$

(d) $1, \dfrac{3}{4}, \dfrac{55}{64}$

$= 1, 0.75, 0.8594, 0.8154, 0.8338, 0.8262$

$L = 2(\sqrt{2} - 1) = 0.8284$

(e) $1, 10, -35, -1430, -2\,053\,475, 4.2168 \times 10^{12}$

(f) $0, 1, 1, 2, 3, 5, 8$

6   (c)   Converges to $-\sqrt{2}$

7   (a) $L = \dfrac{3}{2}$ (not convergent to $L = -1$).

Converges to this value whenever $\dfrac{6}{(3 + 2a_1)^2} < 1$,

i.e. unless $\dfrac{-3 - \sqrt{6}}{2} < a_1 < \dfrac{-3 + \sqrt{6}}{2}$ $(-0.275 < a_1 < 2.725)$

(b) $L = -0.5671$
Converges to this value for all choices of $a_1 > 0$

(c) $L = \dfrac{7 - \sqrt{41}}{2} = 0.2984$

Converges to this value if $\left| \dfrac{1}{4}(-3 + 2a_1) \right| < 1$, i.e. $-\dfrac{1}{2} < a_1 < \dfrac{7}{2}$
Since $g'(\alpha) \sim -0.60$, convergence is slow.

8 $L = \pm 1$

9   (a)   divergent   (b)   divergent   (c)   convergent
(d)   convergent   (e)   divergent   (f)   convergent

10 (c)   Diverges if $k \le 1$

11 (a) If $a_n \to 0$ then $a_n^2 < a_n < 1$ if n is large enough, so $\sum a_n^2$ converges. But if $\sum a_n$ converges, $\sqrt{a_n} > a_n$ for large n, so no conclusion can be drawn, e.g. if

$$a_n = \frac{1}{n^2} \qquad \sqrt{a_n} = \frac{1}{n} \qquad \text{no}$$

$$a_n = \frac{1}{n^4} \qquad \sqrt{a_n} = \frac{1}{n^2} \qquad \text{yes}$$

(b) Argument reversed in every sense.

$$\sum a^2 \quad \text{no conclusion}$$
$$\sum \sqrt{a_n} \quad \text{divergent (examples above)}$$

12 (a) divergent   (b) convergent   (c) convergent

     (d) convergent   (e) convergent   (f) convergent

14 (b) $k > 0$

16 $\dfrac{\pi^2}{12}$

17 (b) (i) Convergent $k > 1$, divergent $k \leq 1$

(ii) Convergent $k > 1$, divergent $k \leq 1$

18 (b) (i) 7.4855   (ii) 12.668

(c) Constant limit $\gamma = 0.5772$ (Euler's constant); convergence slow.

20 (a) 0.278   (b) 0.668   (c) 0.924

     (d) 0.589

Exercise 10.2

1 (a) $S_n = \left(1 + \dfrac{p}{100}\right) S_{n-1}$ (b) $S_3 = \left(1 + \dfrac{p}{100}\right)^3 S_0$

(c) (i) £127.63   (ii) £144.77   (iii) 14.2 years

2 $I_{n+1} = \left(1 + \dfrac{p_n}{36\,500}\right) I_n$

$I_n = \left(1 + \dfrac{p_{n-1}}{36\,500}\right) I_{n-1} = \left(1 + \dfrac{p_{n-1}}{36\,500}\right)\left(1 + \dfrac{p_{n-2}}{36\,500}\right) \cdots \left(1 + \dfrac{p_0}{36\,500}\right) I_0$

(a)　　$D_n = \left(1 + \dfrac{p}{1200}\right)D_{n-1} - R$　　　　(d)　　£18.64

(e)　　11.57%

4　£212.04 (assuming $(7/12)\%$ per month)

5　(a)　　5, 19, 65, 211　　　　　　　　(b)　　$a_n = 3^n - 2^n$

8　$\{0, 1, -2, 9, -50, 481, -2310, \ldots\}$

9　$\{0.3, 0.3960, 0.4698, 0.4973, 0.5000 \ldots\}$

10　$F_n \sim \dfrac{1}{\sqrt{5}}\left(\dfrac{1 + \sqrt{5}}{2}\right)^n,$　　　(n large)

$R_n \sim \dfrac{1 + \sqrt{5}}{2} \sim 1.618,$　　　(n large)

It can be proved this is the limit.

11　$S_0(x) = 0, S_1(x) = x, S_2(x) = 2x\sqrt{1 - x^2}, S_3(x) = 3x - 4x^3, S_n(x)$ is a polynomial of degree n when n is odd only.

13　$P_1(x) = x_1, P_2(x) = \dfrac{1}{2}(3x^2 - 1), P_3(x) = \dfrac{1}{2}(5x^3 - 3x)$

15

n	y_n
0	0.6046
1	0.2470
2	0.1484
3	0.1046
4	0.0803
5	0.0651
6	0.0546

Exercise 10.3

2　Either of n or $n + 1$ is even. $2n + 1$ is odd. Also, either n or $n + 1$ is divisible by 3 or neither is. If not, $2n$ and $2n + 2$ are not divisible by 3 so $2n + 1$ must be. Hence $n(n + 1)(2n + 1)$ is divisible by 6.

4　$n(n + 1)(n + 2)(n + 3)$ is the product of four consecutive integers; 2 and 4 separately are factors and 3 divides at least one, so 24 is always a factor.

6 rth terms

(a) $r(2r+1)$

(b) $(3r-2)(2r-1)^2$

8 $f(n+1) = 9(3^{2n+1} - 2^{2n-1}) + 5 \times 2^{2n-1}$

9 (a) $\dfrac{1}{2}$ (b) $\dfrac{1}{12}$

10 $\ln\dfrac{1}{2} + \ln\dfrac{2}{3} + \ln\dfrac{3}{4} + \cdots + \ln\dfrac{n-1}{n} = \ln\dfrac{1}{n}$

12 e^x

15 (b)   (1) 0   (2) $n!$

(c) $\dfrac{1}{2}\left(-\dfrac{1}{2}\right)\left(-\dfrac{3}{2}\right)\cdots\left(-\dfrac{(2n-3)}{2}\right)x^{1/2-n}$

$= \dfrac{(-1)^{n-1}(2n-3)!}{2^{2(n-1)}(n-2)!}x^{1/2-n}$ $(n \geq 2)$ $(0! = 1)$

16 (a) $x^2\sin\left(x+\dfrac{n\pi}{2}\right) + 2nx\sin\left(x+(n-1)\dfrac{\pi}{2}\right) + n(n-1)\sin\left(x+(n-2)\dfrac{\pi}{2}\right)$

(b) $n!\left\{1 + \dfrac{1}{2}(n+1)(n+2)x^2\right\}$

(c) $e^x\{x^3 + 3nx^2 + 3n(n-1)x + n(n-1)(n-2)\}$

(d) $2(-1)^{n-1}(n-3)!x^{2-n}$

17 (a)   (i) $y^{(n+1)}$ (ii) $y^{(n+5)}$

(b)   (i) $xy^{(n)} + ny^{(n-1)}$ (ii) $(2x+3)y^{(n+1)} + 2ny^{(n)}$

(iii) $x^2y^{(n+2)} + 2nxy^{(n+1)} + n(n-1)y^{(n)}$

18 $y(0) = 0, y'(0) = 0, y^{(2)}(0) = 2, y^{(3)}(0) = 0, y^{(4)}(x) = -8, y^{(5)}(0) = 0, y^{(6)}(0) = 32$

Because $y^{(r+2)}(0) = -4y^{(r)}(0)$, all $y^{(r)} = 0$, r odd because $y'(0) = 0$

19 (b) $B^n = \begin{bmatrix} 1 & 0 \\ 2^n - 1 & 2^n \end{bmatrix}, B^{-1} = \begin{bmatrix} 1 & 0 \\ -\dfrac{1}{2} & \dfrac{1}{2} \end{bmatrix}, B^0 = I$

20 (b) $R(-\theta) = (R(\theta))^{-1}, R^0 = I$

Result holds if n is a negative integer.

Exercise 10.4

1 (a) (i) $-\dfrac{x^2}{6} - \dfrac{x^4}{180} + O(x^6)$ (ii) $-\dfrac{x^2}{2} - \dfrac{x^4}{12} + O(x^6)$

2 (a) $1 + \dfrac{1}{x} + \dfrac{1}{2!x^2} + \dfrac{1}{3!x^3} + O(x^{-4})$

　　(b) Series in (a) must exist for $\dfrac{1}{|x|} < \infty$, i.e. $x \neq 0$

3 $\cos 61° = 0.4848$,　$\cos 59° = 0.51510$

4 (a) $1 + x + \dfrac{x^2}{2} - \dfrac{x^4}{8} - \dfrac{x^5}{15} + O(x^6)$

　　(b) $\cos(1) - \dfrac{x^2}{2}\sin(1) + \dfrac{1}{8}(\sin(1) - \cos(1))x^4 + O(x^6)$

　　(c) $1 + \dfrac{4x}{3} + \dfrac{7x^2}{18} + \dfrac{145x^3}{162} + \dfrac{541x^4}{1944} + \dfrac{19\,033x^5}{29\,160} + O(x^6)$

　　(d) $-3x - 6x^2 - x^3 + x^4 + O(x^5)$

　　(e) $x^{1/2} - \dfrac{x}{2} + \dfrac{11x^{3/2}}{6} - x^2 + \dfrac{63}{40}x^{5/2} - \dfrac{47}{48}x^3 + O(x^{7/2})$

8 (b) Just itself

9 (b) $x - \dfrac{x^3}{3} + \dfrac{x^5}{5} - \dfrac{x^7}{7} + O(x^9)$

$$1 = \dfrac{\pi}{4} - \dfrac{\left(\dfrac{\pi}{4}\right)^3}{3} + \dfrac{\left(\dfrac{\pi}{4}\right)^5}{5} - \dfrac{\left(\dfrac{\pi}{4}\right)^7}{7} + O\left(\dfrac{\pi}{4}\right)^9$$

10 (a) 2　　(b) $\sqrt{2}$　　(c) $\dfrac{1}{\sqrt{6}}$　　(d) $\sqrt{\dfrac{5g}{7(R-r)}}$

11 $1 - 2t - \dfrac{5}{2}t^2$;　as $t \to \infty, x(t) \to 0$

12 $\theta = \dfrac{x\left(3 - \dfrac{b^2}{a^2}\right)}{gt^2}$

13 $-\dfrac{8\pi}{\lambda^4}\left(kT + \dfrac{hc}{2\lambda}\right)$

14 (a) $\sqrt{2} - 1$　　(b) ∞　　(c) $\dfrac{1}{2}$

　　(d) $\dfrac{1}{2}$　　(e) $\dfrac{1}{3}$　　(f) $\dfrac{1}{2}(\sqrt{5} - 1)$

15 (a) $\ln 2$　　(b) '$+\infty$'

11 ORDINARY DIFFERENTIAL EQUATIONS

Introduction

Many phenomena which vary with time can be modelled by differential equations. The most frequently used models are those which involve only first derivatives and those with constant coefficients involving first and second derivatives (first order and second order equations). First-order equations model problems in heat flow and electric circuits among many others. Second order equations model small oscillations and can be applied to Newton's second law of motion and its analogues in circuit theory. When an analytical method of solution cannot be found, a numerical approach may be necessary.

Objectives

After working through this chapter you should be able to

- Follow the development of models describing a range of physical phenomena
- Understand the idea of a family of solutions to a differential equation as well as the role of boundary conditions and initial conditions in selecting the unique solution
- Classify differential equations according to order and degree
- Recognise and solve a variable-separable equation
- Recognise and solve a homogeneous equation
- Recognise and solve an exact equation
- Recognise and solve an equation using an integrating factor
- Solve an equation by the Euler method and by the modified Euler method
- Identify the role of the complementary function and the particular integral in the solution of a linear second-order equation
- Solve a linear second-order equation with constant coefficients
- Interpret solutions in terms of the physical problem.

11.1 BASIC IDEAS

Definitions and solution concepts

Many situations in the physical world are concerned with the rates of change of quantities. Sometimes the mathematical model of the situation is in the form of an equation which involves the derivatives of one quantity with respect to another.

An example is Newton's law of cooling, which relates the rate of change of the temperature of a cooling liquid to the temperature itself. The model is

$$\frac{d\theta}{dt} = -k(\theta - \theta_s) \tag{11.1}$$

where θ is the temperature of the liquid at time t, θ_s is the temperature of the air, assumed constant, and k is a thermal constant, which is a property of the liquid in question.

The model is completed by an **initial condition**:

$$\theta = \theta_0 \quad \text{at} \quad t = 0 \tag{11.2}$$

Equation (11.1) is an example of an **ordinary differential equation**. In this example, t is the **independent variable** and θ is the **dependent variable**. In Section 11.2 the **general solution** of equation (11.1) will be shown to be

$$\theta = \theta_s + Ae^{-kt}$$

where A is a constant. The particular solution which satisfies the initial condition (11.2) is

$$\theta = \theta_s + (\theta_0 - \theta_s)e^{-kt}$$

The quantities k, θ_s and θ_0 are **parameters** of the model. In any particular case they are constant, but their values may change from one case to another; for example, a different liquid will have a different value of k.

In general an **ordinary differential equation** is an equation which relates one or more of the derivatives of one variable quantity (the independent variable) to another variable quantity (the independent variable); the dependent variable may or may not be present explicitly in the equation. In this chapter we shall present examples of such equations and study several common methods of solution.

When several independent variables are involved, the equations are called **partial differential equations**. We will discuss only ordinary differential equations and from now on we will refer often to them merely as differential equations. An accepted abbreviation for ordinary differential equation is o.d.e. and we shall sometimes use it.

When discussing methods of solution we shall normally use y to represent the dependent variable and x to represent the independent variable.

We know that derivatives represent rates of change and that differential equations model a wide variety of situations in which change is involved. They include dynamical systems, fluid motion, stability of systems, chemical reactions, population growth, random processes, economic prediction and genetic evolution.

Systems modelled by ordinary differential equations

Electric circuit

The equation which models the d.c. circuit of Figure 11.1 is

$$L\frac{di}{dt} + Ri = E$$

where the constants L, R, E represent the inductance, resistance and voltage in the d.c. circuit. Physically, this system represents a heavy coiled electric lamp which is battery operated; i is the electric current flowing at time t. When the switch S is closed, the current builds up to a maximum and the lamp brightens. Likewise, when the switch is reopened the lamp dims out.

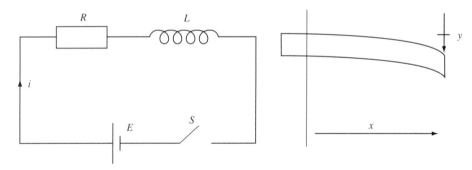

Figure 11.1   Direct current circuit:
$$L\frac{di}{dt} + Ri = E$$

Figure 11.2   Bending of a beam

Bending of a beam

A beam which is supported at one end is deflected by the application of a distributed load (Figure 11.2). The equation describing the deflection is

$$EI\frac{d^2y}{dx^2} = \frac{1}{2}wx(l-x)$$

where y represents the downward displacement of the beam at a distance x along the length. E is Young's modulus, I the second moment of area of the beam cross-section, l the length of the beam and w the magnitude of the distributed load, all constants.

Population stability

The equation governing the change in population in certain circumstances is

$$\frac{dP}{dt} = r(N - P)P$$

where $P = P(t)$ represents the population as a function of time t, which ultimately converges towards a fixed value N: r is a positive constant related to growth per unit time. $P(t)$ tends to follow one of the graphs in Figure 11.3; it converges asymptotically to N.

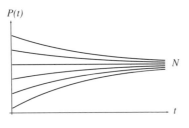

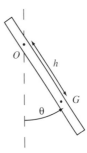

Figure 11.3 Population growth and decay

Figure 11.4 Oscillations of a pendulum:
$$\frac{d^2\theta}{dt^2} = -\frac{Mgh}{I}\theta$$

Small-angle oscillations of a pendulum

The governing equation is

$$\frac{d^2\theta}{dt^2} = -\frac{Mgh}{I}\theta$$

The angle θ, greatly exaggerated in Figure 11.4, represents the small angle of oscillation and t is time. M is the mass of the pendulum, g is the accerlation due to gravity, h is the distance from the pivotal position O to the centre of gravity G, and I is the moment of inertia of the pendulum bar; all are constant.

Classification of differential equations

The **order** of a differential equation is the order of the highest derivatives which occurs; the **degree** of a differential equation is the power to which its highest derivative occurs.

Example

The following differential equations are classified according to order and degree.

(a) $\dfrac{d^2y}{dx^2} + 5\dfrac{dy}{dx} + 2y = e^{-x}\tan x$ second order, first degree

(b) $\left(\dfrac{dy}{dx}\right)^2 + 3y = \cos x$ first order, second degree

(c) $El\dfrac{d^2y}{dx^2} = \dfrac{1}{2}wx(l - x)$ second order, first degree

    $(E, I, w, l$ constants)

(d) $\dfrac{d^3y}{dx^3} + 2\left(\dfrac{d^2y}{dx^2}\right)^2 + \cos\left(\dfrac{dy}{dx}\right) = \dfrac{5x}{y}$ third order, first degree

(e) $\left(\dfrac{d^4y}{dx^4}\right)^2 = 4$ fourth order, second degree

(f) $\left(\dfrac{d^3y}{dx^3}\right)^2 + 2\dfrac{d^2y}{dx} + \cos\left(\dfrac{dy}{dx}\right) = \dfrac{5x}{y}$ third order, second degree

(g) $\dfrac{d^3y}{dx^2} \times \dfrac{dy}{dx} + \left(\dfrac{d^2y}{dx^2}\right)^2 + 15y = 2$ third order, first degree

(h) $y = \dfrac{x^2\sin\left(\dfrac{dy}{dx}\right)}{\dfrac{d^2y}{dx^2} + e^{-x}}$ second order, first degree

(i) $\sin\left(\dfrac{dy}{dx}\right) + \tan\left(\dfrac{dy}{dx}\right) = e^{-xy}$ first order, unclassifiable degree

    The degree cannot be classified as no power of $\dfrac{dy}{dx}$ can be isolated.

(j) $\tan\left(\dfrac{x^2}{y}\right) + e^{3xy^2} = 20$ zeroth order, zeroth degree

This is an **algebraic equation**, not a differential equation, as no derivative is present.

                                                               ■

A differential equation either as a model or in abstract form can often give useful information about the behaviour of the system it represents. Normally we need to **solve** the differential equation, i.e. we need to find $y(x)$ having dispensed with the derivatives $\dfrac{dy}{dx}, \dfrac{d^2y}{dx^2}$, etc. The way to do this is to integrate the differential equation, either directly or indirectly, and shortly we discuss some methods of integration.

Before attempting the solution of simple differential equations by direct integration we first derive some differential equations from their general solutions.

Example   Obtain the simplest differential equation which have the following general solutions; A and B are constants.

(a) $y = Ae^{-x}$ (b) $y = Ae^x + Be^{-x}$ (c) $y = A\sin x + B\cos x$

(d) $y = Ax^B$ (e) $y = \dfrac{A}{x} + 1$

Solution

(a) $\dfrac{dy}{dx} = -Ae^{-x} = -y$ hence $\dfrac{dy}{dx} + y = 0$

(b) Two arbitrary constants imply that we must differentiate twice:

$\dfrac{dy}{dx} = Ae^x - Be^{-x}$ and $\dfrac{d^2y}{dx^2} = Ae^x + Be^{-x} = y$

hence $\dfrac{d^2y}{dx^2} - y = 0$

(c) Again, we need to differentiate twice:

$\dfrac{dy}{dx} = A\cos x - B\sin x$ and $\dfrac{d^2y}{dx^2} = -A\sin x - B\cos x = -y$

hence $\dfrac{d^2y}{dx^2} + y = 0$

(d) Take natural logarithms to obtain $\ln y = \ln A + B\ln x$

Differentiation produces $\dfrac{1}{y}\dfrac{dy}{dx} = \dfrac{B}{x}$ or $\dfrac{x}{y}\dfrac{dy}{dx} = B$

A second differentiation produces $\dfrac{d}{dx}\left(\dfrac{x}{y}\right) \times \dfrac{dy}{dx} + \dfrac{x}{y}\dfrac{d^2y}{dx^2} = 0$

i.e. $\left(\dfrac{y - x\dfrac{dy}{dx}}{y^2}\right) \times \dfrac{dy}{dx} + \dfrac{x}{y}\dfrac{d^2y}{dx^2} = 0$

Multiplying through by y^2 leads to

$$xy\dfrac{d^2y}{dx^2} + y\dfrac{dy}{dx} + x\left(\dfrac{dy}{dx}\right)^2 = 0$$

Using the alternative notation in which y' replaces $\dfrac{dy}{dx}$ and y'' replaces $\dfrac{d^2y}{dx^2}$, we obtain the alternative form of the equation:

$$xyy'' + yy' + x(y')^2 = 0$$

(e) Rearrange to get

$$x(y - 1) = A$$

Then differentiating gives

$$xy' + y - 1 = 0$$ ∎

The reverse process, obtaining the general solution given the o.d.e., is usually harder. In this section we consider only equations of the form

$$\frac{dy}{dx} = f(x)$$

The solution is given by $y = \int f(x)dx$

Examples

1. Find the general solutions of the following differential equations:

(a) $\dfrac{dy}{dx} = 4x + 5$ (b) $\dfrac{dy}{dx} = \ln x$

(c) $\dfrac{dy}{dx} = \dfrac{1}{\sqrt{1 + x^2}}$ (d) $\dfrac{d^2y}{dx^2} = x^2$

And find the particular solutions of the following differential equations with the attached conditions:

(e) $\dfrac{dy}{dx} = \sin x$ $y(\pi) = 2$ (f) $\dfrac{d^2y}{dx^2} = \dfrac{3}{x}$ (i) $y(1) = 0, y(e) = 0$

                                                           (ii) $y(1) = 1, y'(1) = 0$

Solution

(a) $y = 2x^2 + 5x + C$, by straightforward integration, C constant

(b) $y = x \ln x - x + C$, using integration by parts

(c) $y = \sinh^{-1} x + C$, using integration by substitution

(d) $\dfrac{dy}{dx} = \dfrac{x^3}{2} + C$ and $y = \dfrac{x^4}{12} + Cx + D$, D constant

(e) $y = -\cos x + C$; $y(\pi) = -(-1) + C = 2$. Hence $C = 1$ and $y = 1 - \cos x$

(f) $\dfrac{dy}{dx} = 3 \ln x + C$ and $y = 3x \ln x - 3x + Cx + D$

(i) $y(1) = 0$ means $-3 + C + D = 0$ and $y(e) = 0$ means $Ce + D = 0$.

Hence $C = -\dfrac{3}{e-1}$ and $D = \dfrac{3e}{e-1}$

So $y = 3x \ln x - \dfrac{3e}{e-1}x + \dfrac{3e}{e-1}$

(ii) $y'(1) = 0$ means $C = 0$ and $y(1) = 1$ means $D = 3$

Hence $y = 3x \ln x - 3x + 3$

2. A curve is such that its gradient at any point is equal to the distance of that point from the y-axis. Given that the curve passes through the point $(2, -3)$ find its equation. The gradient at a point (x, y) on the curve is $\dfrac{dy}{dx}$ and the distance of the point from the y-axis is x. Hence the curve is defined by the differential equation

$$\frac{dy}{dx} = x$$

This equation can be integrated directly to give $y = \dfrac{1}{2}x^2 + C$

When $x = 2$, $y = \dfrac{1}{2} \times 4 + C = -3$ so that $C = -5$ and $y = \dfrac{1}{2}x^2 - 5$ ∎

Families of solutions

Varying the values of the arbitrary constants in the general solution of a differential equation produces a **family of solutions** as we now see.

Example Determine the family of solutions of the equation $\dfrac{dy}{dx} = 4x$
Integrating directly gives

$$y = 2x^2 + C$$

with the graphs for three different values of C shown in Figure 11.5. Observe that the slope at any x-coordinate must satisfy the condition $y'(x) = 4x$ so that, for example, $y'(1) = 4$, as shown by the graphs.

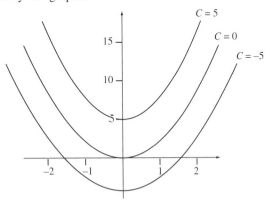

Figure 11.5 Family of solutions to $\dfrac{dy}{dx} = 4x$

Exercise 11.1

1 Identify the order and degree of the following differential equations wherever possible:

(a) $\left(\dfrac{dy}{dx}\right)^2 = 3xy$

(b) $\dfrac{d^3y}{dx^2} + 2\dfrac{d^2y}{dx^2} + 6\dfrac{dy}{dx} + y = \tan x$

(c) $2\left(\dfrac{d^2y}{dx^2}\right)^2 + 6\dfrac{dy}{dx} + \cos\left(\dfrac{x}{y}\right) = \dfrac{d^3y}{dx^3}$

(d) $\dfrac{h^2}{8\pi^2 m}\dfrac{d^2y}{dx^2} + E\dfrac{dy}{dx} = 0$ \qquad (h, E, m constant)

(e) $\dfrac{d^3y}{dx^3} + y\dfrac{d^2y}{dx^2} = 0$

(f) $x^2\dfrac{d^4y}{dx^4}(x+y)\dfrac{dy}{dx} = 0$

(g) $\sin\left(y\dfrac{dy}{dx}\right) = \cos(10xy^2)$

(h) $\left(\dfrac{d^4y}{dx^4}\right)^2 = 4$

(i) $\tan^2\left(\dfrac{dy}{dx}\right) - 3\tan\left(\dfrac{dy}{dx}\right) + 2 = 0$

In cases (h) and (i) find simpler differential equations satisfied by y.

2 Each of the following formulae is the general solution of a differential equation. A and B are arbitrary constants. Differentiate each formula as many times as indicated to eliminate the constants and obtain the differential equations.

(a) $y = Ae^{3x}$ \qquad (once)

(b) $y = Ae^{2x} + Be^{-2x}$ \qquad (twice)

(c) $y = A\cos 3x + B\sin 3x$ \qquad (twice)

(d) $y = Ae^{Bx}$ \qquad (take logarithms, then differentiate twice)

(e) $y = Ax + A^3$ \qquad (once)

3 Given that A, B and C are constants, find the differential equations whose solutions are the following:

(a) $y = Ae^x + Be^{-x}$ (b) $y = Ae^x + Be^{2x} + Ce^{3x}$

(c) $y = e^x(A\cos x + B\sin x)$

4 (a) Prove that for a straight line with equation $lx + my + n = 0$, $\dfrac{d^2y}{dx^2} = 0$

(b) By integrating twice, prove that the general solution of the equation $\dfrac{d^2y}{dx^2} = 0$ can be regarded as the equation of a straight line.

5 (a) The equation describing simple harmonic motion has the form

$$x = A\cos(nt - \alpha)$$

where A is the amplitude, n the frequency and α the phase angle. Prove that

$$\frac{d^2x}{dt^2} + n^2x = 0$$

(b) The given differential equation is the equation of lowest order which has the given solution. Differentiate the differential equation twice to obtain a fourth-order equation which involves x and its fourth derivative only.

6 Find the general solution of the following first-order equations:

(a) $\dfrac{dy}{dx} = 4x$ (b) $\dfrac{dy}{dx} = \dfrac{x}{\sqrt{1-x^2}}$

(c) $\dfrac{dy}{dx} = x\ln x$ (d) $\dfrac{dy}{dx} = xe^{x^2/2}$

7 Using the general solutions of Question 6, write down the particular solutions, where these exist, which satisfy the following conditions:

(a) $y(2) = 3$ (b) $y(2) = 1$

(c) $y(-1) = 0$ (d) $y(-1) = 2\sqrt{e}$

8 By integrating twice find the general solution of each of the following equations. Find also the particular solution, where it exists, which satisfies the given condition.

(a) $\dfrac{d^2y}{dx^2} = 3x$ $\qquad\qquad y(0) = y(1) = 0$

(b) $\dfrac{d^2y}{dx^2} = \sin x$ $\qquad\qquad y(0) = 0, y(\pi) = 1$

(c) $\dfrac{d^2y}{dx^2} = \dfrac{1}{x}$ $\qquad\qquad y(-1) = 1, y(1) = 2$

(d) $\dfrac{d^2y}{dx^2} = \dfrac{4}{\sqrt{1-x^2}}$ $\qquad\qquad y(0) = 1, y(3) = 2$

(e) $\dfrac{d^2y}{dx^2} = x \ln x$ $\qquad\qquad y(1) = 1, y'(1) = 2$

9 For the equations of Question 2, find the particular solutions which satisfy the given conditions.

(a) $y(0) = 1$

(b) (i) $y(0) = 1, \quad y(1) = \cosh 2$ $\qquad$ (ii) $y(0) = y'(0) = 0$

(c) (i) $y(0) = 0, \quad y\left(\dfrac{\pi}{6}\right) = 2$ $\qquad$ (ii) $y\left(\dfrac{\pi}{12}\right) = 1, \quad y\left(\dfrac{\pi}{4}\right) = 0$

(d) (i) $y(0) = 1, \quad y(2) = e$ $\qquad$ (ii) y is constant

(e) (i) $y(2) = 12$ $\qquad\qquad\qquad$ (ii) y is constant

10 The general solution of the simple harmonic oscillator equation

$$\dfrac{d^2x}{dt^2} + x = 0 \quad \text{is} \quad x(t) = A \sin(t + \varepsilon)$$

Find the values of A and ε in each of the following cases, assuming that $A \neq 0$ and $-\dfrac{\pi}{2} \le \varepsilon \le \dfrac{\pi}{2}$

(a) $x(0) = 0, \quad x\left(\dfrac{\pi}{2}\right) = 4$ $\qquad\qquad$ (b) $x(0) = 1, \quad x\left(\dfrac{\pi}{2}\right) = 1$

Show that if the given conditions are $x(0) = 0, x(\pi) = 0$ then we cannot find the value of A.

11 If $Q(t)$ denotes a quantity which varies with time, write down differential equations which model the following situations.

(a) $Q(t)$ decreases with time elapsed; the rate of change of $Q(t)$ is proportional to the amount remaining and invesely proportional to the initial amount, Q_0.

(b)   The rate of change of $Q(t)$ is proportional to the amount present and inversely proportional to the square of the time elapsed since the process started at time $t = t_0$.

12   In the theory of gravitation the force F exerted by one object, mass M, on another object, mass m, a distance r away is given by $F = \dfrac{GMm}{r^2}$, where G is a constant, known as the **constant of gravitation**. This force is the rate of change of the **gravitational potential** $V(r)$ with distance r. Write down the differential equation satisfied by $V(r)$.

If the object of mass m is brought from an infinite distance to a distance r_0 from the other object, find the change in gravitational potential which results.

13   In probability theory the Poisson process can describe a random sequence of events that occur in such a way that the mean number of events in unit time, λ, is constant.

If $P_0(t)$ denotes the probability that no events have occurred up to time t and δt is a small interval of time, it can be shown that $P_0(t + \delta t) = P_0(t)(1 - \lambda \delta t)$.

(a)   Using the defintion

$$P_0'(t) = \lim_{\delta t \to 0} \frac{P_0(t + \delta t) - P_0(t)}{\delta t}$$

obtain a differential equation satisfied by $P_0(t)$ and show that $P_0(t) = Ae^{-\lambda t}$ is a solution. Noting that the probability of an event which is certain to occur is 1, i.e. $P_0(0) = 1$, determine *the* solution.

(b)   Cars on a relatively uncrowded motorway pass under a particular bridge on average once every four seconds. What is the probability of a ten-second gap between two successive cars passing under that bridge?

11.2 First-order equations

We shall confine ourselves to ordinary differential equations of the first order and first degree. The general problem is to solve the equation

$$\frac{dy}{dx} = f(x, y)$$

where $f(x, y)$ is some function of x and y. In many cases we cannot solve the differential equation without resorting to numerical methods but in this section we examine some important special cases which can be solved analytically.

Separation of variables

If $f(x, y) = g(x)h(y)$, i.e. a function of x multiplied by a function of y, we **separate the variables** and write

$$\frac{dy}{dx} = g(x)h(y)$$

so that $\dfrac{1}{h(y)}\dfrac{dy}{dx} = g(x)$

We integrate both sides with respect to x to obtain

$$\int \frac{1}{h(y)}\frac{dy}{dx}dx = \int \frac{1}{h(y)}dy = \int g(x)dx$$

i.e. $H(y) = G(x) + A$, where $H(y)$ is an indefinite integral of $\dfrac{1}{h(y)}$, $G(x)$ is an indefinite integral of $g(x)$, i.e. $H'(y) = \dfrac{1}{h(y)}$, $G'(x) = g(x)$ and A is an arbitrary constant of integration.

We separated the differential equation into an equation where the left-hand side was a function of y and the right-hand side was a function of x. Before discussing the meaning of the solution, let us examine some examples.

Examples

1. Solve the equation $\dfrac{dy}{dx} = y(1 + x^2)$

Separating the variables we obtain the form $\dfrac{1}{y}\dfrac{dy}{dx} = (1 + x^2)$.

Integrating both sides with respect to x we obtain

$$\int \frac{1}{y}\frac{dy}{dx}dx = \int \frac{1}{y}dy = \int (1 + x^2)dx$$

Hence $\ln y = x + \dfrac{1}{3}x^3 + A$, where A is an arbitrary constant.

Note that we did not put an arbitrary constant on each side of the equation after the two indefinite integrations. Since the purpose of the arbitrary constant is to remind us there is an element of uncertainty in the final answer, a single constant is sufficient.

We would not write $\ln y + 4 = x + \frac{1}{3}x^3 + 6$; instead we would write

$$\ln y = x + \frac{1}{3}x^3 + 2$$

In the same way we will not write $\ln y + C = x + \frac{1}{3}x^3 + D$, where C and D are two arbitrary constants.

The formula for y is the general solution of the differential equation. If, however, we write $A = \ln B$, then

$$\ln y = x + \frac{1}{3}x^3 + \ln B$$

hence $\ln \dfrac{y}{B} = x + \dfrac{1}{3}x^3$

Applying exponentials to both sides and multiplying by B gives

$$y = B \exp\left(x + \frac{1}{3}x^3\right)$$

No solutions have been lost, because an arbitrary real number A can be expressed as the natural logarithm of a positive number B, since the range of the logarithmic function is $\mathbb{R}$.

2. If $y(1 + x^2)\dfrac{dy}{dx} = 2x(1 - y^2)$, prove that

$$(1 + x^2)^2(1 - y^2) = D$$

where D is a constant.

Separating the variables and integrating both sides with respect to x we obtain

$$\int \frac{y}{1 - y^2}\frac{dy}{dx}\,dx = \int \frac{y}{1 - y^2}\,dy = \int \frac{2x}{1 + x^2}\,dx$$

i.e. $-\dfrac{1}{2}\displaystyle\int \dfrac{-2y}{1 - y^2}\dfrac{dy}{dx}\,dx = \int \dfrac{2x}{1 + x^2}\,dx$

Hence $-\dfrac{1}{2}\ln(1 - y^2) = \ln(1 + x^2) + \ln C$

where we have written the constant of integration as $\ln C$ without loss of generality.

Then $\ln(1 - y^2) + 2\ln(1 + x^2) + 2\ln C = 0$

i.e. $\ln\{(1 + x^2)^2(1 - y^2)\} = \ln D$

where we have regrouped the logarithms and, without loss of generality, replaced $-2 \ln C$ by $\ln D$. Taking exponentials gives

$$(1 + x^2)^2(1 - y^2) = D$$

To insist that $D > 0$ seems a necessary consequence of having determined the solution in this way via $\ln D$, but if you differentiate the last equation you can re-obtain the differential equation regardless of the sign of D. This case demonstrates that the solution of differential equations may be full of pitfalls.

Some classical problems give rise to differential equations that can be solved using the method of separation of variables.

Examples

1. Newton's law of cooling states:

> *The rate of cooling of a body in a cooler environment is proportional to the temperature difference between the body and the environment.*

Let $\theta(t)$ be temperature of the body and let θ_s be the temperature of the environment, assumed constant. The rate of cooling is therefore $\dfrac{d\theta}{dt}$ and the law is of the form

$$\frac{d\theta}{dt} = -k(\theta - \theta_s)$$

$k > 0$ is a parameter which depends upon the thermal properties of the substance; nevertheless, it can be assumed constant. For cooling, $\dfrac{d\theta}{dt} < 0$ and $\theta > \theta_s$. Separating the variables, we obtain

$$\int \frac{1}{\theta - \theta_s} d\theta = \int kt\, dt$$

Integration produces the equation $\ln(\theta - \theta_s) = kt + \ln A$, and taking exponents gives

$$\theta - \theta_s = Ae^{-kt}$$
i.e. $\qquad \theta = \theta_s + Ae^{-kt}$

Suppose that $t = 0$ the body began to cool from a given temperature, say $\theta = \theta_0$. Putting this information into the solution we find that

$$\theta_0 = \theta_s + Ae^0 = \theta_s + A$$

so that $\quad A = \theta_0 - \theta_s \quad$ and $\quad \theta \equiv \theta(t) = \theta_s + (\theta_0 - \theta_s)e^{-kt}$

As $t \to \infty$, i.e. after a very long time, θ converges to θ_s, the temperature of the environment, and its graph is of the form shown in Figure 11.6.

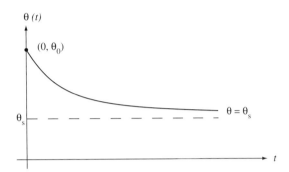

Figure 11.6 Graph to show the temperature of a cooling liquid

2. If a beaker of boiling water, initial temperature $100\,^\circ\mathrm{C}$, is found 10 minutes later to have a temperature of $80\,^\circ\mathrm{C}$ in an environment of temperature $20\,^\circ\mathrm{C}$, we can calculate how long it will take for the temperature of the liquid to fall to $60\,^\circ\mathrm{C}$.

Putting the data into the general solution of the previous example, we obtain the formula

$$\theta = 20 + (100 - 20)e^{-kt} = 20 + 80e^{-kt}$$

When $t = 10$ we have

$$80 = 20 + (100 - 20)e^{-10k}$$

i.e. $\quad e^{-10k} = \dfrac{60}{80} = \dfrac{3}{4}$

and taking logarithms gives

$$k = \frac{1}{10}\ln\frac{4}{3}$$

When $\theta = 60$ we have

$$60 = 20 + 80e^{-kt}$$

so that $\quad e^{-kt} = \dfrac{40}{80} = \dfrac{1}{2}$

and, taking logarithms,

$$kt = \ln 2$$

Substituting for k gives

$$t = \frac{10 \ln 2}{\ln \dfrac{4}{3}} = 24.094$$

Since t is measured from the outset, the temperature falls to $60\,^\circ\mathrm{C}$ in just over 14 minutes from the time that it reaches $80\,^\circ\mathrm{C}$. ∎

Specifying θ_0 in the cooling problem totally defines the solution. This principle operates generally in that a first-order differential equation usually possesses a unique solution if the arbitrary constant of integration is fixed by a starting or **initial condition**, i.e.

$$\frac{dy}{dx} = f(x, y) \quad \text{given} \quad y(0) = y_0$$

usually has a **unique solution**.

This situation parallels the relationship between the indefinite and definite integral. The indefinite integral includes an arbitrary constant but the definite integral has a unique value.

Example   Measured in amps (A), the current $i(t)$ in the circuit of Figure 11.7 satisfies the equation

$$L\frac{di}{dt} + Ri = E$$

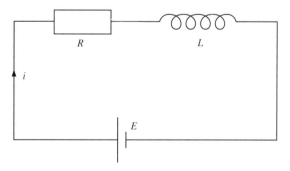

Figure 11.7   Current in an LR circuit

We determine $i(t)$ then find its value when $t = 0.01\,\text{s}$ given that $R = 20\,\Omega$ (ohms), $L = 0.25\,\text{H}$ (henries) and $E = 40\,V$ (volts) with the initial condition $i(0) = 0$. Rearranging the equation, we obtain

$$L\frac{di}{dt} = E - Ri$$

Integrating both sides gives

$$\int \frac{1}{E - Ri}\frac{di}{dt}\,dt = \int \frac{1}{L}\,dt$$

i.e.
$$\int \frac{-R}{E - Ri}\,di = -\int \frac{R}{L}\,dt$$

i.e.
$$\ln(E - Ri) = -\frac{Rt}{L} + C, \text{ on integration}$$

Assuming that $i(0) = 0$ and that $E > Ri$ just afterwards, we are taking the logarithm of a positive quantity. Putting $t = 0$ gives $C = \ln E$, so

$$\ln(E - Ri) - \ln E = -\frac{Rt}{L}$$

hence
$$\ln\left(1 - \frac{Ri}{E}\right) = -\frac{Rt}{L}$$

i.e.
$$1 - \frac{Ri}{E} = e^{-Rt/L}$$

so that
$$i(t) = \frac{E}{R}(1 - e^{-Rt/L})$$

The current i therefore increases from zero to a constant value E/R, usually termed the ohmic voltage (Figure 11.8).

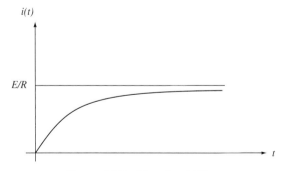

Figure 11.8 Graph of $i(t)$

Substituting the values of the parameters, we obtain

$$i(t) = \frac{40}{20}(1 - e^{-20t/0.25}) = 2(1 - e^{-80t})$$

When $t = 0.01, i = 2(1 - e^{-0.8}) = 1.1013$ A (amps)   ■

Homogeneous equations

Certain first-order and first-degree differential equations which are not separable can be reduced to separable form by a substitution. An important class of this kind has the form

$$\frac{dy}{dx} = \frac{M(x, y)}{N(x, y)}$$

where M and N are polynomial functions of x and y of the same degree, and the sum of the powers of x and y within each term is of the same degree. Such differential equations are termed **homogeneous**.

Example   Consider the case where $M(x, y) = xy$ and $N(x, y) = x^2 + y^2$, both of degree 2. Then

$$\frac{dy}{dx} = \frac{xy}{x^2 + y^2}$$

Let $y = vx$, where $v = v(x)$ is a function of x. Then

$$\frac{dy}{dx} = v + x\frac{dv}{dx}$$

Hence

$$v + x\frac{dv}{dx} = \frac{vx^2}{x^2(1 + v^2)} = \frac{v}{1 + v^2}$$

so that

$$x\frac{dv}{dx} = \frac{v}{1 + v^2} - v = -\frac{v^3}{1 + v^2} \quad \text{or} \quad -\frac{v^3}{1 + v^2} = x\frac{dv}{dx}$$

Separating the variables we obtain

$$\frac{1}{x} = -\frac{1 + v^2}{v^3}\frac{dv}{dx} = -\left(\frac{1}{v^3} + \frac{1}{v}\right)\frac{dv}{dx}$$

Integration produces

$$\int \frac{1}{x} dx = -\int \left(\frac{1}{v^3} + \frac{1}{v} \right) \frac{dv}{dx} dx$$

i.e. $$\int \frac{1}{x} dx = -\int \left(\frac{1}{v^3} + \frac{1}{v} \right) dv$$

then $\ln x + \ln A = \dfrac{1}{2v^2} - \ln v$

so that $\ln vx + \ln A = \dfrac{1}{2v^2}$.

and $\ln Ay = \dfrac{x^2}{2y^2}$ i.e. $x^2 = 2y^2 \ln Ay$ ■

In general, the method proceeds as follows.

If $\dfrac{dy}{dx} = f(v)$ then $v + x\dfrac{dv}{dx} = f(v)$

and $\ln x = \displaystyle\int \frac{dx}{x} = \int \frac{dv}{f(v) - v}$

Provided that the integral on the right-hand side can be found, the procedure is straightforward.

Exact equations

The first-order o.d.e. $\dfrac{dy}{dx} = 3x^2$ can be solved by direct integration to give the general solution $y = x^3 + C$, where C is an arbitrary constant. Now consider the o.d.e.

$$\frac{d}{dx}(xy) = 3x^2 \tag{11.3}$$

Either by substituting $z = xy$ or proceeding straight to the solution, the result of integration is

$$xy = \int 3x^2 \, dx = x^3 + C$$

from which we see that the general solution of (11.3) is

$$y = x^2 + \frac{C}{x}$$

A differential equation, such as (11.3), which can be solved in effect by a simple integration, is said to be **exact**.

Note that $y = x^2$ is a particular solution $(C = 0)$ of (11.3); it is called a **particular integral**. Also, $y = \dfrac{C}{x}$, which contains the arbitrary constant, is called the **complementary** function and it is the general solution of the related differential equation

$$\frac{d}{dx}(xy) = 0$$

Examples

1. Find the general solution of the differential equation $\dfrac{d}{dx}(xy) = 4x^3$
   Integrating gives

   $$xy = \int 4x^3\,dx = x^4 + C$$

   i.e. $$y = x^3 + \frac{C}{x}$$

   Note that the complementary function is the same as before; it is entirely determined by the left-hand side of the o.d.e. The particular integral depends on the right-hand side.

2. Find the general solution of the differential equation $\dfrac{d}{dx}(\sin x \times y) = \cos x$. Integrating gives

   $$\sin x \times y = \int \cos x\,dx = \sin x + C$$

   hence $$y = 1 + C\operatorname{cosec} x \qquad\blacksquare$$

Recognising whether a differential is exact is not always simple; for example, the differential equation

$$x^2\frac{dy}{dx} + xy = 3x^3 \tag{11.4}$$

is equivalent to equation (11.3). The general method of determining whether a differential equation is exact is beyond the scope of this book.

Integrating factor

Consider the differential equation $x^2\dfrac{dy}{dx} + xy = 3x^3$. As it stands, it is not exact, but by dividing it throughout by x we produce equation (11.3), which *is* exact.

A second equation, $\dfrac{dy}{dx} + \dfrac{1}{x}y = 3x$, is also not exact, but by multiplying throughout by x we again recover equation (11.3). The quantity by which we multiply a differential equation to make it exact is called an **integrating factor**. In the first example the integrating factor was $1/x$; in the second example it was x (in the first example, dividing by x is equivalent to multiplying by $1/x$).

In general it is not simple to determine an integrating factor for a differential equation always supposing that the equation has an integrating factor. In this section we look at one special class of differential equation for which a procedure can be followed in order to find its integrating factor. We present the procedure as a set of rules to follow; the derivation of the rules is beyond our scope. The use of this class of equation in modelling physical situations makes the method worth mastering.

First, the equation must be in standard form:

$$\frac{dy}{dx} + P(x)y = Q(x) \tag{11.5}$$

where $P(x)$ and $Q(x)$ are functions of x.

(i)   Take the coefficient of y, including its sign.

(ii) Integrate the coefficient, ignoring the arbitrary constant.

(iii) Take the exponential of the result; this is the integrating factor.

(iv) Multiply the standard form of the differential equation by the integrating factor to give an exact equation.

(v)   As a check, the left-hand side of the current equation can be re-written as the derivative of the product of the integrating factor and the variable y.

(vi) Integrate both sides; this time include the arbitrary constant.

(vii) Divide the resulting equation by the integrating factor to obtain the solution of the original differential equation.

Examples

1. Use the integrating factor method to solve the equation

$$\frac{dy}{dx} + \frac{1}{x}y = 3x$$

The equation is in standard form already.

(i)   The coefficient of y is $+\dfrac{1}{x}$

(ii)   Integration gives $\ln x$

(iii) The integrating factor is $e^{\ln x} \equiv x$

(iv)   Multiplying the standard form by x we obtain the exact equation

$$x\frac{dy}{dx} + y = 3x^2$$

(v)   We can rewrite the exact equation as $\frac{d}{dx}(xy) = 3x^2$

(vi)   Integration gives $xy = \int 3x^2 \, dx = x^3 + C$

(vii) Finally, $y = x^2 + \frac{C}{x}$

2. Find the solution of the differential equation

$$\cos x\frac{dy}{dx} - \sin x \times y = 1$$

which satisfies the condition $y\left(\frac{\pi}{2}\right) = 2$.

First, we divide the equation by $\cos x$ to obtain the standard form

$$\frac{dy}{dx} - \tan x \times y = \sec x$$

(i)   The coefficient of y is $-\tan x$

(ii)   Integration gives $\ln \cos x$

(iii) The integrating factor is $e^{\ln \cos x} \equiv \cos x$

(iv)   Multiplying the standard form by $\cos x$ we obtain the equation

$$\cos x\frac{dy}{dx} - \sin x \times y = 1$$

This is the equation which we were given; it is exact, but we didn't see that—hard luck.

(v)   We can rewrite this equation as $\frac{d}{dx}(\sin x \times y) = 1$

(vi)   Integration gives $\sin x \times y = x + C$

(vii) Finally, $y = (x + C)\operatorname{cosec} x$

Applying the given condition we obtain $\left(\frac{\pi}{2} + C\right) \times 1 = 2$, from which we see that $C = 2 - \frac{\pi}{2}$ so that the solution we require is

$$y = \left(x + 2 - \frac{\pi}{2}\right) \csc x$$

∎

Note that, in applying the method, we have to integrate twice: once to find the integrating factor and once to obtain the penultimate equation. If we cannot carry out both these integrations then the method will fail.

Exercise 11.2

1 The deflection y of a beam of constant cross-section satisfies the differential equation

$$EI\frac{d^2y}{dx^2} = \frac{1}{2}wx(l - x)$$

where E, I, w and l are constant and x is the displacement along the beam. Integrate $\dfrac{d^2y}{dx^2}$ twice and find the general form for the downward displacement $y(x)$.

2 Solve the following differential equations by separation of variables:

(a) $\cos^2 x \dfrac{dy}{dx} = \cos^2 y$

(b) $x^2(1 + y)\dfrac{dy}{dx} + (1 - x)y^2 = 0$

(c) $xy(1 + y^2)\dfrac{dy}{dx} = 1 + y^2$

(d) $x \tan y \dfrac{dy}{dx} = 1$

(e) $\dfrac{dy}{dx} = 2xy$

(f) $\dfrac{dy}{dx} = \dfrac{y}{3y^2 - x}$

(g) $\dfrac{2y}{1 + y^2}\dfrac{dy}{dx} = 1$

(h) $\quad \dfrac{dy}{dx} = x(1 - y)$

(i) $\quad \ln x \dfrac{dy}{dx} + \dfrac{y}{x} = 0, \qquad y(e) = 1$

(j) $\quad (1 + x^2) \dfrac{dy}{dx} = y^2, \qquad y\left(\dfrac{\pi}{4}\right) = 1$

3 In the modelling of conflict by continuous attrition, the following differential equation arises:

$$\frac{dy}{dx} = \frac{bx}{ay}$$

where a and b are constants. Separate the variables to obtain the **Lanchester square law** model

$$ax^2 - by^2 = k$$

as the integral of the differential equation. If x_0 and y_0 denote the initial strengths of the opposing forces, what is the value of k?

4 Determine the solution to the following homogeneous equations:

(a) $\quad \dfrac{dy}{dx} = \dfrac{y(x + 2y)}{x(2x + y)}$

(b) $\quad \dfrac{dy}{dx} = \dfrac{y}{x} + \tan\left(\dfrac{y}{x}\right)$

(c) $\quad \dfrac{dy}{dx} = \dfrac{x + y}{x}$

(d) $\quad \dfrac{dy}{dx} = \dfrac{x^2 + 2y^2}{xy}, \qquad y(1) = 0$

5 For each of the following, determine the general solution to the differential equation then sketch the family of curves which represent the solution.

(a) $\quad \dfrac{dy}{dx} = \dfrac{y}{x}$

(b) $\quad \dfrac{dy}{dx} = \dfrac{x}{y}$

(c) $\quad \dfrac{dy}{dx} = x - y$

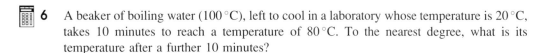

6 A beaker of boiling water ($100\,^{\circ}$C), left to cool in a laboratory whose temperature is $20\,^{\circ}$C, takes 10 minutes to reach a temperature of $80\,^{\circ}$C. To the nearest degree, what is its temperature after a further 10 minutes?

7 The rate of decay of radioactive substance is directly proportional to the amount (m) of substance which remains, i.e.

$$\frac{dm}{dt} = -km$$

(a) If m_0 is the initial mass of the substance, integrate the differential equation to prove that

$$m = m_0 e^{-kt}$$

(b) The **half-life** T of the material is the time for half of the original mass to decay. Prove that

$$T = \frac{1}{k}\ln 2$$

Why is T independent of m_0?

(c) If 10 g of a certain substance decay to 9 g in 100 days, calculate to the nearest day.

(i) the time taken until only 8 g remain

(ii) the half-life.

8 The mass of a rocket at time t is M and its speed is v. M consists of partly explosive material which is assumed to burn at a constant rate $m = -\dfrac{dM}{dt}$ and which is expelled backwards at a constant speed u relative to the rocket. Neglecting air resistance, the equation of motion is therefore

$$\frac{d}{dt}(Mv) - m(u - v) = -Mg$$

Show that this reduces to the equation

$$\frac{dv}{dt} = \frac{mu}{M_0 - mt} - g$$

with $M(t) = M_0 - mt$, where M_0 is the initial mass of the rocket. Find the speed acquired by the rocket at time T.

9 The velocity of a bomb falling vertically, where the air resistance proportional to its velocity, satisfies the differential equation

$$\frac{dv}{dt} = -kv + 10$$

(a)   Given that $v = 0$ when $t = 0$, show that $v = \frac{1}{k}(1 - e^{-kt})$

(b)   If the bomb falls for long enough, e^{-kt} becomes very small and the bomb reaches a **terminal velocity**, which is constant. What is the value of the terminal velocity?

10 A capacitor of capacitance C farads is charged through a resistance R ohms by a battery of constant voltage E_0 volts. The charge Q on the capacitor at time t is given by the equation

$$\frac{Q}{C} + Ri = E_0$$

where i amps is the current flowing in the circuit. Noting that $i = \frac{dQ}{dt}$ and assuming that initially $Q = 0$, find the charge at time t and the value of Q when the capacitor is fully charged.

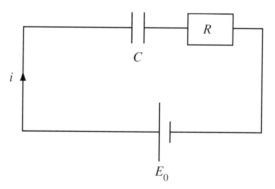

(a)   If $E_0 = 500$ volts (V), $C = 0.1$ farads (F) and $R = 400$ ohms (Ω) determine $i(t)$.

(b)   The **time constant** t_c of the circuit is defined to be the time (in seconds) taken for the capacitor to be 99% fully charged. Prove that $t_c = RC \ln 100$.

11 (a)   Show that the following equations are exact and hence obtain their general solutions.

(i) $x^3 \dfrac{dy}{dx} + 3x^2 y = x$

(ii) $\dfrac{1}{x}\dfrac{dy}{dx} - \dfrac{1}{x^2} y = \dfrac{1}{x}$

(iii) $\tan x \dfrac{dy}{dx} + \sec^2 xy = \cos x$

(iv) $e^{2x} \dfrac{dy}{dx} + 2e^{2x} y = e^x$

(b)   The following equations are not exact; show that the accompanying function is an integrating factor for the equation.

(i) $x^2 \dfrac{dy}{dx} + 3xy = 1$ $\qquad\qquad\qquad x$

(ii) $x \dfrac{dy}{dx} - y = x$ $\qquad\qquad\qquad \dfrac{1}{x^2}$

(iii) $\cos x \sin x \dfrac{dy}{dx} + y = \cos^3 x \qquad \sec^2 x$

(iv) $\dfrac{dy}{dx} + 2y = e^{-x} \qquad\qquad\qquad e^{2x}$

12   For each of the following equations find the integrating factor and use it to solve the equation.

(a) $\dfrac{dy}{dx} + 2xy = 2e^{-x^2}$ $\qquad\qquad$ (b) $\dfrac{dy}{dx} + 3y = e^{2x}$

(c) $\dfrac{dy}{dx} + \cot xy = \sin 2x$ $\qquad\qquad$ (d) $x \dfrac{dy}{dx} - 3y = x^5$

(e)* $x \cos x \dfrac{dy}{dx} + (x \sin x + \cos x)y = 1$

13   A chain is coiled up near the edge of a smooth table and begins to fall over the edge. When a length x has fallen, the equation of motion is

$$\frac{d}{dt}(mxv) = mxg$$

where m is the constant mass per unit length. Show that this reduces to

$$xv\frac{dv}{dx} + v^2 = gx$$

By substituting $y = v^2$, so that $\frac{dy}{dx} = 2v\frac{dv}{dx}$, show that

$$\frac{dy}{dx} = \frac{2(gx - y)}{x}$$

Hence prove that $v = \sqrt{\frac{2gx}{3}}$

14 The family of curves $x^2 - y^2 = C$ are rectangular hyperbolas with asymptotes $y = \pm x$.

(a) Sketch some curves of the family.

(b) Find the value of $\frac{dy}{dx}$ at any point on the curves.

(c) The **orthogonal trajectories** of this family of curves form another family of curves with the property that when a curve of the second family passes through one of the first family it does so at right angles. By finding the gradient at any point on one of the second family of curves, find the differential equation satisfied by the orthogonal trajectories and solve it. Then sketch some members of the second family on the same axes as you used for the first family. Remember that when two straight lines are perpendicular the product of their gradient is -1; hence if one of the gradients is $\frac{dy}{dx}$ then the other must be $-1 \Big/ \frac{dy}{dx}$

15 Find the equation describing the orthogonal trajectories of the following families of curves. Then draw sketches of the two families in each case.

(a) $ay^2 = x^3$ (b) $x^2 + y^2 = a^2$

16 The motion of a mass suspended on a light helical spring is governed by the equation $v\frac{dv}{dx} = -x$, where x is the displacement of the mass from its equilibrium position and v is its velocity. The term $v\frac{dv}{dx}$ is the acceleration of the mass.

(a) Show that if $v = 0$ when $x = 1$ then $v = \sqrt{1 - x^2}$

(b) Writing $v = \frac{dx}{dt}$ find the displacement as a function of time.

17 If the retardation of a particle moving along a straight line is numerically equal to its velocity, i.e. $v\dfrac{dv}{dx} = -v$ show that $x + v = C$, a constant. If $v = 0$ when $x = 4$ and $x = 0$ when $t = 0$ find the displacement as a function of time and describe the motion of the particle.

18 (a)   Write down a differential equation involving the displacement x and the velocity v of a particle moving in a straight line subject to a retardation which is

   (i) proportional to the velocity

   (ii) proportional to the square of the velocity

   (b)   Following the previous exercise, write down in each case an equation for the displacement which involves indefinite integrals.

19 A body radiating in interstellar space obeys **Stefan's law**

$$\frac{dT}{dt} = -k(T^4 - T_0^4)$$

where T is the temperature of the body in kelvins (K) and T_0 is the background temperature of space, namely $3°$K.

   (a)   Prove that the solution of the differential equation is

$$2\tan\left(\frac{T}{3}\right) + \ln\left(\frac{T+3}{T-3}\right) = 112(kt + C)$$

   where C is an arbitrary constant.

   (b)   Find C if the body was initially at the relatively low temperature of $6°$K.

   (c)   Show also that when the temperature of the body is very close to $3°$K then Stefan's law reduces to Newton's law of cooling, i.e.

$$\frac{dT}{dt} = \alpha(T - 3)$$

   where α is constant.

20 A stone falling through the air is subject to a retarding force which is proportional to the square of its velocity. If its acceleration $\dfrac{dv}{dt}$ is measured downwards then the equation of motion is

$$\frac{dv}{dt} = g\left(1 - \frac{v^2}{k^2}\right)$$

where k is a constant.

If the stone is dropped from rest, i.e. $v = 0$ when $t = 0$, show that $v = k \tanh\left(\dfrac{gt}{k}\right)$

If the stone falls for a sufficiently long time, so that we can effectively take the limit of v as $t \to \infty$, what does our model predict will happen to the velocity?

21 In a chemical reaction the rate of conversion of x units of a certain substance at any time is jointly proportional to the amount of substance remaining, i.e. to $x(t)$, and to the amount $N - x$ of other material in the reaction; N is the total amount of material present. Assuming that the constant of proportionality is 1, the reaction is modelled by the differential equation

$$\frac{dx}{dt} = x(N - x)$$

If $N = 10$ and $x(0) = 4$ solve the differential equation. What are the values of $x(0.1)$ and $x(0.2)$? What happens to the reaction?

11.3 Numerical Methods of Solution

Further methods exist for solving first-order and first-degree differential equations analytically, i.e. using the calculus, but in general numerical methods have to be used. In practice we would be determining the solution to an **initial-value problem**, i.e. the unique solution to a differential equation together with an initial condition.

Consider the initial-value problem

$$\frac{dy}{dx} = f(x, y) \quad \text{given} \quad y = y_0 \text{ at } x = x_0 \tag{11.6}$$

The aim of a numerical method is to estimate y at predetermined values of x, namely $x_1, x_2, \ldots, x_n$. Usually, these values are equally spaced so that $x_1 = x_0 + h$,

$x_2 = x_1 + h = x_0 + 2h, \ldots$ and in general $x_r = x_{r-1} + h = x_0 + rh$. The quantity h is called the **step size**. The aim is therefore to produce values $y_1, y_2, \ldots, y_n$ where y_r is the estimated value of y when $x = x_r$.

Euler's method

Euler's method approximates the solution curve in each interval, $x_0 < x < x_1$, $x_1 < x < x_2$, etc., by a straight line (Figure 11.9). At the point $A(x_0 y_0)$ we can evaluate $\dfrac{dy}{dx}$ from equation (11.6). The tangent to the solution curve at A has the value of the function $f(x, y)$, i.e. $f(x_0, y_0)$, as its gradient. The equation of this tangent is therefore

$$y - y_0 = f(x_0, y_0)(x - x_0)$$

At $x = x_1$ the tangent passes through the point C whose y-coordinate is given by

$$y_1 - y_0 = f(x_0, y_0)(x_1 - x_0)$$
i.e. $$y_1 = y_0 + hf(x_0, y_0)$$

Then C is the point (x_1, y_1). Provided that h is small enough in relation to the rate of bending of the solution curve then C will be close to B, the point on the solution curve where $x = x_1$. Starting from C we repeat the procedure to obtain $y_2 = y_1 + hf(x_1, y_1)$.

The point E in Figure 11.9(b) has coordinates (x_2, y_2). From the diagram it seems that E is a poorer approximation to D than C is to B. This will be true for an increasing function;

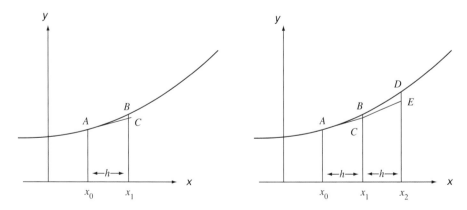

Figure 11.9   Euler's method: (a) in the interval $x_0 < x < x_1$ and (b) in the interval $x_1 < x < x_2$

for a decreasing function things may not be so bad, and in some cases the inaccuracies actually decrease as the solution proceeds.

At every stage there is round-off error, a factor in any numerical process. From the first stage there is a second source of error, known as **formula error**, which arises because we are approximating the solution curve in the interval $x_{r-1} \le x \le x_r$ by a straight line. From the second stage onwards there is a third source of error, **inherited error**, which is the result of starting the stage from a point which is not on the solution curve. It is the interaction between the formula error and the inherited error which determines whether or not the numerical solution gets progressively worse as the solution proceeds.

The Euler formula for the general stage is

$$y_r = y_{r-1} + hf(x_{r-1}, y_{r-1}) \tag{11.7}$$

Examples

1. Use the Euler method with step size $h = 0.1$ to estimate the values of $y(0.1)$, $y(0.2), \ldots, y(0.6)$ for the equation $\dfrac{dy}{dx} = y$ with the initial condition $y(0) = 2$. Compare your results against the analytical solution $y = 2e^x$. Now use a step size $h = 0.2$ to re-estimate the values $y(0.2)$, $y(0.4)$ and $y(0.6)$. Comment on your findings. First we take $h = 0.1$

First $(x_0, y_0) = (0, 2), f(x_0, y_0) = y_0 = 2$ and therefore $y_1 = 2 + 0.1 \times 2 = 2.2$

then $(x_1, y_1) = (0.1, 2.2), f(x_1, y_1) = y_1 = 2.2$ and $y_2 = 2.2 + 0.1 \times 2.2 = 2.42$

Table 11.1 shows the complete set of solutions.

Now we take $h = 0.2$

$(x_0, y_0) = (0, 2), f(x_0, y_0) = y_0 = 2$ and therefore $y_1 = 2 + 0.2 \times 2 = 2.4$

Table 11.1

x	y (Euler, $h=0.1$)	True value (4 d.p.)	Error (4 d.p.)	y (Euler, $h=0.2$)	Error (4 d.p.)
0		2			
0.1	2.2	2.2103	−0.0103		
0.2	2.42	2.4428	−0.0228	2.4	−0.0428
0.3	2.662	2.6997	−0.0377		
0.4	2.928 2	2.9836	−0.0555	2.88	−0.1036
0.5	3.221 02	3.2974	−0.0764		
0.6	3.543 122	3.6442	−0.1011	3.456	− 0.1882

The results are shown in Table 11.1. By halving the step size, the error in the estimates has been approximately halved. Note that

$$y_r = y_{r-1} + hy_{r-1} = (1 + h)y_{r-1} \quad \text{so that} \quad y_r = (1 + h)^r y_0$$

2. Use Euler's method to solve the differential equation

$$\frac{d\theta}{dt} = -0.1(\theta - 16)$$

with the accompanying initial condition $\theta = 80$ at $t = 0$.

(a) Obtain estimates of θ at $t = 2.5$ using a step size $h = 0.5$

(b) Estimate the error at each step given the analytical solution $\theta = 16 + 64e^{-0.1t}$. Comment on your results.

Solution

(a) If we write the equation as

$$\frac{d\theta}{dt} = f(t, \theta)$$

then $f(t, \theta) = -0.1(\theta - 16)$. We start from $(t_0, \theta_0) = (0, 80)$ and calculate

$$f(t_0, \theta_0) = -0.1(80 - 16) = -6.4$$
$$\theta_1 = \theta_0 + hf(t_0, \theta_0) = 80 + 0.5(-6.4) = 76.8$$

The full set of results are shown in Table 11.2.

(b) The errors are shown as part of Table 11.2. Note how the error increases at each step.

Table 11.2

t_r	θ_r (Euler, $h = 0.1$)	True value (4 d.p.)	Error (4 d.p.)	$f(t_r, \theta_r)$
0		80		−6.4
0.5	76.8	76.8787	−0.0787	−6.08
1.0	73.76	73.9096	−0.1496	−5.776
1.5	70.872	71.0853	−0.2133	−5.4872
2	69.3988	68.3988	−0.2704	−5.2184
2.5	65.8433	65.8433	−0.3213	

■

Modified Euler method

Euler's method is very straightforward but very inaccurate. It is clearly possible to travel from A to B in Figure 11.9(a) in a straight line. We chose the wrong gradient for this line. In the case of an increasing function such as the one depicted the slope of the tangent at x_0 is too shallow. There *is* a value of x in the interval $x_0 < x < x_1$ at which the gradient $f'(x)$ is exactly equal to the gradient of the line segment AB; but where it is we cannot say.

As a compromise we could take the average of the gradients at A and B, except that we do not know where B is! As a further compromise we take the average of the gradients at A and C.

The **modified Euler method** is an example of a class of methods known as **predictor-corrector methods**. At each stage the modified Euler method makes two estimates of y_r. The first estimate, using the Euler method, is an initial shot—a predicted value. The second estimate refines, or corrects, the first estimate.

Examples

1. Consider again the equation $\dfrac{dy}{dx} = y$ with initial condition $y(0) = 2$

   We work through one stage to estimate $y(0.1)$, taking $h = 0.1$

   First we use the Euler approximation (11.7) to predict the value $y(0.1)$; we label the predicted value y_1^p, p representing 'prediction'.

   Then, as before, $y_1^p = y_0 + hy_0 = (1 + h)y_0 = 2.2$

   Now we estimate the gradient at (x_1, y_1^p); this is $y_1^p = 2.2$

   The average gradient is $A = \dfrac{1}{2}(2 + 2.2) = 2.1$

   Finally we apply the corrector formula to obtain the revised estimate, $y_1{}^c$. This is
   $y_1{}^c = y_0 + hA = 2 + 0.1 \times 2.1 = 2.21$

   Compared with the simple Euler estimate, the value 2.21 is much closer to the correct value; see Table 11.1. The solution is continued in Table 11.3

Table 11.3

x	y^p	A	y^c	True value (4 d.p.)	Error (4 d.p.)	Euler error (4 d.p.)
0						
0.1	2.2	2.1	2.21	2.210 3	−0.000 3	−0.010 1
0.2	2.431	2.320 5	2.442 05	2.442 8	−0.000 7	−0.027 8
0.3	2.674 1	2.558 075	2.637 857 5	2.699 7	−0.001 8	−0.037 7
0.4	2.967 643 21	2.832 750 375	2.982 132 538	2.983 6	−0.002 5	−0.055 5
0.5	3.179 245 791	3.130 891 64	3.284 154 54	3.297 4	−0.003 2	−0.076 4
0.6	3.623 566 6	3.458 859 027	3.640 037 357	3.644 2	−0.004 2	−0.101 1

2. Keeping $h = 0.1$ continue the solution of the initial-value problem of the previous example until $x = 0.6$. Calculate the error at each stage and comment on the error in relation to the error incurred when using the simple Euler method.

Starting from $(x_1, y_1) = (0.1, 2.21)$ we calculate $f(x_1, y_1) = 2.21$

Then we predict $y_2^p = y_1 + hf(x_1, y_1) = (1 + h)y_1 = 1.1 \times 2.21 = 2.431$

Next we calculate $f(x_2, y_2^p) = 2.431$

Then we find the average gradient $A = \dfrac{1}{2}(2.21 + 2.431) = 2.3205$

We now apply the correction $y_2^c = y_1 + hA = 2.21 + 0.1 \times 2.3205 = 2.44205$

The results are shown in full in Table 11.3

Note that the error using the modified Euler method is $O(h^3)$; using the simpler Euler method the error is $O(h^2)$.   ■

The general formulation of the modified Euler method is

$$y_r^p = y_{r-1} + hf(x_{r-1}, y_{r-1}) \qquad A = \frac{1}{2}(f(x_{r-1}, y_{r-1}) + f(x_r, y_r^p))$$
$$y_r^c = y_{r-1} + hA \tag{11.8}$$

Exercise 11.3

 1 Use Euler's method to solve the following equations with the given initial conditions; use the step size stated and advance five steps.

(a) $\dfrac{dy}{dx} = 2x$ $y(0) = 1;$ $h = 0.2$

(b) $\dfrac{dy}{dx} = 3x^2$ $y(0) = 1;$ $h = 0.2$

(c) $\dfrac{dy}{dx} = -y$ $y(0) = 2;$ $h = 0.1$

(d) $\dfrac{dy}{dx} = x + y$ $y(0) = 0;$ $h = 0.1$

In each case obtain the analytical solution and compare your Euler approximate values with those obtained from the analytical solution. Comment.

 2 Repeat Question 1 using the modified Euler method.

3 For each of the following differential equations and accompanying initial conditions, use Euler's method with the step size given to find the stated value of y.

(a) $\dfrac{dy}{dx} = \sin(xy)$ $\qquad$ $y(1) = 1, h = 0.1;$ $\quad$ find $y(1.6)$

(b) $\dfrac{dy}{dx} = x - y - 1$ $\qquad$ $y(0) = 1, h = 0.2;$ $\quad$ find $y(1)$

(c) $\dfrac{dy}{dx} = x - y - 1$ $\qquad$ $y(0) = 1, h = -0.2;$ $\quad$ find $y(-1)$

(d) $\dfrac{dy}{dx} = \sin(x^2 + y)$ $\qquad$ $y(1) = 0, h = 0.1;$ $\quad$ find $y(1.8)$

(e) $\dfrac{dy}{dx} = x^2 + y^2$ $\qquad$ $y(0) = 2, h = 0.1;$ $\quad$ find $y(0.5)$

(f) $\dfrac{dy}{dx} = y - \dfrac{2}{y}$ $\qquad$ $y(0) = 1, h = 0.1;$ $\quad$ find $y(0.4)$.

In cases (b), (c) and (f) compare the results with the values obtained from the analytical solution.

11.4 Higher-Order Equations

We have so far met first-order differential equations which model population change, cooling, radioactive decay, the charging of electrical capacitors, etc. However, motion in the physical world is governed largly by Newton's laws and the mathematical models that arise from these laws lead commonly to second- and higher-order differential equations. In this section we concentrate on second-order differential equations with constant coefficients.

We start with some basic definitions.

t is time, independent variable, measured in seconds (s), dimensions $[T]$

x is displacement, or distance, measured in metres (m), dimension $[L]$

$v = \dfrac{dx}{dt}$ is velocity, i.e. rate of change of displacement with respect to time, measured in metres per second (m s^{-1}), dimensions $[LT^{-1}]$

$a = \dfrac{dv}{dt} = \dfrac{d^2x}{dt^2}$ is acceleration, i.e rate of change of velocity with respect to time, or second derivative of displacement, measured in metres per second per second (m s^{-2}), dimensions $[LT^{-2}]$

M is mass, measured in kilograms (kg), dimension $[M]$

Newton's second law applied to a particle (i.e. point mass) or body (i.e. mass of finite size) can be stated

Mass $\times$ acceleration $=$ sum of applied forces

In mathematical terms, with forces $F_1, \ldots, F_k$ applied, we have

$$M\frac{d^2x}{dt^2} = F_1 + F_2 + \cdots + F_k = \sum_{i=1}^{k} F_i$$

We will now discuss examples using, where necessary, the notation $\dot{x}$ for $\dfrac{dx}{dt}$ (velocity) and $\ddot{x}$ for $\dfrac{d^2x}{dt^2}$ (acceleration).

Example

Constant acceleration: law and formulae

A body or particle is assumed to be moving with constant acceleration a, perhaps due to gravity. A the start of the motion, $t = 0$, it has velocity u, and at the end of the motion it has velocity v. Meanwhile it covers a distance s. Show that

(i) $v = u + at$ (ii) $s = ut + \dfrac{1}{2}at^2$ (iii) $v^2 = u^2 + 2as$

First, $\ddot{x} = a$, so integrating gives

$$\int_0^t \ddot{x}\, dt = [\dot{x}]_0^t = v - u = \int_0^t a\, dt = at$$

so that $\dot{x}(t) = v = u + at$, which is formula (i).

Now integrate $\dot{x}$ to obtain

$$\int_0^t \dot{x}\, dt = [x]_0^t = x(t) - x(0) = s - 0 = s$$

and

$$\int_0^t (u + at)dt = ut + \frac{1}{2}at^2$$

so that $s = ut + \dfrac{1}{2}at^2$, which is formula (ii).

Eliminating t from (i) and (ii) gives formula (iii). This is left as an exercise. ∎

Motion of a projectile

The motion of a projectile under the action of gravity alone is governed by Newton's laws. The convention is to take axes as shown in Figure 11.10.

A projectile moving through the air follows a path from P to Q, say with gravity directed downwards, i.e. the vertical acceleration is $-g$. Ignoring air resistance, motion in the x and y directions is governed by

$$\ddot{x} = 0 \qquad \text{(no horizontal force)}$$
$$\ddot{y} = -g \qquad \text{(downward constant gravity force)}$$

Integrating twice gives a general solution of the form

$$x = A + Bt, \qquad y = C + Dt - \frac{1}{2}gt^2$$

The variable t can be eliminated from these equations, and the resulting equation shows that the path followed by the projectile, $y(x)$, is a parabola. You are left to prove this. A, B, C and D can be found from the initial conditions.

Example

A shot is fired into the air from the ground with velocity U at an elevation α to the horizontal. Ignoring air resistance, determine

(a) the horizontal and vertical displacements x and y as functions of t

(b) The equation of the parabola of the trajectory, i.e. y as a function of x

(c) the maximum height reached, h

(d) the range, R

(e) the value of α which gives the maximum range.

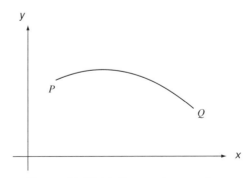

Figure 11.10 Motion under gravity

Solution

(a) At $t = 0$ the components of U in the x and y directions are respectively $U \cos \alpha$ and $U \sin \alpha$ (Figure 11.11).

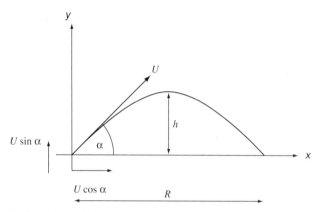

Figure 11.11 Horizontal and vertical motions of a shot fired in the air

Horizontal motion: no forces act horizontally so that

$$\ddot{x} = 0$$

i.e. $\dot{x} = B = U \cos \alpha$, constant.
Integrating again, $x = A + Bt$
Now $B = U \cos \alpha$ and $x = 0$ when $t = 0$, so that $A = 0$, therefore $x = (U \cos \alpha)t$

Vertical motion:

$$\ddot{y} = -g$$

i.e. the acceleration is negative, the force due to gravity acts downwards and is constant.
Integrating, we obtain $\dot{y} = D - gt, D$ a constant.
Now when $t = 0 \ \dot{y} = U \sin \alpha$, the initial upwards velocity, so $D = U \sin \alpha$
Integrating again gives

$$y = C + (U \sin \alpha)t - \frac{1}{2}gt^2$$

Now $y = 0$ when $t = 0$, so $C = 0$ and

$$y = (U \sin \alpha)t - \frac{1}{2}gt^2$$

(b) We can eliminate t from the formulae for x and y to obtain

$$y = x \tan \alpha - \frac{1}{2} \frac{gx^2}{U^2} \sec^2 \alpha$$

which is the equation of a parabola.

(c) The maximum height h is reached when the shot is no longer rising and is about to fall to the ground, i.e. when $\dot{y} = 0$, hence

$$t = \frac{D}{g} = \frac{U \sin \alpha}{g}$$

Therefore

$$y = \frac{(U \sin \alpha)^2}{g} - \frac{1}{2} g \frac{(U \sin \alpha)}{g} = \frac{U^2 \sin^2 \alpha}{2g}$$

(d) The range R is reached when the shot has fallen back to earth, i.e. when $y = 0$ for the second time. We have

$$y = x \left(\tan \alpha - \frac{1}{2} \frac{gx}{U^2} \sec^2 \alpha \right)$$

so $y = 0$ at the start, when $x = 0$, and when

$$x = \frac{2U^2}{g} \frac{\tan \alpha}{\sec^2 \alpha} = \frac{U^2}{g} \sin 2\alpha$$

Therefore $\quad R = \frac{U^2}{g} \sin 2\alpha$

(e) The range is a maximum when $\sin 2\alpha = 1$, i.e. when $\alpha = \frac{\pi}{4}$, giving

$$R_{\text{max}} = \frac{U^2}{g}$$

∎

When the acceleration of a body is a simple function of time its velocity and displacement can be found relatively easily as the following example shows.

Example

The acceleration of a moving body is equal to $3t(t-1)$ where t is the time in seconds.

(a) Determine when the acceleration is a minimum and its value at that time.

(b) Given that the body starts from rest at $t=0$, determine its position and velocity when $t=2$.

Solution

(a) $a = 3t(t-1)$ so that

$$\frac{da}{dt} = 3(t-1) + 3t = 3(2t-1)$$

Hence when $t = \frac{1}{2}, \frac{d^2a}{dt^2} = 6$, and a is a minimum at $t = \frac{1}{2}$

(b) Writing a as $\frac{dv}{dt}$, we obtain

$$\frac{dv}{dt} = 3t(t-1) = 3(t^2-t)$$

so $$\int_0^t \frac{dv}{dt} dt = v(t) - v(0) = 3\left(\frac{t^3}{3} - \frac{t^2}{2}\right)$$

but $v(0) = 0$ therefore $v = t^3 - \frac{3t^2}{2}$. The velocity of the body at $t=2$ is

$$v(2) = 8 - 6 = 2$$

Writing $v = \frac{dx}{dt}$ and integrating gives

$$x(t) - x(0) = 3\left(\frac{t^4}{12} - \frac{t^3}{6}\right) = t^3\left(\frac{t}{4} - \frac{1}{2}\right)$$

We have $x(0) = 0$, which means the position of the body at $t=2$ is

$$x(2) = 2^3\left(\frac{2}{4} - \frac{1}{2}\right) = 0. \qquad \text{So } x(2) \text{ is also } 0 \qquad \blacksquare$$

When $\ddot{x} = f(t)$ and $f(t)$ is easily integrated—maybe $f(t)$ is a polynomial—then $\dot{x}$ and x are easily found. On the other hand, equations of motion are often of the form $\ddot{x} = f(\dot{x}, x, t)$ and are more difficult to integrate. We will examine some of the better-known elementary cases.

Hooke's law

Hooke's law states that, when it is stretched by a small amount from a position of equilibrium, the tension in a light helical spring is proportional to the extension of the spring.

For example, if a weight of mass M, supported by a spring, is pulled by a small amount x from a position of equilibrium (Figure 11.12), the equation governing the motion once the mass is released is

$$M\ddot{x} = -kx$$

$M\ddot{x}$ is the tension (i.e. force) and the extension is x with $-k$ ($k > 0$) as the constant of proportionality. The negative sign indicates the force acts upwards to oppose the extension.

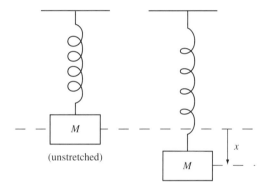

Figure 11.12 Stretching a spring

We have a different equation of the type $\ddot{x} = -\lambda x$, where

$$\lambda = -\frac{k}{M}$$

It is not possible to separate variables, so instead we use the result

$$\frac{d^2x}{dt^2} = \frac{d}{dt}\left(\frac{dx}{dt}\right) = \frac{d}{dx}\left(\frac{dx}{dt}\right)\frac{dx}{dt} = \frac{dv}{dx}v = v\frac{dv}{dx}$$

which was obtained using the chain rule. The differential equation becomes

$$v\frac{dv}{dx} = \lambda x$$

The variables v and x can now be separated to produce

$$\int v \, dv = \int \lambda x \, dx$$

from which we obtain the equation

$$\frac{1}{2}v^2 = \frac{1}{2}\lambda x^2 + C \qquad (C \text{ is a constant})$$

The value of the constant will depend upon initial and/or boundary conditions, but taking the square root of the equation we obtain

$$\frac{dx}{dt} = v = \pm(\lambda x^2 + 2C)^{1/2}$$

In principle, a further separation of variables is possible for x and t, but the integral this time is somewhat more complicated.

Note that the $v\dfrac{dv}{dx}$ form of acceleration applies widely to integrals of motion belonging to the type

$$\ddot{x} = f(x)$$

in which integration is a two-stage process depending on initial and boundary conditions as well as the form of $f(x)$.

Example

A mass M is attached to a helical spring and is extended by an amount a and held fixed. When released it is observed to oscillate up and down. Obtain a formula for the displacement $x(t)$.

Note first of all that $x(0) = a$, and $\dot{x}(0) = 0$. Therefore

$$M\ddot{x} = -kx \quad \text{or} \quad \ddot{x} = -\frac{k}{M}x$$

and by integration

$$\frac{1}{2}\dot{x}^2 = -\frac{k}{2M}x^2 + C$$

Using the initial conditions we see that

$$C = \frac{k}{2M}a^2$$

so $\qquad \dot{x}^2 = \dfrac{k}{M}(a^2 - x^2)$

and therefore

$$\dot{x} = \pm\sqrt{\frac{k}{M}}(a^2 - x^2)^{1/2} = \frac{dx}{dt}$$

i.e. $$\pm\int \frac{1}{(a^2 - x^2)^{1/2}} \, dx = \int \sqrt{\frac{k}{M}} \, dt$$

Since $\dot{x}$ is to be negative initially, i.e. the spring contracts but towards the equilibrium position, the negative sign makes physical sense. Therefore

$$-\int \frac{1}{(a^2 - x^2)^{1/2}} \, dx = \cos^{-1}\left(\frac{x}{a}\right)$$

i.e. $$\cos^{-1}\left(\frac{x}{a}\right) = \sqrt{\frac{k}{M}}t + C$$

When $t = 0$, $x = a$ and $\cos^{-1}(1) = 0$, so $C = 0$. This gives

$$x = a\cos\sqrt{\frac{k}{M}}t$$

The spring therefore oscillates with amplitude a and period $T = 2\pi\sqrt{\dfrac{M}{k}}$ ■

The solution of the differential equation, which was derived from Hooke's law, leads to a solution $x(t)$ which is a sinusoidal function of time. Solutions of this type form the essence of the theory behind mechanical vibration where M is mass and k stiffness. Shock absorbers, for example, vibrate when jolted into motion but viscous damping terms proportional to $\dot{x}$ ensure that the vibrations decay to zero exponentially with time.

Example   Assume a solution of the form $y = Ae^{mx}$ for the equation

$$\frac{d^2y}{dx^2} + 3\frac{dy}{dx} + 2y = 0$$

Hence obtain the general solution of the equation.

Assuming $y = Ae^{mx}$, then

$$\frac{dy}{dx} = mAe^{mx}, \qquad \frac{d^2y}{dx^2} = m^2Ae^{mx}$$

Substituting these formulae into the differential equation, we obtain

$$Ae^{mx}(m^2 + 3m + 2) = 0$$

Obviously, $A \neq 0$ for a non-trivial solution, and $e^{mx} \neq 0$, so that

$$m^2 + 3m + 2 = 0$$

which has solutions $m = -1$ or $m = -2$.

The general solution of the differential equation is

$$y = Ae^{-x} + Be^{-2x}$$

where A and B are arbitrary constants. ∎

Differential equations of the type where each derivative is of degree 1 and where the coefficients are constants, are called **linear**. The quadratic equation $m^2 + 3m + 2 = 0$ is called the **auxiliary equation**. It can always be solved irrespective of whether the roots are real and distinct, real and equal, or complex. A systematic consideration is beyond our scope.

We must now explore another important property in determining a solution to a second-order equation. Two constants of integration usually emerge so that two specified values are needed to determine a unique solution. If both of these values are given at one point, we have an **initial-value problem**; if the values are given at two distinct points, we have a **boundary value problem**.

Mathematically this means that for an initial-value problem we are given

$$y(a) \quad \text{and} \quad y'(a)$$

e.g., $y(0) = 0$, $y'(0) = \tan \alpha$ for a projectile fired from the ground $x = y = 0$, at an angle α, to the horizontal.

For a boundary-value problem we are given

$$y(a) \quad \text{and} \quad y(b)$$

e.g. $y(0) = 0$, $y(R) = 0$ for a projectile fired from the ground, $x = y = 0$, with specified range R, i.e. when $x = R$, $y = 0$.

Example Solve the equation $\dfrac{d^2y}{dx^2} + 3\dfrac{dy}{dx} + 2y = 0$ and sketch the solution curve for

(a) the initial conditions $y(0) = 1, y'(0) = 0$

(b) the boundary conditions $y(0) = 1, \ y(1) = 0$

The general solution is of the form

$$y(x) = Ae^{-x} + Be^{-2x}$$

therefore

$$y'(x) = -Ae^{-x} + 2Be^{-2x}$$

(a) $y(0) = A + B = 1$ and $y'(0) = -A - 2B = 0$ hence $A = 2, B = -1$
This leads to the solution $y = 2e^{-x} - e^{-2x}$. The curve is shown in Figure 11.13(a)

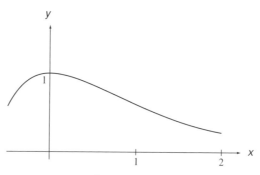

Figure 11.13(a) Solutions to $\dfrac{d^2y}{dx^2} + 3\dfrac{dy}{dx} + 2y = 0$: under $y(0) = 1, y'(0) = 0$

(b) The conditions are

$$y(0) = A + B = 1 \quad \text{and} \quad y(1) = \frac{A}{e} + \frac{B}{e^2} = 0$$

therefore $Ae + B = 0$ or $B = Ae$
Solving these equations gives

$$A = -\frac{1}{e - 1} \qquad B = \frac{e}{e - 1}$$

The solution is

$$y = \frac{1}{e - 1}(-e^{-x} + e^{1-2x})$$

and the curve is sketched in Figure 11.13(b)   ■

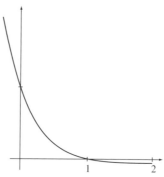

Figure 11.13(b) Solutions to $\dfrac{d^2y}{dx^2} + 3\dfrac{dy}{dx} + 2y = 0$: under $y(0) = 1, y(1) = 0$

Exercise 11.4

1   From the equations of straight-line motion with constant acceleration

$$v = u + at \qquad s = ut + \frac{1}{2}at^2$$

prove that

$$v^2 = u^2 + 2as$$

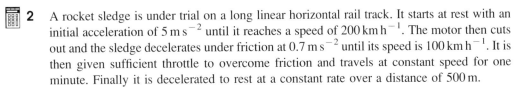 2   A rocket sledge is under trial on a long linear horizontal rail track. It starts at rest with an initial acceleration of $5\,\mathrm{m\,s}^{-2}$ until it reaches a speed of $200\,\mathrm{km\,h}^{-1}$. The motor then cuts out and the sledge decelerates under friction at $0.7\,\mathrm{m\,s}^{-2}$ until its speed is $100\,\mathrm{km\,h}^{-1}$. It is then given sufficient throttle to overcome friction and travels at constant speed for one minute. Finally it is decelerated to rest at a constant rate over a distance of $500\,\mathrm{m}$.

(a)   Determine the total distance covered and the time taken.

(b)   Sketch the graph of velocity versus time. How do you estimate the distance covered from the graph?

3   A rough shooting chuck is fired from a catapult at $30\,\mathrm{m\,s}^{-1}$ upwards at $40°$ to the horizontal. Determine the maximum height reached and its horizontal range, assuming there is no air resistance.

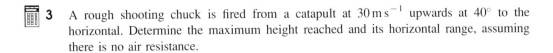 4   A stone is thrown from a cliff top of height $70\,\mathrm{m}$ above the sea with velocity $25\,\mathrm{m\,s}^{-1}$ at an upward elevation of $45°$. If $g = 9.81\,\mathrm{m\,s}^{-2}$ determine

(a)   the time of flight

(b)   the distance from the bottom of the cliff when the stone hits the sea.

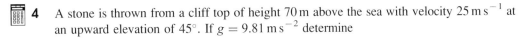 5*   A man throws a stone from $1.5\,\mathrm{m}$ above the ground at a velocity of $25\,\mathrm{m\,s}^{-1}$ at an angle of elevation α. The stone just clears a wall $5\,\mathrm{m}$ high a distance of $20\,\mathrm{m}$ away.

(a)   Draw the two possible trajectories.

(b)   Determine for each flight path

(i) the angle of elevation

(ii) the maximum height attained and the range in each case.

6 A bullet is fired vertically into the air. It is at a height h after t_1 seconds on the way up and after t_2 seconds on the way down. Show that

(a) $h = \dfrac{1}{2} g t_1 t_2$

(b) the initial velocity is $\dfrac{1}{2} g(t_1 + t_2)$

7* A projectile is fired with velocity μ and elevation angle α above a plane surface of angle β, as illustrated.

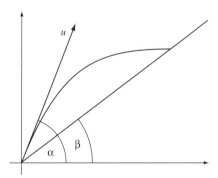

(a) Show that the projectile lands upon the plane when the horizontal distance travelled is

$$x = \frac{2u^2 \cos^2 \alpha}{g}(\tan \alpha - \tan \beta)$$

(b) Hence show that the range on the plane is

$$\frac{u^2}{g \cos^2 \beta}(\sin(2\alpha - \beta) - \sin \beta)$$

(c) Deduce that the range is a maximum when $\alpha = \dfrac{1}{2}\left(\dfrac{\pi}{2} + \beta\right)$ and that its value is

$$\frac{u^2}{g(1 + \sin \beta)}$$

8   A particle is moving such that its acceleration is given by

$$\ddot{x} = 19 - te^{-t}$$

(a)   Write down the general form of the equation for

(i) the velocity

(ii) the displacement

(b)   If the velocity is zero when $t = 0$ and the displacement is $-\dfrac{19}{2}$ when $t = 1$, write down the solutions for the velocity and for the displacement.

9   A particle is oscillating under a restoring force $-n^2 x$ and its displacement is governed by an equation of motion which satisfies Hooke's law, i.e. $\ddot{x} = -n^2 x$. By substituting into the differential equation, verify that

$$x(t) = A \cos nt + B \sin nt$$

If at $t = 0$ the initial displacement and velocity are x_0 and u_0, replace A and B in terms of them. What is the maximum amplitude of $x(t)$?

10   The equation of motion of a simple pendulum of length l satisfies the equation $m\ddot{\theta} = -mg \sin \theta$.

(a)   If $\sin \theta \approx \theta$, where θ is a small angle, write down the solution for $\theta(t)$.

(b)   Show that the period of oscillation for the simple pendulum is $2\pi \sqrt{l/g}$. What is the period when $l = 1$ m?

11 A particle moves under a repulsive force away from the origin; the equation of motion is $\ddot{x} = x$. If $v = 0$ when $x = a$, prove that $v^2 = x^2 - a^2$. If the general solution is of the form $Ae^t + Be^{-t}$, determine A and B if $x(0) = a$.

12 For the following constant-coefficient homogeneous linear differential equations write down the auxiliary quadratic equation and obtain the general solution.

(a) $y'' + 6y' + 5y = 0$ (b) $y'' - 9y' + 18y = 0$

(c) $3y'' + 10y' + 3y = 0$ (d) $10y'' - 37y' + 30y = 0$

(e) $y'' + 4y' + y = 0$ (f) $3y'' + y' - y = 0$

13 A differential equation of the type

$$y'' + 2ny' + n^2 y = 0$$

has an auxiliary quadratic equation with equal roots $-n$. In this case the general solution is of the form

$$y = (A + Bx)e^{-nx}$$

(a) Write down the general solution of the equation $y'' + 6y' + 9y = 0$

(b) Obtain the solution when $y(0) = 1, y'(0) = 0$

14 Show that the auxiliary quadratic associated with the differential equation $y'' + 2y' + 2y = 0$ does not have real roots. Verify, however, that the general solution is $y(x) = e^{-x}(A \cos x + B \sin x)$.

15 Write down the general solution of the equation

$$y'' + y' - 2y = 0$$

Hence obtain the specific solutions in the following initial-value problem:

(a) $y(0) = 0, \qquad y'(0) = 1$

and the following boundary-value problems:

(b) $y(0) = 0, \qquad y(1) = 1$

(c) $y'(0) = 0, \qquad y(1) = 1$

(d) $y(0) + y'(0) = 0, \qquad y(1) + y'(1) = 1$

16 The downward displacement x of a particle falling vertically satisfies the differential equation $\ddot{x} = g - k\dot{x}$. The air resistance is assumed to be proportional to the velocity of the particle. If the particle is dropped from a position of rest, its velocity increases until it reaches a value $V = \dfrac{g}{k}$, known as the **terminal velocity**.

(a)   Why does the velocity reach this terminal value?

(b)   Writing the differential equation as $\dfrac{dv}{dt} = g - kv$ and adding the initial condition $v(0) = 0$, prove that

(i) $v = V(1 - e^{-gt/V})$

(ii) $x = V\left(t - \dfrac{V}{g}(1 - e^{-gt/V})\right)$

where the downward displacement is measured from the point of release.

(c)   If a second particle is thrown downwards from the same point at the same time with velocity $v_0(<V)$ show that its downward displacement is given by

$$x = V\left(t - \frac{(V - v_0)}{g}(1 - e^{-gt/V})\right)$$

(d)   If both particles fall from a sufficient height, and therefore for a long time, what will be their ultimate distance apart?

17 If a paricle is thrown vertically upwards according to the same law of motion as in the previous problem, determine

(a)   the time taken to reach its greatest height

(b)   the height of this point above the point of projection.

18 A particle is dropped from rest. If we assume that the air resistance is proportional to the square of the velocity then the equation of motion is $\ddot{x} = g - kv^2$. Prove that

(a) $v = V \tanh\left(\dfrac{gt}{V}\right)$

(b) $x = \dfrac{V^2}{g} \ln \cosh\left(\dfrac{gt}{V}\right)$

where $V = \sqrt{\dfrac{g}{k}}$ is the terminal velocity of the particle.

19 A closed-loop circuit containing a capacitance C and an inductance L is isolated from a main electric circuit. The current in the circuit, I, satisfies the differential equation $L\dfrac{dI}{dt} + \dfrac{Q}{C} = 0$ where $Q = \displaystyle\int I\,dt$ is the charge on the capacitance.

Differentiating this equation, we obtain

$$L\frac{d^2I}{dt^2} + \frac{I}{C} = 0$$

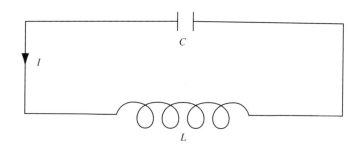

(a) If no current flows in the circuit until time $t = 0$, prove that I is oscillatory. If I_0 is the maximum value of the current after $t = 0$, find I as a function of time.

(b) If $L = 2\,\text{H}$, $C = 200 \times 10^{-6}\,\text{F}$ and $I_0 = 100 \times 10^{-3}\,\text{A}$, determine the value of $I(0.1)$ and find the period of the oscillations.

20 An electric circuit contains inductance L, resistance R and capacitance C in series with an applied alternating voltage. The current I satisfies the differential equation

$$L\frac{dI}{dt} + RI + \frac{Q}{C} = E_0 \sin \omega t$$

where $Q = \displaystyle\int I\,dt$ is the charge stored by the capacitor and $E_0 \sin \omega t$ is the applied voltage.

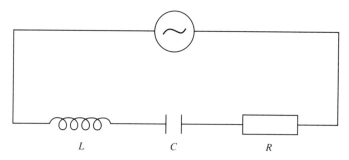

The solution of the differential equation includes terms with a decaying exponential component, called **transients**, which become negligible for large values of t. The other part of the solution has the form

$$I = A \cos \omega t + B \sin \omega t$$

where A and B are constants.

(a)   By differentiating the given differential equation with respect to t, obtain a second-order differential equation for I.

(b)   Substitute the formula for I into the differential equation, equate the coefficients of $\sin \omega t$, and $\cos \omega t$ and show that

$$A = \frac{E_0 \left(\dfrac{1}{\omega C} - \omega L \right)}{R^2 + \left(\dfrac{1}{\omega C} - \omega L \right)^2} \quad \text{and} \quad B = \frac{E_0 R}{R^2 + \left(\dfrac{1}{\omega C} - \omega L \right)^2}$$

(c)   Write I in the form $I = \dfrac{E_0}{Z} \sin(\omega t - \phi)$ and determine ϕ and the impedance of the circuit, Z, in terms of the other parameters.

(d)   Find the impedance in ohms (Ω) of a series LCR circuit in which $L = 2\,\text{H}$, $C = 100 \times 10^{-6}\,\text{F}$, $R = 200\,\Omega$ and $\omega = 50\,\text{Hz}$.

SUMMARY

- **General solution and boundary conditions**: for an ordinary differential equation (o.d.e.), containing derivatives of a function, the general solution represents a family of curves; the boundary conditions specify a unique solution

- **Order and degree**: the order is the number of differentiations to reach the highest derivative and the degree is the power to which the highest derivative is raised

- **Separation of variables**

$$\frac{dy}{dx} = g(x)h(y) \quad \text{leads to} \quad \int \frac{1}{h(y)} dy = \int g(x)dx$$

- **Homogeneous equations**: if $y = vx$ is substituted on $\dfrac{dy}{dx} = f(v)$ then

$$v + x\frac{dv}{dx} = f(v)$$

- **Exact equations**

$$\frac{d}{dx}(f(x) \times y) = g(x)$$

- **Integrating factor**: a factor by which an o.d.e. is multiplied to make it exact; the integrating factor for

$$\frac{dy}{dx} + p(x) \times y = q(x) \quad \text{is} \quad e^{\int p(x)dx}$$

- **Euler's method**

$$y_r = y_{r-1} + hf(x_{r-1}, y_{r-1})$$

- **Modified Euler method**

$$y_r^p = y_{r-1} + hf(x_{r-1}, y_{r-1})$$

$$A = \frac{1}{2}(f(x_{r-1}, y_{r-1}) + f(x_r, y_r^p))$$

$$y_r^c = y_{r-1} + hA$$

- **Linear second-order, constant coefficients**: the o.d.e.

$$\frac{d^2y}{dx^2} + a\frac{dy}{dx} + by = 0$$

has **auxiliary equation** $m^2 + am + b = 0$; if this has roots m_1, m_2 then the general solution of the differential equation is $y = Ae^{m_1 x} + Be^{m_2 x}$, where A and B are constants.

Answers

Exercise 11.1

1

	Order	Degree
(a)	1	2
(b)	3	1
(c)	3	1
(d)	2	1
(e)	3	1
(f)	4	1
(g)	1	1, using $\sin^{-1}$
(h)	4	2
(i)	1	does not exist

Writing (h) as $\dfrac{d^4y}{dx^4} = \pm 2$ yields two fourth-order differential equations each of degree 1.

Solving (i) gives

$$\frac{dy}{dx} = \tan^{-1}(1) = \frac{\pi}{4} \quad \text{and} \quad \frac{dy}{dx} = \tan^{-1}(2)$$

two first-order, first-degree equations.

2 (a) $y' - 3y = 0$ (b) $y'' - 4y = 0$
 (c) $y'' + 9y = 0$ (d) $yy'' = (y')^2$
 (e) $y = xy' + (y')^3$

3 (a) $y''' = y'$ (b) $y''' - 6y'' + 11y' - 6y = 0$
 (c) $y'' - 2y' + 2y = 0$

5 (a) (i) $\dfrac{d^3x}{dt^3} + n^2\dfrac{dx}{dt} = 0$ (ii) $\dfrac{d^4x}{dt^4} = n^4x$

6 (a) $2x^2 + C$ (b) $-(1 - x^2)^{1/2} + C$
 (c) $\dfrac{x^2}{2}\left(\ln x - \dfrac{1}{2}\right) + C$ (d) $e^{x^2/2} + C$

7 (a) $2x^2 - 5$ (b, c) does not exist (d) $e^{x^2/2} + \sqrt{e}$

8 (a) $\dfrac{x}{2}(x^2 - 1)$ (b) $\dfrac{x}{\pi} - \sin x$

  (c) $x(\ln|x| - 1) + \dfrac{3}{2}(x + 1)$ (d) cannot be defined

  (e) $\dfrac{x^3}{36}(6\ln|x| - 5) + \dfrac{9x}{4} + \dfrac{5}{36}$

9 (a) e^{3x}

  (b) (i) $\cosh 2x$ (ii) 0

  (c) (i) $4\sin 3x$ (ii) $\dfrac{1}{\sqrt{2}}(\cos 3x + \sin 3x)$

  (d) (i) $e^{x/2}$ (ii) y is constant, $B = 0$

  (e) (i) $2(x + 4)$ (ii) not possible

10 (a) $4\sin t$ (b) $\sqrt{2}\sin\left(t + \dfrac{\pi}{4}\right)$

11 (a) $\dfrac{dQ}{dt} = -\dfrac{Q}{Q_0}$ (minus sign since $Q, Q_0 > 0$)

  (b) $\dfrac{dQ}{dt} = \dfrac{KQ}{(t - t_0)^2}$

12 $\dfrac{dV}{dr} = -\dfrac{GMm}{r^2}, \; V = \dfrac{GMm}{r} + C; \; G$ is the constant of gravitation

13 (a) $P_0(0) = 1; A = 1$ (b) 0.082

Exercise 11.2

1 $y(x) = \dfrac{wx^2}{24EI}(2l - x) + Ax + B$, where A and B are arbitrary constants

2 (a) $\tan y = \tan x + A$ (b) $y = Ax\exp\left(\dfrac{1}{x} + \dfrac{1}{y}\right)$

  (c) $1 + y^2 = \dfrac{Ax^2}{1 + x^2}$ (d) $x\cos y = A$

  (e) $y = Ae^{x^2}$ (f) $xy = y^3 + A$

  (g) $x = \ln(1 + y^2) + A$ (h) $y = 1 + Ae^{-x^2/2}$

  (i) $y = 1/\ln x$ (j) $y = \dfrac{-1}{\tan^{-1} x}$

3 $k = ax_0^2 - by_0^2$

4 (a) $\quad x^2 y^2 = A(y - x)^3$ (b) $\quad x = A \sin \left(\dfrac{y}{x} \right)$

  (c) $\quad y = x \ln x + A$ (d) $\quad y^2 = x^4 - x^2$

5 (a) $\quad y = Ax$

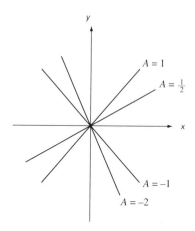

Families of straight lines, with any slope passing through the origin

  (b) $\quad x^2 - y^2 = A$

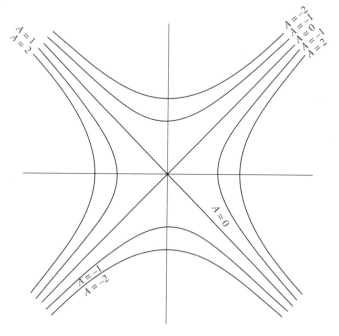

Families of rectangular hyperbolas with asymptotes $y = \pm x (A = 0)$

(c)

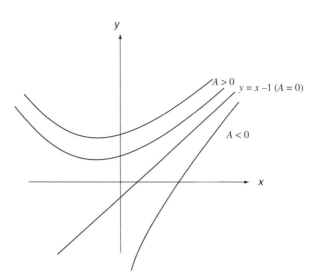

6 $65\,^\circ\text{C}$

7   (c)   (i) 212 days

(ii) 657 days

8 $u\ln\left(\dfrac{M_0}{M_0 - mT}\right) - gT$

9   (b) $\dfrac{10}{k}$

10   (a) $Q = E_0 C(1 - e^{-t/RC})$. Final value of Q is $E_0 C$

$$i(t) = -\frac{1}{8}e^{-t/40}$$

11   (a) (i) $y = \dfrac{1}{2x} + \dfrac{C}{x^3}$ (ii) $y = x\ln|x| + C$

(iii) $y = \cos x + C\cot x$ (iv) $y = e^{-x} + Ce^{-2x}$

12   (a) $(2x + C)e^{-x^2}$ (b) $\dfrac{1}{5}e^{2x} + Ce^{-3x}$

(c) $\dfrac{2}{5}\sin^2 x + C\operatorname{cosec} x$ (d) $\dfrac{1}{2}x^5 + Cx^3$

(e) $\dfrac{1}{x}(\sin x + C\cos x)$

14 (a)

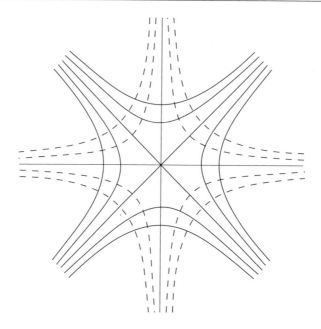

(b) $\dfrac{dy}{dx} = \dfrac{x}{y}$

(c) Orthogonal trajectories satisfy $\dfrac{dy}{dx} = -\dfrac{y}{x}$; these are rectangular hyperbolas,

$xy = $ constant, whose asymptotoes are the coordinate axes.

15 (a) $y = \pm\sqrt{\dfrac{x^3}{a}}$ so that either $x > 0$ and $a > 0$ or $x < 0$ and $a < 0$; orthogonal

trajectories are $2x^2 + 3y^2 = C$, which are similar ellipses, centred at the origin.

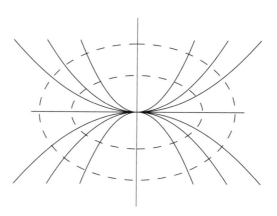

(b)   Orthogonal trajectories are $y = Kx$, i.e. straight lines through the origin.

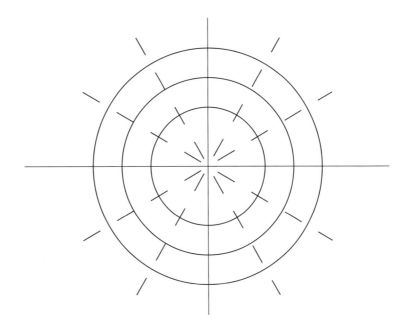

16 (b) $x = \sin t$

17 $x(t) = 4(1 - e^{-t})$; motion starts with $v = 4$ at $x = 0$ and proceeds towards $x = 4$, slowing to zero speed when it reaches that point.

18 (a)   (i) $v\dfrac{dv}{dx} = -kv$ or $\dfrac{dv}{dx} = -k$ (ii) $v\dfrac{dv}{dx} = -kv^2$ or $\dfrac{dv}{dx} = -kv$

(b)   (i) $v + kx = C$ and $\displaystyle\int \dfrac{1}{C - kx}\,dx = \int dt$ (ii) $v = Ae^{-kx}$ and $\displaystyle\int e^{kx}\,dx = A\int dt$

19 (b) -0.0292

20 $v \to k$ as $t \to \infty$; this is the terminal velocity of the stone.

21 $x(t) = \dfrac{10}{1 + 1.5e^{-10t}}$ so that $x(0.1) = 6.444$, $x(0.2) = 8.313$

$x(t) \to 10$ very quickly. The reaction ceases when there is only substance 'x' remaining (10 units).

Exercise 11.3

1 (a)

x	y_r	$f(x_r, y_r)$	y	Error
0	1	0	1	–
0.2	1	0.4	1.04	−0.04
0.4	1.08	0.8	1.16	−0.08
0.6	1.24	1.2	1.30	−0.12
0.8	1.48	1.6	1.64	−0.16
1	1.80	2	2	−0.28

$y_r = y_{r-1} + 0.4x_{r-1}$
Error $= -h^2$ at each step

(b)

x	y_r	$f(x_r, y_r)$	y	Error
0	1	0	1	–
0.2	1	0.12	1.008	−0.008
0.4	1.024	0.48	1.064	−0.046
0.6	1.120	1.08	1.216	−0.096
0.8	1.336	1.92	1.512	−0.176
1	1.720	3	2	−0.280

(c)

x	y_r	$f(x_r, y_r)$	y	Error (4 d.p.)
0	2	−2	2	–
0.1	1.8	−1.8	1.809 67	−0.0097
0.2	1.62	−1.62	1.637 46	−0.1075
0.3	1.458	−1.458	1.481 64	−0.0236
0.4	1.312 32	−1.312 2	1.340 64	−0.0284
0.5	1.180 98	–	1.213 00	−0.0321

(d)

x	y_r	$f(x_r, y_r)$	y (5 d.p.)	Error (4 d.p.)
0	0	0	0	–
0.1	0	0.1	0.005 17	−0.0052
0.2	0.01	0.21	0.021 40	−0.6114
0.3	0.031	0.331	0.049 86	−0.0189
0.4	0.064 1	0.464 1	0.091 82	−0.0277
0.5	0.110 51	–	0.148 72	−0.0382

2 (a)

x	y_c	y_p	y	Error
0	–	–	1	–
0.2	1.04	1.08	1.04	0
0.4	1.16	1.32	1.16	0
0.6	1.36	1.608	1.36	0
0.8	1.64	1.962	1.64	0
1	2	–	2	0

(b)

x	y_p	y_c	y	Error
0	–	–	1	–
0.2	1	1.012	1.008	0.004
0.4	1.030	1.072	1.064	0.008
0.6	1.168	1.228	1.216	0.012
0.8	1.444	1.528	1.512	0.016
1	1.824	2.020	2 (5 d.p.)	0.020

(c)

x	y_p	y_c	y	Error (4 d.p.)
0	–	–	2	–
0.1	1.8	1.81	1.809 67	−0.0003
0.2	1.629	1.638	1.637 46	−0.0006
0.3	1.4742	1.4824	1.481 64	−0.0008
0.4	1.3342	1.3416	1.340 04	−0.0010
0.5	1.2074	1.2142	1.213 06	−0.0011

(d)

x	y^p (5 d.p.)	y^c (5 d.p.)	y (5 d.p.)	Error (4 d.p.)
0	–	–	–	–
0.1	0	0.005	0.005 17	−0.0002
0.2	0.015 5	0.021 03	0.021 40	−0.0004
0.3	0.043 13	0.049 23	0.049 86	−0.0006
0.4	0.084 16	0.090 90	0.091 82	−0.0009
0.5	0.140 00	0.147 45	0.148 72	−0.0013

3 (a)

x	y
1	0
1.1	1.084
1.2	1.177
1.3	1.276
1.4	1.376
1.5	1.470
1.6	1.555

(b)

x	y (Euler)	y (exact)
0	1	1
0.2	1	1.0214
0.4	1.0400	1.0918
0.6	1.1120	1.2221
0.8	1.2776	1.4255
1.0	1.4883	1.7183

(c)

x	y (Euler)	y (exact)
0	1	0
−0.2	1	1.0187
−0.4	1.0400	1.0703
−0.6	1.1120	1.1488
−0.8	1.2096	1.2493
−1.0	1.3277	1.3679

(d)	x	y
	1	0
	1.1	0.0841
	1.2	0.1803
	1.3	0.2802
	1.4	0.3723
	1.5	0.4447
	1.6	0.4879
	1.7	0.4973
	1.8	0.4730

(e)	x	y
	0	2.000
	0.1	1.600
	0.2	1.345
	0.3	1.168
	0.4	1.041
	0.5	0.949

(f)	x	y (Euler)	y (exact)
	0	1.000	1.000
	0.1	1.100	1.100
	0.2	1.192	1.183
	0.3	1.278	1.265
	0.4	1.359	1.342

Exercise 11.4

2 (a)   4129 m

(b)

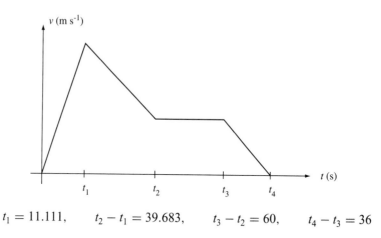

$t_1 = 11.111,$ $t_2 - t_1 = 39.683,$ $t_3 - t_2 = 60,$ $t_4 - t_3 = 36$

Distance covered is equal to the area under the curve.

3 $h_{max} = 18.95$ m, $R = 90.35$ m

4 (a)   5.4850 s   (b)   77.57 m

5 (a)

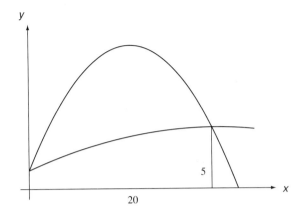

(b)　(i)　$19.36°, 80.57°$

(ii)　$h = 5.001$ m, $R = 43.73$ m
　　　$h = 32.50$ m, $R = 20.84$ m

8　(a)　(i)　$\dot{x} = 19t + e^{-t}(t+1) + a$

(ii)　$\dot{x} = \dfrac{19t^2}{2} - e^{-t}(t+2) + at + b$

(b)　$v = 19t - 1 + e^{-t}(t+1), \qquad x = \dfrac{19t^2}{2} - 19t - e^{-t}(t+2) + \dfrac{3}{e}$

9　$A = x_0, B = \dfrac{u_0}{n}, \left(x_0^2 + \left(\dfrac{u_0}{n} \right)^2 \right)^{1/2}$

10　2.006 s

11　$A = B = \dfrac{a}{2}, x(t) = a\cosh t$

12　(a)　$Ae^{-x} + Be^{-5x}$ (b)　$Ae^{3x} + Be^{6x}$

(c)　$Ae^{-x/3} + Be^{-3x}$ (d)　$Ae^{6x/5} + Be^{5x/2}$

(e)　$e^{-x}(Ae^{\sqrt{3}x} + Be^{-\sqrt{3}x})$ (f)　$e^{-x/6}(Ae^{\sqrt{13}x/6} + Be^{-\sqrt{13}x/6}$

13　(a)　$(A + Bx)e^{-3x}$ (b)　$(1 + 3x)e^{-3x}$

15　(a)　$\dfrac{1}{3}(e^x - e^{-2x})$ (b)　$\dfrac{e^2}{e^3 - 1}(e^x - e^{-2x})$

(c)　$\dfrac{e^2}{2e^3 + 1}(2e^x + e^{-2x})$ (d)　$\dfrac{e^2}{e^3 - 1}\left(\dfrac{1}{2}e^x + e^{-2x} \right)$

16 (d) $\dfrac{V v_0}{g}$

17 (a) $t = \dfrac{V}{g} \ln\left(\dfrac{V + v_0}{V}\right)$ (b) $h = \dfrac{V}{g}\left(v_0 - V \ln\left(\dfrac{V + v_0}{g}\right)\right)$

19 (a) $I = I_0 \sin\left(\dfrac{t}{\sqrt{LC}}\right)$

(b) $I(0.1) = -95.89 \times 10^{-3}$ A, period $= 2\pi\sqrt{LC} = 0.889$ s

20 (a) $L\dfrac{d^2 I}{dt^2} + R\dfrac{dI}{dt} + \dfrac{I}{C} = \omega E_0 \cos \omega t$

(c) $Z = \left(R^2 + \left(\dfrac{1}{\omega C} - \omega L\right)^2\right)^{1/2}$

$\phi = \tan^{-1}\left(\dfrac{R}{\omega L - \dfrac{1}{\omega C}}\right)$

(d) $223.6\,\Omega$

12 COMPLEX NUMBERS

Introduction

Many problems lead to the square root of a negative number. To accommodate this mathematical quantity we extend the set of real numbers to create the set of complex numbers. Using complex numbers we can still obtain real solutions to important mathematical problems. Nowhere is this more apparent than in studying lightly damped or undamped oscillating systems.

Objectives

After working through this chapter you should be able to

- Define a complex number
- Represent a complex number on an Argand diagram
- Write down the conjugate of a complex number
- Write a complex number in Cartesian, polar or exponential form
- Find the sum, difference, product and quotient of two complex numbers
- Understand De Moivre's formula
- Find the roots of a complex number
- Define loci in the plane using complex numbers
- Use complex numbers to solve problems involving differential equations
- Use complex numbers to solve problems in wave interaction and electric circuits.

12.1 COMPLEX ALGEBRA

In Section 11.4 we looked at linear second-order differential equations with constant coefficients. The first stage in their solution involved a trial solution of the form $x = e^{mt}$, which led to an auxiliary quadratic equation. In the case where there is no damping, the equation will be of the form $m^2 + k^2 = 0$, which cannot be solved in terms of real numbers. An example of a system with light damping is one with auxiliary equation $m^2 + 4m + 13 = 0$; attempting to solve this by completing the square we obtain $(m + 2)^2 = -9$ and we cannot progress. Yet these systems exist and solutions must be found.

Mathematicians long believed it was impossible to find the square root of a negative number, and only as late as the eighteenth century did this belief become a real stumbling-block. We have said that the equation $q(x) = x^2 + 1 = 0$ does not possess roots because the graph of $q(x)$ does not cross the x-axis; in fact, $q(x) \geq 1$ for all x, as in Figure 12.1.

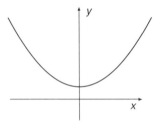
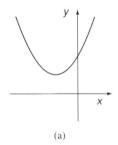
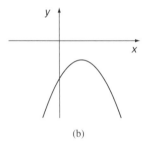

(a)   (b)

Figure 12.1 The graph of $q(x) = x^2 + 1$

Figure 12.2 Quadratic curves which do not cross the x-axis

In general, the quadratic equation

$$ax^2 + bx + c = 0$$

cannot be solved when $b^2 < 4ac$ if a **real** solution is required. The graphs shown in Figure 12.2 correspond to cases where $b^2 < 4ac$ and do not cross the x-axis.

To order to make the quadratic equation solvable it is necessary to introduce a new concept. If $x^2 + 1 = 0$ then formally we write $x = \pm\sqrt{-1}$. The quantity $\sqrt{-1}$, usually denoted j or i, is called the **imaginary unit**. Engineers and applied physicists use j to avoid confusion with the symbol for electric current. In this book we shall use the symbol j. In calculations we replace j^2 by -1 wherever it occurs.

We will assume that j acts like an algebraic constant and obeys all the conventional rules of algebra, for example

$$-3j = -3 \times j$$
$$(-3j) \times (2j) = -6j^2 = 6$$
$$5j^7 = 5j \times j^6 = 5j \times (-1)^3 = -5j$$
$$(-j)^2 = +j^2 = -1$$

To solve the equation $ax^2 + bx + c = 0$ where $b^2 < 4ac$, we first note that

$$x = \frac{-b \pm \sqrt{-1(4ac - b^2)}}{2a} = \frac{-b \pm \sqrt{-1} \times \sqrt{(4ac - b^2)}}{2a} = \frac{-b \pm \sqrt{4ac - b^2}\, j}{2a}$$

Then $x = \beta \pm \gamma j$ where

$$\beta = -\frac{b}{2a} \quad \text{and} \quad \gamma = \frac{\sqrt{4ac - b^2}}{2a}$$

are **real numbers**, i.e. conventional numbers.

A number which is a (real) multiple of j, such as $3j$ is called a **purely imaginary number**. A number of the form $x + yj$ where x and y are real numbers is called a **complex number**. We write a general complex number as

$$z = x + yj$$

x is called the **real part** of z, and y is called the **imaginary part** of z. In symbols

$$x = \mathrm{Re}(z) \quad \text{and} \quad y = \mathrm{Im}(z)$$

It follows that the set of real numbers and the set of imaginary numbers are both subsets of the set of complex numbers. If $y = 0$, z is real and if $x = 0$, z is imaginary.

Using the symbol $\Leftrightarrow$ to represent the words 'if and only if' we can state that

$$z = 0 \Leftrightarrow x = 0 \quad \text{and} \quad y = 0$$

i.e. the two statements on either side of the symbol $\Leftrightarrow$ are equivalent.

Furthermore, if $z_1 = z_2$ i.e. $x_1 + y_1 j = x_2 + y_2 j \Leftrightarrow x_1 = x_2$ and $y_1 = y_2$. Note that the equality of two complex numbers implies two separate equalities in real numbers. For example

$$3u - v + (u + v)j \equiv 1 + 3j \Rightarrow \quad 3u - v = 1 \quad \text{and} \quad u + v = 3$$

so that $u = 1$, $v = 2$.

Before combining complex numbers we make one more definition. The **complex conjugate** of a number $z = x + yj$ is $\bar{z} = x - yj$.

Algebra of Complex Numbers

Complex numbers obey all the usual laws of algebra and the quantity j behaves like an algebraic constant. Hence, if z_1, z_2 and z_3 are complex numbers and k is a real number, then

$$z_1 + z_2 = z_2 + z_1 \qquad \text{(commutative law)}$$
$$(z_1 + z_2) + z_3 = z_1 + (z_2 + z_3) \quad \text{(associative law)}$$
$$k(z_1 + z_2) = kz_1 + kz_2 \qquad \text{(distributive law)}$$

The complex numbers form a **complete** set with respect to conventional algebra. In other words, the sum, difference, product and ratio of two complex numbers are also complex numbers.

Let $z_1 = x_1 + y_1 j$ and $z_2 = x_2 + y_2 j$ be two complex numbers; then any function of z_1 and z_2 can be expressed as a complex number $a + bj$. In particular

$$z_1 + z_2 \quad \text{is defined to be} \quad (x_1 + x_2) + (y_1 + y_2)j$$
$$z_1 - z_2 \quad \text{is defined to be} \quad (x_1 - x_2) + (y_1 - y_2)j$$

The product $z_1 \times z_2$ and the ratio $z_1 \div z_2$ can be expressed as complex numbers in the form $a + bj$ by using the usual laws of algebra. First the product:

$$z_1 \times z_2 = (x_1 + y_1 j)(x_2 + y_2 j) = x_1x_2 + y_1y_2 j^2 + (y_1x_2 + x_1y_2)j$$
$$= (x_1x_2 - y_1y_2) + (x_1y_2 + x_2y_1)j$$

Note that when a complex number is multiplied by its conjugate, the result is a real number:

$$z\bar{z} = (x + yj)(x - yj) = x^2 - y^2 j^2 = x^2 + y^2$$

Also
$$\frac{z_1}{z_2} = \frac{(x_1 + y_1 j)}{(x_2 + y_2 j)} = \frac{(x_1 + y_1 j)}{(x_2 + y_2 j)} \times \frac{(x_2 - y_2 j)}{(x_2 - y_2 j)} = \frac{(x_1x_2 + y_1y_2)}{x_2^2 + y_2^2} + \frac{(x_2y_1 - x_1y_2)}{x_2^2 + y_2^2} j$$

The ratio has been found by multiplying top and bottom by the conjugate of the bottom.

Examples

1. Express the following quantities in the form $a + bj$.

(a) $(2 + j)(1 + 2j)$ 　　(b) $(1 + j)^3$ 　　(c) $\dfrac{3 - j}{3 + j}$

(d) $\dfrac{2j}{(6 - j)^2}$ 　　(e) $\left(\dfrac{c + dj}{c - dj}\right)^2$ 　　$(c, d \text{ real})$

Solution

(a) $(2+j)(1+2j) = 2 + 5j + 2j^2 = 2 + 5j - 2 = 5j$

(b) $(1-j)^3 = 1 - 3j + 3j^2 - j^3 = 1 - 3j - 3 - (-j) = -2 - 2j$

(c) $\dfrac{3-j}{3+j} = \dfrac{(3-j)(3-j)}{(3+j)(3-j)} = \dfrac{9 - 6j + j^2}{3^2 + 1^2} = \dfrac{8 - 6j}{10} = \dfrac{4 - 3j}{5}$

(d) $\dfrac{2j}{(6-j)^2} = \dfrac{2j}{(6-j)^2} \times \dfrac{(6+j)^2}{(6+j)^2} = \dfrac{2j(6+j)^2}{(6^2+1^2)^2}$

$\qquad = \dfrac{2j}{(37)^2}(36 + 12j + j^2) = \dfrac{2j}{1369}(35 + 12j) = \dfrac{2}{1369}(-12 + 35j)$

(e) $\left(\dfrac{c+dj}{c-dj}\right)^2 = \left(\dfrac{c^2 - d^2 + 2cdj}{c^2 + d^2}\right)^2$

$\qquad = \dfrac{1}{(c^2+d^2)^2}\{(c^2 - d^2)^2 - 4c^2d^2 + (4cd)(c^2 - d^2)j\}$

2. $f(z) = az^2 + bz + c$ is a **real quadratic** if a, b, c are real. Also z is a root of the equation if $f(z) = 0$.

(a) If $f(z) = 0$, show that $\bar{f}(z) = f(\bar{z}) = 0$

(b) What are the roots of $z^2 + z + 20 = 0$?

Solution

(a) Let $z = x + yj$ and $\bar{z} = x - yj$ be its conjugate, then $(\bar{z})^2 = \overline{z^2}$. This follows because $(x - yj)^2 = x^2 - y^2 - 2xyj$ and

$$\overline{(x + yj)^2} = \overline{(x^2 - y^2 + 2xyj)} = x^2 - y^2 - 2xyj$$

Noting also that $\overline{z_1 + z_2} = \bar{z}_1 + \bar{z}_2$, i.e. the conjugate of the sum of the two complex numbers is the sum of their conjugates, then if a, b, c are real, $\overline{az^2 + bz + c} = \overline{az^2} + \overline{bz} + \bar{c} = a\bar{z}^2 + b\bar{z} + c$ then $\overline{f(z)} = f(\bar{z})$. This means that if $f(z) = 0, (= \bar{0})$, then

$$az^2 + bz + c = 0 \Leftrightarrow \overline{az^2 + bz + c} = 0 \Leftrightarrow a\bar{z}^2 + b\bar{z} + c = 0$$

In other words, if z is a complex root of the quadratic equation, then so is $\bar{z}$. In fact the equation has roots

$$\dfrac{-b \pm \sqrt{4ac - b^2}j}{2a} = \beta \pm \gamma j$$

so that $z = \beta + \gamma j$ and $\bar{z} = \beta - \gamma j$ are roots.

(b) For $z^2 + z + 20 = 0$ we obtain

$$z = \frac{-1 \pm \sqrt{80 - 1}\,j}{2} = \frac{1}{2}(1 \pm \sqrt{79}\,j)$$

3. Consider $w = \dfrac{1}{2 - z}$

    (a) If w is real prove that $y = 0$

    (b) If the real and imaginary parts of w are equal, prove that $x + y = 2$

First $\dfrac{1}{2 - z} = \dfrac{1}{2 - (x + yj)} = \dfrac{2 - x + yj}{(2 - x)^2 + y^2} = \dfrac{2 - x}{(2 - x)^2 + y^2} + \dfrac{y}{(2 - x)^2 + y^2}\,j$

    (a) If w is purely real then $\dfrac{y}{(2 - x)^2 + y^2} = 0$, i.e. $y = 0$

    (b) If the real and imaginary parts of w are equal, then $2 - x = y$, i.e. $x + y = 2$

4. The quartic polynomial equation

$$z^4 - 3z^3 + 3z^2 - 37z + 36 = 0$$

has two roots which are small positive integers and another two which are complex. Determine them all.

    The remainder theorem easily shows that $z = 1$ and $z = 4$ are the real roots and that the factors of the polynomial are

$$(z - 1)(z - 4)(z^2 + 2z + 9)$$

To find the complex roots we set the quadratic factor equal to zero and we find that the remaining roots of the quartic are $z = -1 \pm 2\sqrt{2}\,j$

Note that in the fourth example the complex roots were conjugates of one another. This is always the case when the coefficients of the polynomial are real. It is left as an exercise for you to verify that in general $\overline{P_n(z)} = P_n(\bar{z})$ ∎

> A polynomial $P_n(z)$ whose coefficients are real has n factors, each counted according to their multiplicity. These are either real or occur in complex conjugate pairs.

Exercise 12.1

1 Express in the form $a + bj$

(a) $(2 + 3j)(3 - 2j)$ (b) $(1 - j)^3$ (c) $\dfrac{2 + 3j}{3 + 2j}$

(d) $\dfrac{1}{(3 + 2j)^2}$ (e) $\left(\dfrac{2 - 3j}{3 - 2j}\right)^2$ (f) $\dfrac{(1 - j)^3}{(1 + j)^4}$

(g) $j + (2 + 3j) + (4 + 5j) + \cdots + \{2n - 2 + (2n - 1)j\}$

(h) $\displaystyle\sum_{n=0}^{N} j^n$

2 If $z = x + yj$, express the quotient $\dfrac{z - 1}{z + 1}$ in the form $a + bj$

(a) If this quotient is real, prove that $y = 0$

(b) It is a purely imaginary, prove that $x^2 + y^2 = 1$

(c) If $a = b$ prove that $x^2 + y^2 - 2y - 1 = 0$

3 Solve for z the equations

(a) $(2 + j)z + j = 3$

(b) $\dfrac{z - 1}{z - j} = \dfrac{2}{3}$

(c) $\dfrac{z - 5j}{2z + 3j} = \dfrac{z + j + 1}{2z - j}$

4 Prove by multiplication that

(a) $(\cos\alpha + \sin\alpha j)(\cos\beta + \sin\beta j) \equiv [\cos(\alpha + \beta) + \sin(\alpha + \beta)\,j]$

(b) $\dfrac{1}{\cos\alpha + \sin\alpha\ j} \equiv \cos\alpha - \sin\alpha\ j$

(c) $\dfrac{\cos\alpha + \sin\alpha\ j}{\cos\beta + \sin\beta\ j} \equiv \cos(\alpha - \beta) + \sin(\alpha - \beta)\ j$

5 Solve the equation $z^2 + z + 1 = 0$. Hence prove that

$$z^3 - 1 = (z - 1)(z - \lambda)(z - \mu) \quad \text{where} \quad \lambda = \frac{-1 + \sqrt{3}\,j}{2}, \mu = \frac{-1 - \sqrt{3}\,j}{2}$$

Show that λ and μ are the square of the other.

6   Write down all the roots of the polynomial equation

$$(z + 5)(z - 7)^2(z^2 + 4z + 5)(z^2 + 6z + 10) = 0$$

7   (a)   Show that $z + \bar{z}$ is real and that $z - \bar{z}$ is purely imaginary.

(b)   The **modulus** of a complex number $|z|$, is defined as

$$|z| = \sqrt{z\bar{z}} = (x^2 + y^2)^{1/2}$$

Find the modulus of

(i) $3 + 4j$ (ii) $12 - 5j$ (iii) $\dfrac{1}{5 - 7j}$

8   If z and w are any two complex numbers prove that

(a) $|z + w|^2 + |z - w|^2 = 2\{|z|^2 + |w|^2\}$

(b) $\left|\dfrac{z}{w}\right| = \dfrac{|z|}{|w|}$

Take $z = x + yj$, $w = u + vj$

(c)   Determine the modulus of $\dfrac{(3 + 4j)^{10}}{(4 - 3j)^8}$

9   The square root of a complex number z, written $z^{1/2}$ or $\sqrt{z}$, produces z when squared, e.g. $(3 - 2j)^2 = 5 - 12j$.
To find possible square roots of $5 - 12j$, write

$$5 - 12j = (\alpha + \beta j)^2$$

then equate real and imaginary parts, i.e.

$$\begin{aligned} \alpha^2 - \beta^2 &= 5 &&\text{(i)} \\ 2\alpha\beta &= -12 &&\text{(ii)} \end{aligned}$$

But $\alpha^2 + \beta^2 = \sqrt{169} = 13$ (iii)

Add equations (i) and (iii) to obtain α^2, giving two values of α. Substitute in (ii) to find the values of β.

10   (a)   If $w = \alpha + \beta j$ is a square root of $z = x + yj$, prove that $-w$ is also a square root.

(b)   Prove that $\alpha^2 = \dfrac{1}{2}(x + |z|) \geq 0$ and hence that any complex number can only possess two square roots at most.

11   Evaluate

(a) $\sqrt{j}$ (b) $(2 + 2j)^{1/2}$ (c) $(3 + 4j)^{-1/2}$

12 Determine the roots of the following quadratic equations:

(a) $z^2 - j = 0$ (b) $z^2 - 2jz + 8 = 0$

(c) $z^2 - (2 + j)z - 6j - 8 = 0$ (d) $(2 + j)z^2 - 5z + (2 - j) = 0$

(e) $3z^2 - (1 + j)z + (3 - j) = 0$

13 Solve completely the equation

$$z^4 - 2z^2 + 4 = 0$$

14 Express in the form of partial fractions

$$\frac{1}{z^2 - 4jz + 21} = \frac{A}{z - \alpha} + \frac{B}{z - \beta}$$

15 (a) Prove by induction

(i) $\overline{z_1 + z_2 + \cdots + z_n} = \bar{z}_1 + \bar{z}_2 + \cdots + \bar{z}_n$

(ii) $\overline{z_1 z_2 \ldots z_n} = \bar{z}_1 \bar{z}_2 \ldots \bar{z}_n$

(b) A polynomial of degree n in z can be, written as

$$P_n(z) \equiv a_0 + a_1 z + \cdots + a_n z^n$$

If the coefficients $a_0, a_1, \ldots, a_n$ are all real, establish that

(i) $\overline{P_n(z)} \equiv P_n(\bar{z})$

(ii) If α is a root of $P_n(z)$ then so is $\bar{\alpha}$.

(Hints are given with the answers.)

16 We are given that $1, j, 1 + j$ are simple roots and -2 is a double root of a seventh-order polynomial. Write out the polynomial as a product of linear and quadratic factors and expand it completely.

12.2 Complex Numbers and Geometry

The Argand diagram

A complex number $x + yj$ can be considered as an ordered pair of real numbers (x, y); it can be represented as a point P in the x–y plane called the z-**plane** or **Argand diagram**, named after the French mathematician Argand. As there is a one-to-one correspondence between the set of complex numbers and the seet of points in the z-plane, the complex number z is often referred to as the point $P(x, y)$. The x- and y-axes are referred to as the real and imaginary axes respectively. Refer to Figure 12.3 where four complex numbers have been plotted. Note that z_1 and z_4 are a conjugate pair and that the points which represent them are mirror images of each other in the real axis.

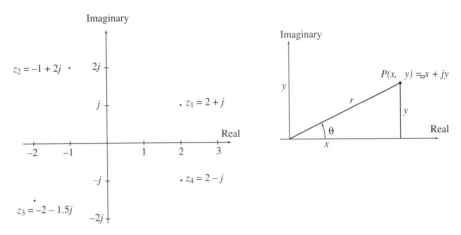

Figure 12.3 Points on an Argand diagram

Figure 12.4 Polar representation of a complex number

The point P also has a **polar representation** (Figure 12.4). Its distance from the origin is r and the **polar angle** is θ. We call $r = \sqrt{x^2 + y^2}$ the **modulus** of z, often written $|z|$, and θ the **argument** of z or **arg** z; θ is measured anticlockwise from the positive x-axis.

Note that θ is **multi-valued**, i.e. any multiple of 2π can be added to it or subtracted from it without changing its position, and the convention is to restrict the argument of a complex number to either $0 \le \arg z < 2\pi$ or $-\pi \le \arg z < \pi$. We therefore write

$$\theta = \arg z = \tan^{-1}\left(\frac{y}{x}\right) \quad \text{and} \quad z = r(\cos\theta + \sin\theta\ j)$$

Note that since $x = r\cos\theta$, $y = r\sin\theta$ then $\bar{z} = x - yj = r\cos\theta - r\sin\theta\ j$; alternatively, $\bar{z} = r[\cos(-\theta) + \sin(-\theta)\ j] = r(\cos - \sin\theta\ j)$

The exponential form of a complex number

If we define e^z to be the infinite series $1 + z + \dfrac{z^2}{2!} + \dfrac{z^3}{3!} + \dfrac{z^4}{4!} + \dfrac{z^5}{5!} + \ldots$ and substitute $z = \theta j$, we obtain the series

$$e^{\theta j} = 1 + \theta j + \frac{\theta^2 j^2}{2!} + \frac{\theta^3 j^3}{3!} + \frac{\theta^4 j^4}{4!} + \frac{\theta^5 j^5}{5!} + \ldots$$

$$= \left(1 - \frac{\theta^2}{2!} + \frac{\theta^4}{4!} - \ldots\right) + \left(\theta - \frac{\theta^3}{3!} + \frac{\theta^5}{5!} - \ldots\right) j$$

Therefore we may write $e^{\theta j} \equiv \cos\theta + \sin\theta \; j$ because the two series in brackets above are just those of $\cos\theta$ and $\sin\theta$ respectively. Hence z may now be written as

$$z = r(\cos\theta + \sin\theta \; j) = re^{\theta j}$$

Note that if $z = re^{\theta j}$ then $\bar{z} = re^{-\theta j}$

Multiplication and division are very straightforward for complex numbers in exponential form. Begin with two complex numbers, z_1 and z_2:

$$z_1 = r_1 e^{\theta_1 j} \qquad z_2 = r_2 e^{\theta_2 j}$$

$$z_1 \times z_2 = r_1 r_2 e^{\theta_1 j} e^{\theta_2 j} = r_1 r_2 e^{(\theta_1 + \theta_2) j}$$

so

$$|z_1 z_2| = |z_1| \times |z_2| \quad \text{and} \quad \arg(z_1 z_2) = \arg z_1 + \arg z_2 \tag{12.1}$$

We also have that

$$\frac{z_1}{z_2} = \frac{r_1 e^{\theta_1 j}}{r_2 e^{\theta_2 j}} = \frac{r_1}{r_2} e^{\theta_1 j - \theta_2 j} = \frac{r_1}{r_2} e^{(\theta_1 - \theta_2) j}$$

therefore

$$\left|\frac{z_1}{z_2}\right| = \frac{r_1}{r_2} = \frac{|z_1|}{|z_2|} \quad \text{and} \quad \arg\left(\frac{z_1}{z_2}\right) = \arg z_1 - \arg z_2 \tag{12.2}$$

Example

Write the following complex numbers in exponential form:

(a) $12 + 35j$ (b) $\sqrt{5} - 2j$ (c) $-1.22 + 3.861j$

Solution

(a) modulus $= 37$, argument $= \tan^{-1}\dfrac{35}{12} = 1.240^c$; $z = 37e^{1.240j}$

(b) modulus $= 3$, argument $= 2\pi - \tan^{-1}\dfrac{2}{\sqrt{5}} = 5.553^c$; $z = 3e^{5.553j}$

(c) modulus $= 4.021$, argument $= \pi - \tan^{-1}\left(\dfrac{3.861}{1.122}\right)$; $z = 4.021e^{1.854j}$ ∎

Complex numbers as vectors

There is a direct equivalence between complex numbers and two-dimensional vectors; see Figure 12.5. We make use of the fact that the point $P(x, y)$ can be associated uniquely

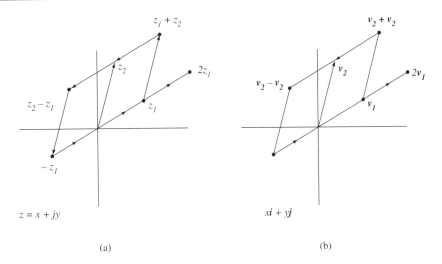

Figure 12.5   Complex numbers and vectors

with a line segment joining that point to the origin and hence can be associated uniquely with the vector $\overrightarrow{OP}$. This means, for example, that the addition of two complex numbers is equivalent to the parallelogram law for adding vectors in two dimensions, as illustrated. And note that the simple equivalence between $2z_1$ and $2\mathbf{v}_1$, between $z_1 - z_2$ and $\mathbf{v}_1 - \mathbf{v}_2$, and between $-z_1$ and $-\mathbf{v}_1$.

The points representing the complex numbers $z_1, z_1 + z_2, z_2 - z_1$ and $-z_1$ form the vertices of a parallelogram, as is the case for the vectors $\mathbf{v}_1, \mathbf{v}_1 + \mathbf{v}_2, \mathbf{v}_2 - \mathbf{v}_1$ and $-\mathbf{v}_1$.

Loci in the complex plane

We can draw loci in the Argand Diagram, or **complex plane**. The concept is best illustrated by means of examples.

Examples

1. Determine the curve represented by $|z| = a$ and by $|z - z_1| = a$. Writing $z = x + yj$ and squaring both sides we obtain

$$|z|^2 = z\bar{z} = x^2 + y^2 = a^2$$

which is the equation of a circle, centre the origin, radius a (Figure 12.6). On the circle $|z| = a$, whereas inside the circle $|z| < a$ and outside the circle $|z| > a$.

In the context of the complex plane, the modulus function represents **distance**. Therefore the equation $|z - z_1| = a$ represents the locus of a point z whose distance from a fixed point, z_1, is constant (Figure 12.7).

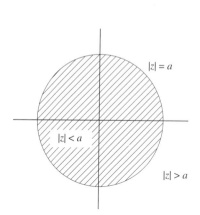

Figure 12.6 The locus $|z| = a$

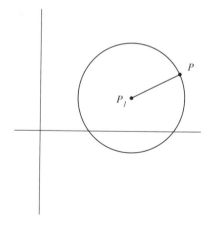

Figure 12.7 The locus $|z - z_1| = a$

P traces out a circle of radius a centred on P_1. It is left as an exercise for you to verify that its Cartesian equation is

$$(x - x_1)^2 + (y - y_1)^2 = a^2$$

where $z_1 = z_1 + y_1 j$.

2. Find the locus of the point which satisfies the equation $|z - a| + |z - b| = 2k$ where k is positive. The locus is an ellipse as shown in Figure 12.8. The foci are at the points a and b, as illustrated. The defining equation expresses the property of the ellipse, which states that the sum of the distances from any point on the ellipse to the foci is constant.

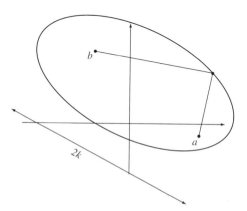

Figure 12.8 The locus $|z - a| + |z - b| = 2k$

3. Find the locus specified by $\left|\dfrac{z-a}{z-b}\right| = k$ where k is positive. The locus is a circle as shown in Figure 12.9. The proof is the subject of Question 7 in Exercise 12.2.

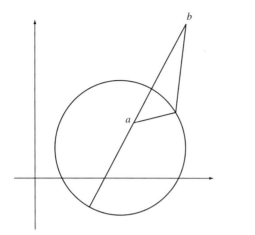

Figure 12.9 The locus $\left|\dfrac{z-a}{z-b}\right| = k$ **Figure 12.10** The locus $\arg(z-a) = \theta$

4. Find the locus defined by the equation $\arg(z-a) = \theta$. The locus is a semi-infinite straight line as shown in Figure 12.10.

5. Find the locus specified by the equation $\arg\left|\dfrac{z-a}{z-b}\right| = \theta$, where θ is an acute angle. The locus is the major segment of the circle subtended from A to B in Figure 12.11. This follows from the chord–angle theorem.

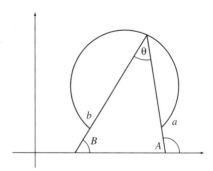

Figure 12.11 The locus $\arg\left|\dfrac{z-a}{z-b}\right| = \theta$

Exercise 12.2

 1 Determine the modulus and argument of the following complex numbers then write them in exponential form.

(a) $3 + 4j$ (b) $3 - 4j$ (c) $-1 + j$

(d) $1 + \sqrt{3}j$ (e) $2.7015 + 4.2074j$

2 If the argument of z is θ determine the argument of the following complex numbers:

(a) jz (b) $-z$ (c) $(1 + j)z$

(d) z^3 (e) $2z$

3 Find z_3 given that $z_1 = 2 + j$, $z_2 = -2 + 4j$ and $\dfrac{1}{z_3} = \dfrac{1}{z_1} + \dfrac{1}{z_2}$. If z_1, z_2, z_3 are represented on an Argand diagram by the points P_1, P_2, P_3 show that

$$z_2 = 2z_1 j \quad \text{and} \quad z_3 = \frac{2}{5}(z_1 - z_2)j$$

Remembering that multiplication of a complex number by j is equivalent to rotating the corresponding vector anticlockwise through a right angle, prove that P_3 is the foot of the perpendicular from O to the line P_1P_2.

 4 Define the modulus and argument of a complex number $z = x + yj$. If the points P_1 and P_2 represent the complex numbers z_1 and z_2 on an Argand diagram, give geometrical constructions to find the points representing $z_1 + z_2$ and $z_1 - z_2$.

(a) If $z = 1 + j$, mark on an Argand diagram the four points A, B, C, D representing z, z^2, z^3, z^4 respectively.

(b) Find, by calculation or from your diagram, the modulus and arguments of the complex numbers $(z^3 - 1)$ and $(z + z^4)$.

(c) Show that $\arg[(z^3 - z^4)/(z^2 - z^4)]$ is given by the angle $B\hat{D}C$; hence or otherwise show that the angles $B\hat{D}C$ and $A\hat{C}B$ are equal.

5 Begin by squaring up, and prove that the equation of a circle if radius a, centred on z_1, i.e.

$$|z - z_1| = a$$

is equivalent to the Cartesian equation

$$(x - x_1)^2 + (y - y_1)^2 = a^2$$

where $z_1 = x_1 + y_1 j$

6   Give geometric interpretations of the following loci in the complex plane; c is real and positive.

(a) $|z - c| + |z + c| = 6c$

(b) $|z - cj| - |z + cj| = c$

(c) $|z - c|^2 + |z + c|^2 = 4c^2$

(d) $\arg\left(\dfrac{z - c}{z + c}\right) = \dfrac{\pi}{2}$

In cases (a), (c) and (d), determine the equivalent Cartesian equation.

7   Prove that the locus of the point z in the complex plane such that

$$\left|\frac{z - 1}{z + 1}\right| = K \qquad (K \text{ a positive constant})$$

is a circle if $K \neq 1$. Draw a sketch. What is the locus when $K = 1$?

8   Sketch the following loci:

(a) $\arg(z - 1 - j) = \dfrac{\pi}{4}$

(b) $\arg(z + 1) = \arg(z - 1) + \dfrac{\pi}{6}$

(c) $\left|\dfrac{z - 3j}{z + 1 + j}\right| = 2$

(d) $|z + 3| - |z + 3j| = 7$

(e) $\mathrm{Im}\left(z - 1 + \dfrac{1}{2}\right) = 0$

(f) $\mathrm{Re}\left(\dfrac{2z - 1}{z}\right) = 1$

9   Identify the loci in the following sketches

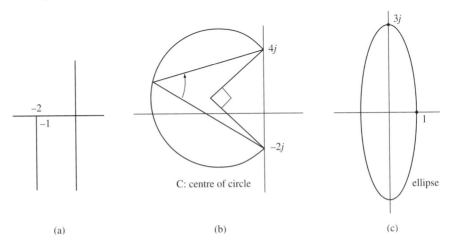

(a)

(b)   C: centre of circle

(c)   ellipse

10 (a)   Shade the following regions on an Argand diagram:

        (i) $|z - 1 - j| < \sqrt{2}$, $0 < \arg(z - 1 - j) < \dfrac{\pi}{2}$

        (ii) $\dfrac{\pi}{6} < \arg z < \dfrac{\pi}{3}$, $\dfrac{2\pi}{3} < \arg(z - 4) < \dfrac{5\pi}{6}$

        (iii) $|zj - 5| < 3$, $\text{Re}(z) > 0$

    (b)   Which values of z satisfy the following conditions?

        (i) $|z - 1 - 2j| = 3$, $\text{Re}(z - 1) = 0$

        (ii) $|z| = 5$, $\arg z = \pi + \tan^{-1}\left(\dfrac{4}{3}\right)$?

    (c)   Shade in those values of z in the complex plane for which $\text{Re}(z^2) > 0$, $0 \leq \arg z < 2\pi$

11   The triangle inequality $|z_1 - z_2| \leq |z_1| + |z_2|$ seems to hold in accordance with the diagram. To prove it start by squaring up the left-hand side and observe that $\text{Re}(z_1 \bar{z}_2) = \text{Re}(\bar{z}_1 z_2) < |z_1||z_2|$

Now prove that

    (a) $||z_1| - |z_2|| \leq |z_1 - z_2|$ and $||z_1| - |z_2|| \leq |z_1 + z_2|$

    (b) $|z_1 - z_2| \leq |z_1| + |z_2|$ and $|z_1 + z_2| \leq |z_1| + |z_2|$

Replace z_2 by $-z_2$ in (a) to prove the inequality when signs are reversed.

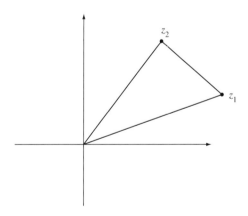

12   Prove by induction that

$$|z_1 + z_2 + \cdots + z_n| \leq |z_1| + |z_2| + \cdots + |z_n|$$

13 If $|z|$ is very large and $f(z) = P_n(z)$ is a polynomial of degree n, use the results from the previous questions to prove that

$$|P_n(z)| = |a_n z^n + a_{n-1} z^{n-1} + \cdots + a_0|$$

$$\geq \left| |a_n z^n| - \left| \sum_{r=0}^{n-1} a_r z^r \right| \right|$$

$$\geq |T_1 - T_2|, \text{ say.}$$

14 As $|z| \to \infty$, $T_1 \geq T_2$ in Question 13 so $|f(z)|$ is strictly positive, because the $|z^n|$ term dominates. For the following polynomials determine the smallest positive integer for which $|f(z)| > 0$

(a) $z^4 - 3z^3 + 2z^2 + z + 1$ (b) $z^5 - 11z^3 + 5z$

(c) $4z^6 - 30z^5 j + 21z^4 - 100z + 5000(1+j)$

Interpret your results.

15 $P_6(z) = z^6 + z^4 + z^2 + 1$

(a) Let $z = x$ (real) and prove that there are no real roots. ($P_6(x) \geq 1$, why?)

(b) Let $z = yj$ and prove that there are two purely imaginary roots.

(c) Obtain a **circle of containment**, inside which all the roots must lie.

12.3 De Moivre's Formula

Given the relationships

$$e^{\theta j} \equiv \cos\theta + \sin\theta \ j \quad \text{and} \quad (e^{\theta j})^n \equiv e^{n\theta j}$$

we can write

$$(\cos\theta + \sin\theta \ j)^n \equiv \cos n\theta + \sin n\theta \ j \tag{12.3}$$

This result is known as **de Moivre's formula** and we look at some of its applications.

Example   Use de Moivre's formula with $n = 3$ to find formulae for $\cos 3\theta$ and $\sin 3\theta$. First

$$\cos 3\theta + \sin 3\theta \ j \equiv (\cos\theta + \sin\theta \ j)^3$$

Expanding the right-hand side, we obtain

$$\cos^3 \theta + 3 \cos^2 \theta \sin \theta \; j - 3 \cos \theta \sin^2 \theta - \sin^3 \theta \; j$$

Equating the real part and the imaginary part on either side of the identity, we obtain the identities

$$\cos 3\theta \equiv \cos^3 \theta - 3 \cos \theta \sin^2 \theta \equiv 4 \cos^3 \theta - 3 \cos \theta$$

$$\sin 3\theta \equiv 3 \cos^2 \theta \sin \theta - \sin^3 \theta \equiv 3 \cos \theta - 4 \sin^3 \theta$$

∎

You will recall that if an equation involves complex quantities then both the real and imaginary parts must equate separately, i.e. two real equations are produced. Similar identities hold for other multiple angles. Two further results will be useful. If

$$z = e^{\theta j} \quad \text{then} \quad |z|^2 = 1 \quad \text{and} \quad \bar{z} = \frac{1}{z} = e^{-\theta j}$$

Also $\quad z^{-n} = e^{-n\theta j} = \cos n\theta - \sin n\theta \; j$

Example

Express $\cos^6 \theta$ in terms of cosines of multiples of θ. First

$$z^n + z^{-n} \equiv \cos n\theta + \sin n\theta \; j + \cos n\theta - \sin n\theta \; j \equiv 2 \cos n\theta$$

then $\quad (z + z^{-1})^6 = z^6 + 6z^5 z^{-1} + 15 z^4 z^{-2} + 20 z^3 z^{-3} + 15 z^2 z^{-4} + 6zz^{-5} + z^{-6}$

using the binomial theorem.

Combining powers and regrouping positive and negative powers together, we obtain

$$(z + z^{-1})^6 \equiv z^6 + z^{-6} + 6(z^4 + z^{-4}) + 15(z^2 + z^{-2}) + 20$$

i.e. $\quad (2 \cos \theta)^6 \equiv 2 \cos 6\theta + 6(2 \cos 4\theta) + 15(2 \cos \theta) + 20$

so $\quad \cos^6 \theta \equiv \left(\dfrac{\cos 6\theta + 6 \cos 4\theta + 15 \cos 2\theta + 10}{32} \right)$

The nth roots of unity

When we wrote $z = re^{\theta j}$ we restricted the argument of z to a range of 2π, e.g.

$$-\pi < \arg z \le \pi \quad \text{or} \quad 0 \le \arg z \le 2\pi$$

However, $\arg z$ is an angle which could be assigned any value. If therefore we add to θ the quantity $2k\pi$, where k is an integer, we make no change to z because

$$z = re^{(\theta + 2k\pi)j} = re^{\theta j} e^{2k\pi j} = re^{\theta j}$$

This follows because $e^{2k\pi j} \equiv \cos 2k\pi + \sin 2k\pi \; j = 1$

Example

Solve the equation $z^6 = 1$.
Now $1 = e^{2k\pi j}$ for integer k, so that

$$z^6 = e^{2k\pi j}$$

Therefore, taking the sixth root, $z = e^{k\pi j/3}$
Remember that k is any integer, but taking in turn $k = 0, 1, 2, 3, 4$ and 5 generates six distinct solutions.

$$k = 0 \qquad z_0 = \cos 0 + j\sin 0 = 1$$

$$k = 1 \qquad z_1 = \cos\frac{\pi}{3} + \sin\frac{\pi}{3}\, j = \frac{1}{2} + \frac{\sqrt{3}}{2}j$$

$$k = 2 \qquad z_2 = \cos\frac{2\pi}{3} + \sin\frac{2\pi}{3}\, j = -\frac{1}{2} + \frac{\sqrt{3}}{2}j$$

$$k = 3 \qquad z_3 = \cos\pi + \sin\pi\, j = -1$$

$$k = 4 \qquad z_4 = \cos\frac{4\pi}{3} + \sin\frac{4\pi}{3}\, j = -\frac{1}{2} + \frac{\sqrt{3}}{2}j$$

$$k = 5 \qquad z_5 = \cos\frac{5\pi}{3} + \sin\frac{5\pi}{3}\, j = \frac{1}{2} - \frac{\sqrt{3}}{2}j$$

If other integer values of k are taken, the results obtained are merely a repetition of those above. For example

$$z_6 = \cos\frac{6\pi}{3} + \sin\frac{6\pi}{3}\, j = \cos 2\pi + \sin 2\pi\, j = 1$$

The six roots lie on a circle of radius 1,
centred at the origin, at the vertices of a regular hexagon (Figure 12.12). ∎

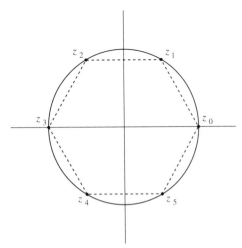

Figure 12.12 The six roots of unity

By the fundamental theorem of algebra we would expect the equation $z^6 = 1$ to possess six roots, so the result is no surprise. The coefficient of z^5 is zero, so the sum of the roots is 0. Note that $z_5 = \overline{z_1}$ and $z_4 = \overline{z_2}$.

Factorising the polynomial gives

$$z^6 - 1 \equiv (z^3 - 1)(z^3 + 1)$$
$$\equiv (z - 1)(z^2 + z + 1)(z + 1)(z^2 - z + 1)$$

the complex roots being those of the quadratics. ∎

Circular and exponential functions

The formula

$$e^{\theta j} \equiv \cos \theta + \sin \theta \; j$$

links the trigonometric and hyperbolic functions. For example

$$\cos \theta = \frac{1}{2}(e^{\theta j} + e^{-\theta j}) = \cosh(\theta j) \tag{12.4}$$

$$\sin \theta = \frac{1}{2j}(e^{\theta j} - e^{-\theta j}) = \frac{1}{j}\sinh(\theta j) \tag{12.5}$$

$$\cos \theta j = \frac{1}{2}(e^{-\theta} + e^{\theta}) = \cosh \theta \tag{12.6}$$

$$\sin \theta j = \frac{1}{2j}(e^{-\theta} - e^{\theta}) = -\frac{1}{j} \times \sinh \theta = \sinh \theta j \tag{12.7}$$

We can now find trigonometric and hyperbolic functions of a complex quantity. For example

$$\sin z \equiv \sin(x + yj) \equiv \sin x \cos(yj) + \cos x \sin(yj) \equiv \sin x \cosh y + \cos x \sinh yj$$
$$\cos z \equiv \cos(x + yj) \equiv \cos x \cos(yj) - \sin x \sin(yj) \equiv \cos x \cosh y - \sin x \sinh yj$$
$$\sinh z \equiv \sinh(x + yj) \equiv \sinh x \cosh(yj) + \cosh x \sinh(yj)$$
$$\equiv \sinh x \cos y - \cosh x \sin yj$$

Example   Determine all possible solutions to the equation $\sin z = 2$.

No purely real value of z can satisfy this equation, so set $z = x + yj$, whence

$$\sin(x + yj) = 2$$

i.e. $\sin x \cosh y + \cos x \sinh y \; j = 2 + 0j = 2$

Equating the real and imaginary parts separately gives two equations

$$\sin x \cosh y = 2 \quad \text{(i)}$$
$$\cos x \sinh y = 0 \quad \text{(ii)}$$

Tackling (ii) first yields

$$\cos x = 0 \quad \text{or} \quad \sinh y = 0$$

If $\cos x = 0$, then $x = (2k+1)\dfrac{\pi}{2}$, k integer, and $\sin x = \pm 1$. Substituting in (i) gives $\cosh y = \pm 2$. As $\cosh y \geq 0$, only the positive option is acceptable, so

$$\sin x = 1, \qquad \cosh y = 2 \qquad \text{(iii)}$$

therefore $\cos x = 0, \qquad \sinh y = \pm\sqrt{3} \qquad \text{(iv)}$

Hence $x = (4k+1)\dfrac{\pi}{2}, \qquad (k \text{ integer})$

and $y = \ln(2 \pm \sqrt{3}), \qquad \text{(from (iii))}$

Is $\sinh y = 0$ possible? Yes it is. If so, $\cosh y = 1$ and $\sin x = 2$. This cannot happen for x real, so there are no solutions for this option.

The roots lie in matching rows, equally spaced at an interval of 2π (Figure 12.13).

■

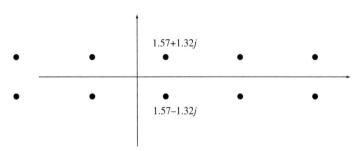

Figure 12.13   Some of the y roots of $\sin z = 2$

Multiple-valued functions

If w is an arbitrary complex number then the solution of the equation

$$e^w = z$$

may seem straightforward, but this is not the case.
    Taking logarithms we write

$$w = \ln z = \ln(re^{(\theta+2k\pi)j}) = \ln r + \theta j + 2k\pi j$$

where θ is a restricted argument, e.g. $-\pi < \theta \le \pi$ and k is an integer.

The function $\ln z$ has infinitely many values. In practice we deal with a function called Ln z, with θ restricted as $k = 0$. This is called the **principal value** of the logarithmic function.

Example

Write down values of the following:

(a) $\ln(3 + 4j)$

(b) $\mathrm{Ln}(-2), -\pi < \arg z \le \pi$

Solution

(a) $\ln(3 + 4j) = \dfrac{1}{2}\ln 5 + \left(\tan^{-1}\dfrac{4}{3} + 2k\pi\right) j$

(b) $\mathrm{Ln}(-2) = \ln 2 + \pi j$ ∎

Exercise 12.3

1   Express the following as a simple complex number or quantity:

(a) $\left(\cos\dfrac{\pi}{12} + \sin\dfrac{\pi}{12}j\right)^{15}$

(b) $(\sin\theta + \cos\theta\ j)^2$

(c) $e^{\pi j}$

2   Determine $\dfrac{\sin 6\theta}{\sin\theta}$ in terms of powers of $\cos\theta$

3   Prove that $\displaystyle\int_0^{\pi/2} \cos^6\theta\ d\theta = \dfrac{5\pi}{32}$

4   Given

$$C = \cos(\alpha + \theta) + \cos(\alpha + 2\theta) + \cdots + \cos(\alpha + n\theta)$$
$$S = \sin(\alpha + \theta) + \sin(\alpha + 2\theta) + \cdots + \sin(\alpha + n\theta)$$

determine an exponential form for $C + Sj$. Hence prove that

$$C = \cos\left(\alpha + \frac{n+1}{2}\theta\right)\frac{\sin\dfrac{n\theta}{2}}{\sin\dfrac{\theta}{2}}$$

and find S.

5   Find the six roots of the equation

$$(z - 1)^6 = 1$$

giving each in the form $a + bj$

6   Use de Moivre's theorem to find the roots of the equation

$$(1 - z)^5 = z^5 \qquad \left(\text{set } w = \frac{1 - z}{z}, w^5 = 1\right)$$

7*   Consider the solution to an equation of the form

$$z^n = c$$

where c is a general complex number, and $c = |c|e^{\phi j}, -\pi < \phi \le \pi$
De Moivre's theorem gives

$$z = |c|^{1/n}e^{(\theta + 2k\pi)j/n} \qquad k = 0, 1, \ldots, n - 1$$

i.e. n roots of magnitude $|c|^{1/n}$, the nth real root of $|c|$.
Obtain the following:

(a)   the values of $j^{1/3}$
(b)   the solutions of the equation $z^5 = -1 + 4j$, giving the argument in degrees.

8   Determine forms for

(a) $\cosh z$
(b) $\tan z$ (involving tan, tanh, sec, sech only)

9   Express the following in the form $a + bj$:

(a) $\sin(1 + 2j)$ (b) $\cosh(3 - 2j)$
(c) $\tanh(2 + j)$

10 Solve the following equations for z:

(a) $\sin z = 0$ (b) $\cos z = 2$

(c) $\cosh z = j$ (plot the solutions on an Argand diagram)

(d) $\ln z = -2$

11 Resolve $z^5 + 1$ into linear and quadratic factors with real coefficients. Deduce that
$$4 \sin \frac{\pi}{10} \cos \frac{\pi}{5} = 1$$

12* The six roots of
$$z^6 + z^3 + 1 = 0$$

lie at the vertices of a regular **nonagon**, a nine-sided figure, on $|z| = 1$ as shown.

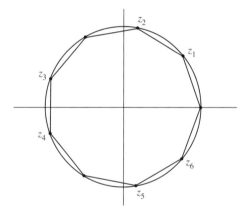

(a)   Identify the missing three vertices as complex numbers and identify the cubic equation whose roots they are.

(b)   Explain why the sixth-order equation must have the roots given above.

12.4 Applications

Complex numbers are necessary to expand and extend ideas and developments in mathematics that could only be partly completed using real numbers. Our aim in this section is to develop insight into some of these extensions.

Solving differential equations

In Chapter 11 we looked at second-order linear differential equations with constant coefficients. The simplest form is

$$\frac{d^2y}{dx^2} + \lambda y = 0, \qquad \text{where } \lambda \text{ is a constant}$$

A trial solution of the type Ae^{mx} yields the equation

$$Ae^{mx}(m^2 + \lambda) = 0$$

For a non-trivial solution the auxiliary quadratic must be zero, i.e.

$$m^2 + \lambda = 0$$

The sign of λ is all-important as this characterises the type of solution

(i) $\lambda = 0$ i.e. $m = 0$
(ii) $\lambda = k^2 > 0$ i.e. $m = \pm k$
(iii) $\lambda = -k^2 < 0$ i.e. $m = \pm kj$

Case (i) is simplest, the differential equation has the elementary solution $y = Ax + B$, where A and B are arbitrary constants.

Case (ii) gives an exponential solution of the form $y = Ae^{kx} + Be^{-kx}$, i.e. growth or decay.

Case (iii) gives a trigonometric solution

$$y = C \sin kx + D \cos kx \tag{a}$$

i.e. oscillation or simple harmonic motion, for Hooke's law perhaps.

If we allow $m = kj$ and solutions of the type Ae^{kxj}, for case (iii) we obtain

$$y = Ae^{kxj} + Be^{-kxj} \tag{b}$$

Forms (a) and (b) for y are identical because y is real, as are C and D, so A and B are complex

$$A = \frac{1}{2}(C - Dj), \qquad B = \frac{1}{2}(C + Dj)$$

By allowing complex roots of the auxiliary quadratic and using a solution of the form (a) we can determine the general solution to any second-order linear differential equation with constant coefficients.

Examples

1. Determine the general solutions of

(a) $\dfrac{d^2y}{dx^2} + 6\dfrac{dy}{dx} + 8y = 0$ (b) $\dfrac{d^2y}{dx^2} + 6\dfrac{dy}{dx} + 9y = 0$

(c) $\dfrac{d^2y}{dx^2} + 6\dfrac{dy}{dx} + 10y = 0$

Setting $y = Ae^{mx}$ in each case gives the auxiliary quadratic equations, roots and general solutions as follows

(a) $m^2 + 6m + 8 = 0$, i.e.

$$m = -2, -4$$

so $y = Ae^{-2x} + Be^{-4x}$

(b) $m^2 + 6m + 9 = 0$, i.e.

$$m = -3 \quad \text{(equal roots)}$$

so $y = (A + Bx)e^{-3x}$

(c) $m^2 + 6m + 10 = 0$, i.e.

$$m = -3 \pm j$$

so $y = e^{-3x}(A\cos x + B\sin x)$

This example demonstrates that very similar-looking differential equations have very different general solutions. It also emphasises the importance of quadratic equations and complex numbers.

2. The spring–mass system with damper is modelled by the oscillation of a mass at the end of a helical spring subject to Hooke's law and a resistive force which is linearly proportional to the velocity of the mass. The shock absorber of a car can be modelled in this way (Figure 12.14).

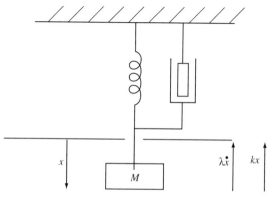

Figure 12.14 Modelling a spring–mass system and damper subject to Hooke's law

The equation of motion is

$$M\ddot{x} = -\lambda\dot{x} - kx$$

or $M\ddot{x} + \lambda\dot{x} + kx = 0$

where λ is the damping constant and k is the spring constant.
It is usually the case that the damping force controls vibration, rather than eliminating it, i.e. λ is relatively small, so that assuming a solution of the form Ae^{mt} gives the auxiliary quadratic equation

$$Mm^2 + \lambda m + k = 0$$

This has complex roots

$$m = \frac{-\lambda \pm (4kM - \lambda^2)^{1/2}j}{2M} \qquad \text{(if } \lambda < 2\sqrt{kM})$$

Writing

$$\beta = -\frac{\lambda}{2M} \qquad \text{(the } damping \text{ term)}$$

and $$\gamma = \frac{(4kM - \lambda^2)^{1/2}}{2M} \qquad \text{(the } oscillatory \text{ term)}$$

The general solution is therefore of the form

$$x(t) = e^{-\beta t}(Ae^{\gamma tj} + Be^{-\gamma tj})$$

or $x(t) = e^{-\beta t}(C\cos\gamma t + D\sin\gamma t)$

or $x(t) = Ee^{-\beta t}\sin(\gamma t + \varepsilon)$

3. A car's shock absorber is jerked into motion from a stationary position. With $x(0) = 0, \dot{x}(0) = 1, M = \frac{1}{2}, K = 148, \lambda = 10$, find the subsequent displacement $x(t)$.

   (a) What is $\dot{x}(0.1)$?
   (b) Sketch the graphs of $x(t), \dot{x}(t)$ for $t < 0.4$

First we calculate $\beta = -10, \gamma = 14$. Then

$$x(t) = Ee^{-10t}\sin(14t + \varepsilon)$$

The condition

$$x(0) = 0 \Rightarrow \varepsilon = 0$$

Also $\dot{x}(t) = E(-10e^{-10t}\sin 14t + 14e^{-10t}\cos 14t)$

so $\dot{x}(0) = 14E \quad \text{and} \quad E = \frac{1}{14}$

Finally

$$x(t) = \frac{(e^{-10t} \sin 14t)}{14}$$

(a) $\dot{x}(0.1) = e^{-1} \dfrac{(-10 \sin 1.4 + 14 \cos 1.4)}{14} = -0.196$

(b) The graphs of $x(t)$ and $\dot{x}(t)$ are shown in Figure 12.15.

For the spring–mass system with damper the graphs for $x(t)$ when $x(0) > 0$, $\dot{x}(0) > 0$, is bounded by $\pm e^{-\beta t}$ and $x(t) \to 0$ as $t \to \infty$ (Figure 12.16). ■

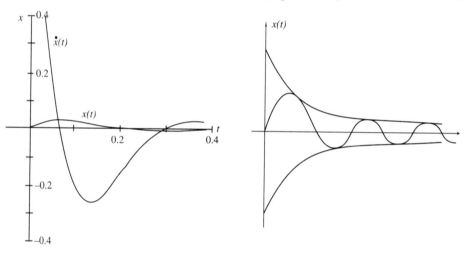

Figure 12.15   Oscillations of a shock absorber   **Figure 12.16**   Damped oscillation

The introduction of a **forcing term** on the right-hand side of a differential equation leads to a general solution which is the sum of two parts. The general form of the equation is

$$a\frac{d^2y}{dx^2} + b\frac{dy}{dx} + cy = f(x)$$

We state without proof that

General solution = complementary function + particular integral

The **complementary function** (CF) is the solution of the unforced equation, i.e. the equation with 0 on the right-hand side in place of $f(x)$. The **particular integral** (PI) is *any* solution of the full equation and needs to be found separately. It is beyond our scope to find particular integrals systematically but we will look at a straightforward example of an initial-value problem in which the particular integral can be found by the method of trial solution.

Example

A spring–mass system with damper has unit mass and is subject to the following balance of forces:

$$\text{Mass} \times \text{acceleration} = \text{stiffness force} + \text{damping force} + \text{applied force}$$

Consider the case where

$$\ddot{x} = -2x - 2\dot{x} + \cos t \quad \text{i.e.} \quad \ddot{x} + 2\dot{x} + 2x = \cos t$$

Describe the motion if the mass started from rest at $t = 0$. What form does its motion take after a long time has elapsed?

The auxiliary quadratic equation is

$$m^2 + 2m + 2 = 0$$

which has roots $m = -1 \pm j$, so

$$x_{\text{CF}} = e^{-t}(A \cos t + B \sin t)$$

Assume that the particular integral has the form $C \cos t + D \sin t$, and substitute into the differential equation to obtain

$$-(C \cos t + D \sin t) + 2(-C \sin t + D \cos t) + 2(C \cos t + D \sin t) \equiv \cos t$$

Comparing coefficients gives

$$C = \frac{1}{5}, D = \frac{2}{5},$$

hence

$$x_{\text{PI}} = \frac{1}{5}(-\cos t + 2 \sin t)$$

and the general solution is

$$x(t) = e^{-t}(A \cos t + B \sin t) + \frac{1}{5}(\cos t + 2 \sin t)$$

Now $x(0) = 0$, so

$$A + \frac{1}{5} = 0 \quad \text{i.e.} \quad A = -\frac{1}{5}$$

Differentiating the general solution gives

$$\dot{x}(t) = -e^{-t}(A \cos t + B \sin t) + e^{-t}(-A \sin t + B \cos t) + \frac{1}{5}(-\sin t + 2 \cos t)$$

And $\dot{x}(0) = 0$, therefore

$$-A + B + \frac{2}{5} = 0 \quad \text{so} \quad B = -\frac{3}{5}$$

Finally

$$x(t) = -\frac{e^{-t}}{5}(\cos t + 3 \sin t) + \frac{1}{5}(\cos t + 2 \sin t)$$

After a long time the natural damping term decays to a very small value and the system responds only to the forcing term, i.e.

$$x \sim \frac{1}{5}(\cos t + 2 \sin t)$$

∎

Electric circuits

In an electric circuit with alternating current of frequency ω the flow of the current is determined by the **reactance**, a real or complex quantity composed of resistance (R), capacitance (C) and inductance (L). It is aggregated in a form known as **impedance**. Reactance is in essence phased resistance has has components

Resistance (R ohms) $\qquad R$

Capacitance(C farads) $\qquad \dfrac{1}{\omega Cj}$

Inductance (L henries) $\qquad \omega Lj$

Capacitance leads by $90°$ in phase and inductance lags by $90°$. Impedance Z is the sum of these components, as determined in the following example.

Example

Consider the LCR circuit shown in Figure 12.17. The resistance and the inductance are in series and the impedance of that part of the circuit is the sum $R + \omega Lj$. This part of the circuit is in parallel with the capacitance, so for the complete circuit, the impedance Z is given by

$$\frac{1}{Z} = \frac{1}{\dfrac{1}{\omega Cj}} + \frac{1}{R + \omega Lj}$$

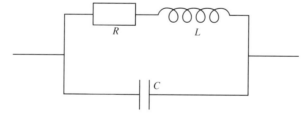

Figure 12.17 Series LCR circuit

Now

$$\frac{1}{Z} = \omega Cj + \frac{1}{R + \omega Lj} = \frac{1 - \omega^2 LC + \omega CRj}{r + \omega Lj}$$

so that

$$Z = \frac{r + \omega Lj}{1 - \omega^2 LC + \omega CRj}$$

$$= \frac{R + \omega Lj}{1 - \omega^2 LC + \omega CRj} \times \frac{1 - \omega^2 LC - \omega CRj}{1 - \omega^2 LC - \omega CRj}$$

$$= \frac{[R(1 - \omega^2 LC) + \omega^2 LCR] + (-\omega CR^2 + \omega L - \omega^3 L^2 C)j}{(1 - \omega^2 LC)^2 + \omega^2 C^2 R^2}$$

$$= \frac{R + (-\omega CR^2 + \omega L - \omega^3 L^2 C)j}{(1 - \omega^2 LC)^2 + \omega^2 C^2 R^2}$$

Therefore, after some manipulation, the impedance has modulus

$$|Z| = \frac{[R^2 + (-\omega CR^2 + \omega L - \omega^3 L^2 C)^2]^{1/2}}{(1 - \omega^2 LC)^2 + \omega^2 C^2 R^2}$$

and its phase angle is

$$\phi = \tan^{-1}\left[\frac{-\omega CR^2 + \omega L - \omega^3 L^2 C}{R}\right]$$

Suppose $\omega = 50$ Hz, $L = 2$ H, $C = 50$ μF and $R = 200\,\Omega$, Then

$$|Z| = 248.07\,\Omega \quad \text{and} \quad \phi = \tan^{-1}\left(-\frac{1}{8}\right) = -7.125° \qquad \blacksquare$$

Solving difference equations

As with differential equations, the auxiliary equation of a linear difference equation can have complex roots. We are now able to handle such cases.

Example   Solve the difference equation $y_{n+2} - 2y_{n+1} + 2y_n = 0$

We put $y_n = Ab^n$ to obtain the auxiliary equation $b^2 - 2b + 2 = 0$, which has roots

$$b_1 = 1 + j = \sqrt{2}\left(\cos\left(\frac{\pi}{4}\right) + \sin\left(\frac{\pi}{4}\right)j\right) = \sqrt{2}\,e^{\pi j/4}$$

$$b_2 = 1 - j = \sqrt{2}\left(\cos\left(\frac{\pi}{4}\right) - \sin\left(\frac{\pi}{4}\right)j\right) = \sqrt{2}\,e^{-\pi j/4}$$

The general solution of the difference equation is $y_n = (\sqrt{2})^n (Ae^{n\pi j/4} + Be^{-n\pi j/4})$; A, B complex.

However, y_n is real, so we can rewrite this as

$$y_n = (\sqrt{2})^n \left(C \cos \frac{n\pi}{4} + D \sin \frac{n\pi}{4} \right); \; C, \; D \text{ real}$$

where $C = A + B, D = j(A - B)$ ∎

In general $y_n = r^n (C \cos n\theta + D \sin n\theta)$, where r is the modulus of the roots and θ the argument.

Exercise 12.4

1 Determine general solutions to the following homogeneous constant-coefficient differential equations:

(a) $\dfrac{d^2y}{dx^2} + 2\dfrac{dy}{dx} + 2y = 0$

(b) $\dfrac{d^2y}{dx^2} - \dfrac{dy}{dx} + 2y = 0$

(c) $\dfrac{d^2y}{dx^2} + 8\dfrac{dy}{dx} + 16y = 0$

(d) $3\dfrac{d^2y}{dx^2} + 11\dfrac{dy}{dx} + 12y = 0$

2 For the given values of a and b, determine the solution to the initial-value problem

$$\ddot{x} + a\dot{x} + bx = 0; \qquad x_0 = 2, \dot{x}_0 = 0$$

(a) $a = 4, b = 13$

(b) $a = b = 4$

3 The current in an electric circuit with an inductor (L), resistor (R), and capacitor (C) in series without applied voltage satisfies the differential equation

$$L\frac{d^2i}{dt^2} + R\frac{di}{dt} + \frac{i}{C} = 0$$

Given that L, R and C are constant, determine the condition that needs to be fulfilled for the current to be oscillatory.

4 Determine the solution to the boundary value problem

$$\frac{d^2y}{dx^2} + 4\frac{dy}{dx} + 13y = 0 \qquad y(0) = 1 \quad \text{and} \quad y\left(\frac{\pi}{6}\right) = 0$$

5 A spring–mass system with damper is forced into motion by an applied vibration. The equation of motion is

$$M\ddot{x} + \lambda\dot{x} + Kx = F\cos pt$$

Solve the equation when $M = 1$, $\lambda = 3$, $K = 2$, $F = 10$, $p = 1$; $x(0) = \dot{x}(0) = 0$

6 A simple harmonic oscillation is executed by an applied undamped force of exactly the same period. The equation of motion is

$$\ddot{x} + x = \sin t$$

Because the applied force is resonant with the natural frequency $\sin t$, a modified trial solution to the response is necessary, i.e. $x = t(C\cos t + D\sin t)$. Obtain the general solution. Sketch the graph of the solution given $x(0) = 0$, $\dot{x}(0) = 10.5$.

7 The complex impedance Z for each of the following circuits is given. Determine $|Z|$ in each case symbolically.

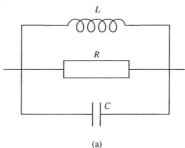

(a)

(a) C and R in parallel, connected in series with L:

$$Z = \omega Lj + \left(\frac{1}{(\omega Cj)} + \frac{1}{R}\right)^{-1}$$

(b) L and R in parallel, connected in series with C:

$$\frac{1}{Z} = \frac{1}{\omega Lj} + \frac{1}{R} + \frac{1}{(\omega Cj)}$$

(c)   Evaluate $|Z|$ in each case where

$$\omega = 50 \text{ Hz} \qquad L = 2 \text{ H} \qquad R = 100 \text{ }\Omega \qquad C = 100 \text{ }\mu F$$

8 Solve the following difference equations:

(a) $y_{n+2} + y_n = 0$

(b) $3y_{n+2} - 6y_{n+1} + 4y_n = 0$

SUMMARY

- **Complex number**: $z = x + yj$, x and y real numbers, $j^2 = -1$; conjugate $\bar{z} = x - yj$

- **Arithmetic**

$$z_1 + z_2 = (x_1 + x_2) + (y_1 + y_2)j$$

$$z_1 \times z_2 = (x_1 x_2 - y_1 y_2) + (x_1 y_2 + x_2 y_1)j$$

$$\frac{z_1}{z_2} = \frac{(x_1 + y_1 j)}{(x_2 + y_2 j)} = \frac{(x_1 + y_1 j)}{(x_2 + y_2 j)} \times \frac{(x_2 - y_2 j)}{(x_2 - y_2 j)}$$

$$= \frac{(x_1 x_2 + y_1 y_2)}{x_2^2 + y_2^2} + \frac{(x_2 y_1 - x_1 y_2)}{x_2^2 + y_2^2} j$$

- **Argand diagram**: $z = x + yj$ can be considered as an ordered pair of real numbers (x, y); it can be represented as a point P

- **Polar and exponential forms**: the complex number $z = r(\cos\theta + j\sin\theta) = re^{\theta j}$ has modulus r, $|z| = r$, and argument θ, $\arg z = \theta$; its complex conjugate is $\bar{z} = re^{-\theta j}$; other useful results are

$$|z_1 z_2| = |z_1| \times |z_2| \qquad \arg(z_1 z_2) = \arg z_1 + \arg z_2$$

$$\left|\frac{z_1}{z_2}\right| = \frac{r_1}{r_2} = \frac{|z_1|}{|z_2|} \qquad \arg\left(\frac{z_1}{z_2}\right) = \arg z_1 - \arg z_2$$

- **Loci in the plane**: $|z - z_1| = a$ represents a circle with centre z_1 and radius a; $\arg z = \alpha$ represents a half-line making an angle α with the positive x-axis.

- **De Moivre's theorem**

$$(\cos\theta + \sin\theta\, j)^n \equiv \cos n\theta + \sin n\theta\, j$$

- **Roots of a complex number**: the nth roots of a complex number all have modulus equal to the nth root of its modulus. One root has argument equal to its argument divided by n, the others are equally spaced around the circle with radius equal to the new modulus

- **Circular functions**

$$\sin z \equiv \sin x \cosh y + \cos x \sinh y \; j$$
$$\cos z \equiv \cos x \cosh y - \sin x \sinh y \; j$$

- **Multiple-valued functions**: $\ln z = \ln r + \theta j + 2k\pi j$, k integer; when $k = 0$ we have the principal value, Ln z.

Answers

Exercise 12.1

1   (a) $12 + 5j$ (b) $-2 + 2j$ (c) $\dfrac{12}{13} + \dfrac{5}{13}j$

(d) $\dfrac{5}{169} - \dfrac{12}{169}j$ (e) $\dfrac{119}{169} - \dfrac{120}{169}j$ (f) $\dfrac{1}{2}(1 + j)$

(g) $n(n - 1 + nj)$

(h) $1 + j - 1 - j + 1 + j - \cdots = \dfrac{1 - j^N}{1 - j}$

$$= \begin{cases} 1 & N = 4k \\ 1 + j & N = 4k + 1 \\ j & N = 4k + 2 \\ 0 & N = 4k + 3, \; k \text{ non-negative integer} \end{cases}$$

3   (a) $1 - j$ (b) $3 - 2j$ (c) $\dfrac{1}{10}(-2 + j)$

6 $-5, 7$ (double), $-2 + j, -2 - j, -3 - j$

7   (b)   (i) 5 (ii) 13 (iii) $\dfrac{1}{\sqrt{74}}$

8   (c) 25

9 $5 - 12j = (3 - 2j)^2 = (-3 + 2j)^2$

10   (b) $|z| = (x^2 + y^2)^{1/2} > |x|$; i.e. $x, -x$

$x = \alpha^2 - \beta^2$ or $(-\alpha)^2 - (-\beta)^2$

$y = 2\alpha\beta$ or $2(-\alpha)(-\beta)$

Roots $\pm(\alpha + \beta j)$; also $|z| = \alpha^2 + \beta^2$

11   (a) $\pm\dfrac{1}{\sqrt{2}}(1 + j)$ (b) $\pm((\sqrt{2} + 1)^{1/2} + (\sqrt{2} - 1)^{1/2}j)$

(c) $\pm\dfrac{1}{5}(2 - j)$

12   (a) $\pm\dfrac{1}{\sqrt{2}}(1 + j)$

(b) $-2j, 4j$

(c) $-2 - j, 2(2 + j)$

(d) $\left(1 + \dfrac{1}{\sqrt{5}}\right)\left(1 - \dfrac{j}{2}\right), \left(1 - \dfrac{1}{\sqrt{5}}\right)\left(1 - \dfrac{j}{2}\right)$

(e) $\left(r_1 + \dfrac{1}{6}\right) + \left(r_2 + \dfrac{1}{6}\right)j = 0.3577 + 1.1847j$

$\left(-r_1 + \dfrac{1}{6}\right) + \left(-r_2 + \dfrac{1}{6}\right)j = -0.0243 - 0.8514j$

$r_1 = \left(\dfrac{\sqrt{373}}{36} - \dfrac{1}{2}\right)^{1/2}, \quad r_2 = \left(\dfrac{\sqrt{373}}{36} + \dfrac{1}{2}\right)^{1/2}$

13 $\dfrac{1}{2}(\pm\sqrt{6} \pm \sqrt{2}j)$, four roots

14 $\dfrac{-j}{10}\left(\dfrac{1}{z - 7j} - \dfrac{1}{z + 3j}\right)$

15   (b)   (i) $\overline{P_n(z)} = \overline{a_0 + a_1 z + \cdots + a_n z^n}$

$= a_0 + a_1\bar{z} + \cdots + a_n\bar{z}^n = P_n(\bar{z})$

(ii) If $P_n(\alpha) = 0$ then $\overline{P_n(\alpha)} = \bar{0} = 0 = P_n(\bar{\alpha})$

16 $(z - 1)(z^2 + 1)(z^2 - 2z + 2)(z + 2)^2 \equiv z^7 + z^6 - 3z^5 + 3z^4 - 4z^3 - 6z^2 + 8z - 8$

Exercise 12.2

1   (a) $5e^{0.9273j}$ (b) $5e^{5.3559j}$

(c) $1.4142e^{2.3562j}(\sqrt{2}e^{3\pi j/4})$ (d) $2e^{\pi j/3}$

(e) $5e^j$

2   (a) $\theta + \dfrac{\pi}{2}$ (b) $\theta + \pi$ (c) $\theta + \dfrac{\pi}{2}$

(d) 3θ (e) θ

3 $\dfrac{6}{5} + \dfrac{8}{5}j$

4 $\sqrt{13}; \ 146.32° = \tan^{-1}\left(-\dfrac{2}{3}\right); \ \sqrt{10}; \ 161.57° = \tan^{-1}\left(-\dfrac{1}{3}\right)$

6   (a)   Ellipse with $z = \pm c$ as foci, semi-major axis $3c$, semi-minor axis $2\sqrt{2}c$, i.e.
$\dfrac{x^2}{9c^2} + \dfrac{y^2}{8c^2} = 1$

(b)   Hyperbola with $z = \pm cj$ as foci, semi-major axis $\dfrac{c}{2}$

(c)   Circle $x^2 + y^2 = c^2$

(d)   Semicircle $x^2 + y^2 = c^2, y \geq 0$

7   Families of circles with centres on the real axis; $K = 1 \Rightarrow \text{Re}(z) = 0$, imaginary axis.

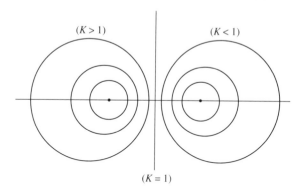

$(K > 1)$ $(K < 1)$

$(K = 1)$

8   (a)   (b)   (c)

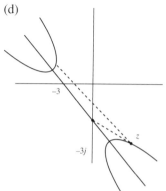

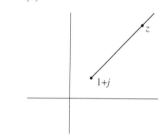

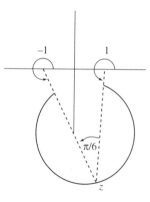

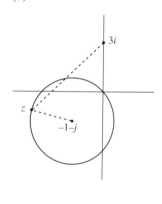

$1+j$

-1 1

$\pi/6$

z

$3i$

z

$-1-j$

(d)   (e)   (f)

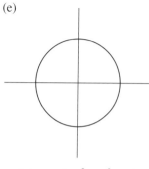

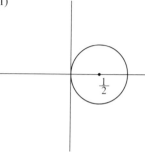

-3

$-3j$

z

$\frac{1}{2}$

(e) $y = 0, x^2 + y^2 = 1$

(f) $\left(x - \dfrac{1}{2}\right)^2 + y^2 = \left(\dfrac{1}{2}\right)^2$

9 (a) $\quad \arg(z + 2) = \dfrac{3\pi}{2}$

(b) $\arg\left(\dfrac{z - 4j}{z + 2j}\right) = \dfrac{\pi}{4}$

(c) $\quad |z - 2\sqrt{2}j| + |z + 2\sqrt{2}j| = 6$

10 (a) (i)

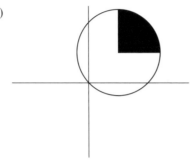

(ii)

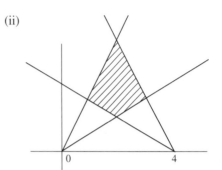

(iii)

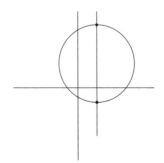

(b) (1) $\ 1 - j, 1 + 5j$

(2) $-3 - 4j$

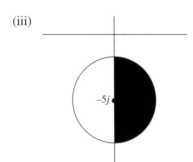

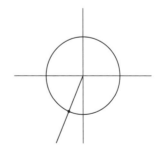

(c) $z^2 = x^2 - y^2 + 2xyj$, so $x^2 > y^2$

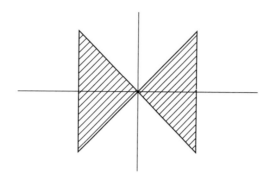

14   (a)   4; all zeros lie inside $|z| = 4$ (actually $|z| = 2$)

         (b)   4; all zeros lie inside $|z| = 4$ (actually all real, 0, 689, ± 3.244)

         (c)   8; all zeros lie inside $|z| = 8$

15   (b) $\pm j$ roots

         (c)   Roots must lie inside $|z| = 2$ (actually they all have modulus 1)

Exercise 12.3

1   (a) $-\dfrac{1}{\sqrt{2}}(1 + j)$ (b) $-\cos 2\theta + \sin 2\theta\ j$ (c) -1

2 $6\cos\theta - 32\ \cos^3\theta + 32\cos^5\theta$

4 $\sin\left(\alpha + \dfrac{n+1}{2}\theta\right) \times \dfrac{\sin\dfrac{n\theta}{2}}{\sin\dfrac{\theta}{2}}$

5 $2, \dfrac{3}{2} \pm \dfrac{\sqrt{3}}{2}j, \dfrac{1}{2} \pm \dfrac{\sqrt{3}}{2}j, 0$

6 $\dfrac{1}{2}\left(1 \pm \tan\dfrac{n\pi}{5}j\right), n = 0, 1, 2, 3, 4$

7   (a) $\dfrac{\sqrt{3}}{2} + \dfrac{j}{2}, -\dfrac{\sqrt{3}}{2} + j, -j$

         (b) $|z| = 1.3275$

               $\arg z = 20.80° + 72n°, n = 0, 1, 2, 3, 4$

8 $\cosh z = \cosh x \cos y + \sinh x \sin y \; j$

$$\tan z = \frac{\tan x + \tanh y \; j}{1 - \tan x \tanh y \; j} = \frac{\tan x \; \mathrm{sech}^2 y + \sec^2 x \tanh y \; j}{1 + \tan^2 x \tanh^2 y}$$

9   (a) $\sin(1)\cosh(2) + \cos(1)\sinh(2) \; j = 3.1658 + 1.9596 \; j$

(b) $\cosh(3)\cos(2) + \sinh(3)\sin(2) \; j = -4.1896 - 9.1092 \; j$

(c) $1.014\,79 + 0.0338j$

10   (a) $z = n\pi,\; n$ integer

(b) $z = 2n\pi \pm \ln(2 + \sqrt{3}) \; j$

(c) $\ln(1 + \sqrt{2}) + (4n + 1)\dfrac{\pi}{2}j$ or $-\ln(1 + \sqrt{2}) + (4n + 1)\dfrac{\pi}{2}j$

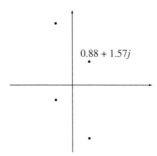

0.88 + 1.57j

(d) 0.1353

11 $(z + 1)\left(z^2 + \dfrac{1}{2}(\sqrt{5} - 1)z + 1\right)\left(z^2 - \dfrac{1}{2}(\sqrt{5} + 1)z + 1\right)$

12   (a)   Roots of $z^3 = 1$

(b) $z^6 + z^3 + 1 = \dfrac{z^9 - 1}{z^3 - 1}$

Therefore the roots are those of $z^9 - 1$ apart from those of $z^3 - 1$.

Exercise 12.4

1   (a) $e^{-x}(A \cos x + B \sin x)$

(b) $e^{x/2}\left(A \cos \dfrac{\sqrt{7}}{2}x + B \sin \dfrac{\sqrt{7}}{2}x\right)$

(c) $e^{-4x}(Ax + B)$

(d) $\quad e^{-11x/6}\left(A\cos\dfrac{\sqrt{23}}{6}x + B\sin\dfrac{\sqrt{23}}{6}x\right)$

2 (a) $\quad e^{-2t}\left(2\cos 3t + \dfrac{4}{3}\sin 3t\right)$

   (b) $\quad e^{-2t}(2 + 4t)$

3 $\quad R < 2\sqrt{LC}$

4 $\quad e^{-2x}\cos 3x$

5 $\quad -5e^{-t} + 4e^{-2t} + \cos t + 3\sin t$

6 $\quad A\cos t + B\sin t - \dfrac{t}{2}\cos t$

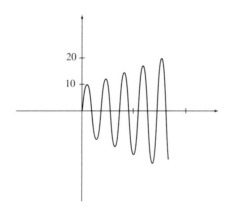

Amplitude ~ 10 when $t = 0$, rising to 20 when $t = 4$

7 (a) $\quad \dfrac{[C^2(R^2 + L^2\omega^2) + 2CLR^2 + R^2L^2]^{1/2}}{[\omega^2 C^2 + R^2]^{1/2}}$

   (b) $\quad \dfrac{[L^2C^2R^2\omega^4 - 2LCR^2 + \omega^2 L^2 + R^2]^{1/2}}{\omega LR}$

   (c)   (i) 100.005

           (ii) 89.443

8 (a) $\quad y_n = C\cos\left(\dfrac{n\pi}{2}\right) + D\sin\left(\dfrac{n\pi}{2}\right)$

   (b) $\quad y_n = \left(\dfrac{2}{\sqrt{3}}\right)^n \left(C\cos\left(\dfrac{n\pi}{6}\right) + D\sin\left(\dfrac{n\pi}{6}\right)\right)$

Appendix: An Introduction to Mathematical Modelling

A.1 CASE STUDY: VENTILATING A LECTURE ROOM

Suppose that a lecture room is ventilated by pumping fresh air into it and removing stale air. The purpose of this form of ventilation is to prevent the concentration of carbon dioxide rising above an acceptable level. Given a particular pumping capacity, how many people can a room safely accommodate to meet this criterion, bearing in mind that people breathe out carbon dioxide?

It is always useful to draw a diagram that summarises the information. Figure A.1 is a schematic diagram of the ventilation process. The room has volume $V \, \text{m}^3$ and air is pumped into the room at a rate of $Q \, \text{m}^3 \, \text{s}^{-1}$ and extracted at the same rate. The (constant) concentration of carbon dioxide in the fresh air is c_f and the concentration in the room air at time t is $c = c(t)$.

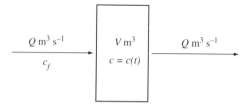

Figure A.1   Schematic diagram of the room

Already we have made an **assumption**: at any time t the concentration of carbon dioxide is uniform throughout the room. The notation $c(t)$ indicates that, in the room, the concentration depends on the time of observation, not on the position where the observation was made.

We now make a second assumption: the concentration of carbon dioxide in the air leaving the room at any instant is the same as the concentration in the room at the same instant.

A final introductory point is that the concentration in these situations is often quoted in parts per $10\,000$, so when we say $c = 3$ it means the concentration of carbon dioxide in the air is 3 parts per $10\,000$.

First model: empty room

It is usually wise to begin by modelling a simplified system instead of the actual situation. After all, if the model of the simpler situation produces predictions which are not borne out in practice, it is not worthwhile to generalise the model.

Before we embark on the mathematical model, we ask the question, what do you expect will happen to the concentration of carbon dioxide in the room after the lecture room has been left empty for the night? We would expect the concentration to decrease, but it cannot decrease below c_f, the concentration in the incoming air. Figure A.2 shows a possible graph of c against t; c_0 is the initial concentration.

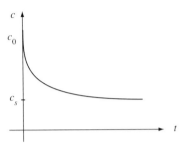

Figure A.2 Concentration against time for empty room

The graph suggests an exponential decay for the concentration towards a **steady-state value** of c_f.

The key element in building the model is the general statement for a system with input and output

$$\text{Rate of inflow} - \text{rate of outflow} = \text{rate of increase} \tag{A.1}$$

We know that the rate of inflow of air to the room is $Q\,\text{m}^3\,\text{s}^{-1}$ and the fraction of carbon dioxide entering the room is $\dfrac{c_f}{10\,000}$. Hence the rate at which carbon dioxide enters the room is $\dfrac{Qc_f}{10\,000}\,\text{m}^3\,\text{s}^{-1}$, which is a constant.

The rate at which air leaves the room is $Q\,\text{m}^3\,\text{s}^{-1}$ and the fraction of carbon dioxide is $\dfrac{c}{10\,000}$ at time t. Hence the rate at which carbon dioxide leaves the room is $Q\dfrac{c}{10\,000}\,\text{m}^3\,\text{s}^{-1}$, which is a function of time. At time t the fraction of air in the room which is carbon dioxide is $\dfrac{c}{10\,000}$, so the volume of carbon dioxide in the room is $(Vc/10\,000)\,\text{m}^3$. The rate at which this changes $\dfrac{d}{dt}\left(\dfrac{Vc}{10\,000}\right) \equiv \left(\dfrac{V}{10\,000}\right)\dfrac{dc}{dt}$, since V is a constant. Putting these three elements into (A.1) gives

$$Q\left(\frac{c_f}{10\,000}\right) - Q\left(\frac{c}{10\,000}\right) = \left(\frac{V}{10\,000}\right)\frac{dc}{dt}$$

or

$$\frac{dc}{dt} = \left(\frac{Q}{V}\right)c_f - \left(\frac{Q}{V}\right)c \tag{A.2}$$

Note that if we wish to determine the steady-state concentration, we simply put $\dfrac{dc}{dt}$ equal to zero in (A.2). This gives

$$\left(\frac{Q}{V}\right)c_f = \left(\frac{Q}{V}\right)c \quad \text{i.e.} \quad c = c_f$$

as predicted.

Equation (A.2) can be rearranged into the form

$$\frac{dc}{dt} + \left(\frac{Q}{V}\right)c = \left(\frac{Q}{V}\right)c_f \tag{A.3}$$

which is the standard form for solution by an integrating factor.
   The integrating factor is $e^{Qt/V}$ and the general solution is

$$c = c_f + Ae^{Qt/v}$$

where A is a constant.
   When $t = 0$ let $c = c_0$, so we can deduce that

$$c = c_f + (c_0 - c_f)e^{-Qt/V} \tag{A.4}$$

The graph of c against t is of the form shown in Figure A.2.

Second model: room with occupants

With the confidence we have gained from the simple model, we now consider the case when the room is occupied. Suppose that after a long period of non-occupation, say overnight, the room beings a period of almost continuous occupation. What shape might we expect for the graph of carbon dioxide concentration against time?
   Given that the purpose of pumping fresh air into the room is to keep that concentration below a recommended level, a potential candidate is the graph in Figure A.3. The curve suggests an exponential behaviour. The concentration tends to a steady-state value c_s; but unlike the previous model, we cannot predict what the value might be.

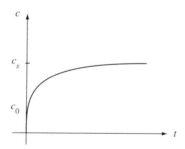

Figure A.3   Concentration against time for occupied room

We assume that the occupants between them breathe out carbon dioxide at a rate $Q_p\, \text{m}^3\, \text{s}^{-1}$. We now have two inputs to the system: the external input provided by the incoming air and the internal input provided by the people in the room.

Equation (A.1) now takes the form

$$Q_p + Q\left(\frac{c_f}{10\ 000}\right) - Q\left(\frac{c}{10\ 000}\right) = \left(\frac{V}{10\ 000}\right)\frac{dc}{dt}$$

which rearranges to

$$\frac{dc}{dt} + \left(\frac{Q}{V}\right)c = \frac{10\ 000Q_p + Qc_f}{V} \tag{A.5}$$

The steady-state concentration is found by putting $\frac{dc}{dt} = 0$, so that

$$\left(\frac{Q}{V}\right)c = \frac{10\ 000Q_p + Qc_f}{V}$$

Hence $\qquad c = c_s\left(\frac{10\ 000Q_p}{Q}\right) + c_f \tag{A.6}$

It would have been difficult to anticipate this result. We cannot guarantee it is correct but we can verify that it gives correct predictions in some special cases.

If the room is empty then $Q_p = 0$ and (A.6) reduces to $c = c_f$, as before. Therefore, the case of the empty room is covered by our more general model.

If we increase the speed at which fresh air is pumped into the room, i.e. Q is increased, then the fraction $\frac{10\ 000Q_p}{Q}$ decreases and so does c_s. This shows that the effect is to lower the steady-state value, which is what we might have anticipated.

If more people are allowed into the room, then Q_p increases and so does the fraction $\frac{10\ 000Q_p}{Q}$, hence c_s assumes a greater value, something else we might have expected.

Most surprising, perhaps, equation (A.6) does not depend on the volume of the room.

The equation (A.5) can be solved by the same integrating factor as (A.2). The solution is

$$c = \left(\frac{10\ 000Q_p}{Q}\right) + c_f + \left[c_0 - \left(\frac{10\ 000Q_p}{Q} + c_f\right)\right]e^{-Qt/V} \tag{A.7}$$

If we assume that $c_0 = c_f$ then (A.7) reduces to

$$c = \left(\frac{10\ 000Q_p}{Q} + c_f\right) - \left(\frac{10\ 000Q_p}{Q}\right)e^{-Qt/V} \tag{A.8}$$

Note that if

$$c_0 < \left(\frac{10\,000 Q_p}{Q}\right) + c_f$$

then equation (A.6) is of the form $c = A - Be^{-Qt/V}$ where A and B are constants greater than 0. The graph of this relationship conforms to Figure A.3.

So far the model has behaved well, but it makes no allowance for additional air currents through the room, e.g. by doors opening and shutting. We made several assumptions when building the model:

1. The concentration of carbon dioxide in the fresh air is constant.
2. The number of people in the room is constant, so Q_p is constant.
3. At any instant the concentration of carbon dioxide is uniform throughout the room.

The third assumption is the most worrying, since it is unlikely to be true. But if we had not made this assumption it is difficult to see how we would have progressed.

Exercise A.1

1 Show that the general solution of (A.3) is $c = c_f + (c_0 - c_f)e^{-Qt/V}$.

2 Show that the general solution of (A.7) is given by (A.8)

3 Draw graphs of solution (A.7) in the cases

    (a) $c_0 > \left(\dfrac{10\,000 Q_p}{Q}\right) + c_f$

    (b) $c_0 < \left(\dfrac{10\,000 Q_p}{Q}\right) + c_f$

4 Assume that $V = 170\,\text{m}^3$, $Q = 2$ air changes per hour (equivalent to $340/3600\,\text{m}^3\,\text{s}^{-1}$), $c_f = 0.03\%$ by volume (equivalent to 3 parts per 10 000, i.e. $c_f = 3$) and $c_0 = c_f$.

    Assume also that the average rate at which people breathe out carbon dioxide is $4.7 \times 10^{-6}\,\text{m}^3\,\text{s}^{-1}$ per person. Suppose that the 'safe level' of carbon dioxide is specified as 10 parts per 10 000. How many people, say N, may be allowed into the room if the safe level is not to be exceeded? Make a prediction based on common sense before carrying out the calculations.

5 If in Question 4 it is desired to increase the room capacity by 25%, what should be the airflow setting of the pump in cubic metres per second.

A.2 THE MODELLING PROCESS

To illustrate the wider context of mathematics for science and engineering, Secion A.1 developed a mathematical model for the ventilation of a room. Many books are devoted to the subject of mathematical modelling, so our treatment is comparatively brief. In this section we sketch out the general process of modelling; its eight stages are flowcharted in Figure A.4.

Understanding the problem

The first task is to clarify ideas about the problem under consideration and to define as precisely as we can those aspects of the general problem we wish to investigate.

Simplification of the problem

Most problems cannot be tackled in their entirety. We need to decide on the relevant factors and whether we can measure them. Then we must be ruthless and reduce the number of factors to a set we can sensibly handle. We also make some assumptions to ease our calculations; for example, certain variables may hardly change so we may assume they are constant. Relationships between the problem variables are determined by theory or by empirical (experimental) laws.

Mathematical formulation

We now convert the simplified problem into mathematical statements. For example, Newton's law of cooling can be stated as: the rate at which a heated liquid cools is proportional to the temperature difference between the liquid and its environment (i.e. the air). The mathematical statement of the law is

$$\frac{d\theta}{dt} = -k(\theta - \theta_s)$$

where θ is the temperature of the liquid at time t, θ_s is the air temperature and k is a thermal constant. Mathematically we know that the problem is not complete without an initial condition, e.g. $\theta = \theta_0$ at $t = 0$.

Solving the mathematical problem

The ideal way to solve the mathematical problem is often an analytical technique using a mix of algebra and calculus to obtain formulae which express the dependent variables in terms of the independent variables. However, an analytical solution may not be possible, so instead of a formula, we may have to settle for a computer-based solution, perhaps in the form of a table of (approximate) values. Sometimes the 'solution' may appear as a set of graphs or charts from which appropriate information can be taken.

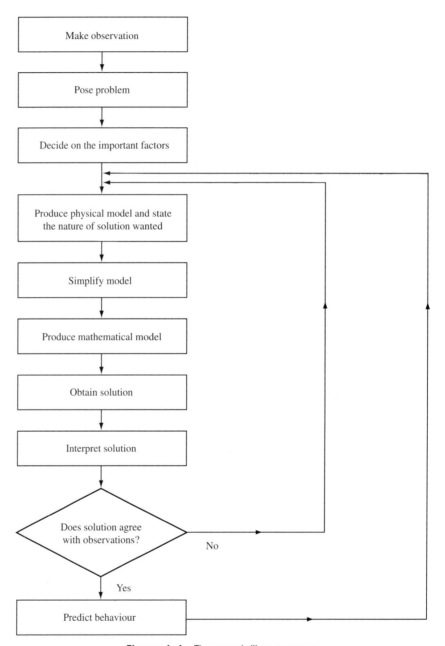

Figure A.4 The modelling process

We may also need to collect information on the values of parameters in the problem formulation. These may be available in published tables or we may have to carry out experimental measurements.

Interpreting the mathematical solution

What distinguishes a model from a mere problem-solving exercise is the essential step of linking the mathematical solution back to the real-world problem with which we started. Have we answered the questions we set out to resolve? Do the results from the mathematical model conform to the experimental results we have obtained?

It may be that, in attempting to solve the mathematical model, we could not obtain an analytical solution and made further simplifications in order to produce a model we could solve analytically. So long as the results of our simplified model are borne out by experiment, further simplification was justified. But often we obtain an analytical solution that is *not* consistent with experiment, so the model from which it was obtained is inadequate.

Revising the Model

If we have made too many simplifications, the model may not be realistic; we have to make fewer assumptions at the formulation stage. This might mean that an analytical solution becomes impossible, but if we want to obtain a realistic solution then this is the price we have to pay.

For many years Newton's laws of motion were perfectly adequate, but they could not model speeds close to the speed of light; eventually the theory of special relativity evolved to cope with these cases.

As computers grow more powerful it becomes possible to solve problems which hitherto could not be tackled, albeit using numerical methods for approximate solutions. Models can be made more sophisticated so they describe physical situations more comprehensively.

Developments

If the problem changes—either by extending the situation being considered or because external circumstances have altered—we may need to rebuild the model.

Reporting on the findings of the model

In any practical situation it is important to make a full report on the modelling process. A report should include an explanation of the model, why is was chosen, what assumptions were made and why they were made, an outline of the solution, comparison with experimental evidence and an estimate of the accuracy it could achieve.

Exercise A.2

1 Follow steps 1 to 8 for the problem of cooling in a hot liquid. Use Newton's law of cooling as the basis.

2 The figure shows a schematic diagram of a soldering iron. We wish to know how the temperature of the copper bit varies with time after the power has been switched on. This is a prelude to determining how long it takes for the temperature of the bit to reach the operating temperature at which it can melt the solder. We assume that all heat produced in the barrel goes directly to the bit and none is lost to the air. We also assume there is no temperature variation along the bit, i.e. its temperature, $\theta = \theta(t)$, depends only on time.
   We use three physical laws:

(a) The rate of energy storage in the bit is the product of the mass m of copper, the specific heat capacity C of copper and the rate of change of temperature in the bit.

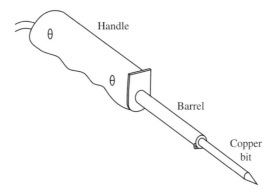

(b) The rate of loss of heat from the bit to the air has the form $kA(\theta - \theta_a)$ where θ_a is the temperature of the air, A is the (constant) cross-sectional area of the bit and k is a constant.

(c) The heat coming from the barrel to the bit is equal to the sum of the heat loss from the bit and the heat stored in the bit (conservation of energy).

Let W be the heat supplied to the soldering iron per second. Produce an equation for W. What is the steady-state temperature of the bit? If $\theta = \theta_0$ at $t = 0$ find the solution for θ as a function of time. Assume that $\theta_0 = \theta_a$ then sketch the graph of θ against t.

3 In the RC circuit illustrated we wish to find the voltage across the capacitor as a function of time. Assume there are no strong inductive or capacitive effects, there is no temperature variation and there is no leakage of charge between the plates of the capacitor. Then we may use the ideal law that the voltage across the resistor is $V_R = iR$ where R is the resistance and i is the current in the circuit. The voltage V_c across the capacitor satisfies

$$i = C\frac{d}{dt}(V_C)$$ where C is the capacitance.

Use Kirchhoff's law $V = V_R + V_C$ where V is the applied voltage to build a differential equation model for V_C. Solve the equation and sketch a graph of V_C against time. Compare the differential equation with the equation in Question 2. What conclusion do you draw?

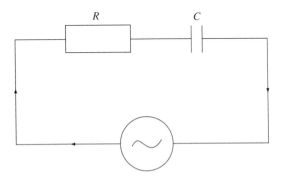

4 A model for the run-off of water over land, following a rainstorm, and into a river is provided by a cylindrical tank from which water, initial depth h_0, is allowed to leak at time $t = 0$. Plot a graph of the depth of water (the head) in the tank against time. Does this model provide a realistic representation of the situation if the observed depth of water $h(t)$ in the river is as illustrated?

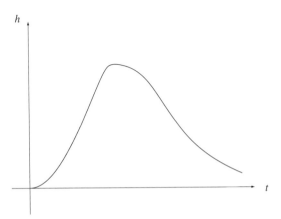

Now try a model in which the output from the first tank leaks into an identical tank that is initially empty. Obtain a formula for the head of water in the second tank as a function of time. Plot the graph of this function and comment.

What does the use of a leaking tank imply about the nature of the flow of water into the river? How could you refine the model further?

 5   The flow rate f of vehicles in a stream of traffic is given by

$$f = \frac{v}{d}$$

where v is the velocity of the vehicles (assumed constant) and d the separation distance between successive vehicles (also assumed constant). What does this model imply about the flow rate if v is increased?

A second model suggests that $d = 3v\tau + l$ where l is the length of a car and τ is the thinking distance. To what does the formula approximate when the speeds are high and when they are low? Sketch the graph f as a function of v.

A third model suggests that d should be the stopping distance, i.e.

$$d = v\tau + \frac{v^2}{2a} + l$$

where a is the deceleration. Again, consider the resulting formula for f in the cases of high speed and low speed. Sketch the graph of f as a function of v. Find the value of v at which the flow rate is a maximum. You may use the values $\tau = 0.7\,\mathrm{s}$, $l = 4\,\mathrm{m}$ and $a = 65\,\mathrm{m\,s}^{-2}$.

Answers

Exercise A.1

3 (a)

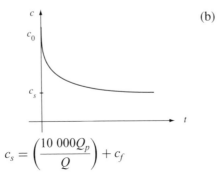

$$c_s = \left(\frac{10\,000Q_p}{Q}\right) + c_f$$

(b)

4 $N = 14$

5 $2\frac{1}{2}$ air changes per hour, i.e. a 25% increase

Exercise A.2

2 $\dfrac{d\theta}{dt} = \dfrac{W}{mc} - \dfrac{kA}{mc}(\theta - \theta_a);$ $\qquad \theta = \dfrac{W}{kA} + \theta_a - \dfrac{W}{kA}e^{-(hAt/mc)}$

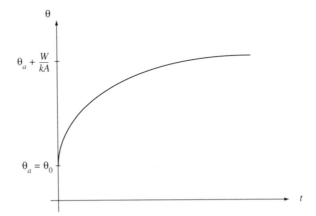

3 $\dfrac{dV_c}{dt} + \dfrac{V_c}{RC} = \dfrac{V}{RC}$

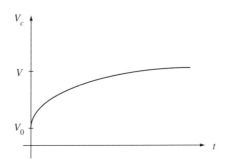

The voltage in the circuit behaves like the temperature of the copper bit.

4 One tank: $h = h_0 e^{-kt}$

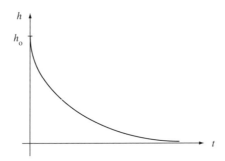

Second tank: $h = kh_0 t e^{-kt}$, which gives better agreement.

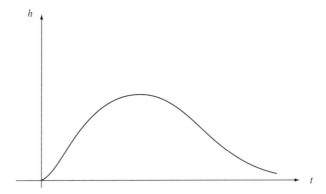

Water close to the river gets to it more quickly. More tanks in sequence give greater scope for modelling.

5 First model: the faster you travel, the greater the flow, without limit.
Second model: at high speeds $3v\tau \gg l$, implying f is constant; at low speeds $f = \dfrac{v}{l}$, suggesting f increases with speed.

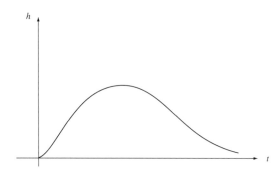

Third model: at low speeds $d = l$, so $f = \dfrac{v}{l}$; at high speeds $f = \dfrac{v}{v^2/2a} = \dfrac{2a}{v}$ and decreases to zero.

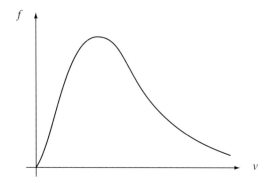

f is a maximum where $\dfrac{1}{f} = l + \dfrac{v}{2a} + \dfrac{l}{v}$ is a maximum. Differentiating $\dfrac{1}{f}$ with respect to v we obtain $\dfrac{1}{2a} - \dfrac{l}{v^2}$. This expression is zero where $v^2 = 2al$. Therefore $v = \sqrt{4 \times 13} \approx 7\,\mathrm{m\,s^{-1}}$ (15 mph).

INDEX